AF714004

ELECTROMAGNETICS OF BODY-AREA NETWORKS

IEEE Press
445 Hoes Lane
Piscataway, NJ 08854

ELECTROMAGNETICS OF BODY-AREA NETWORKS

Antennas, Propagation, and RF Systems

Edited by

DOUGLAS H. WERNER
ZHI HAO JIANG

IEEE PRESS

WILEY

Library of Congress Cataloging-in-Publication Data is available.

ISBN: 978-1-119-02946-5

Printed in the United States of America

10 9 8 7 6 5 4 3 2 1

To my wife Pingjuan L. Werner
Douglas H. Werner

To my parents Yujun Jiang and Qiuping Jin
Zhi Hao Jiang

CONTENTS

LIST OF CONTRIBUTORS

Kush Agarwal, Department of Electrical and Computer Engineering, National University of Singapore, Singapore

Sara Amendola, Department of Civil Engineering and Informatics, University of Rome Tor Vergata, Rome, Italy

Joonsung Bae, Department of Electrical Engineering, Korea Advanced Institute of Science and Technology, Daejeon, Republic of Korea

Nacer Chahat, NASA Jet Propulsion Laboratory, California Institute of Technology, Pasadena, CA, USA

Zhi Ning Chen, Institute for Infocomm Research, A*STAR, Singapore, and Department of Electrical and Computer Engineering, National University of Singapore, Singapore

Christian Enz, Integrated Circuits Laboratory (ICLAB), IMT Institute, École Polytechnique Fédérale de Lausanne, Neuchâtel, Switzerland

Yong-Xin Guo, Department of Electrical and Computer Engineering, National University of Singapore, Singapore

Anda Guraliuc, Institute of Electronics and Telecommunications of Rennes, UMR CNRS 6164, University of Rennes 1, Rennes, France

Yang Hao, School of Electronic Engineering and Computer Science, Queen Mary University of London, London, UK

Aravind Prasad Heragu, Semtech Neuchâtel Sàrl, Neuchâtel, Switzerland

Koichi Ito, Center for Frontier Medical Engineering, Chiba University, Chiba-shi, Japan

Zhi Hao Jiang, Department of Electrical Engineering, Pennsylvania State University, University Park, PA, USA

John Kimionis, School of Electrical and Computer Engineering, Georgia Institute of Technology, Atlanta, GA, USA

Gianluca Lazzi, Department of Electrical and Computer Engineering, University of Utah, Salt Lake City, UT, USA

Gaetano Marrocco, Department of Civil Engineering and Informatics, University of Rome Tor Vergata, Rome, Italy

Max O. Munoz, School of Electronic Engineering and Computer Science, Queen Mary University of London, London, UK

Cecilia Occhiuzzi, Department of Civil Engineering and Informatics, University of Rome Tor Vergata, Rome, Italy

Xianming Qing, Institute for Infocomm Research, A*STAR, Singapore

Yahya Rahmat-Samii, Department of Electrical Engineering, University of California, Los Angeles, CA, USA

Harish Rajagopalan, Department of Electrical Engineering, University of California, Los Angeles, CA, USA

Anil K. RamRakhyani, Department of Electrical and Computer Engineering, University of Utah, Salt Lake City, UT, USA

Hendrik Rogier, Department of Information Technology, Ghent University, Ghent, Belgium, and iMinds, Ghent, Belgium

Ronan Sauleau, Institute of Electronics and Telecommunications of Rennes, UMR CNRS 6164, University of Rennes 1, Rennes, France

Terence Shie Ping See, Institute for Infocomm Research, A*STAR, Singapore

Ping Jack Soh, Advanced Communication Engineering Centre, School of Computer and Communication Engineering, Universiti Malaysia Perlis, Arau, Perlis, Malaysia

Adrian Tang, NASA Jet Propulsion Laboratory, California Institute of Technology, Pasadena, CA, USA

Manos (Emmanouil) M. Tentzeris, School of Electrical and Computer Engineering, Georgia Institute of Technology, Atlanta, GA, USA

Raghavasimhan Thirunarayanan, Integrated Circuits Laboratory (ICLAB), IMT Institute, École Polytechnique Fédérale de Lausanne, Neuchâtel, Switzerland

Patrick Van Torre, Department of Information Technology, Ghent University, Ghent, Belgium, and iMinds, Ghent, Belgium

Guido Valerio, Laboratoire d'Electronique et Electromagnétisme, Sorbonne Universités – Université Pierre et Marie Curie, Paris, France

Luigi Vallozzi, Department of Information Technology, Ghent University, Ghent, Belgium, and iMinds, Ghent, Belgium

Guy A. E. Vandenbosch, ESAT-TELEMIC Research Division, Department of Electrical Engineering, KU Leuven, Leuven, Belgium

Douglas H. Werner, Department of Electrical Engineering, Pennsylvania State University, University Park, PA, USA

Hoi-Jun Yoo, Department of Electrical Engineering, Korea Advanced Institute of Science and Technology, Daejeon, Republic of Korea

Taiwei Yue, Department of Electrical Engineering, Pennsylvania State University, University Park, PA, USA

Maxim Zhadobov, Institute of Electronics and Telecommunications of Rennes, UMR CNRS 6164, University of Rennes 1, Rennes, France

PREFACE

The subject of this book is the burgeoning area of research and development on the electromagnetic aspects of body area networks (BANs) including antennas and propagation, as well as the associated RF systems and sensor technology. BANs, sometimes also referred to as body sensor networks or body-centric networks, are wireless communication networks of wearable devices that transmit information within, on, near, or around the human body. In these systems, distributed wearable and/or implanted sensors are required to communicate with either other on-body computing devices such as smart phones, watches or glasses, or more distant off-body systems like wireless routers. In such a network, the sensors are responsible for collecting various critical physiological signals and/or environmental parameters, and effectively relaying this data to a smart wearable processing unit. The smart wearable processing unit, which is capable of storing the data temporarily and performing fundamental signal processing, is primarily used to communicate with a wireless router or a smart phone such that the data can then be sent via internet to a remote center (e.g., the cloud) for further processing and analysis by specialists. These wearable devices are either implanted within or placed in extremely close proximity to the body, which leads to a number of unconventional challenges for wireless system and RF designers. Over the past few years, the field of BAN electromagnetics has witnessed tremendous advancements in many of its associated sub-areas, which have jointly enabled systems with revolutionary functionalities for a wide variety of applications including health monitoring, patient tracking, physiotherapy, wearable computing, battlefield survival, the Internet of Things (IoT), and so on. As such, BAN technology holds great promise for revolutionizing many aspects of the daily life of human beings.

The main goal of this book is to provide an overview of the many recent and significant accomplishments regarding the various electromagnetic topics related to BAN systems. In this book, a comprehensive treatment of the field has been compiled from a group of leading scholars and researchers throughout the world on subjects ranging from the fundamental theoretical principles and new technological advancements to the state-of-the-art device design and demonstration examples encompassing a wide range of related sub-areas. In particular, topics that will be covered include recent developments on novel wearable and implantable antenna concepts and designs, advanced wireless propagation models for on-body, in-body (or through-body), and off-body modes, efficient numerical human body modeling techniques, artificial phantom synthesis and fabrication, as well as low-power RF systems and related sensor technology. Newly emerging flexible materials and fabrication methodologies are also highlighted for the key role they play as enabling technologies for wearable and implantable devices. Moreover, an in-depth discussion of wearable and implantable RF transmitter/receiver design and implementation using low-power integrated circuits has also been included, providing a valuable perspective on BANs from the systems viewpoint. Throughout the book, a plethora of application examples have been included, encompassing wireless capsule endoscopy, physiological data acquisition, wireless implantable neuroprobe microsystems for motor prosthetics, retina prosthetics, medical compliance sensing, temperature sensing, wearable wireless energy harvesting, fire-fighter tracking, exposimeters, security and identification, and others. These topics have been carefully selected for their transformational impact on the next generation of BAN systems and beyond. Each chapter was contributed by an internationally recognized author or group of authors covering the latest research results. We hope that this book will be an indispensable resource for graduate students and researchers in academia as well as professionals in the consumer electronics and telecommunication industries.

This book contains 14 invited chapters contributed by leading experts in the field of electromagnetics of BANs. A brief summary of each chapter is provided as follows. Chapter 1 discusses the materials, design, fabrication, and characterization methods for textile wearable antennas. Various antenna performance evaluation techniques are reviewed, including numerical simulations based on the HUGO human body model, specific absorption rate measurements using a DASY-4 system, and antenna efficiency measurements using a reverberation chamber. Chapter 2 introduces several new antenna designs for off-body communications that provide more compact form factors and/or multiple functionalities, including metasurface-enabled linear-polarized antennas and circularly polarized integrated filtering antennas. The performance of these antennas is evaluated numerically and experimentally when mounted on different positions of a human body. Chapter 3 presents several novel wearable antenna designs for on-body communications based on miniaturized Yagi-Uda antennas backed by artificial magnetic conducting ground planes. Numerical and experimental evaluations of antenna prototypes fabricated using flexible latex substrate materials are reported. Chapter 4 presents the design and characterization methodologies of implantable antennas operating at different frequency bands. These antenna examples are exploited for applications including wireless capsule endoscopy diagnostics, medical compliance sensing, and patient monitoring. Chapter 5 considers the basic characteristics of wearable antennas and their associated wave propagation over a wide frequency range from 3 MHz to 3 GHz. The impacts of human body movement and the surrounding environment on the antenna properties as well as wave propagation are investigated. Chapter 6 presents a study of the channel properties for through-body communication modes excited using different basic antenna structures. In addition, more advanced implantable antenna designs for wireless capsule endoscopy and wireless implantable neuroprobe microsystems as well as the corresponding experimental results are presented. Chapter 7 discusses the use of phantoms, wearable antenna design, and the associated material and fabrication considerations important for millimeter wave applications. Several wearable antennas are demonstrated at the 60 GHz band, which are used to study the channel properties of on-body propagation modes. Chapter 8 investigates the multiple-in multiple-out (MIMO) channel characteristics in a BAN for off-body communications. The design and experimental results for active wearable antennas are also presented, which demonstrate a scenario for MIMO off-body communication between fire fighters. Chapter 9 reviews the performance of antennas when they are mounted on the human body and proposes epidermal antennas for radio frequency identification (RFID) applications. By exploiting the epidermal membranes, devices that perform wireless measurement of body temperature and which monitor fluid change at the skin interface are demonstrated. Chapter 10 provides an overview of inkjet printing technology for implementing wearable devices including antennas, sensors, microfludic-tunable filters, and RF receivers. A specific example is included, which achieves wireless energy harvesting at an ultra-high frequency band. Chapter 11 presents the concepts and system design of wearable RF integrated circuits. In particular, a BAN system is demonstrated that includes an energy efficient dual-band wearable hub receiver, a compact implantable temperature sensor node, and a wearable electrocardiogram sensor node. Chapter 12 proposes the use of MEMS thin-film bulk acoustic resonators for ultra low power wearable and implantable chips that can be employed in a BAN system. Moreover, the design and implementation of a transmitter and a receiver in the ISM band are presented. Chapter 13 describes different non-invasive RF technologies that have the potential for remote monitoring of vital signs. To this end, the received signal strength-based method, continuous wave method, and the wideband pulse radar method are explained in detail. Chapter 14 discusses the various methods of supplying power for implantable and wearable devices including non-regenerative and regenerative approaches. In addition, a methodology to achieve simultaneous power/data telemetry for BAN devices is proposed, including an illustrative example for a retinal prosthesis.

We are indebted to the authors of the 14 chapters for their outstanding contributions to this book, each reporting on cutting-edge research in the field. The cooperation of the contributing authors during the course of the book preparation as well as their dedication to delivering a high-quality product is deeply appreciated. We would also like to gratefully acknowledge the editorial assistance provided by Jeremy A. Bossard, Kenneth L. Morgan, Zachary C. Morgan, and John A. Easum of the Pennsylvania State University Computational Electromagnetics and Antenna Research Lab.

DOUGLAS H. WERNER
ZHI HAO JIANG

State College, PA, USA

ACKNOWLEDGMENTS

Ping Jack Soh and Guy A. E. Vandenbosch would like to acknowledge Ms. Nor Farahiyah Mohd Aun of Universiti Malaysia Perlis for her contributions to Chapter 1. The work in Chapter 1 is also supported in part by the Malaysian Ministry of Higher Education, COST IC1102 VISTA – Versatile Integrated and Signal-Aware Technologies for Antennas Short Term Scientific Mission (STSM) and the FP7 CARE (Coordinating Antenna Research in Europe) Secondment Programme. The work reported in Chapter 2, contributed by Zhi Hao Jiang, Taiwei Yue, and Douglas H. Werner, was supported by the National Science Foundation ASSIST Nanosystems ERC under Award EEC-1160483. The work summarized in Chapter 3, contributed by Kush Agarwal and Yong-Xin Guo, was supported in part by the National Research Foundation, Prime Minister's Office, Singapore, under Grant NRF-CRP8-2011-01, and in part by the National Natural Science Foundation of China under Grant 61372012. Koichi Ito would like to acknowledge Dr. Nozomi Haga of Gunma University, Dr. Masaharu Takahashi and Dr. Kazuyuki Saito of Chiba University, for their contributions to Chapter 5. Finally, the work reported in Chapter 7 was partly supported by the European Union Seventh Framework Program under the grant no 619563 (MiWaveS project) and by Labex CominLabs under the French National Research Agency program "Investing for the Future" ANR-10-LABX-07-01 (ResCor/BoWi project). The editors would also like to thank Jeremy A. Bossard and Mario Pantoja for providing ideas and assistance during the preparation of the cover art for this book. The sample shown in the photograph on the front book cover of the flexible artificial metasurface for wearable antenna applications was fabricated by Prof. Yong Zhu and Zheng Cui at North Carolina State University.

1

TEXTILE ANTENNAS FOR BODY AREA NETWORKS: DESIGN STRATEGIES AND EVALUATION METHODS

PING JACK SOH[1] AND GUY A. E. VANDENBOSCH[2]

[1] *Advanced Communication Engineering Centre, School of Computer and Communication Engineering, Universiti Malaysia Perlis, Arau, Perlis, Malaysia*
[2] *ESAT-TELEMIC Research Division, Department of Electrical Engineering, KU Leuven, Leuven, Belgium*

1.1 INTRODUCTION

Due to the increasing demand for multi-functional, multi-band wireless operation and consumer-centric technology, textile antennas have been receiving growing attention [1]. Future wearable systems should be unobtrusive, flexible, and operating with minimal degradation in proximity to the human body. These antennas have to meet the bandwidth, efficiency, and safety requirements, while being consistent with low-cost manufacturing techniques. Moreover, in wearable applications, flat surfaces cannot be guaranteed. Thus, an important antenna requirement is its ability to work with good robustness against environmental, positional, and location changes when being worn, besides complying with medical and safety regulations.

Body-centric antennas are crucial in catering for various current and future wireless standards. Among others, wearable antennas could assist medical monitoring for hospitalized, homebound, or outpatients [2, 3]. They could be applied in emergency service communication and public safety support (e.g., firefighters) [4–6]. They could

Electromagnetics of Body-Area Networks: Antennas, Propagation, and RF Systems, First Edition.
Edited by Douglas H. Werner and Zhi Hao Jiang.

also provide flexibility in assisting communication in search, rescue, and location-tracking alerts, especially in hazardous environments [7, 8]. There is also the possibility that they will become popular in consumer electronics in the near future, applied for communication [9], positioning, and navigation for recreational purposes [10].

Wearable antennas are electrical radiators being made flexible enough to be worn and to work in the proximity of a user's body. Since it is ergonomically more suitable that a wearable antenna for Wireless Body Area Networks (WBAN) applications is flexible and made to conform to the body, it is only natural that textiles be used to achieve these requirements compared to conventional metallic structures, for example, rigid copper plates or tapes which are worn. However, degradation of the antenna performance when worn on the human body has been one of the major deterrents in its successful implementation, be it in terms of frequency detuning, bandwidth reduction, and efficiency degradation or radiation distortion [11]. In other words, ideally, a wearable antenna must be designed to be immune enough for an on-body operation. Moreover, a flexible antenna made from textile is regarded as a realistic candidate due to the advancements in conductive textiles and the ergonomic properties that it is able to offer. Since these textiles are either newly introduced or have been traditionally used for other purposes, for example, electromagnetic interference (EMI) shielding or grounding, one of the important and yet challenging aspects of this work is to properly characterize their electrical properties at the intended frequencies.

In this chapter, firstly, a brief overview of the types of textiles (conductive and non-conductive) is given. Next, the characterization procedure using a commercial setup is explained prior to the proposal of a systematic antenna fabrication procedure. Finally, this chapter also describes the evaluation methods used for the fabricated antenna prototypes, that is, in terms of reflection coefficient, radiation characteristics, efficiency, and specific absorption rates (SAR).

1.2 TEXTILE MATERIALS AND ANTENNA FABRICATION PROCEDURE

Textile antenna prototyping materials generally consist of two textile types, conducting and non-conducting. The former is typically used to form antenna conductive elements (radiator, ground plane, shorting wall, etc.), whereas the latter is used to form the substrate, spacer, etc. For example, in the case of a Planar Inverted-F Antenna (PIFA) topology, conductive textiles are used as its radiator, ground plane, and shorting wall, whereas felt or fleece is used as the substrate. The properties of several popular commercial off-the-shelf textiles can be found in [4] and [12–14], and will be explained in the following sections.

1.2.1 Conductive Textiles/Foils

Initially, flexible antennas are prototyped using copper foil, as this is flexible and a rough representation of a textile antenna. However, the introduction of electrically conductive acrylic adhesives which are reasonably homogeneous in terms of surface

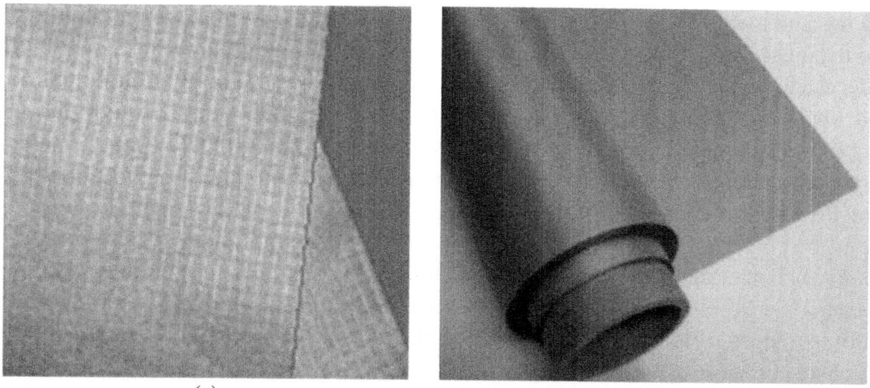

(a) (b)

FIGURE 1.1 Conductive textiles: (a) ShieldIt Super [15] and (b) PCPTF [16]. Used with permission from LessEMF.com.

resistivity/conductivity and mechanically stable has eased the fabrication of antennas made using commercial off-the-shelf textiles. These materials can be used as the radiating or grounding element for a textile antenna, and are required to be highly conductive, with surface resistivities (R_s) of less than 0.05 Ω/sq.

Three of the more popular conducting materials for wearable antennas are described as follows and are shown in Figure 1.1

(a) Copper foil tape: a 0.035 mm thick foil, coated with a 0.03 mm thickness.
(b) ShieldIt Super [15]: a ripstop, woven polyester textile coated with copper and nickel. Its thickness is 0.17 mm and its estimated weight is 230 g/m^2.
(c) Pure copper polyester taffeta fabric (PCPTF) [16]: plain woven and coated using pure copper. It has a thickness of 0.08 mm and an estimated weight of 80 g/m^2.

The parameter of prime importance is the equivalent conductivity of the textile used. This parameter ultimately determines the losses, and consequently, the efficiency and gain of the antennas. The homogenized conductivities were calculated based on the surface resistances provided by the manufacturer. The thickness of the conductive textile can be chosen depending on the application and location of the on-body deployment. For example, if the antennas are needed for health monitoring of the aged, thinner and low-cost materials may be chosen as they might be less exposed to harsh environments compared to the use of such antennas in military applications.

1.2.2 Non-conductive Textiles

The substrates used to fabricate wearable textile antennas are generally chosen to enable ease of integration onto the users' clothing. Upon the selection of a suitable textile substrate, it is important that its permittivity and loss tangent are characterized via measurements. Note that material properties for the same substrate might differ due to (small) variations in the fabrication process. For example, it is found that for felt [17], values given in literature are within the following ranges: ε_r in between 1.18 [18] and 1.45 [19], and δ_{tan} in between 0.004 [18] and 0.025 [14]. This illustrates that a proper measurement of felt characteristics is still necessary. On the other hand, measurements using an in-house developed technique based on the cavity method yielded a relative permittivity of $\varepsilon_r = 1.45$, and $\delta_{tan} = 0.044$ at 2.45 GHz. Another example is the polyester fleece [20], a soft napped insulating fabric made from synthetic fibers. Such material is typically used to manufacture loss tangent jackets, hats, pants, sweaters, gym clothes, and high-performance outdoor clothing [21]. Its reported permittivity ranges from 1.04 [7, 22] to 1.25 [23], and none of the publications reported the characterization of its loss tangent.

1.2.3 Textile Antenna Fabrication Procedure Using Commercial Textiles

Fabrication starts by cutting a single piece of textile or copper foil according to the designed and optimized shape. Then, the textile is folded over and fastened on the felt fabric to form the antennas. Depending on the antenna topology, a sufficiently large hole for the insertion of a 50 Ω SMA connector is needed. This is mainly to avoid shorting contacts between the radiator and the ground plane, once soldering or conductive epoxy is applied. Fastening copper foil and ShieldIt textile to the substrate is simpler, as both materials come with adhesive reverse sides. ShieldIt comes with a heat-activated reverse side adhesive, which is firm and typically enables it to be secured accurately onto substrates. On the other hand, it should be noted that the contact between the copper foil adhesive and the non-woven felt substrate degrades with time, thus the use of additional non-conductive adhesive is highly recommended. Although this might slightly affect the overall material thickness and its electrical properties, more adverse effects are expected as the original adhesive degrades and as the copper foil edges peel off and lift up, allowing the existence of air gaps. PCPTF, on the other hand, has to be stitched onto the substrate along its perimeter. This almost always results in an uneven conductive surface around the antenna perimeter, at locations where the conductive surface is sewn to the substrate layer. Besides, manual stitching, if not carefully done, could also result in misalignments. Thus it is also recommended that a small amount of non-conductive adhesive be applied prior to the stitching process. Meanwhile, to ensure proper electrical connection between PCPTF textile and the SMA connector, a silver conductive epoxy model 8331 from MG Chemical is used. Fabrication using this simple method provides an estimated accuracy of about 0.2 mm [12].

To realize a full textile fabrication, several high-level requirements are imposed.

(a) The whole structure or structure on the same layer has to be fabricated out of a single piece of textile. This avoids additional interconnections using glue/epoxy and maintains a consistent surface conductivity.

(b) The structure has to be simple and as symmetrical as possible, to ease fabrication.

(c) Hard-to-cut shapes/corners should be avoided to minimize fabrication inconsistencies.

1.3 DESIGN STRATEGIES AND EVALUATION METHODS

1.3.1 Antenna Simulation and Evaluation in Free Space

Similarly as for a conventional antenna, it is crucial that the performance of a textile antenna is first evaluated using a numerical electromagnetic solver. Such antenna is typically designed using a series of design equations, resulting in a set of initial dimensions. The general input parameters to these equations are the operational frequency, bandwidth, and permittivities and loss tangents of the materials. For this reason, it is important that the material properties are determined using the most accurate method available to the antenna designer. This is to ensure that the performance predicted by the antenna is not affected by a mischaracterization of the materials.

Textile substrates can be inserted in simulators in exactly the same way as conventional substrates: in most cases, the conductive textiles are defined as a lossy metal. This simplifies the analysis. For example, the two conductive textiles introduced earlier are defined with an estimated conductivity of $\sigma_S = 1.18 \times 10^5$ S/m for ShieldIt and $\sigma_P = 2.5 \times 10^5$ S/m for PCPTF, using the approximate equation proposed in [24], as follows:

$$\sigma = \frac{1}{R_s \cdot t} \tag{1.1}$$

where σ is the conductivity (S/m), R_s is the surface resistance (Ω/sq), and t is the thickness of the conductive textile (m). The copper foil conductivity is simplified as $\sigma_C = 5.88 \times 10^7$ S/m, assuming that the conductivity of the adhesive is negligible.

Further, it is also important to properly understand how the simulator operates to ensure the most accurate results. Most simulators are categorized according to the method used to solve Maxwell's equations, either in time or frequency domain. Several numerical techniques exist: the finite integration technique (FIT) [25, 26], the finite-difference time-domain (FDTD) [27–30], the transmission line matrix (TLM) method [31, 32], the finite element method (FEM) [33, 34], and the method of moments (MoM) [35]. All methods involve a discretization of the entire analysis space, except for the MoM, which only discretizes the antenna itself. The computational effort for volume-based methods depends on the volumetric size, which must always be made finite. To this goal, boundary conditions are introduced. Absorbing boundary conditions are used to represent the continuous flow of energy in free space; electrically conducting walls are used for closed boxes, or periodic boundary conditions are used to represent periodic structures. Symmetries can be taken advantage of in many simulators.

The optimal simulation strategy works as follows. First, the antennas are simulated using coarse meshes across a larger frequency range. In this way computational resources are more efficiently used, as the initial antenna performance is already predicted with a limited simulation time. Several optimization iterations are then necessary to ensure that the antenna is operating at the targeted frequency. The simulated frequency range can then be narrowed down, and a more detailed mesh can then be defined. For more exotic antenna topologies, it is important to understand the structure well via parameter sweeps.

In experiments, measurement of the reflection coefficient is performed using a vector network analyzer (VNA), whereas radiation characteristics, that is, patterns, gain, and efficiency are evaluated in an anechoic chamber. For reflection coefficient measurements, the VNA is first calibrated using the well-known short–open–load method to remove any systematic errors (losses, reflections, and phase delays) resulting from connectors, cables, and transitions. Measurement repeatability is first validated, and at least three measurements are averaged to deliver the final reflection coefficient for each antenna.

Radiation patterns are measured with the antenna under test (AUT) as the receiving antenna. A transmitting broadband antenna is placed in the far-field region of the AUT. Received power from two orthogonal AUT axes is evaluated from $-180°$ to $+180°$, before the transmitting antenna is turned to the other orthogonal polarization. This measurement sequence is then repeated, resulting in four main readings, co- and cross-polarization in *E*- and *H*-plane for a linearly polarized textile antenna. Antenna gains are measured using the widely accepted gain transfer method [36], based on the Friis transmission equation. In this method, a two-step procedure is performed. Received power is first evaluated using a standard gain antenna as the receiving antenna. This results in the following equation

$$P_{\mathrm{rREF}} = P_{\mathrm{T}} + G_{\mathrm{T}} + G_{\mathrm{rREF}} - L_{\mathrm{S}} \quad (\mathrm{dB}) \tag{1.2}$$

where P_{rREF} is the received power at the reference antenna, G_{T} is the gain of the transmitting antenna, G_{rREF} is the known gain of the receiving reference antenna and L_{S} is the free space path loss. The receiving reference antenna is then replaced with the AUT, and the received power is measured, resulting in

$$P_{\mathrm{rAUT}} = P_{\mathrm{T}} + G_{\mathrm{T}} + G_{\mathrm{rAUT}} - L_{\mathrm{S}} \quad (\mathrm{dB}) \tag{1.3}$$

where P_{rAUT} is the received power at the AUT and G_{rAUT} is the gain of the receiving AUT. Subtracting (1.1) from (1.2), the desired G_{rAUT} can then be calculated as

$$G_{\mathrm{rAUT}} = P_{\mathrm{rAUT}} - P_{\mathrm{rREF}} + G_{\mathrm{rREF}} \quad (\mathrm{dB}) \tag{1.4}$$

1.3.2 On-Body Co-simulations and Experimental Evaluations

Since the proposed antennas are designed to be worn, it is crucial that they are evaluated when mounted on the human body. For wearable antennas, the focus mounting area is on the upper human torso as it is the largest and most suitable location. This is because antennas placed in this area are less prone to bending or other deformations.

Moreover, in health monitoring applications, for example, this is the location where many physiological parameters can be gathered in a non-obtrusive way. Nonetheless, in general, on-body investigations should consider several aspects such as the effect of on-body mounting locations, antenna orientations, and distances from the body.

The simplest step to predict on-body detuning is to co-simulate and further optimize the antenna in proximity of a human phantom. This will enable the antenna performance changes to be forecasted, and the designer can ensure that the antenna remains operational when placed on the body despite the coupling with the body. As the body is composed of different tissues with different material properties, the choice of a proper body phantom is critical in ensuring a good tradeoff between simulation accuracy and complexity. Moreover, tissue properties vary with frequency, while the tissue composition also depends on the body part where the antenna is to be placed. In general, body models can be categorized as either homogeneous or heterogeneous. The former generally indicates that the model is filled using a simplified, single tissue property, or an average of several tissue properties, whereas a heterogeneous model consists of a large number of tissues and a voxel resolution. Examples are found in [37] and [38] where the phantom is defined from a two-third muscle equivalent material. The dielectric properties of the tissues involved are obtained via *in vivo* measurements on human or autopsy materials. A compromise between the homogeneous and heterogeneous models is the three-layer model as illustrated in Figure 1.2 [39].

The main parameter of concern in determining the level of on-body detuning is the distance between the antenna and the body. In simulations, this can be controlled quite accurately. On the contrary, this is not the case in experiments, especially when the antenna is very flexible. Foams with different thicknesses are typically used to consistently space the antenna at a certain distance from the user's body. To maintain consistent antenna mounting location and cable routing during the on-body experiments, a custom-modified fleece jacket routed with sewn-in RF cables can be utilized. For example, in [40], two 40 cm long cables were mounted vertically on the jacket on the front and the back, connected to the AUT at one end, and to the VNA or radiation pattern measurement system at the other end. Throughout the course of this work, four placement locations were evaluated, that is, the chest, back, shoulder, and on the elbow in bent conditions. However, only two antenna mounting locations

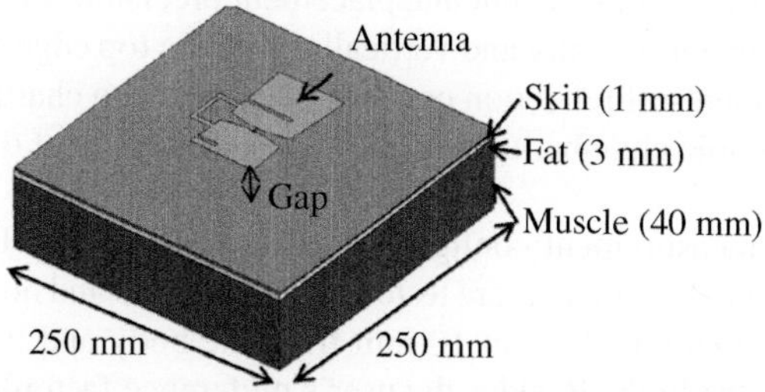

FIGURE 1.2 The three-layered human body model as defined in Wang et al. 2014 [39]. Reproduced with permission of IEEE.

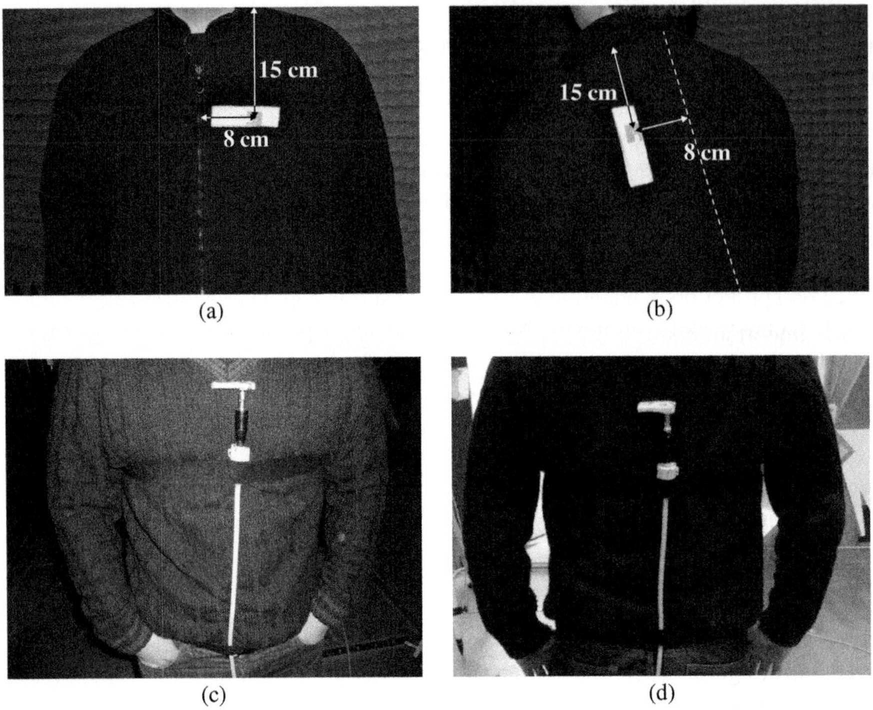

FIGURE 1.3 Examples of on-body mounting locations, horizontally and vertically: (a) on the chest at 10 mm distance, (b) on the back at 10 mm distance, (c) on the chest at 0 mm distance, and (d) on the back at 0 mm distance. *Source:* Soh et al. 2012 [40]. Reproduced courtesy of The Electromagnetics Academy.

on the upper human torso (i.e., chest and back) were tested for all antennas due to the difficulty in ensuring a consistent antenna bending condition and placement position for two other locations: on the shoulder and elbow. Moreover, these two locations are less likely to be favorable for use in practice, since the antenna performances are inconsistent when operating in a bent condition. Furthermore, the substrate thickness of at least 3 mm and the felt material properties enable bent antennas to quickly regain its original planar form. To ensure antenna placement precision, two holes were created at a fixed location horizontally and vertically from the top edge of the jacket, see Figure 1.3. Measurements of reflection coefficient or radiation characteristics can be performed in an anechoic chamber using real human volunteers of different genders, heights, and weights.

Several on-body measurement configurations are illustrated in Figures 1.3a and 1.3b. For each location, the antennas are tested in both vertical and horizontal settings, with its ground plane mounted about 10 mm from the body and its radiator facing outward, away from the body. Besides the user's preference factor—whether he/she is more comfortable in tighter/looser clothing—the choice of the chest and back for this evaluation defines two locations with distinct gap sizes. In other words, clothes are relatively closer to the chest and slightly distant from the back when a user is standing or walking, for instance. With the additional clothing worn under a jacket, it is reasoned that a 10 mm gap would be a realistic implementation. Taking into account the possibility of a simultaneous change in orientation and spacing distance caused by the users' movement, two additional configurations can be introduced to simulate a worst-case scenario: the antenna facing upward, with the substrate touching the user's clothing, as shown in Figures 1.3c and 1.3d.

1.3.3 Deformation Study

Another important component of textile antenna evaluation is studying its deformation. Such situations easily occur when a flexible or thin textile antenna is worn on the body. Deformations can be generally decomposed into

- Bending
- Crumpling
- Stretching
- Displacement

It is important for a textile antenna designer to model and evaluate these deformations in the design stage to be able to predict the practical applicability of the antenna. These investigations aim to capture the impact of instability in the antenna parameters because of variations in, for instance, the amount of antenna bending. Bending can be modeled by most electromagnetic solvers available today. They are mainly influenced by the following factors:

1. the movements or activities of users wearing the antennas
2. the variations in body morphology of the users

In a typical bending study, the textile antenna is first evaluated in its flat form prior to the evaluation of a number of bending radii. These radii need to be realistic and representative of the actual on-body placement location: on thighs, arms, neck, etc. The study has to be performed for two orthogonal bending axes, for example, the y-axis and x-axis, of a z-pointing planar antenna; see Figure 1.4. This is to ensure that the effect of bending in either of the two orthogonal planes is fully predicted.

Antennas which are easily deformed must also be experimentally evaluated. This can be done by placing them either onto a human phantom or directly onto a real human arm; see Figure 1.5. For the former, a cylindrically shaped container is typically required. This container is filled with proper tissue simulating liquid. While these liquids are commercially available, it is also possible to fabricate them using a series of chemicals. A sample of the fabricated phantom at 2.45 GHz is shown in Figure 1.6.

Due to the randomness of a user's activity and the various possible on-body placement locations, any textile antenna integrated on a worn garment may also suffer

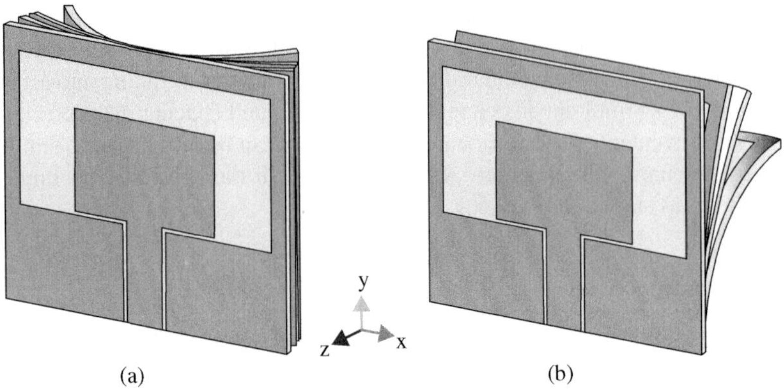

FIGURE 1.4 Modeling bending in the two orthogonal planes. *Source:* Bai and Langley 2009 [41]. Reproduced with permission of IEEE.

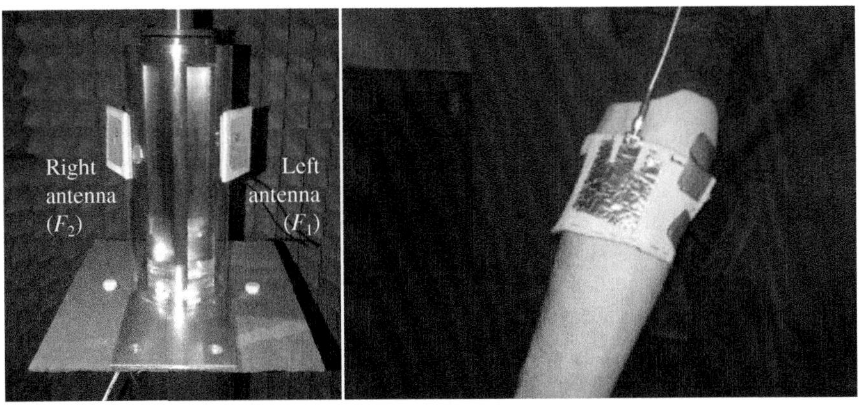

FIGURE 1.5 On-body evaluation of bending on a human phantom and on a human volunteer. *Source:* Ivsic et al. 2013 [42] and Soh et al. 2013 [43]. Reproduced with permission of IEEE.

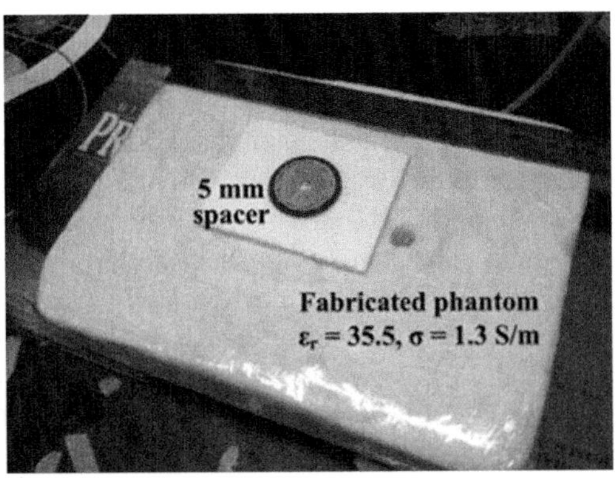

FIGURE 1.6 Fabricated sample of a semi-solid phantom. *Source:* Choi et al. 2015 [44]. Reproduced with permission of IEEE.

TABLE 1.1 The effect of crumpling on the dimensions of a sample dipole antenna [45].

	Antenna length (mm)	*h* (mm)	*l* (mm)
Flat	140	N/A	N/A
Crumpling case 1 (CC1)	133.24	11	40
Crumpling case 2 (CC2)	115.19	11	18
Crumpling case 3 (CC3)	90.23	11	10
Crumpling case 4 (CC 4)	79.28	11	8

from crumpling. A series of crumpling models is illustrated in Figure 1.7. Crumpling can be most detrimental since it reduces the electrical dimensions of the antenna. It generally results in an upward shift of the operational frequency band, and leads to an increased mismatch. This is illustrated in Table 1.1. When placed on body, it is noticed that the frequency shift is dependent on the level of the crumpling. For example, in [45], for antenna CC4, the frequency shifted upward by the largest rate, followed by CC3, CC2, and CC1. Remarkable is that matching for all crumpling cases improved at the detuned frequency; see Figure 1.8.

1.3.4 Antenna Efficiency Evaluation

Generally, since it is the easiest, the efficiency of an antenna is evaluated using numerical simulators. However, there are several methods for experimental efficiency assessment. One of the more popular methods is using a reverberation chamber (RC). Thanks to the larger size of such a chamber, a real human volunteer can be located inside the chamber while the AUT is placed on the body, either in a flat or a bent position. A reverberation chamber is typically constructed from galvanized steel and many are fit with two principle sets of mechanical stirring paddles. They are configured to be able to stir fields orthogonally throughout the volume of the chamber. One set can be mounted about a central rotational shaft from the floor to the roof and

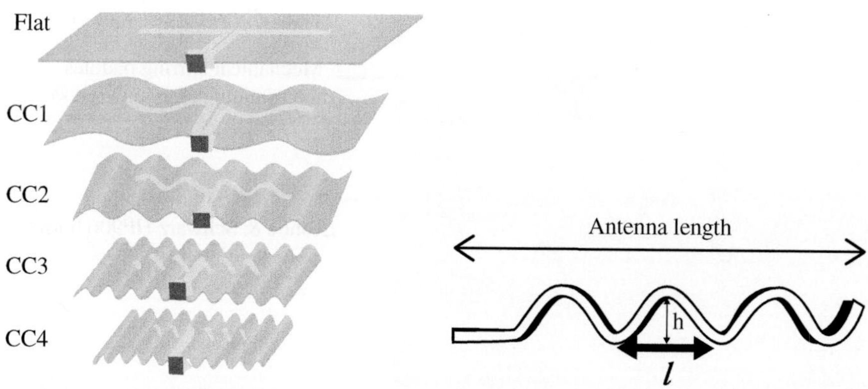

FIGURE 1.7 Modeling crumpling on a sample dipole antenna. *Source:* Elias et al. 2013 [45]. Reproduced with permission of IEEE.

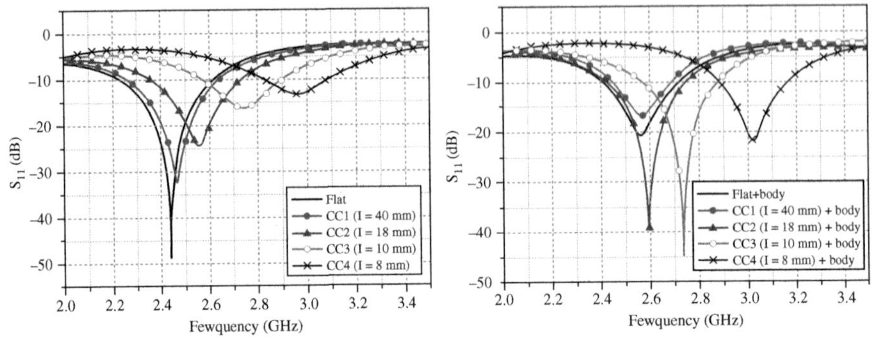

FIGURE 1.8 The effect of crumpling in free space and on the body for the case of a dipole. *Source:* Elias et al. 2013 [45]. Reproduced with permission of IEEE.

situated toward one corner of the chamber, the other set can be mounted about a rotational shaft from the front to the back wall at ceiling height. Figure 1.9 illustrates this chamber configuration. Further, as a representative example, the RC at the University of Liverpool as used in the investigations of [46–48] is described.

The physical interpretation of each stirring sequence can be briefly summarized as follows [49–51]:

(a) Mechanical stirring is the rotation or movement of electrically large paddles situated inside the chamber which change the boundary conditions of the electromagnetic field inside upon each increment.

(b) Polarization stirring requires the variation of the source (transmit) polarization to encompass both vertical and horizontal linear polarizations (in this case). It is required to ensure that no bias exists in the amount of TM and TE modes which are excited. This means that the average power level in the

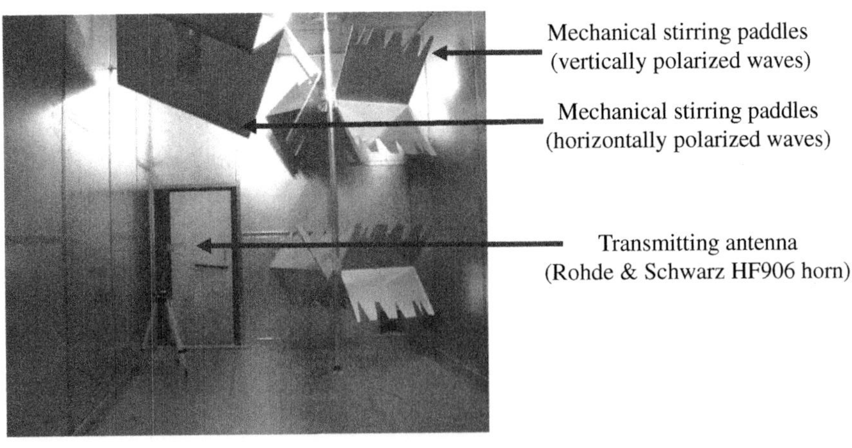

FIGURE 1.9 The RC at the University of Liverpool dimensioned at 3.6 m (width), 5.8 m (length), and 4 m (height). *Source:* Boyes et al. 2012 [46], Boyes et al. 2012 [47], and Boyes et al. 2013 [48]. Reproduced with permission of IEEE.

TABLE 1.2 RC measurement parameters at the University of Liverpool [46–48].

Parameter	Description
Frequency range	2–7 GHz
Frequency points	1601
Stirring sequences	5° mechanical stirring 2 × (orthogonal) polarization stirring 5 × position stirring 100 MHz frequency stirring
No. of measured samples	710 per frequency sample
Transmitting antenna (fixed)	Rohde & Schwarz HF906 horn
Reference antenna	Satimo SH2000 horn

measurements can be more accurately formed irrespective of the receive antennas' polarization.

(c) Position stirring is the physical movement of the AUT to various locations inside the chamber, movement distances are typically of the order of $\lambda/2$ in order to create independent samples.

(d) Frequency or electronic stirring is performed by taking a further average to effectively 'smooth' any measured results. This is usually applied in the post processing stage.

The measurement parameters used in the investigation are detailed in Table 1.2. The stirring sequences detailed in the table have been configured because

(a) a large number of measured samples (a percentage of which should be statistically independent) are required to keep the uncertainty at an acceptable level;

(b) the sequences have to be organized in such a way that the human subject does not have to spend prolonged amounts of time in the chamber to complete each measured sequence (i.e., the subject could take regular breaks if required).

All these stirring sequences (from a field perspective) are used in order to create a statistically isotropic and homogeneous field distribution in the cavity as long as the test location is situated approximately $\lambda/2$ from the walls [46–48]. The distance from the walls advocates the establishment of a "calibrated area" in which the statistically isotropic and homogeneous field distribution should be demonstrated.

As with any efficiency measurement in the RC, a reference measurement is required for calibration purposes. This was performed using a reference antenna with known efficiency values and with the preferred unidirectional pattern. Throughout the reference measurements, a human subject was located inside the chamber with the AUT located on the chest via a rigid cable terminated with a 50 Ω load. The human subject in the chamber during the reference measurement is expected to significantly load the chamber, and serves as the dominant contributor to the loss mechanisms that exist. Therefore any calibration without the human loading would simply not be accurate. Figure 1.10 illustrates the measurement setup used throughout the

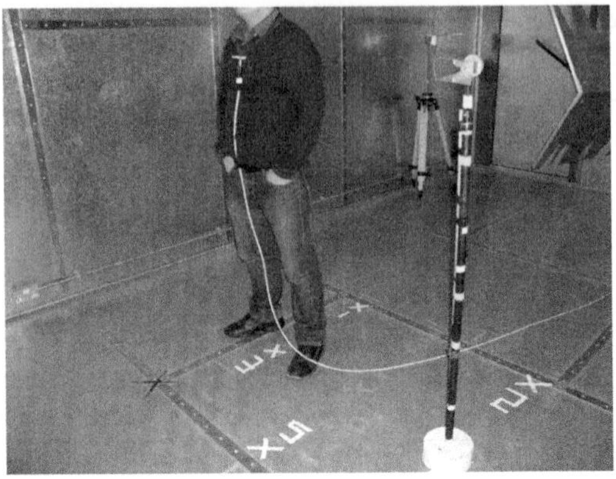

FIGURE 1.10 Measurement setup of Table 1.2. *Source:* Boyes et al. 2013 [48]. Reproduced with permission of IEEE.

efficiency investigation, including reference measurements. Standard Velcro straps have been employed to hold the antenna against the human test subject. The straps themselves have not been directly touching the antenna element, but have been only used to secure the attached cable against the human subject, sufficiently holding the antenna in place to prevent any movement. The contents inside the human subject's pockets were kept constant (without mobile phone, coins, etc.) to ensure consistency. The distance between the consecutive locations shown in Table 1.3 was such that each location was positioned further than the correlation distance (of about $\lambda/2$) to create independent samples. Further, throughout all position stirring locations (for both the reference and AUT measurements), the separation between the human subject and the reference antenna was always kept a prescribed distance apart ($\gg\lambda/2$) to avoid any coupling issues that could potentially corrupt any measurement results.

As defined by the IEEE, the radiation efficiency ($\eta_{\rm rad}$) is the "Ratio of total radiated power to the net power accepted by the antenna at its terminals." Mathematically, when measured in an RC, the radiation efficiency can be deduced from (1).

$$\eta_{\rm rad} = \left\{ \frac{\langle |S_{21\rm AUT}|^2 \rangle}{\langle |S_{21\rm REF}|^2 \rangle} \times \frac{(1-(S_{22\rm REF})^2)}{(1-(S_{22\rm AUT})^2)} \right\} \times \eta_{\rm REF} \tag{1.5}$$

TABLE 1.3 Position stirring locations in the RC with a minimum distance of 8 λ (at 2 GHz) from walls or other objects [48].

Position	Location
1	1.2 m from left wall
2	1.15 m from right wall, 1.1 m from stirrers
3	1.2 m from left wall, 0.9 m from position 1
4	1.15 m from right wall, 0.9 m from position 2
5	1.8 m from left wall, 1.2 m from back wall, 1 m diagonally from positions 3 and 4

where $\langle \rangle$ = average of the scattering parameter over the total number of measured samples (defined in Table 1.2), S_{21} = transmission coefficient, η_{REF} = known reference efficiency. One of the ports was always connected to the transmitting antenna, while another port was connected to the receiving antenna. The reflection coefficients, denoted as S_{22} in (1.5) are not signified as being an average because these parameters were acquired in an anechoic chamber.

Total efficiency (η_{tot}) is defined as the *"Ratio of total radiated power to the power incident on the antenna port."* Mathematically, this is a product of the radiation efficiency and mismatch efficiency, and can be deduced from (1.6).

$$\eta_{tot} = \eta_{rad} \times (1 - (S_{22AUT})^2) \tag{1.6}$$

Note that the reflection coefficient quantities were acquired in an anechoic chamber, hence the omission of the averaging.

It is shown in [47] that efficiency measurements performed using textile antennas on live human beings can be performed in an accurate and controlled manner, with a repeatability sometimes being as close as 2%. The magnitude of on-body losses for a given textile antenna with a small ground plane is seen in this case to be a function of the material properties of that antenna.

1.3.5 Specific Absorption Rate Evaluation

Due to the often very costly measurements caused by the scarcity of certified equipment, the most crucial parameter often investigated with the aid of simulators is the on-body operational safety level, defined in terms of SAR. SAR characterizes the absorption of the radiated electromagnetic energy by the human tissues, and is defined as follows [24]:

$$\text{SAR} = \frac{\sigma \cdot E^2}{\rho} \tag{1.7}$$

where σ (S/m) is the conductivity of the biological material, ρ (kg/m^3) is the material density, and E (V/m) is the electric field strength.

SAR simulations are typically performed in a numerical simulator using a detailed human body model [52]. A very popular model is the Hugo voxel model, which is normally truncated to reduce simulation time. Consistent with practical applications, the antennas are placed at a specific distance from this voxel model to emulate realistic worn placements. Depending on the model's resolution and truncation size, each simulation can easily generate as much as 18–20 million mesh cells, which means that the problem can only be solved using high-performance workstations or graphic processing units (GPUs). A typical simulation setup is shown in Figure 1.11.

An investigation providing information about measurement repeatability and simulation accuracy compared with measurements in various practical scenarios is presented in [55]. A good agreement is obtained, overall of the order of 17%, which is well below the published equipment uncertainty. Measurements are also

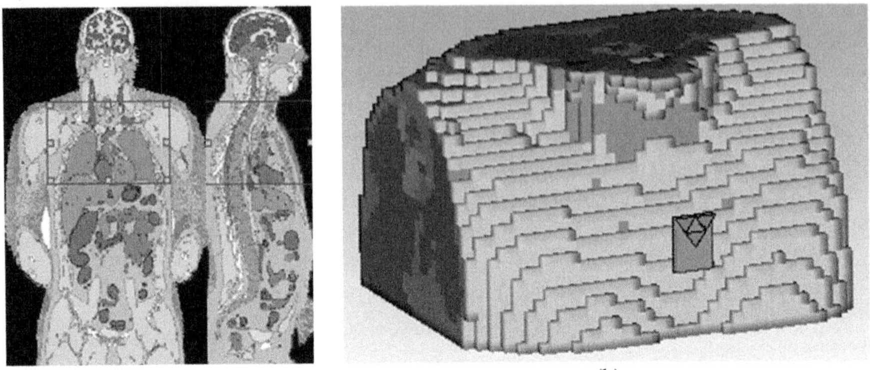

(a) (b)

FIGURE 1.11 SAR simulation setup: (a) Hugo voxel model and (b) AUT placement on the truncated model. *Source:* Soh et al. 2013 [53]. Reproduced with permission of IEEE.

proven to be reproducible with a maximum uncertainty of about 10%. On the other hand, a comparison between the SAR assessed for a flat antenna and several cases of crumpled antennas on a voxel model is presented in Figure 1.12. It is noticed that a higher level of bending results in a higher SAR due to the deeper penetration of the resulting radiation in general. It is worth to note that the resulting average SAR also depends on where the maximum point SAR is located when a detailed model is used in simulations. Lower average SAR can be produced when the antenna is deformed and the radiation penetrates deeper into the body. These lower values are typically located in the fat or muscle layers, which have a lower conductivity. On the contrary, a flat antenna may produce higher SARs, but mainly in the skin layer due to the skin layer's relatively higher conductivity [45].

A commercial DASY-4 system from Schmid & Partner Engineering AG (SPEAG) [56] shown in Figure 1.13 is often utilized to perform SAR measurements. It consists of a measurement server, robot controller, computer, near-field probe, probe alignment sensor, and a specific anthropomorphic mannequin (SAM). The SAM is filled using two tissue-emulating liquids for AUT evaluations at the frequencies of interest. For example, their properties for 2.45 and 5.2 GHz measurements are $\varepsilon_{r2.45} = 39.2$, $\sigma_{2.45} = 1.8$ S/m and $\varepsilon_{r5.2} = 36.0$, $\sigma_{5.2} = 4.66$ S/m, respectively. Prior to AUT measurements, calibrations are performed near the SAM at both frequencies using standard SAR validation dipoles to ensure that the evaluated SAR values are within a specified level. For the cases of 2.45 and 5.2 GHz, they must be below $SAR_{2.45\ GHz} < 6.15$ W/kg and $SAR_{5.2\ GHz} < 5.93$ W/kg [56].

To initialize the measurement, the lineup to the AUT is first connected to a power meter to ascertain that a 100 mW CW signal is fed into the AUTs during subsequent evaluations. The AUT feed cable is secured onto the SAM to ensure a fixed measurement location and to avoid variations due to cable movements. AUTs are placed at a specified distance from the phantom to emulate a realistic practical worn spacing, for instance, a 5 mm thick fleece jacket and another 5 mm air gap modeling the jacket-to-body distance. This additional distance depends on the AUT on-body location and

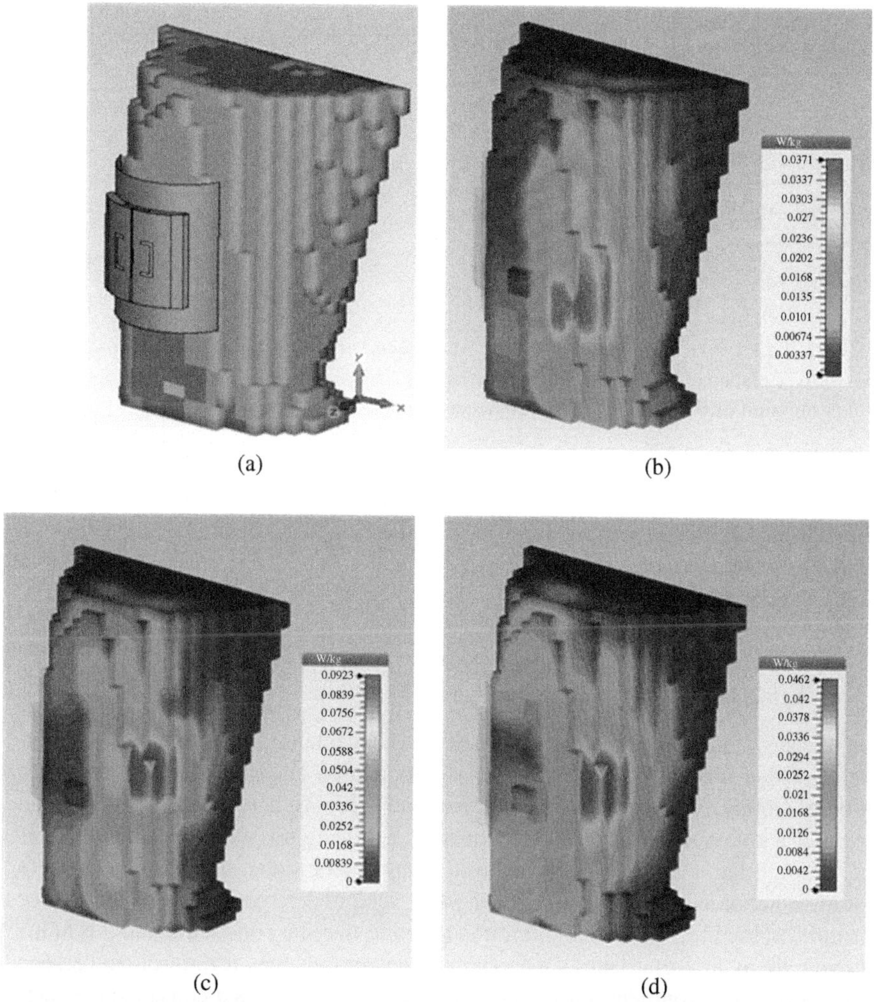

FIGURE 1.12 The SAR of a (a) bent dual-band antenna integrated with an artificial magnetic conductor (AMC) at (b) 2.45 GHz, (c) 5.2 GHz, and (d) 5.8 GHz evaluated on a voxel model. *Source:* Yan et al. 2015 [54]. Reproduced with permission of IEEE. For a color version of this figure, see the color plate section.

the jacket's fitting against the user's body. It is also important to note that the SARs for antennas placed on the body trunk must be evaluated at less than 20 mm distance from a flat phantom [24].

Higher SARs are generally observed when measuring antennas fabricated using conductive textiles with lower conductivities. Further, SAR values are dependent on the distance between the antenna and the phantom. However, these SAR values remain unchanged when the antenna-to-phantom distance approaches 20 mm, along with an increase of the measurement uncertainty. The antenna topology and the size of the ground plane that is shielding the user from the radiation are critical in ensuring

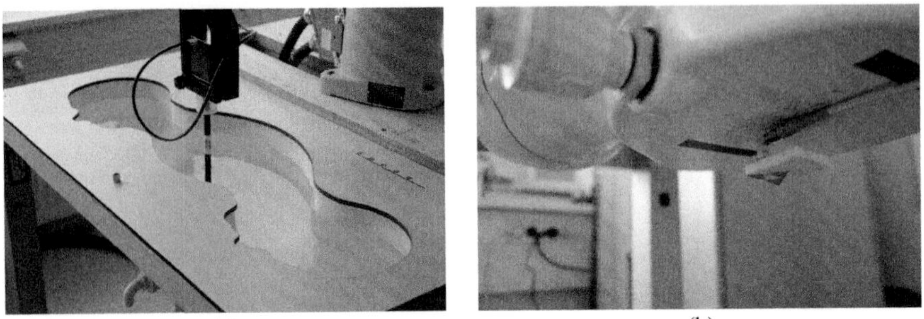

(a) (b)

FIGURE 1.13 SAR measurement setup: (a) top view of the SAM on the DASY-4 system and (b) AUT placement at the measurement location. *Source:* Soh et al. 2013 [53]. Reproduced with permission of IEEE.

that the antenna operates well below the safety limit. Most measured SARs are well below their respective simulated equivalent, indicating that a numerical solver can be used to estimate the worst-case on-body SAR, which is useful to enable a reasonable safety margin in the design phase.

1.3.6 Aging and Varying Environmental Conditions

Just like any conductors, conductive textiles used in antennas are subjected to degradations, either mechanical, electrical, or a combination of both. When implemented on daily clothing, the exposure to washing and cleaning has to be considered. On the other hand, garments for more specialized applications, such as search and rescue, fire-fighting, and the military, require specialized evaluation methods to ensure that the antenna is still fully functional when used under harsh circumstances. For example, in the case of a fire fighter, it is required that the conductive textiles in his garment are as fire retardant as the original apparel, whereas for search and rescue in a maritime environment, they need to be water resistant. As a consequence, dedicated methods to embed or protect the antennas need to be introduced so that they meet their requirements.

In [57], the authors performed an evaluation of the possible factors in the fabrication process that may degrade the usage life of the antenna. The epoxy curing process used and environmental factors may cause water to be trapped in the antenna substrate, resulting in resonant frequency and bandwidth changes. Besides that, textile antennas also experience a decrease in conductivity due to oxidation. To assess both factors, the reflection coefficients of the antennas were measured in a predefined time schedule: just after the time of fabrication, and after a few days, a month, and 2 months. Two fabrication process types were also evaluated: Flectron electrotextile was either stitched and glued, or secured onto the substrate using adhesive sheets. In both cases, it was noticed that a slight resonant frequency shift occurred, without compromising however, the operation in the intended design band, see Figure 1.14.

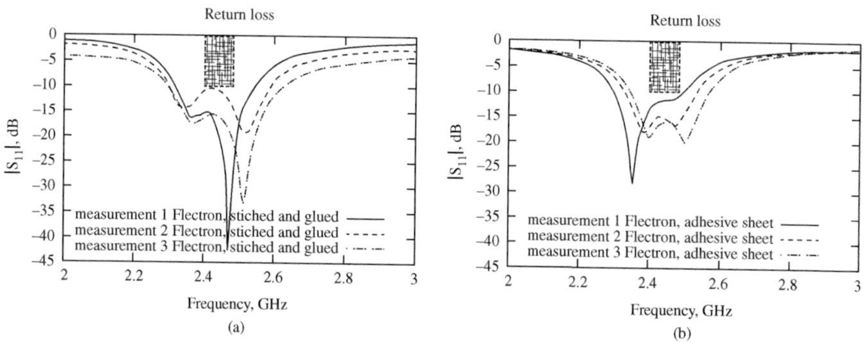

FIGURE 1.14 Reflection coefficients of the Flectron prototypes evaluated over time: (a) prototype fabricated using glueing and stitching, and (b) prototype with adhesive sheet. *Source:* Tronquo et al. 2006 [57]. Reproduced with permission of IEEE.

Several other investigations were also performed to evaluate the effect of different fabrication steps. In [58], the accuracy of manually fabricating a 0.2 mm gap in a co-planar waveguide (CPW) line is investigated. For the first prototype, the CPW line-radiator structure and the ground plane structure are cut individually before being secured onto the fleece. The best effort is put in accurately forming the required 0.2 mm gap by means of manual measurement. The alternative method is to cut an overall CPW–radiator–ground plane structure and securing it first onto the substrate. The gap between the feed line and the ground plane is then cut away after fastening to create the CPW line. The fabrication steps used for the second antenna prototype resulted in a better performance in terms of similarity with simulations.

The investigation in [42] assesses the effects of moisture and freezing a wearable antenna, in addition to the usual bending and on-body coupling investigations. The antenna is made of copper tape and fleece as the conductor and substrate, respectively. The influence of moisture was quantified as the percentage of water in the total mass of the antenna, which was measured using a 0.01 g technical balance. Water was applied to the antenna using a sprinkler and the antenna was gradually dried using a hairdryer. The reflection coefficient was measured after each minute of drying, and it was found that a relative humidity of 10% was sufficient to considerably shift the resonant frequency downward. As a consequence, it is suggested that such antennas need to be water-proofed by using coating for extreme weather conditions. On the other hand, antenna freezing can be expected when such antennas are to be operated by search and rescue teams. In the related investigation, the antenna was first covered using nylon superstrate to avoid moisture. The antenna was then immersed into water and frozen, prior to reflection coefficient measurements. Afterward, it was immersed into a pot of boiling water for 2 minutes before being measured for the second time. It was noticed that the water still penetrated into the substrate despite having the nylon superstrate, thus rendering the antenna unusable.

A comprehensive evaluation of non-conductive textiles is performed in [59]. First, a transmission line method is used to extract the measured scattering parameters of two copper microstrip lines. They are made of a stack of Cordura sheets attached using a pressure sensitive adhesive. The temperature is varied from −20°C to 40°C and the relative humidity from 10% to 95%. Similarly as in the previous investigation, it is noticed that the observed permittivity and loss tangent behavior follow the changes in relative humidity, which then relates to the absorbed water in the fabric mass. Another aspect of assessment in this work is related to a protective cover for the antenna. The investigation found that surface contamination may be a cause of failure of fabric water-proofing. The two possible factors are explained as follows. First, if the water repellence is chemically added to the fabric, a strong salt solution can wash the added layer away. Secondly, user comfort includes the breathability factor; thus, small holes are introduced to enable water evaporation. These holes are impenetrable by water due to the surface tension. However, salt lowers the surface tension of water rendering salt water to be penetrable into many waterproof fabrics.

In [60], the effect of several washing and drying cycles based on the ISO 6330:2000 and ISO 139 standards, respectively, was investigated for an inset-fed microstrip patch antenna. The washing temperature was set at 30 ± 3°C, and the samples were placed in a room at a temperature of 23 ± 2°C and humidity of 50 ± 4 % upon drying. The textile was considered dried when weighing performed in intervals of 2 hours indicated no change in mass greater than 0.25%. Measurements of the antenna made using Flectron and Electrodag and covered with a breathable thermoplastic polyurethane (TPU) layer were performed after one, three, and six washing cycles. With the increase in the number of washing cycles, the antenna based on Flectron was observed to degrade in terms of radiation efficiency and matching, thus resulting in resonance shifting. This is believed to be due to mechanical delamination and detachment from the substrate. Besides that, the Flectron layer itself suffered deterioration in terms of resistivity to a value of about 1.2 Ω/sq, resulting in the observed efficiency decrease despite still being within acceptable limits. On the contrary, the antennas based on the Electrodag displayed a more stable performance in terms of matching and efficiency.

1.4 CONCLUSION

In this chapter, the design, development, and evaluation procedures for textile antennas have been presented. Since these materials are either newly introduced or have been traditionally used for other purposes, for example, EMI shielding or grounding, one of the important and yet challenging aspects in the field of textile antennas is to properly characterize the electrical properties of these materials at the intended frequencies. Basically, two types of materials to realize textile antennas are discussed, conductive and non-conductive textiles.

First, a simple and pragmatic, but at the same time, efficient and result-oriented procedure to design and fabricate textile antennas using commercially available textiles and flexible conductive layers has been proposed and described. This simple procedure utilizes manual fabrication tools. The accuracy of the produced prototypes, verified across the various antennas operating in distinct frequency bands is satisfactory, with an accuracy of 0.2 mm. A general top-level design guideline and its limitations are also presented. Since a classical soldering technique, as used in fabricating conventional antennas, might not always be applicable for conductive textiles, there is a need for alternatives such as the use of conductive epoxy.

Next, procedures to verify the performance of flexible/textile antennas have been presented. Most of the procedures are adapted versions of the evaluation procedures used for conventional, rigid antennas. In contrast to the general on-head or in-hand evaluations performed for handset antennas, which are already widely available in the literature, evaluation of textile antennas has been mainly focusing on the upper human torso or the arms. Additional tools to characterize and properly validate these antennas and measures to keep the performance stable have been outlined in this chapter. Since the major practical issues when using textile antennas are the bandwidth detuning and power absorption when operating in the vicinity of a user, a crucial aspect is the use of on-body evaluation methods. Antenna performance is assessed in the proximity of either a human user or a phantom and must lead to sufficient design information to guarantee this performance in the simulation-based design process, where the proximity of the human body also has to be taken into account. Minor steps such as the use of a fleece jacket and spacers, properly applied, aim at providing a better measurement accuracy and consistency when evaluating textile antennas. The typical small antenna-to-body distance also causes safety concerns. This chapter has also addressed this issue, with a fundamental study on SAR.

REFERENCES

[1] P. S. Hall and Y. Hao, *Antennas and Propagation for Body-Centric Wireless Communications*, 2nd ed. Artech House, Norwood, MA, 2012.

[2] P. J. Soh, G. A. E. Vandenbosch, X. Chen, P.-S. Kildal, S. L. Ooi, and H. Aliakbarian, "Wearable Textile Antenna Efficiency Characterization using a Reverberation Chamber," in IEEE International Symposium on Antennas and Propagation, New York, July 3–8, 2011, pp. 810–813.

[3] M. A. R. Osman, M. K. A. Rahim, M. Azfar, N. A. Samsuri, F. Zubir, and K. Kamardin, "Design, implementation and performance of ultra-wideband textile antenna," *Prog. Electromagnetics Res. B*, vol. 27, pp. 307–325, 2011.

[4] C. Hertleer, H. Rogier, L. Vallozzi, and L. V. Langenhove, "A textile antenna for off-body communication integrated into protective clothing for firefighters," *IEEE Trans. Antennas Propag.*, vol. 57, no. 4, pp. 919–925, April 2009.

[5] E. C. Lee, P. J. Soh, N. B. M. Hashim, G. A. E. Vandenbosch, H. Mirza, I. Adam, and S. L. Ooi, "Design of a Flexible Minkowski-like Pre-fractal (MLPF) Antenna with Different Ground Planes for VHF LMR," in 2011 International Workshop on Antenna Technology (iWAT), Hong Kong, March 7–9, 2011, pp. 298–301.

[6] E. C. Lee, P. J. Soh, N. B. M. Hashim, G. A. E. Vandenbosch, V. Volski, I. Adam, and M. Z. A. A. Aziz, "Design and Fabrication of a Flexible Minkowski Fractal Antenna for VHF Applications," in Proceedings of the 5th European Conference on Antennas and Propagation, Rome, April 11–15, 2011, pp. 521–524.

[7] P. Salonen and L. Hurme, "A Novel Fabric WLAN Antenna for Wearable Applications," in IEEE Antennas and Propagation Society International Symposium, 2003, Columbus, OH, June 22–27, 2003, pp. 700–703.

[8] "Introduction to the Cospas-Sarsat System," International Cospas-Sarsat Programme, 2009.

[9] S. Sankaralingam and B. Gupta, "Use of electro-textiles for development of wibro antennas," *Prog. Electromagnetics Res. C*, vol. 16, pp. 183–193, 2010.

[10] N. H. M. Rais, P. J. Soh, F. Malek, R. B. Ahmad, N. B. M. Hashim, and P. S. Hall, "A Review of Wearable Antennas," in Loughborough Antennas and Propagation Conference (LAPC), Loughborough, November 16–17, 2009, pp. 225–228.

[11] P. J. Soh, G. A. E. Vandenbosch, S. L. Ooi, and H. M. R. Nurul, "Characterization of a plain broadband textile PIFA," *Radioeng.*, vol. 20, no. 4, pp. 718–725, December 2011.

[12] P. J. Soh, G. A. E. Vandenbosch, S. L. Ooi, and N. H. M. Rais, "Design of a broadband all-textile slotted PIFA," *IEEE Trans. Antennas Propag.*, vol. 60, no. 1, pp. 379–384, January 2012.

[13] P. J. Soh, G. A. E. Vandenbosch, V. Volski, and H. M. R. Nurul, "Characterization of a Simple Broadband Textile Planar Inverted-F Antenna (PIFA) for On Body Communications," in International Conference on Applied Electromagnetics and Communication, Dubrovnik, Croatia, September 20–23, 2010, pp. 1–4.

[14] C. Hertleer, A. Tronquo, H. Rogier, L. Vallozzi, and L. V. Langenhove, "Aperture-coupled patch antenna for integration into wearable textile systems," *IEEE Antennas Wireless Propag. Lett.*, vol. 6, pp. 392–395, 2007.

[15] "Specification Sheet—ShieldIt Super," LessEMF Inc., 2015.

[16] "Specification Sheet—Pure Copper Polyester Taffeta Fabric," LessEMF Inc., 2015.

[17] "Specification Sheet—Felt Sheet 6 mm," RS Components Inc., 2015.

[18] T. F. Kennedy, P. W. Fink, A. W. Chu, N. J. Champagne, G. Y. Lin, and M. A. Khayat, "Body-worn E-textile antennas: the good, the low-mass, and the conformal," *IEEE Trans. Antennas Propag.*, vol. 57, no. 4, pp. 910–918, April 2009.

[19] I. Locher, M. Klemm, T. Kirstein, and G. Troster, "Design and characterization of purely textile patch antennas," *IEEE Trans. Adv. Packag.*, vol. 29, no. 4, pp. 777–788, November 2006.

[20] "Polar Fleece," Wikipedia, 2015.

[21] "Specification Sheet—Fleece Jacket," RS Components Inc., 2015.

[22] P. Salonen, Y. Rahmat-Samii, and M. Kivikoski, "Wearable Antennas in the Vicinity of Human Body," in IEEE Antennas and Propagation Society International Symposium, June 20–25, 2004, pp. 467–470.

[23] A. Tronquo, H. Rogier, C. Hertleer, and L. V. Langenhove, "Robust planar textile antenna for wireless body LANs operating in 2.45 GHz ISM band," *Electron. Lett.*, vol. 42, no. 3, pp. 142–143, February 2006.

[24] "Human Exposure to Radio Frequency Fields from Hand-held and Body-mounted Wireless Communication Devices–Human Models, Instrumentation and Procedures IEC 62209 Standard," The International Electrotechnical Commission (IEC), 2010.

[25] T. Weiland, "A discretization method for the solution of Maxwell's equations for six-component fields," *AEÜ - Int. J. Electron. Commun.*, vol. 31, no. 3, p. 116, 1977.

[26] T. Weiland, "Time domain electromagnetic field computation with finite difference methods," *Int. J. Numer. Model., Electron. Netw. Devices Fields*, vol. 9, no. 4, pp. 295–319, 1996.

[27] K. Yee, "Numerical solution of initial boundary value problems involving Maxwell's equations in isotropic media," *IEEE Trans. Antennas Propag.*, vol. 14, no. 3, pp. 302–307, May 1996.

[28] A. Taflove and S. C. Hagness, *Computational Electrodynamics: The Finite-Difference Time-Domain Method*, 3rd ed. Artech House, Norwood, MA, 2005.

[29] F. Zheng, Z. Chen, and J. Zhang, "A finite-difference time-domain method without the Courant stability conditions," *IEEE Microw. Guid. Wave Lett.*, vol. 9, no. 11, pp. 441–443, August 1999.

[30] E. Li, I. Ahmed, and R. Vahldieck, "Numerical dispersion analysis with an improved LOD–FDTD method," *IEEE Microw. Guided Wave Lett.*, vol. 17, no. 5, pp. 319–321, April 2007.

[31] P. B. Johns and R. L. Beurle, "Numerical solution of 2-dimensional scattering problems using a transmission-line matrix," *Proc. Inst. Electrical Eng.*, vol. 11, no. 9, pp. 1203–1208, September 1971.

[32] C. Christopoulos, *The Transmission-Line Modeling Method: TLM*. Wiley-IEEE Press, Piscataway, NJ, 1996.

[33] J.-M. Jin, *The Finite Element Method in Electromagnetics*, 2nd ed. Wiley-IEEE Press, Piscataway, NJ, 2002.

[34] A. Bossavit, *Computational Electromagnetism: Variational Formulations, Complementarity, Edge Elements*. Academic Press, New York, 1997.

[35] R. F. Harrington, *Field Computation by Moment Methods*. Wiley-IEEE Press, Piscataway, NJ, 1993.

[36] C. A. Balanis, *Antenna Theory Analysis and Design*, 3rd ed. John Wiley & Sons, New York, 2005.

[37] N. Chahat, M. Zhadobov, R. Sauleau, and K. Ito, "A compact UWB antenna for on-body applications," *IEEE Trans. Antennas Propag.*, vol. 59, no. 4, pp. 1123–1131, April 2011.

[38] P. Samal, P. J. Soh, and G. A. E. Vandenbosch, "UWB all-textile antenna with full ground plane for off-body WBAN communications," *IEEE Trans. Antennas Propag.*, vol. 62, no. 1, pp. 102–108, January 2014.

[39] Z. Wang, L. Z. Lee, D. Psychoudakis, and J. L. Volakis, "Embroidered multiband body-worn antenna for GSM/PCS/WLAN communications," *IEEE Trans. Antennas Propag.*, vol. 62, no. 6, pp. 3321–3329, March 2014.

[40] P. J. Soh, S. J. Boyes, G. A. E. Vandenbosch, Y. Huang, and S. L. Ooi, "On-body characterization of dual-band all-textile PIFAs," *Prog. Electromagn. Res.*, vol. 129, pp. 517–539, July 2012.

[41] Q. Bai and R. Langley, "Wearable EBG Antenna Bending and Crumpling," in Loughborough Antennas and Propagation Conference (LAPC), Loughborough, November 16–17, 2009, pp. 201–204.

[42] B. Ivsic, G. Golemac, and D. Bonefacic, "Performance of Wearable Antenna Exposed to Adverse Environmental Conditions," in 21st International Conference on Applied Electromagnetics and Communication (ICECom), Dubrovnik, Croatia, October 14–16, 2013, pp. 1–5.

[43] P. J. Soh, B. van den Bergh, H. Xu, H. Aliakbarian, S. Farsi, P. Samal, G. A. E. Vandenbosch, D. M. M.P. Schreurs, and B. K. J. C. Nauwelaers, "A smart wearable textile antenna system for biomedical telemetry system," *IEEE Trans. Microw. Theory Tech.*, vol. 61, no. 5, pp. 2253–2261, May 2013.

[44] J. Choi, J. Tak, and S. Lee, "All-textile higher order mode circular patch antenna for on-body to on-body communications," *IET Microw. Antennas Propag.*, vol. 9, no. 6, pp. 576–584, April 2015.

[45] N. A. Elias, N. A. Samsuri, M. K. A. Rahim, and N. Othman, "Investigation of Crumpling Effects on EM Absorption and Antenna Performance at 2.4 GHz," in 2013 IEEE International RF and Microwave Conference (RFM), Penang, Malaysia, December 9–11, 2013, pp. 395–399.

[46] S. Boyes, Y. Huang, N. Khiabani, P. J. Soh, and G. A. E. Vandenbosch, "Repeatability and Uncertainty Evaluations of On-body Textile Antenna Efficiency Measurements in a Reverberation Chamber," in 2012 Loughborough Antennas and Propagation Conference (LAPC), Loughborough, UK, November 12–13, 2012, pp. 1–5.

[47] S. J. Boyes, P. J. Soh, Y. Huang, G. A. E. Vandenbosch, and N. Khiabani, "On-body performance of dual-band textile antennas," *IET Microw. Antennas Propag.*, vol. 6, no. 15, pp. 1696–1703, December 2012.

[48] S. J. Boyes, P. J. Soh, Y. Huang, G. A. E. Vandenbosch, and N. Khiabani, "Measurement and performance of textile antenna efficiency on a human body in a reverberation chamber," *IEEE Trans. Antennas Propag.*, vol. 6, no. 2, pp. 871–881, February 2013.

[49] K. Rosengren and P.-S. Kildal, "Study of distributions of modes and plane waves in reverberation chambers for the characterization of antennas in a multipath environment," *Microw. Opt. Technol. Lett.*, vol. 30, no. 6, pp. 386–391, September 2001.

[50] P.-S. Kildal and C. Carlsson, "Detection of a polarization imbalance in reverberation chambers and how to remove it by polarization stirring when measuring antenna efficiencies," *Microw. Opt. Technol. Lett.*, vol. 34, no. 2, pp. 145–149, July 2002.

[51] P.-S. Kildal, X. Chen, C. Orlenius, M. Franzen, and C. S. L. Patane, "Characterization of reverberation chambers for OTA measurements of wireless devices: physical formulations of channel matrix and new uncertainty formula," *IEEE Trans. Antennas Propag.*, vol. 60, no. 8, pp. 3875–3891, August 2012.

[52] "CST Microwave Studio," Darmstadt, Germany: CST-Computer Simulation Technology AG, 2015.

[53] P. J. Soh, G. A. E. Vandenbosch, F. H. Wee, A. van den Bosch, M. Martinez-Vazquez, and D. Schreurs, "Specific Absorption Rate (SAR) Evaluation of Biomedical Telemetry Textile Antennas," in 2013 IEEE MTT-S International Microwave Symposium Digest (IMS), Seattle, WA, June 2–7, 2013, pp. 1–4.

[54] S. Yan, P. J. Soh, and G. A. E. Vandenbosch, "Dual-band textile MIMO antenna based on substrate integrated waveguide (SIW) technology," *IEEE Trans. Antennas Propag.*, vol. 63, no. 11, pp. 4640–4647, November 2015.

[55] P. J. Soh, G. A. E. Vandenbosch, F. H. Wee, A. van den Bosch, M. Martinez-Vazquez, and D. Schreurs, "Specific absorption rate (SAR) evaluation of textile antennas," *IEEE Antennas Propag. Mag.*, vol. 57, no. 2, pp. 229–240, April 2015.

[56] "DASY-4 V4.7 System Handbook," Schmid & Partner Engineering AG, 2008.

[57] A. Tronquo, H. Rogier, C. Hertleer, and L. V. Langenhove, "Applying Textile Materials for the Design of Antennas for Wireless Body Area Networks," in First European Conference on Antennas and Propagation (EuCAP), Nice, France, 2006, pp. 1–5.

[58] P. J. Soh, G. A. E. Vandenbosch, and J. Higuera-Oro, "Design and Evaluation of Flexible CPW-fed Ultra Wideband (UWB) Textile Antennas," in 2011 IEEE International RF and Microwave Conference (RFM), Seremban, Negeri Sembilan, December 12–14, 2011, pp. 133–136.

[59] J. Lilja, P. Salonen, T. Kaija, and P. d. Maagt, "Design and manufacturing of robust textile antennas for harsh environments," *IEEE Trans. Antennas Propag.*, vol. 60, no. 9, pp. 4130–4140, September 2012.

[60] M. L. Scarpello, I. Kazani, C. Hertleer, H. Rogier, and D. V. Ginste, "Stability and efficiency of screen-printed wearable and washable antennas," *IEEE Antennas Wireless Propag. Lett.*, vol. 11, pp. 838–841, July 2012.

2

METAMATERIAL-ENABLED AND MICROWAVE CIRCUIT INTEGRATED WEARABLE ANTENNAS FOR OFF-BODY COMMUNICATIONS

Zhi Hao Jiang, Taiwei Yue, and Douglas H. Werner

Department of Electrical Engineering, Pennsylvania State University, University Park, PA, USA

2.1 INTRODUCTION

Throughout the first decade of the twenty-first century, the field of body area networks (BANs) has witnessed rapid advancements including the development of related network theory, component implementation, and low-power system architectures [1–4]. Due to these developments, wireless BAN systems are poised to become an enabling technology for numerous applications in a wide variety of fields, such as physical training, battlefield survival, health monitoring, emergency rescue, human tracking, and so on. Typically, the type of communication within a BAN can be classified into three modes, including on-body, in-body (or sometimes called through-body), and off-body communications. Several frequency bands have been previously utilized for research and commercialization of BAN communication systems, such as the 402–405 MHz Medical Implant Communication Services (MICS) band [5] which is primarily used for on-body and through-body communications, the 2.4–2.48 GHz Industrial, Scientific, and Medical (ISM) band [6] as well as the 3.1–10.76 GHz Ultra-Wide Band (UWB) range [7] which are primary used for off-body communication. More recently, a new Medical BAN (MBAN) band, which operates from 2.36 GHz to

Electromagnetics of Body-Area Networks: Antennas, Propagation, and RF Systems, First Edition.
Edited by Douglas H. Werner and Zhi Hao Jiang.

2.4 GHz, has been issued by the Federal Communication Commission (FCC) for its clean spectrum and low-interference sources [8].

Because wearable antennas operate in close proximity to the human body, the loading effect due to the lossy tissue makes the design of a high-performance antenna extremely challenging, especially when the antenna is also required to possess characteristics such as lightweight, low profile, and small form factor [9]. At the same time, the impact of such wearable antennas on human tissue also needs to be considered, usually characterized by the maximum allowed specific absorption rate (SAR). SAR is a standard measure often used to evaluate the absorption of electromagnetic power in human tissue. According to the specification provided by the FCC, SAR values must be no greater than 1.6 W/kg, averaged over 1 g of tissue [10]. So far, several configurations have been investigated for their suitability as wearable antennas, including the early works on vertical monopoles [9, 11, 12], planar microstrip monopoles [13–16], various inverted-F antennas [17–19], microstrip patch antennas [20], and more recent efforts on surface wave antennas [21], cavity-backed slot antennas [22,23], reflector patch radiators [24], and artificial magnetic conducting surface backed antennas [25–27]. The monopoles and inverted-F antennas are suitable for on-body communications but they protrude from the body and are therefore not conformal. The planar monopole antennas and planar inverted-F antennas have smaller footprint. However, they radiate a significant amount of energy directly into the human body below them due to their near-omnidirectional radiation patterns. Surface wave antennas promise to be much better candidates for on-body communications due to their lower profiles and vertically polarized radiated fields. For off-body communications, microstrip patch antennas and cavity-backed slot antennas have been considered because of their broadside radiation patterns, but they exhibit a narrow bandwidth and ground plane size dependent front-to-back (FB) ratios. Substrate-integrated waveguide cavities can reduce the profile (i.e., thickness) of slot antennas, but they still retain a large lateral size. More recently, isotropic artificial magnetic conducting surface ground planes were utilized to provide a high degree of isolation between the antenna and human tissue while maintaining a reasonably low overall profile. However, they still suffer from relatively large footprints, frequency shifts due to bending, and/or low FB ratios. There are also many innovative antenna designs that have been reported for wearable devices which operate at other frequency bands, such as the low-frequency MICS band [28–31] and the high-frequency UWB band [32–35], but these do not fall under the topic of this chapter. The focus here will be on highly efficient, low-profile, and small form-factor unidirectional antenna designs primarily for off-body communications.

In wearable systems, sensitive receivers along with low-power circuitry are often utilized which allow for a longer battery life [36, 37]. Due to the numerous existing wireless systems operating within the 1–6 GHz spectrum, interference and crosstalk may degrade the performance of these wearable receivers. Therefore, a bandpass filter (BPF) is necessary at the front-end to discriminate the desired signal from out-of-band extraneous signals [38, 39]. Conventionally, the BPF and the antenna are designed separately and then directly cascaded by sharing a common reference impedance or both are connected to a two-port matching circuit, resulting in increased device footprint and possibly degraded performance. A mutual synthesis method for designing the BPF and antenna as a single module, that is, a filtering antenna (sometimes also referred to as a filtenna), appears to be a more attractive solution [40]. Thus far, filtering antennas in various forms have been investigated by integrating different types of filter structures and antennas together, including frequency selective surface (FSS) coverings for monopoles [41] and horn antennas [42, 43], coupled rectangular and substrate-integrated waveguide cavity filters cascaded behind slot antennas [44–46], a planar coaxial collinear antenna [47], as well as planar monopole antennas combined with hairpin [48] and coupled-line filters [49]. These reported works show that co-designed filtennas indeed have smaller overall device form factor and superior band selectivity. However, these proposed filtenna structures are not suitable candidates for wearable applications due to their large footprints, high profile, and linear polarization. In addition, most of the demonstrated planar filtennas possess near-omnidirectional radiation patterns in the E- or H-plane that are undesirable for off-body communications and/or require fabrication of complicated three dimensional (3D) structures.

In this chapter, we present two new types of recently developed wearable antennas. First, transformational designs for lightweight conformal wearable antennas are presented which are highly efficient while, at the same time, having a compact footprint and a low profile in the MBAN band. These designs are enabled by placing a planar monopole antenna on top of an extremely truncated metasurface, which yields a front-to-back ratio in the radiation pattern that is remarkably high for its size and effectively isolates the antenna from the loading effects of the human body. The second part of the chapter focuses on compact, planar circularly polarized (CP) wearable filtennas with both narrow and wide operational bandwidths. The antenna design strategy, simulations, and experimental results, as well as on-body performance evaluations of these wearable antennas will be discussed.

2.2 A METASURFACE-ENABLED COMPACT WEARABLE ANTENNA

In this section, the design methodology and performance analysis of a truncated metasurface integrated antenna are presented. Moreover, the principles governing its unique radiation properties are discussed and verified by the uniform geometrical theory of diffraction (GTD). Bandwidth broadening techniques are also illustrated by introducing various interdigital capacitor loading schemes into the metasurface. Experiments were carried out to confirm the proposed metasurface-enabled antenna designs.

2.2.1 Antenna Design

The configuration of the integrated antenna is shown in Figure 2.1, which consists of two components, a planar monopole fed by a microstrip on the top and a highly truncated anisotropic metasurface backed by a ground plane on the bottom. The top and bottom sections are separated by a thin foam spacer. The metasurface is composed of

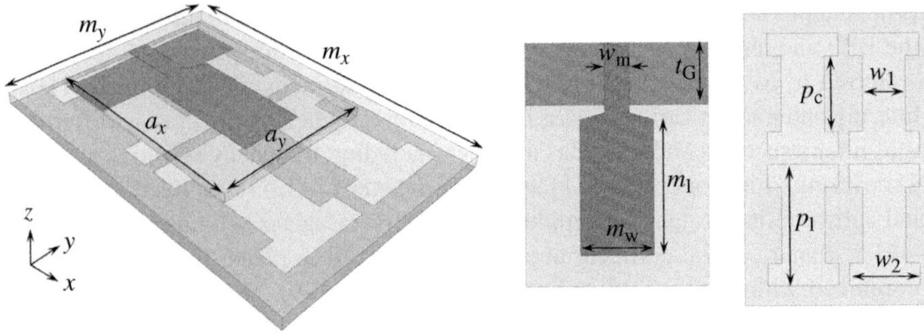

FIGURE 2.1 Configuration of the integrated antenna. The optimized dimensions are m_x = 62, m_y = 42, a_x = 39, a_y = 30, m_w = 12, m_l = 21.8, w_m = 4, t_G = 10, p_c = 16.86, p_l = 27.06, w_1 = 9.75, and w_2 = 16, all in millimeters. The substrate material is Rogers RO3003 ($\varepsilon_r = 3$, $\delta_{tan} = 0.0013$). The thickness values for the top and bottom layers are d_1 = 1.5 mm and d_2 = 2 mm. The foam spacer is 0.5 mm thick.

an array of two by two I-shaped patches [50]. The ground plane backed anisotropic metasurface was first designed as a doubly infinite periodic structure with a zero reflection phase at around 2.5 GHz, which is slightly higher than the targeted band. This enables the response of the metasurface to behave inductively in the MBAN band. A planar monopole was then positioned at a distance d above the metasurface. The coupling between the metasurface and the monopole is controlled by the position of the monopole. The geometrical dimensions of the monopole, as well as its lateral position with respect to the underlying metasurface, were tuned to obtain the optimal impedance match. The final integrated antenna has a form factor of $0.028\lambda_0$ by $0.15\lambda_0^2$, which is more compact than many previously reported wearable antenna designs.

The simulated impedance performance of the monopole and the integrated antenna in free space are shown in Figure 2.2a. At the targeted MBAN band, a poor impedance match is observed for the monopole case, which exhibits an S_{11} higher than −6 dB. In contrast, the integrated antenna achieves an S_{11} lower than −10 dB from 2.32 GHz to 2.43 GHz, that is, its bandwidth is around 4.7%. Because the monopole alone operates at a frequency below its fundamental mode, its input impedance is capacitive. When it is loaded by the inductive impedance surface which stores more magnetic energy, the reactance of the integrated antenna can be suppressed, thus achieving a good impedance match. This also corroborates a previous theoretical study on this topic [51]. The simulated gain patterns of both the monopole and the integrated antenna are displayed in Figure 2.2b. The monopole alone in free space has a dipole like pattern in its E-plane and an omnidirectional pattern in its H-plane, with a gain of about 2 dBi. In contrast, the integrated antenna has the majority of its energy radiated toward the upper half space with a half-power beamwidth (HPBW) of about 81° and 102° in the E-plane and H-plane, respectively, with a gain of more than 6 dBi and an FB ratio greater than 28 dB. The weak backlobes indicate that very little energy would be radiated into the tissue when this antenna is placed directly on the human body. Such a

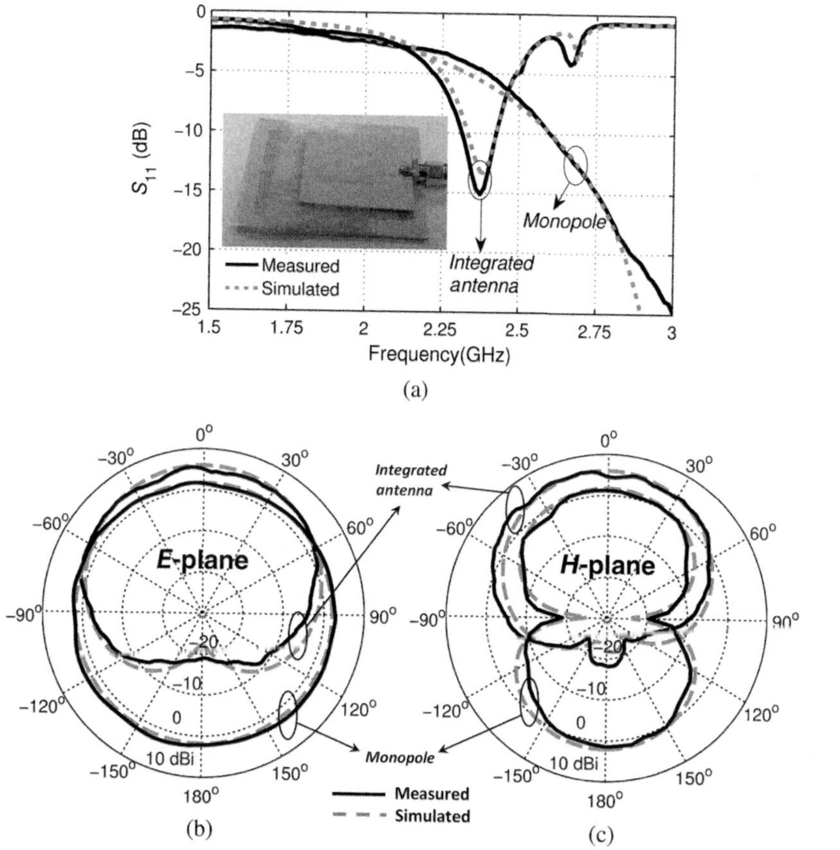

FIGURE 2.2 Simulated and measured (a) S_{11}, (b) E-plane, and (c) H-plane gain patterns of the monopole alone and the integrated antenna. The inset of (a) shows the fabricated integrated antenna.

property helps to minimize the SAR value and also makes the antenna more robust to human body loading, both of which are highly desirable characteristics for wearable applications. This value is much higher than any previously demonstrated antennas with artificial ground plane backings, especially when considering its remarkably compact footprint of $0.15\lambda_0^2$ [25, 26, 52, 53].

2.2.2 Radiation Mechanism

At resonance, the electric fields are mainly concentrated on the periphery of the planar monopole and in the capacitive gaps between the I-shaped patches along the x direction of the metasurface, as illustrated in Figure 2.3a. The waves are radiated by the slots between the edges of the I-shaped resonators and the ground plane, that is,

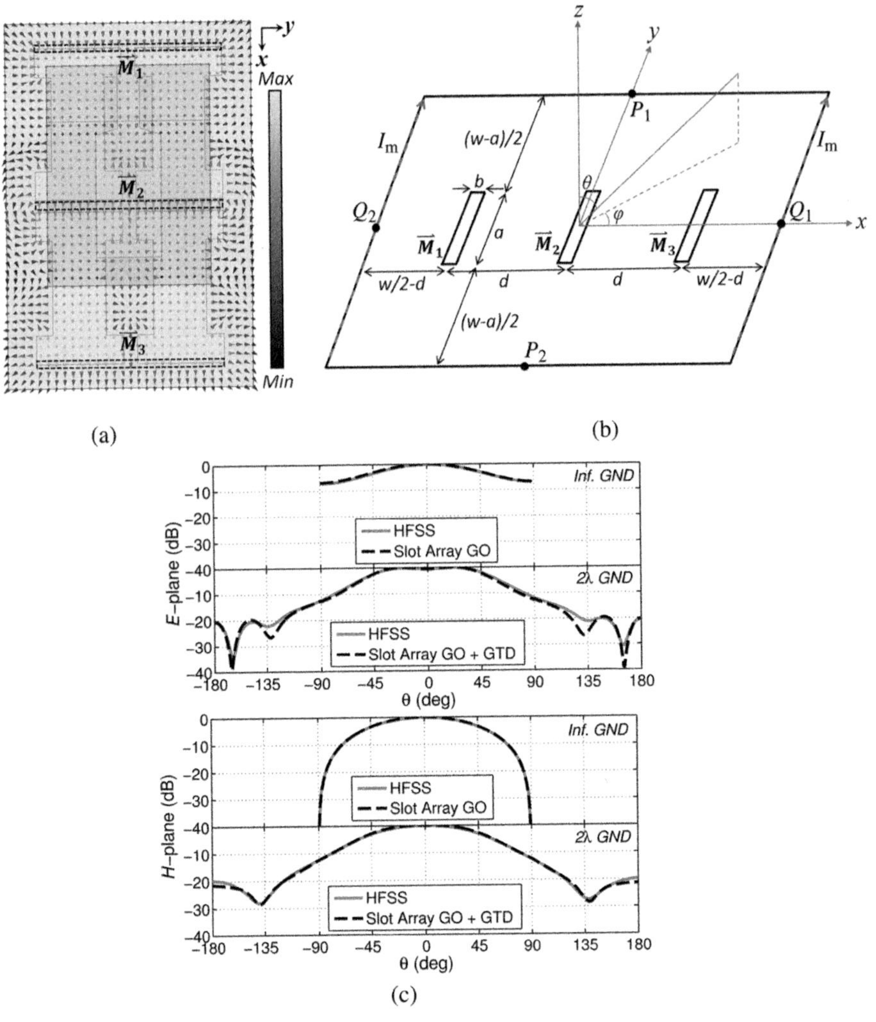

FIGURE 2.3 (a) Simulated electric field distribution of the integrated metasurface-enabled antenna at 2.38 GHz. (b) Array model consists of three magnetic current sources ($\vec{M}_1, \vec{M}_2, \vec{M}_3$) on an infinite- or finite-sized ground plane. (c) Normalized far-field patterns based on HFSS full-wave simulations of the integrated antenna compared to the three-element magnetic current source array on an infinite and a finite $2\lambda_0$ by $2\lambda_0$ ground plane in both the *E*-plane and the *H*-plane.

in a patch mode. In contrast to the previous designs where the metallic backed metasurface acts only as a high-impedance AMC reflector [25, 26, 52, 53], the finite-sized metasurface proposed in our design also acts as the primary radiator. More specifically, the antenna operates like a three-element slot array with amplitude tapering, which gives rise to a remarkably high FB ratio when considering the compactness of its overall footprint.

To validate the model, the radiation pattern was calculated for an array of three uniform equivalent magnetic current densities ($\vec{M}_1, \vec{M}_2, \vec{M}_3$) on both an infinite- and a finite-size ground plane (see Figure 2.3b). The central element is assumed to have twice the power as that of the magnetic current sources on each side. In the E-plane (x-z plane), the far-field provided by the direct geometrical optic (GO) fields resulting from each of the three magnetic current sources can be expressed as

$$\left.\begin{aligned} E_{1\theta}^{\mathrm{GO}} &= M_1 \frac{\sin(k_0 b \sin(\theta)/2)}{k_0 b \sin(\theta)/2} \frac{e^{-jk_0 r}}{r} e^{-jk_0 d \sin(\theta)} \\ E_{2\theta}^{\mathrm{GO}} &= M_2 \frac{\sin(k_0 b \sin(\theta)/2)}{k_0 b \sin(\theta)/2} \frac{e^{-jk_0 r}}{r} \\ E_{3\theta}^{\mathrm{GO}} &= M_3 \frac{\sin(k_0 b \sin(\theta)/2)}{k_0 b \sin(\theta)/2} \frac{e^{-jk_0 r}}{r} e^{jk_0 d \sin(\theta)} \end{aligned}\right\} \tag{2.1}$$

where $-\pi/2 < \theta < \pi/2$. The far-field expression for the singly diffracted GTD field from each of the two straight edges at $x = \pm w/2$ is given by

$$\begin{aligned} E_{\theta}^{\mathrm{di}} = &\left[E_{1\theta}^{\mathrm{GO}}(Q_i) \frac{e^{-jk_0(w/2 \pm d)}}{\sqrt{w/2 \pm d}} D_{1\perp} + E_{2\theta}^{\mathrm{GO}}(Q_i) \frac{e^{-jk_0(w/2)}}{\sqrt{w/2}} D_{2\perp} \right. \\ &\left. + E_{3\theta}^{\mathrm{GO}}(Q_i) \frac{e^{-jk_0(w/2 \mp d)}}{\sqrt{w/2 \mp d}} D_{3\perp} \right] \frac{e^{-jk_0 r}}{r} e^{\pm jk_0 \sin(\theta) w/2} \quad i = 1,\ 2 \end{aligned} \tag{2.2}$$

where $D_{1\perp}$, $D_{2\perp}$, and $D_{3\perp}$ are the hard boundary diffraction coefficients for each slot without the dielectric effect, which have been reported in [54]. In order to obtain more accurate results, the doubly diffracted fields were taken into account where the singly diffracted fields serve as the incident fields which are subsequently edge diffracted for a second time. In the H-plane (y-z plane), the far-field contribution provided by the direct GO fields is the same from each magnetic current source and can be expressed as

$$E_{\varphi}^{\mathrm{GO}} = \cos(\theta) \frac{\sin(k_0 a \sin(\theta)/2)}{k_0 a \sin(\theta)/2} \frac{e^{-jk_0 r}}{r} \tag{2.3}$$

where $-\pi/2 < \theta < \pi/2$. Because the electric field at the edge vanishes for the polarization of a grazing incident wave which is parallel to the surface, that is, the soft boundary condition, the first-order diffraction field from each edge at $y = \pm w/2$ is zero. However, the second-order diffraction, that is, the slope diffraction, is not zero, which can be viewed as the result of the changing incident field at the point of diffraction. This diffracted field can be expressed as

$$\mathrm{E}_{\varphi}^{\mathrm{sdi}} = \frac{1}{jk_0 w/2} \frac{\partial E_{\varphi}^{\mathrm{GO}}}{\partial \theta} |P_i D_{\mathrm{isp}} \frac{e^{-jk_0 r}}{r} e^{\pm jk_0 (w/2) \sin(\theta)}, \quad i = 1,\ 2 \tag{2.4}$$

where D_{1sp} and D_{2sp} are the slope diffraction coefficients as given in [55]. In the backlobe region of the H-plane, the contribution from the E-plane edge diffraction needs to be accounted for by using the equivalent edge current technique. The radiated field produced by the equivalent current on the edges at $x = \pm w/2$ is expressed as

$$E_{\varphi}^{\mathrm{eq}} = \frac{jk_0}{4\pi}\frac{e^{-jk_0 r}}{r}\int_{-w/2}^{w/2} I_{\mathrm{m}}(y')\cos(\theta)\, e^{\pm jk_0 y' \sin(\theta)} dy' \tag{2.5}$$

Hence, the H-plane pattern is the vectorial sum of the GO field ($E_{\varphi}^{\mathrm{GO}}$), the slope diffraction field ($E_{\varphi}^{\mathrm{sdi}}$), and the field radiated from the equivalent edge currents ($E_{\varphi}^{\mathrm{eq}}$).

The normalized radiation patterns predicted by the full-wave simulations of the integrated antenna and those calculated using the GTD method are compared in Figure 2.3c, for both the infinite- and finite-sized ground plane cases. It can be observed that good agreement is achieved for the infinite and finite ground plane sizes in both the E-plane and the H-plane. Since this model does not take the radiated fields from the currents on the monopole into account, small deviations can be found between the simplified analytical model and the full-wave simulations of the actual integrated antenna, especially in the angular region near the first null in the E-plane. Nevertheless, the overall good correspondence verifies that the radiation of this integrated antenna originates from the artificial metallic sheet backed metasurface rather than the monopole, whose radiation is largely cancelled by the imaging source.

2.2.3 Free-Space Experiments

The components of the integrated antenna were fabricated and assembled (see inset of Figure 2.2a). The measured S_{11} curves of both the monopole and the integrated antenna are shown in Figure 2.2a. Good agreement was obtained between simulated and measured responses. The measured S_{11} achieves a 5.5%, −10 dB impedance bandwidth ranging from 2.30 GHz to 2.43 GHz, slightly broader than the simulation prediction. The measured gain patterns at 2.38 GHz are displayed in Figures 2.2b and 2.2c, agreeing well with the simulated results, confirming the predicted behavior of the proposed antenna design. In the E-plane, the measured HPBW of the integrated metasurface-enabled antenna reaches 89°, slightly broader than the simulated value. A stronger backlobe is also observed primarily due to fabrication imperfections. In the H-plane, the measured HPBW agrees well with that obtained from the simulation, which is around 101°. The measured gain values for both the monopole and the integrated metasurface-enabled antenna are slightly smaller than the simulated results, but with a difference of less than 0.4 dB. The measured FB ratio exhibits a value greater than 23 dB.

2.2.4 Bandwidth Extended Design

The concept of using the truncated metasurface as the radiator enables a low mutual impact between the antenna and human body (i.e., the antenna is highly isolated from

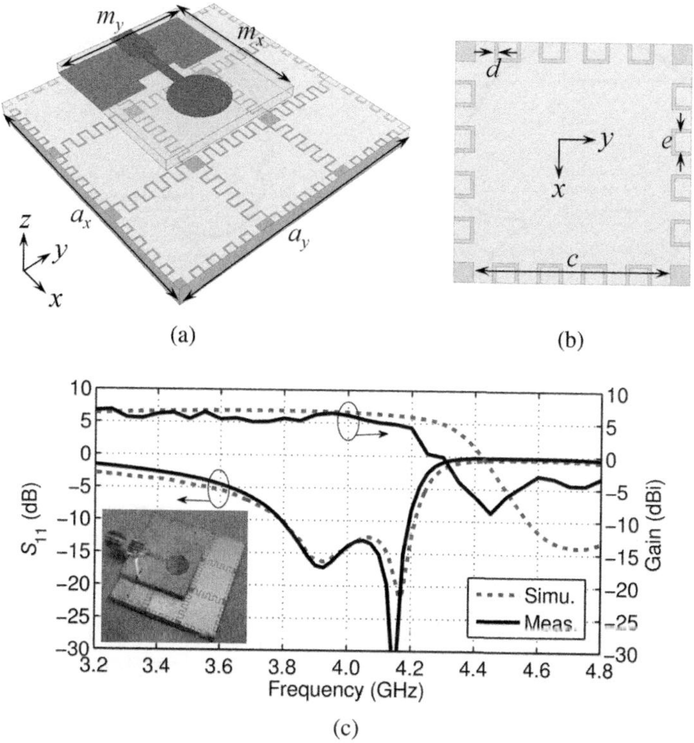

FIGURE 2.4 (a) Configurations of the wideband integrated antenna and (b) the metasurface unit cell. The optimized dimensions are $m_x = 22.9$, $m_y = 20$, $a_x = 39$, $a_y = 36$, $m_w = 36$, $d = 0.18$, $c = 10$, and $e = 0.95$, all in millimeters. The substrate material is Rogers RO3003 ($\varepsilon_r = 3$, $\delta_{tan} = 0.0013$). The thickness values for the top and bottom layers are $d_1 = 1.5$ mm and $d_2 = 2.5$ mm. The foam spacer is 1.5 mm thick. (c) Simulated and measured S_{11} and gain. The inset shows a photograph of the fabricated prototype.

any loading effects of the human body) when they are located close to each other. To extend the operational bandwidth, interdigital capacitors can be incorporated into the metasurface design. As Figure 2.4a shows, instead of using flat edges of the metasurface elements to form the capacitive gap, an interdigital capacitance can be created. This results in a larger shunt capacitance which lowers the Q factor of the metasurface [50]. Moreover, the doubly loaded interdigital capacitors modify the dispersion of the metasurface in both the x and y directions, which makes possible two standing wave resonances, thereby further broadening the bandwidth of the radiator. As a proof-of-concept, here we target an operational band around 4 GHz. Following a similar design procedure, the top monopole feed, which is non-resonant at the targeted frequencies, and the bottom-grounded metasurface are optimized. The simulated S_{11} shows a wide -10 dB bandwidth from 3.8 to 4.2 GHz, within which two resonances are present. Throughout this 10% fractional bandwidth, the gain profile is flat, slowly varying from 6.8 to 6 dBi. The E-plane and H-plane normalized radiation patterns

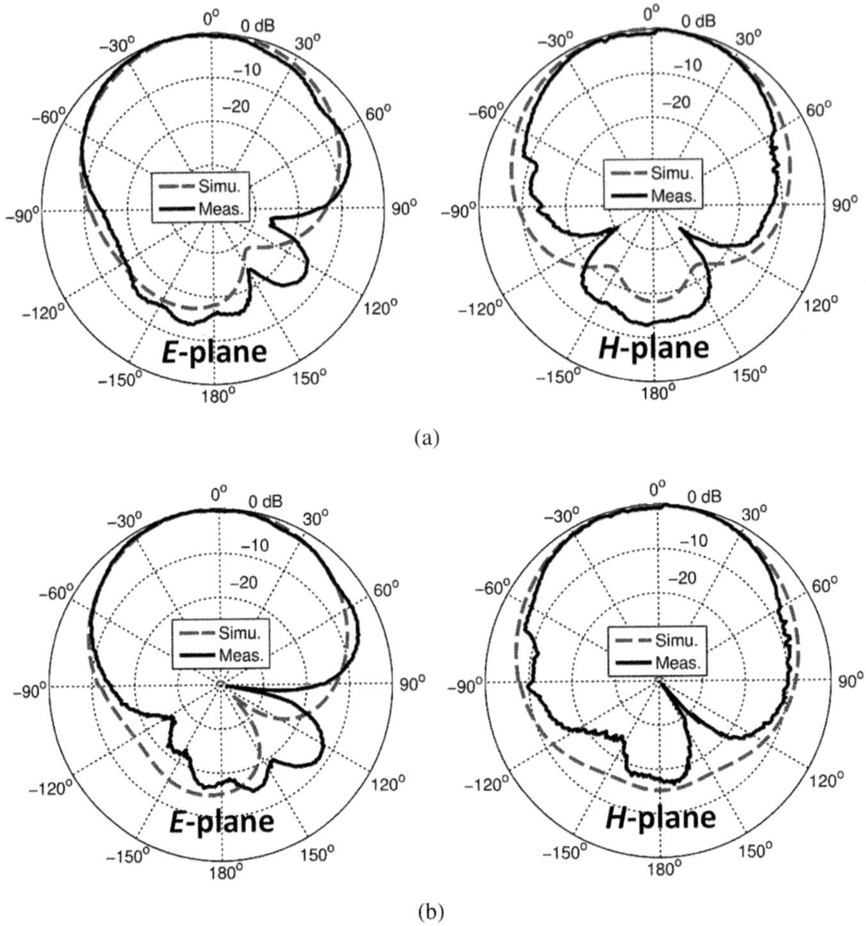

FIGURE 2.5 Simulated and measured *E*-plane and *H*-plane normalized radiation patterns at (a) 3.8 GHz and (b) 4.0 GHz.

at 3.8 and 4.0 GHz are shown in Figure 2.5. Broadside radiation is achieved, with a HPBW of around 74° in the *E*-plane and 86° in the *H*-plane. The FB ratio is 18.5 and 16.1 dB at 3.8 and 4.0 GHz, respectively.

The wideband metasurface-enabled antenna was fabricated, as shown in the inset of Figure 2.4c. The measured S_{11} exhibits a −10 dB band from 3.80 to 4.19 GHz, showing very good agreement with simulated results. Within this band, the measured antenna gain profile experiences a slight redshift, but remains greater than 5 dBi in the intended operational frequency band. The measured *E*-plane and *H*-plane normalized radiation patterns at 3.8 and 4.0 GHz are displayed in Figure 2.5. Broadside radiation is confirmed from the experimental measurements, with good correspondence in terms of the HPBW values. Slightly higher FB ratio can be observed due to the impact of the connecting cable and measurement noise. In all, the experimental results validate the concepts of extending the bandwidth of the metasurface-enabled antenna, where the metasurface functions as the radiator rather than a reflector.

2.3 MICROWAVE CIRCUIT INTEGRATED WEARABLE FILTERING ANTENNAS

In this section, the design methodologies for both narrowband and wideband filtering antennas are presented, based on patch radiator and stripline-coupled resonator circuits. The integrated band selective functionality of the radiating modules is verified by experimental measurements.

2.3.1 Narrowband CP Filtering Antenna Synthesis

When the required bandwidth is narrow, the radiator can be considered simply as a non-resonating complex load on the filter structure [56]. The targeted frequency range for the filtenna design is the 2.4–2.48 ISM band, with stop bands elsewhere. The configuration of the wearable CP filtering antenna is illustrated in Figure 2.6a, which consists of a rectangular patch radiator with truncated corners on the top and a planar stripline microwave BPF circuit on the bottom. The filter and the patch are connected by a metallic pin that passes through the top ground plane of the stripline circuit. The feeding position of the pin on the patch is offset from the center in order to obtain a low CP axial ratio (AR). The top ground plane of the filter also serves as the bottom ground plane of the patch radiator. The BPF circuit consists of four edge-coupled stripline open-loop resonators (SOLRs) [57], where the signal passes sequentially from resonator 1 to resonator 4. The filtering antenna is fed by an SMA connector from the side, which enables it to be placed very close to the surface of a human body compared to the bottom-fed topologies that have been considered in previous CP wearable patch antennas [58, 59].

The equivalent circuit of the integrated filtering antenna is shown in Figure 2.6b. The CP patch radiator is modeled as two series-connected parallel RLC resonators (R_x, L_x, C_x and R_y, L_y, C_y) to account for the orthogonal modes in the x and y directions, respectively. A series inductor (L_p) is used to model the feeding pin while a shunt C_s is employed to represent the parasitic capacitor of the circular stripline open-end. The pin-fed patch and the BPF are connected by a transmission line (TL) with a propagation constant β_s, a characteristic impedance Z_s, and a length l_s. Within the narrow ISM band, the input impedance of the transmission line (TL) connected pin-fed patch (Z_{ANT}) behaves like a simple load with a slowly varying complex impedance. The four parallel RLC resonators represent the four SOLRs, which can be synchronously or asynchronously tuned. The direct- and cross-coupling between the resonators are modeled as admittance inverters (J_{ij}, $i \neq j$) [61].

Using the external quality factors $Q_{e,i}$ and $Q_{e,o}$, and the coupling matrix $\boldsymbol{M}$, the scattering parameters of the filter can be readily calculated [57]. A fourth-order Chebyshev BPF with a 0.1 dB in-band ripple and a fractional band-width (FBW)

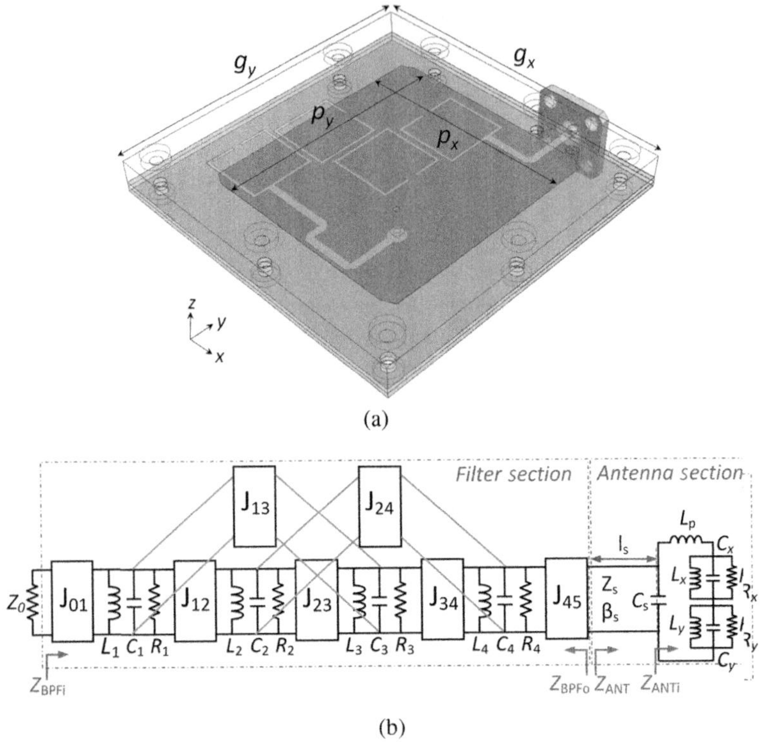

FIGURE 2.6 (a) Configuration of the narrowband CP filtering antenna. The geometrical dimensions are $g_x = 55$, $g_y = 55$, $p_x = 36.2$, and $p_y = 38.9$, all in millimeters. The substrate material is Rogers RT/duroid 5880 ($\varepsilon_r = 2.2$, $\delta_{tan} = 0.0009$). The thickness values for the three layers from top to bottom are 3.68 mm, 0.79 mm, and 0.79 mm, respectively. (b) Equivalent circuit model of the narrowband CP filtering antenna. *Source:* Jiang et al. 2015 [60]. Reproduced with permission of IEEE.

of 4.09% centered at 2.445 GHz was employed as the base design, which has $M_{12} = M_{34} = 0.0326$, $M_{23} = 0.0262$, $Q_{e,i} = Q_{e,o} = 29.725$, and $M_{13} = 0$, and where all the diagonal elements are zero [60]. When loaded by the dispersive CP patch with a transmission line, the coupling matrix, the external quality factors, the transmission-line length, and the circuit elements of the patch are all optimized using the covariance matrix adaptation evolutionary strategy (CMA-ES) [62,63] to produce a passband in the ISM band. The optimized coupling matrix becomes

$$M = \begin{bmatrix} -0.003 & 0.029 & -0.004 & 0 \\ 0.029 & 0.006 & 0.029 & 0.003 \\ -0.004 & 0.029 & 0 & 0.042 \\ 0 & 0.003 & 0.042 & 0 \end{bmatrix} \tag{2.6}$$

indicating that the resonators are asynchronously tuned. The external quality factors are $Q_{e,i} = 27.1$ and $Q_{e,o} = 26.5$, while the optimized circuit values for the patch radiator are $R_x = 42.5\ \Omega$, $C_x = 25.5$ pF, $L_x = 15.6$ nH, $R_y = 47.7\ \Omega$, $C_y = 28.5$ pF, $L_y = 15.7$ nH, $L_p = 15.73$ nH, and $C_s = 0.01$ pF. In addition, a value of 30.5 mm has been chosen for the transmission line connecting the patch and the filter. The cross-coupling parameter between resonators 1 and 3 and between resonators 2 and 4 have opposite signs due to the fact that one is primarily a magnetic coupling while the other is electric.

Once the values of the circuit elements for each component are obtained, the coupled resonators and the patch radiator that comprise the filtenna can be designed separately. The patch radiator has a size of 36.2 mm by 38.9 mm in the *x* and *y* directions, respectively, with 1.98 mm triangular corner cuts. The feeding pin is located at a distance of 5.48 mm offset from the center in both the *x* and *y* directions. The AR of the patch radiator alone is below 3 dB between 2.42 and 2.48 GHz. The resonant frequency for each resonator in the BPF was calculated using the HFSS eigenmode solver. Figure 2.7a shows the resonant frequency as a function of the width of the SOLR where the gap and length are kept the same. The SOLRs with a slightly different value of W_r all have similar quality factors of around 300–320. The external quality factor can be controlled by a properly positioned tapped stripline directly connecting to the SOLR at a distance *t* away from the center, as shown in Figure 2.7b. The coupling parameter between two directly coupled SOLRs as a function of their distance can be found in Figures 2.7c–2.7d, where two different edge coupling configurations are considered. It is noted that the offset between the two resonators has a minor impact for the direct-coupling case but a more significant effect on the cross-coupling parameter between two diagonally coupled resonators.

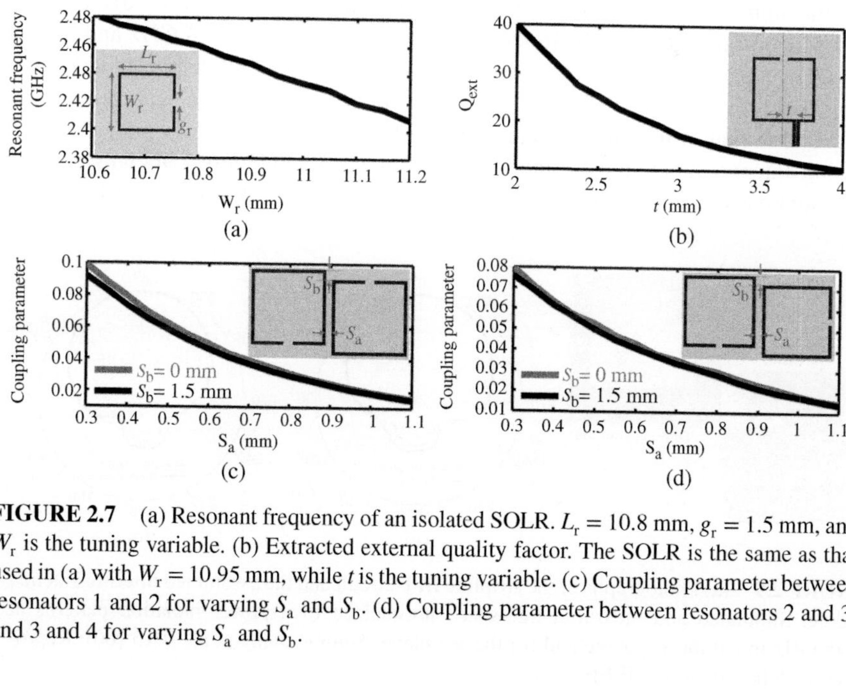

FIGURE 2.7 (a) Resonant frequency of an isolated SOLR. $L_r = 10.8$ mm, $g_r = 1.5$ mm, and W_r is the tuning variable. (b) Extracted external quality factor. The SOLR is the same as that used in (a) with $W_r = 10.95$ mm, while *t* is the tuning variable. (c) Coupling parameter between resonators 1 and 2 for varying S_a and S_b. (d) Coupling parameter between resonators 2 and 3, and 3 and 4 for varying S_a and S_b.

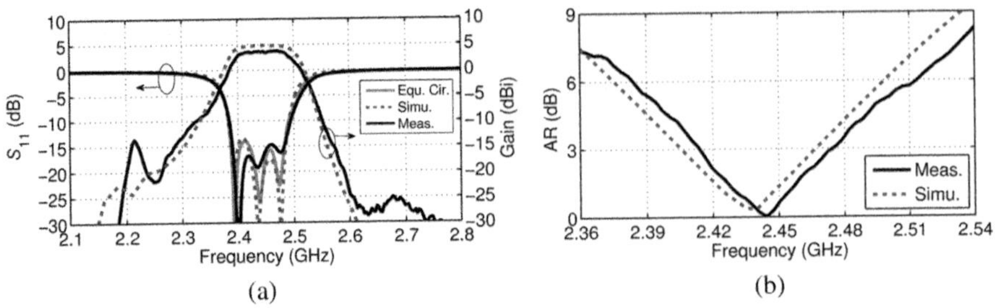

FIGURE 2.8 (a) The S_{11} and gain obtained from full-wave simulation, equivalent circuit model (Equ. Cir.), and free-space measurement of the prototype. (b) Simulated and measured AR in free space. *Source:* Jiang et al. 2015 [60]. Reproduced with permission of IEEE.

The initial CP filtering antenna configuration is obtained by properly selecting the geometrical parameters for both the patch radiator and the BPF structure from the design curves. However, when all the parts are integrated, coupling among adjacent resonators and between resonators and the patch may cause minor frequency shifts and/or coupling parameter variations, resulting in a degraded overall performance of the module. A MATLAB-based CMA-ES optimization code was employed and coupled with HFSS to fine-tune the geometrical dimensions for the integrated module in order to achieve a performance resembling that obtained from the circuit model. As shown in Figure 2.8a, the S_{11} obtained from full-wave simulations and the circuit model are in good agreement with a maximum value of −13.7 dB and a 3.9% fractional bandwidth in the passband centered at 2.44 GHz. Its broadside gain has a profile similar to the transmission of the BPF with a peak gain of 4.7 dBi and in-band gain variation of 1.1 dB. In the frequency regimes below 2.3 GHz and above 2.56 GHz, the gain is at least 20 dB lower than the peak. The AR displayed in Figure 2.8b is below 3 dB from 2.409 to 2.467 GHz, which exhibits only a very minor shift compared to that of the pin-fed patch alone. The total radiated power pattern of the filtering antenna in both the *x-z* and *y-z* planes are shown in Figure 2.9b, with a maximum at broadside and HPBWs of 86° and 87°, respectively. A front-to-back ratio of

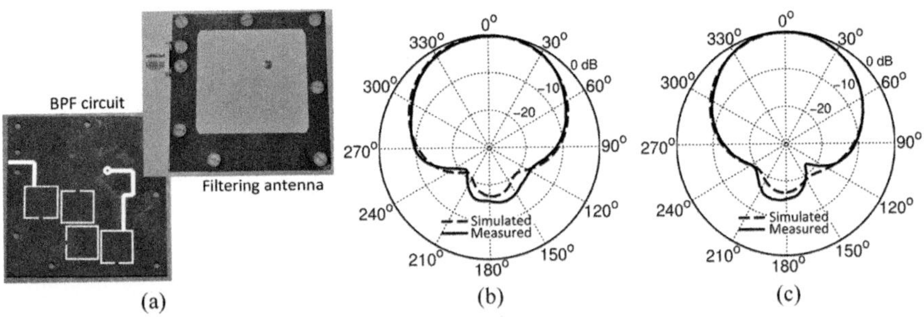

FIGURE 2.9 (a) Photographs of the stripline BPF circuit and the assembled filtering antenna module. Simulated and measured free-space normalized total power radiation patterns at 2.445 GHz in (b) the *x-z* plane and (c) the *y-z* plane. *Source:* Jiang et al. 2015 [60]. Reproduced with permission of IEEE.

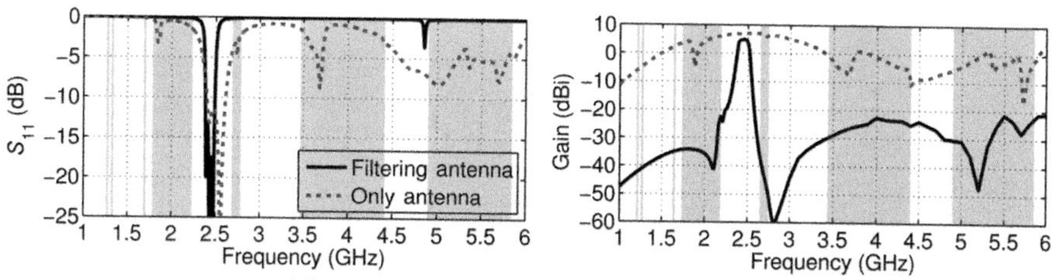

FIGURE 2.10 Simulated (a) S_{11} and (b) gain of the integrated filtering antenna and the antenna without the BPF. The highlighted regions denote some major wireless bands for existing systems. *Source:* Jiang et al. 2015 [60]. Reproduced with permission of IEEE.

about 17.5 dB is achieved, indicating a small amount of radiation into the hemisphere behind the antenna.

The performance of the CP filtering antenna was further compared to a CP patch antenna without the filter within an ultra-wide frequency range of 1–6 GHz, as presented in Figure 2.10. The filtering antenna has a single passband within the targeted frequency range and stopbands elsewhere, except for a narrow shallow resonance around 4.8 GHz, due to a harmonic resonance of the BPF. However, the S_{11} drops only down to −4 dB because the impedances of the TL-loaded pin-fed patch and the BPF are no longer conjugated to each other. Similarly, its gain response also has only one narrow peak in the targeted band, with an out-of-band rejection of more than 25 dB. In contrast, the CP patch antenna without the BPF has a wide S_{11} <-10 dB band around the targeted range from 2.4 to 2.6 GHz with a higher reflection loss than that of the filtering antenna in the ISM band. Other spurious bands can also be observed at around 1.8 GHz, 3.7 GHz, and 4.5–5.9 GHz. Its gain profile is much flatter than that of the filtering antenna, well above −10 dBi within all of the spurious bands, and even higher than 0 dBi at some frequencies. The poor selectivity of both the S_{11} and gain makes the CP patch antenna subject to interference and crosstalk caused by other existing wireless systems such as the various GPS bands between 1 and 2 GHz, the 1.71–1.98 GHz GSM band, the 1.71–2.17 GHz UMTS band, the 2.1 and 2.6 GHz LTE bands, the 3.4–3.6 WiMax band, the 3.6–3.7 GHz and 4.9–5.8 GHz WLAN bands, the 4.2–4.4 GHz aeronautical radio band, and so on, as highlighted in Figure 2.10. This comparison shows that the filtering antenna achieves a much better spectrum selectivity than the patch antenna alone, consequently yielding greatly reduced interference and crosstalk which is especially vital for sensitive low-power wearable receivers. Particularly important is the fact that even with all of the aforementioned performance enhancements, the device form factor remains the same.

2.3.2 Wideband CP Filtering Antenna Synthesis

For the narrowband CP filtering antenna design, the radiator behaves as a complex load with a nearly constant impedance for the BPF and provides the circular polarization by a properly positioned single pin feed. This approach, however, cannot

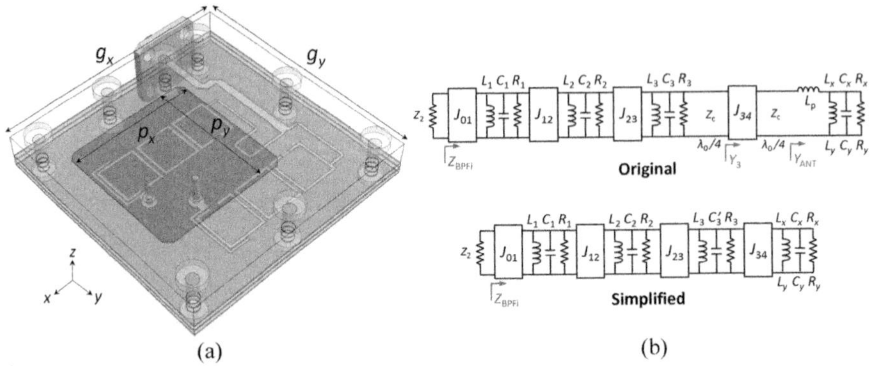

FIGURE 2.11 (a) Configuration of the wideband filtering antenna. The geometrical dimensions are $g_x = 40$, $g_y = 40$, $p_x = 22.5$, $p_y = 22.5$, all in millimeters. The substrate material is Rogers RT/duroid 5880 ($\varepsilon_r = 2.2$, $\delta_{tan} = 0.0009$). The thickness values for the three layers from top to bottom are 3.68 mm, 0.79 mm, and 0.79 mm, respectively. (b) Equivalent circuit models for each branch of the wideband CP filtering antenna. *Source:* Jiang and Werner 2015 [64]. Reproduced with permission of IEEE.

be used for achieving wideband CP filtering antennas due to the intrinsic narrow AR bandwidth and the dispersive impedance of the patch with a single feed. A seamless co-design technique should therefore be adopted [44–46], which utilizes the patch simultaneously as the radiator and the last resonator of the BPF. As shown in Figure 2.11a, the wideband CP filtering antenna consists of four metal layers, similar to the narrowband design due to the employment of the stripline structure. A square patch radiator with cut corners resides on the top layer, which is fed by two pins. The two pins are located on the two orthogonal symmetry lines of the patch in the x and y directions, respectively, but are offset from the center. One branch of a cascaded stripline BPF circuit, which contains three edge-coupled SOLRs, is located behind each pin. Power dividing and impedance matching sections connect the two BPF branches to the SMA connector on the side of the module. The top ground plane of the filter also serves as the bottom ground plane of the patch radiator. The form factor of the filtering antenna is 40 mm × 40 mm × 5.3 mm ($0.53\lambda_0 \times 0.53\lambda_0 \times 0.07\lambda_0$).

Since the same BPF structure is used for each branch, only the equivalent circuit for a single branch of the integrated CP filtering antenna is illustrated in Figure 2.11b. A parallel RLC circuit (R_x, L_x, C_x or R_y, L_y, C_y) was used to model the orthogonal mode of the patch radiator in the x or y direction, respectively. A series inductor (L_p) was employed to represent the inductance of the feeding pin. The three parallel RLC resonators represent the three SOLRs, which can be either synchronously or asynchronously tuned, with the coupling between them modeled as admittance inverters (J_{ij}, $i \neq j$). The stripline that capacitively couples the third resonator to the patch radiator in each filter branch is represented as an admittance inverter with a section of transmission line on each side [39]. At the center frequency, the equivalent circuit model can be further simplified to the one shown in Figure 2.11b. By incorporating the series inductor L_p into the capacitor of the third resonator, the shunt capacitor of the third resonator in the simplified circuit model can thus be modified to be $C_3' = C_3 + L_p J_{34}^2$. The simplified circuit can then be used to synthesize a typical Chebyshev BPF, in which the antenna becomes the last stage resonator of the filter, in addition to its role as the radiator. The two branches are connected in parallel at a T-junction with a 90° phase difference between them to achieve circular polarization. It should be noted that the BPF circuit can also be considered as a lossless two-port network that greatly enhances the bandwidth of the conventional patch antenna [65], which provides simultaneous filtering and impedance matching functionalities.

As a proof-of-concept example, we consider a filtering antenna with a fourth-order Chebyshev equal-ripple broadside antenna gain response. The passband is centered at 4 GHz with a fractional bandwidth of 12.5%, within which the axial ratio (AR) should be maintained below 3 dB. An LP filtering antenna with the same spectrum requirement will be designed first, which will then be used to construct the final CP filtering antenna. As before, the coupling matrix technique is employed here for synthesizing the filtering antenna [57]. To meet the spectral performance requirements, the normalized Chebyshev low-pass filter prototype element values with a ripple level of 0.2 dB were determined to be $g_0 = 1$, $g_1 = 1.303$, $g_2 = 1.284$, $g_3 = 1.976$, $g_4 = 0.847$, and $g_5 = 1.539$, leading to $Q_{e,i} = Q_{e,o} = 10.423$, $M_{12} = M_{34} = 0.097$, and $M_{23} = 0.078$. By further considering the dispersive transmission line sections with an electrical length of a quarter wavelength at the center frequency and a characteristic impedance of $Z_c = 80\ \Omega$ due to the coupled stripline, the external quality factors and the coupling matrix were optimized to be $Q_{e,i} = 9.28$, $Q_{e,o} = 10.02$, and

$$M = \begin{bmatrix} 0 & 0.106 & 0 & 0 \\ 0.106 & 0 & 0.086 & 0 \\ 0 & 0.086 & 0 & 0.107 \\ 0 & 0 & 0.107 & 0 \end{bmatrix} \tag{2.7}$$

Once the values of the circuit elements for each component have been obtained, the physical structures required to realize the desired functionality were designed separately. For the patch radiators, the equivalent circuit elements were extracted from its input impedance. When the edge length of the patch is 22.5 mm with a 1 mm triangular cut at the four corners and it is fed by a pin 5 mm away from the center, the values of the equivalent circuit elements are $L_p = 1.95$ nH, $R_x = 98\ \Omega$, $L_x = 0.34$ nH, and $C_x = 4.6$ pF. Similar to the design procedure of the BPF structure for the narrowband CP filtering antenna, the resonant frequency, coupling parameters, and the external quality factor of the SOLRs are determined by performing a set of parametric studies. The coupled stripline shares one of its transmission line sections with the third SOLR. The length of the coupled stripline should be around a quarter wavelength at the center frequency 4 GHz, which suggests that a value of 12.6 mm should be used. The separation distance (s) is determined to be 0.21 mm, by following the design equations provided in [61]. A wideband CP filtering antenna can be constructed using two LP filtering antennas, each connected to a branch of the T-junction with a stripline. The two striplines have a length difference of around

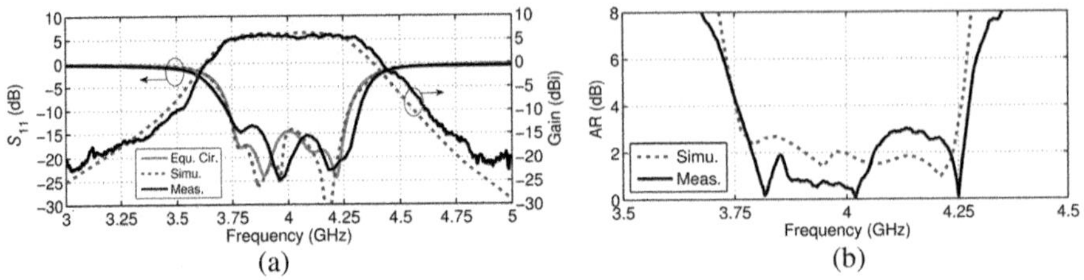

FIGURE 2.12 (a) The S_{11} and gain obtained from full-wave simulation, equivalent circuit model (Equ. Cir.), and free-space measurement of the wideband CP prototype. (b) Simulated and measured AR in free space. *Source:* Jiang and Werner 2015 [64]. Reproduced with permission of IEEE.

12.6 mm, providing the desired 90° phase shift over the 12.5% bandwidth with a phase deviation of less than ±7°. Within the passband, the input impedance at the T-junction is around 40 Ω, which can be readily matched to 50 Ω using a quarter-wave transformer with a characteristic impedance of 44.72 Ω.

Similar to the narrowband filtering antenna, the geometrical dimensions of the module were further tuned by a MATLAB-based CMA-ES optimizer linked together with HFSS to account for the coupling among different waveguiding or resonating components. The S_{11} of the optimized design is shown in Figure 2.12a, which is below −14.5 dB in a 12.5% fractional band centered at 4 GHz with sharp roll-offs on the band edges. The S_{11} calculated using the full circuit model, including the two LP filtering antenna branches, phase shifter, power dividing and matching circuits, is also presented in Figure 2.12a, showing good correspondence with the full-wave simulation result. The filtering antenna has a broadside gain profile (see Figure 2.12a) similar to the transmission of the Chebyshev BPF with a peak gain of 6.8 dBi and an in-band gain variation of 0.8 dB. The AR displayed in Figure 2.12b is below 3 dB within the same window from 3.76 to 4.24 GHz, that is, a fractional bandwidth of 12.4%. The total radiated power patterns of the filtering antenna at 4 GHz in both the *x-z* and *y-z* planes are shown in Figure 2.13b, with a maximum at broadside and a HPBW of 85° and 83°, respectively. A front-to-back ratio of 19.5 dB is achieved, indicating a small amount of radiation in the backward direction (i.e., behind the antenna). The radiation patterns within the 12.5% passband are well maintained with a FB ratio greater than 19 dB.

To appreciate the advantage of the co-designed wideband CP filtering antenna, a CP patch antenna with and without a directly connected BPF was also evaluated. The CP radiation, also centered at 4 GHz, is achieved by a single pin-fed rectangular patch, similar to the narrowband CP filtering antenna presented in Section 4.3.1. A BPF is cascaded behind the CP patch antenna via a 50 Ω transmission line. The BPF was designed to have a Chebyshev response with a passband from 3.75 to 4.25 GHz. As shown in Figure 2.14a, the S_{11} of the CP patch antenna alone is below −10 dB from 3.87 to 4.45 GHz, with poor frequency selectivity at the band edges and spurious bands at frequencies higher than 5 GHz due to the feeding structure. When directly

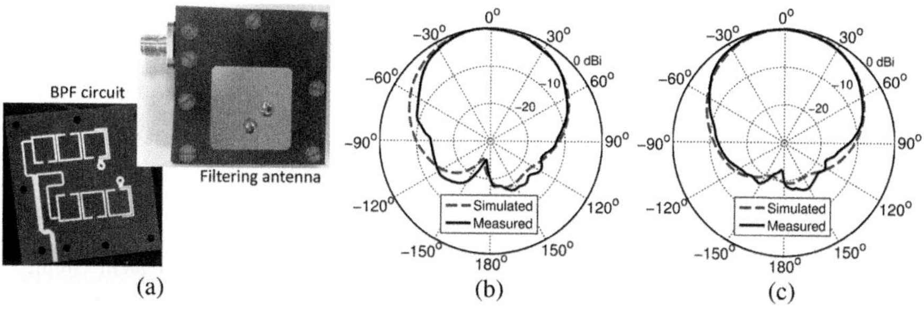

FIGURE 2.13 (a) Photographs of the stripline BPF circuit and the assembled filtering antenna module. Simulated and measured free-space normalized total power radiation patterns at 2.445 GHz in (b) the *x*-*z* plane and (c) the *y*-*z* plane. *Source:* Jiang and Werner 2015 [64]. Reproduced with permission of IEEE.

cascading a BPF after the CP patch antenna, the spurious bands can be eliminated. However, since the antenna resistance is not maintained at 50 Ω throughout the wide band of interest, such a mismatch leads to the degradation of band-edge selectivity and in-band impedance matching, resulting in an S_{11} higher than −10 dB in the targeted band with a maximum value of only −6.9 dB at 3.85 GHz. In addition, the AR values of both the CP patches, with and without the BPF, are only below 3 dB within a narrow band from 3.94 to 4.06 GHz, as shown in Figure 2.14b, which is much narrower than the targeted bandwidth. The broadside gain of the response of the CP patch antenna has a flat profile with a maximum value of 7.25 dBi. For the CP patch with a directly connected BPF, the frequency selectivity of its gain response is improved, but still worse than that of the integrated co-designed filtering antenna. The out-of-band gain values at 3 and 5 GHz for the CP patch with a directly connected BPF are −17.5 and −19 dBi, while those of the proposed integrated filtering antenna are −25 and −29 dBi, respectively. Moreover, the gain variation in the passband is increased to 2.3 dB, which is higher than the 0.8 dB gain variation possessed by the proposed design. This comparison shows that the co-designed filtering antenna achieves a much better spectrum selectivity and in-band impedance match than the

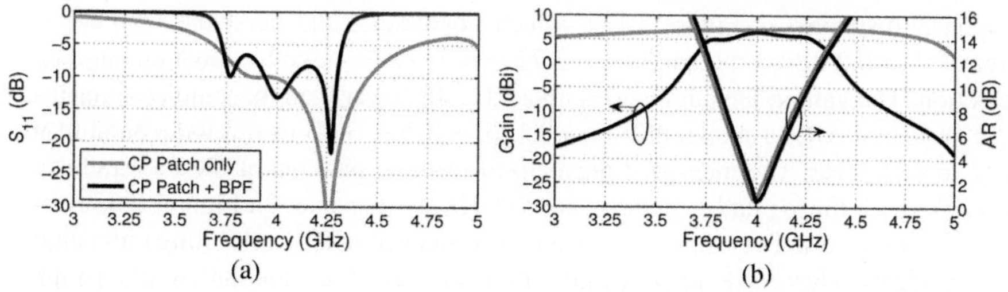

FIGURE 2.14 Simulated (a) S_{11}, (b) gain, and AR of the integrated filtering antenna and a conventional CP patch antenna with and without the directly connected BPF.

patch antenna alone or even with a directly connected BPF, consequently yielding greatly reduced interference and crosstalk. These properties are especially beneficial for the protection of sensitive wireless transceivers.

2.3.3 Free-Space Experiments

The radiator and the BPF circuit for the narrowband and wideband CP filtering antennas were fabricated and assembled to form the final modules. Nylon screws were used on the outer periphery of the filtering antennas in order to achieve a tight bonding of the layers. SMA connectors were soldered to the stripline feed and corresponding ground layers. Figures 2.9a and 2.13a show photographs of the fabricated stripline BPF structures and the two final assembled modules. The impedance of the filtering antennas was measured using an Agilent E8364B network analyzer and the radiation patterns were measured in an anechoic chamber.

As shown in Figure 2.8a, the measured S_{11} of the narrowband CP filtering antenna exhibits good filtering properties with $S_{11} < -14$ dB from 2.385 to 2.485 GHz and stopbands elsewhere. Good correspondence is achieved between simulated and measured results in terms of the positions of reflection zeros, out-of-band rejection levels, and passband bandwidth. The minor difference in the in-band S_{11} and slightly decreased roll-offs on the band edge are attributed to the fabrication and assembly inaccuracy. The measured broadside gain, displayed in Figure 2.8a, has a flat passband with values higher than 3.5 dBi and a high rejection outside of the targeted band. The gain in the passband is about 1.1 dB lower than the simulated value primarily due to the loss caused by the relatively low conductivity solder near the SMA and the vertical pin. The measured AR is smaller than 3 dB from 2.417 to 2.473 GHz, exhibiting a slight blueshift compared to the simulated value. The normalized far-field patterns of the total radiated power in both the x-z and y-z planes are shown in Figures 2.9b and 2.9c, respectively, which were measured in an anechoic chamber. They correspond well to simulated predictions with only a slightly larger backlobe of about 1.5 dB higher.

For the wideband CP filtering antenna, as shown in Figure 2.12a, the measured S_{11} demonstrates good filtering properties with $S_{11} < -13.5$ dB from 3.76 to 4.30 GHz, although slightly shifted to higher frequencies as well as broadened. Such a minor blueshift and band broadening can also be observed in the measured broadside gain response, displayed in Figure 2.12a, which possesses a flat passband with values greater than 5.2 dBi in the band from 3.77 to 4.33 GHz as well as high out-of-band rejection. The gain in the passband is about 1.2 dB lower than the simulated results. The measured AR, as shown in Figure 2.12b, is below 3 dB over a wide band from 3.77 to 4.26 GHz. The measured normalized far-field patterns of the total radiated power in both the x-z and y-z planes at 4.0 GHz are displayed in Figures 2.13b and 2.13c, respectively. They agree well with the simulated results exhibiting only slight discrepancies. They have nearly equal HPBWs of about 83° and 84° in the x-z and y-z planes, respectively, and front-to-back ratios higher than 18 dB. Above all, the measurements confirm a fairly high fidelity implementation of both the proposed narrowband and wideband CP integrated filtering antenna designs.

2.4 INVESTIGATION OF PERFORMANCE FOR WEARABLE APPLICATIONS

In this section, the suitability of the various antenna designs introduced in the previous sections for wearable applications are investigated. In particular, the effects of structural deformation and human body loading on the properties of the metasurface-enabled antenna are presented. On-body simulations employing a full-scale human body model and SAR simulations using a HUGO model for the CP filtering antennas are also carried out.

2.4.1 Metasurface-Enabled Antenna

In order to evaluate the performance of the metasurface-enabled antenna when it is placed on the human body, numerical simulations and experimental measurements were carried out by comparing the integrated antenna and the patch antenna. A cylindrical multilayer human tissue model with an outer radius of 40 mm was used to mimic the human arm, as shown in Figure 2.15a. The model has four layers, including skin ($\varepsilon_r = 37.95$, $\sigma = 1.49$ S/m), fat ($\varepsilon_r = 5.27$, $\sigma = 0.11$ S/m), muscle ($\varepsilon_r = 52.67$, $\sigma = 1.77$ S/m), and bone ($\varepsilon_r = 18.49$, $\sigma = 0.82$ S/m). The permittivity and conductivity values are chosen from measured values reported in the literature [66]. Several different distances between the antenna and the tissue model were considered in the full-wave simulations in order to investigate the impact of the anisotropic metasurface.

Figures 2.15b–2.15d show the simulated S_{11}, gain, and FB ratio values of the integrated metasurface-enabled antenna and a conventional patch antenna with the same footprint. It can be seen that the integrated antenna has a robust reflection coefficient even when it is placed in extremely close proximity (e.g., $d_a = 1$ mm) to the tissue model. A slight bandwidth broadening is observed as compared to the case when it is in free space, showing a −10 dB bandwidth extending from 2.31 to 2.47 GHz. This is due to the decreased quality factor of the radiator caused by the loading effect of the lossy tissue model on the antenna [67]. Similar effects can be observed on the S_{11} of the reference patch in that it has a ground plane to block the backward radiation. In terms of gain, the integrated antenna has a very stable value, only decreasing from 5.9 to 5.8 dBi within the band of interest, whereas the gain of the patch antenna is much lower and undergoes a more severe drop from 4.5 to 3.8 dBi. The FB ratio of the integrated antenna is maintained above 18.6 dB at the MBAN frequencies, while the patch antenna has a FB ratio about 4.5 dB lower. It should be noted that the monopole alone has a poor performance when it is located in close proximity to a human body, since the majority of the radiated power is absorbed by the underlying tissue [68]. To verify the robust impedance property of the integrated metasurface-enabled antenna, experiments were performed with the antenna placed on different parts of a human body. As shown in Figure 2.16, a very stable S_{11} is maintained for the case where the antenna is directly placed on different parts of the human body, including the chest, leg, and arm. Similar to what was observed in the numerical simulations, for the antenna on-arm case as displayed in the inset of Figure 2.16a, the

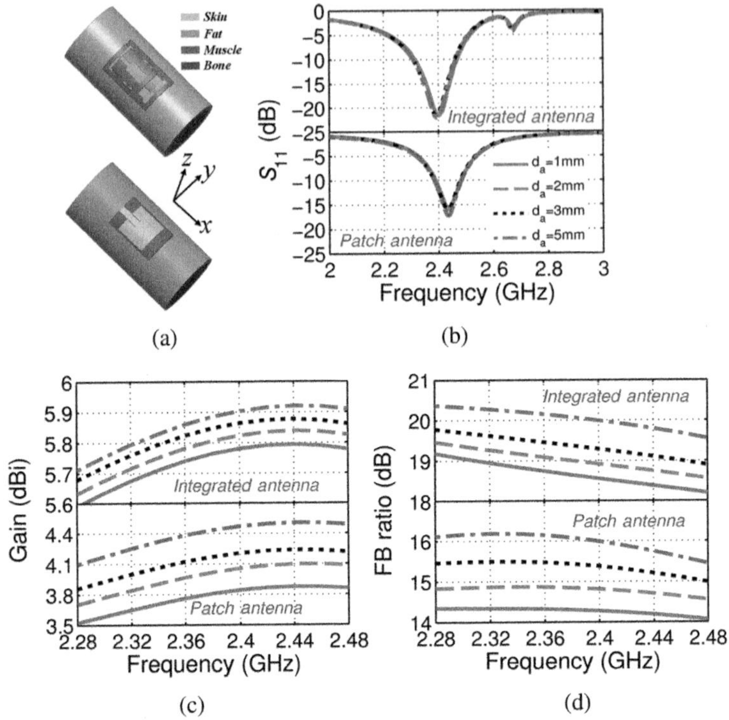

FIGURE 2.15 (a) Structurally deformed integrated metasurface-enabled antenna and patch antenna when placed on a cylindrical multilayer tissue model representing human arms. Simulated (b) S_{11}, (c) gain, and (d) FB ratio of the deformed (40 mm radius of curvature) integrated metasurface-enabled antenna and patch antenna conformal to the multilayer tissue model with different separation distances (d_a) between the antenna and the arm.

−10 dB bandwidth of the integrated metasurface-enabled antenna is broadened from 2.30 to 2.43 GHz in free space to around 2.32–2.46 GHz, which corresponds well to the full-wave predictions.

Having studied the impact of the tissue model on the antenna performance, we now investigate the effect of the antenna on the tissue by evaluating the SAR. The SAR value is related to the applied input power by

$$\text{SAR} = \sigma \left|E\right|^2 / \rho, \tag{2.8}$$

where σ is the conductivity of the tissue in S/m, E is the electric field in V/m, and ρ is the mass density of the tissue in kg/m^3. Figures 2.16b and 2.16c compare the calculated 1 g averaged SAR distribution corresponding to the integrated antenna and the patch antenna at 1 mm above the tissue model, with a 100 mW accepted power. Both antennas have a peak SAR value in the region near the two edges of the structure in the H-plane. The patch antenna has a maximum 1 g averaged SAR around 3.98 W/kg,

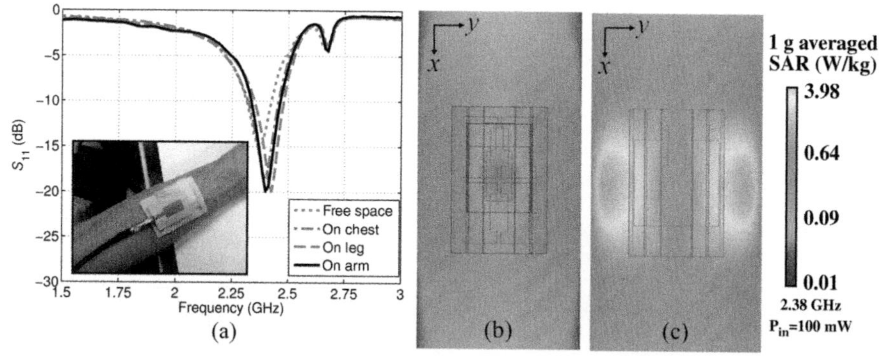

FIGURE 2.16 (a) Measured S_{11} of the structurally deformed integrated metasurface-enabled antenna placed on different parts of the human body. The inset shows a photograph of the integrated metasurface-enabled antenna conformal to a human arm. Simulated 1 g averaged SAR values for (b) the integrated metasurface-enabled antenna and (c) a conventional patch antenna with the same footprint at 1 mm away from the multilayer tissue model. *Source:* Jiang et al. 2014 [68]. Reproduced with permission of IEEE. For a color version of this figure, see the color plate section.

whereas the integrated antenna has a significantly suppressed peak 1 g averaged SAR of only about 0.66 W/kg, indicating an 83.4% reduction. This comparison demonstrates the superiority of the integrated antenna for operation in close proximity to the human body. It can be understood from the fact that the radiating slots of the conventional patch antenna are located near the edges of the ground plane, whereas the proposed antenna has the strongest radiating slot at the center of the ground plane along with the tapered amplitude corresponding to the three-element array configuration, which gives rise to a lower SAR as well as a higher gain and FB ratio.

2.4.2 Narrowband CP Filtering Antenna

The impact of the human body on the input impedance and the radiation properties of the narrowband CP filtering antenna were studied by employing a homogenous full-scale human body model with a height of 174 cm and a chest width of 46 cm [69]. The HFSS integral equation (IE) solver based on the method of moments was utilized, where the human body model and the filtering antenna were assigned to the IE region and the default finite element region, respectively. The permittivity of the homogeneous human body model was set to be 2/3 of the permittivity of muscle, as suggested in the literature [1, 70]. Such a homogeneous model allows for a numerically economical way to provide reasonably accurate evaluation of the impedance and radiation performance of wearable antennas [16].

The narrowband CP filtering antenna was placed at two different locations including the arm and the chest with an antenna to body distance of approximately 2 mm, as shown in Figure 2.17. Because the stripline BPF structure is buried in between the two ground planes, bringing the filtering antenna in close proximity to the human body does not have a significant impact on the electromagnetic properties of either the patch

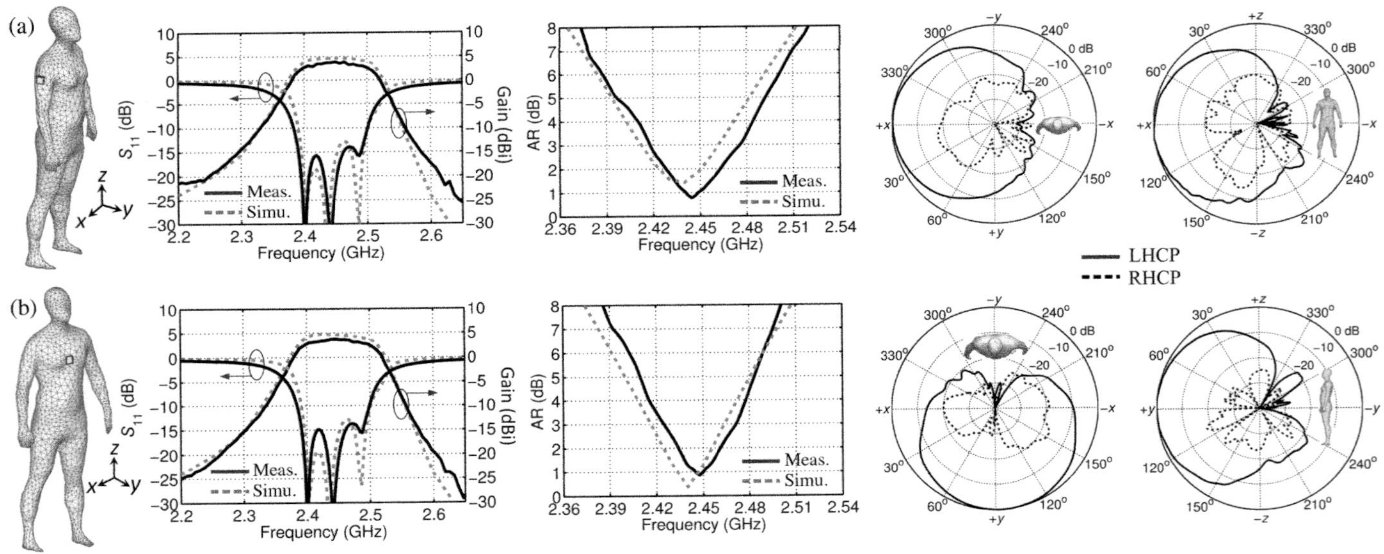

FIGURE 2.17 Simulated and measured S_{11}, gain, AR as well as simulated LHCP and RHCP radiation patterns in both the horizontal *x-y* and vertical (*x-z* or *y-z*) planes for the narrowband CP filtering antenna mounted on (a) the arm and (b) the chest of a full-scale human body model.

antenna or the integrated BPF circuit, thus the S_{11} remains below −12.5 dB in the targeted band. Only a slight variation in the S_{11} can be observed between the CP filtering antenna in free space and when it is placed on-body. The gain profiles for both cases are also well maintained with a passband at the targeted frequencies and high signal rejection elsewhere. In the passband, a peak gain of 4.65 and 4.69 dBi is achieved for the on-arm and on-chest cases, respectively. The in-band variation for both cases is smaller than 1.2 dB. Moreover, a 60 MHz AR < 3 dB is maintained within the targeted band. Due to the shadowing effect, a band shift of around 5 MHz occurs with respect to the AR when the filtering antenna is located at different positions on the human body. The simulated left-hand (LH) and right-hand (RH) CP radiation patterns in both the horizontal (x-y) and vertical (x-z or y-z) planes at 2.44 GHz are also displayed in Figure 2.17. It can be seen that the majority of the radiated energy, which is LHCP, is directed into the hemisphere away from the human body. Slight differences exist between the radiation patterns for the antenna in free space versus when it is mounted on-body. This includes pattern asymmetry and multiple minor lobes in the backward direction which are primarily attributed to the human body shadowing effect. The horizontal/vertical plane HPBWs for the filtering antenna on the arm and chest are 87°/86° and 89°/88°, respectively. By comparing the co-pol (LHCP) and x-pol (RHCP) radiation patterns, it is found that relatively wide angular ranges are achieved within which the AR is below 3 dB for both cases. Specifically, the radiation patterns exhibit a high degree of CP beam coverage, within which the LHCP wave is at least 15 dB stronger than the RHCP wave, of 108°/119° and 123°/110° in the vertical/horizontal plane for the on-arm and on-chest cases, respectively. Importantly, the simultaneous wide angular power and high-purity CP coverage indicates that wireless links can be obtained for off-body communications which are more robust and tolerant to human body movement than their LP counterparts.

Corresponding measurements were carried out where the fabricated filtering antenna was mounted on the arm and chest of an actual subject. The measured S_{11} curves agree well with the simulated results, showing a value below −12.5 dB within the targeted passband and a slightly decreased quality factor. The measured gain curves, shown in Figure 2.17, possess similar profiles compared to those of the filtering antenna in free space, though with a peak that is around 1–1.5 dB lower than the simulated results. The AR curves all show a minimum at 2.443 GHz, with an AR < 3 dB bandwidth of around 60 MHz lying within the targeted ISM band. These experiments verify that the proposed filtering antenna maintains its performance when it is mounted on different positions of an actual human body. Compared to commonly used backside microstrip feeding techniques, the stripline configuration provides better shielding for the high-Q coupled resonator circuits.

Having evaluated the impact of the human body on the performance of the filtering antenna, we now investigate the effect of the antenna on the human body by calculating the SAR. A numerical 3D volumetric anatomical HUGO human body model produced by the National Library of Medicine was employed for the purposes of this study, which contains 40 distinct types of tissues at a resolution of 2 mm × 2 mm × 2 mm mesh size [71]. The HUGO human body model was incorporated into the CST Microwave Studio (MWS) software package for performing the SAR calculations.

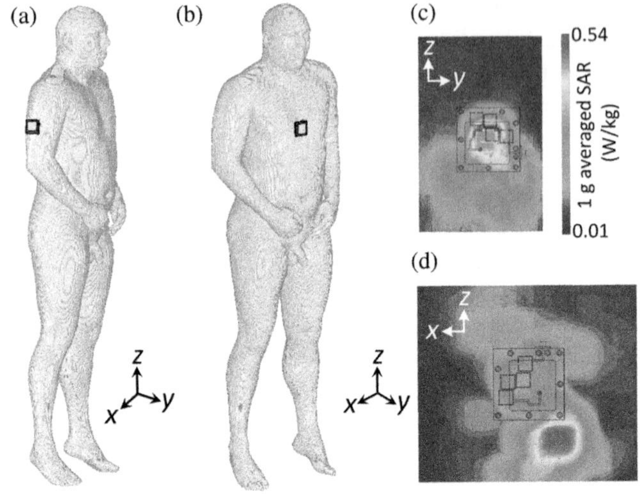

FIGURE 2.18 Configuration of the filtering antenna mounted on (a) the arm and (b) the chest of the HUGO human body model in CST MWS. Simulated 1g averaged SAR values for the filtering antenna when placed on (c) the arm and (d) the chest. For a color version of this figure, see the color plate section.

Similar to Figure 2.17, the filtering antenna was placed at two different positions on the HUGO human body, including the arm and the chest, respectively, as shown in Figure 2.18a and 2.18b. As a benchmark, an assumed 100 mW of power accepted by the antenna is chosen to evaluate the SAR performance. To reduce the simulation time, only a portion of the arm, chest, and back were included in the model (i.e., a sectional area of larger than 10 times the antenna size) since the SAR is a near-field effect. Figures 2.18c and 2.18d show the simulated 1g averaged SAR values for the two cases. It can be seen that, for both cases, the peak 1g averaged SAR value is less than 0.54 W/kg, well below the 1.6 W/kg specification required by the FCC [10] and also smaller than most of the previously proposed compact wearable antennas.

2.4.3 Wideband CP Filtering Antenna

The impact of the human body on the input impedance and the radiation properties of the wideband CP filtering antenna were also investigated using the same full-scale human body model. As shown in Figure 2.19, the filtering antenna was placed at two different locations on the body, including the arm and the chest, with an antenna to body distance of approximately 2 mm. The S_{11}, gain, and AR of the CP filtering antenna under the two different circumstances remain similar to the values obtained when the antenna is in free space. A slight resonance shift in the S_{11} can be observed, but it is maintained below −13.5 dB in the targeted band. The broadside gain profiles are also well maintained with a peak gain of 6.4 and 6.7 dBi for the on-arm and on-chest cases, respectively, with an in-band variation of less than 1.2 dB. The

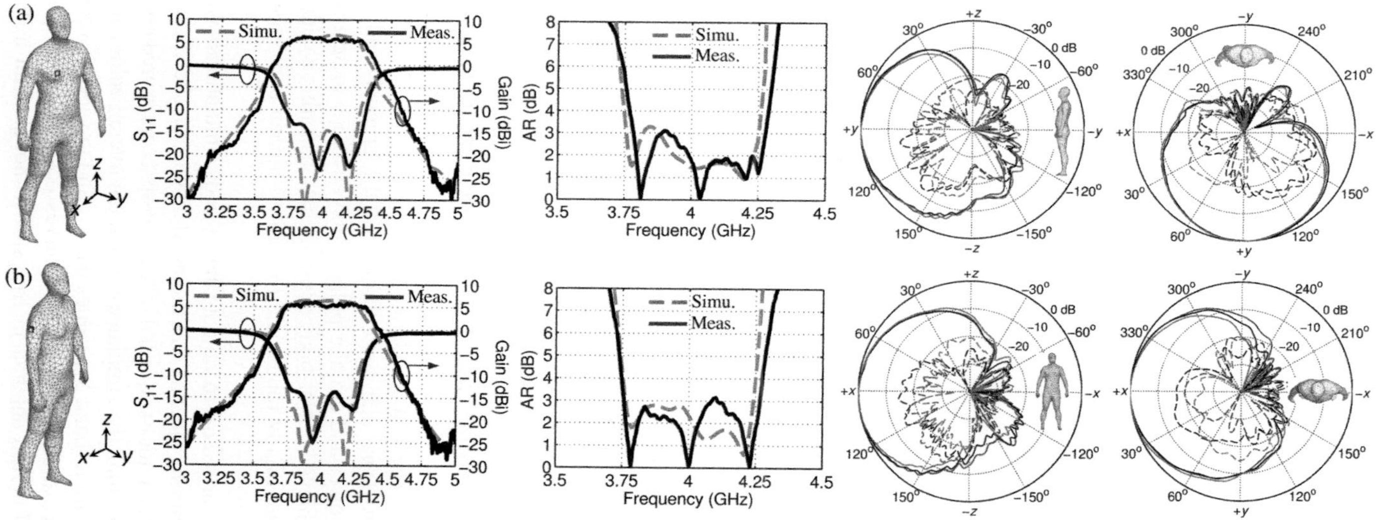

FIGURE 2.19 Simulated and measured S_{11}, gain, AR as well as simulated LHCP (solid lines) and RHCP (dashed lines) radiation patterns at 3.8, 4.0, and 4.2 GHz in both the horizontal *x-y* and vertical (*x-z* or *y-z*) planes of the wideband CP filtering antenna mounted on (a) the arm and (b) the chest of a full-scale human body model. *Source:* Jiang and Werner 2015 [64]. Reproduced with permission of IEEE.

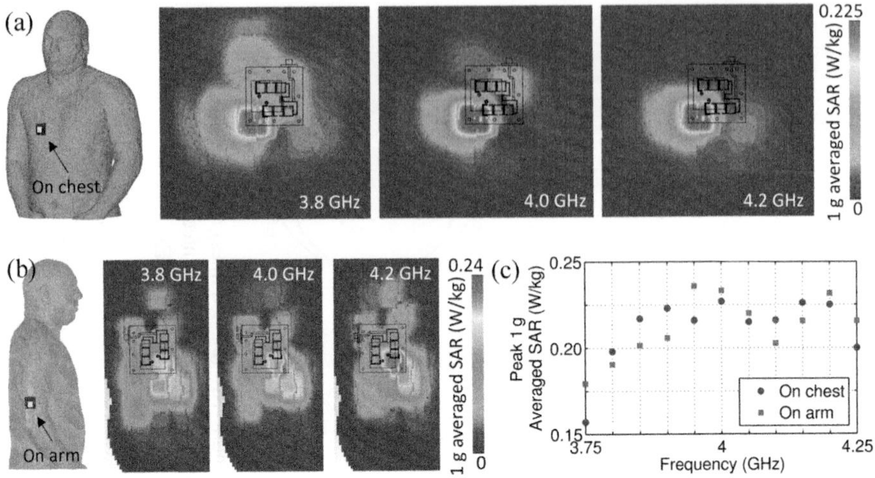

FIGURE 2.20 Simulated 1g averaged SAR values of the filtering antenna mounted on (a) the chest and (b) the arm of the HUGO human body model in CST MWS at 3.8, 4.0, and 4.2 GHz. (c) Simulated peak 1g averaged SAR values in the targeted band as a function of frequency. *Source:* Jiang and Werner 2015 [64]. Reproduced with permission of IEEE. For a color version of this figure, see the color plate section.

ARs for both the cases are below 3 dB throughout the majority of the targeted band, except for small peaks that reach 3.3 dB. The AR band is slightly shifted lower, in frequency, by about 20 MHz, due to the loading and shadowing effect of the human body. The simulated normalized LHCP and RHCP radiation patterns in both the horizontal and vertical planes at 3.8, 4.0, and 4.2 GHz are also displayed in Figure 2.19. It can be seen that the majority of the energy is directed into the hemisphere away from the human body (i.e., into free space). The vertical/horizontal plane HPBWs of the filtering antenna are 80°/83° and 78°/79°, respectively, for the two cases. The beamwidths are slightly reduced from those of the antenna in free space in both the vertical/horizontal plane. The LHCP radiated wave is more than 15 dB stronger than the RHCP wave over a wide angular range throughout the targeted band. Such simultaneous wide angular power and circular polarization coverage indicates that stable wireless links can be achieved for off-body communications within a wide bandwidth of ~500 MHz, which are highly robust and tolerant to human body movement and multipath interference.

Measurements were carried out to validate the on-body simulations, where the fabricated wideband CP filtering antenna was mounted on the arm and chest of a human subject. Similar to the behavior manifested in the free-space measured results, the S_{11} curves are slightly wider than the simulated results, but overall achieving very good agreement. The maximum in-band S_{11} values were found to vary between −12.9 dB and −13.6 dB. The measured broadside gain curves possess similar profiles compared to those of the filtering antenna in free space, though with a 0.5 dB lower peak value and slightly higher in-band fluctuation. The AR curves are smaller than

3 dB within the majority of the targeted band, with a slight shift to higher frequencies and some bandwidth broadening, which corresponds to the behavior that was also exhibited in the S_{11} and gain curves. These experimental measurements verify that the proposed antenna module maintains its simultaneous wideband filtering and CP radiating functionalities when it is mounted on different positions of an actual human body.

Using the HUGO human body model, the SAR were simulated in CST MWS. Figures 2.20a and 2.20b show the simulated 1g averaged SAR values for the on-arm and on-chest cases at 3.8, 4.0, and 4.2 GHz. It can be observed that the SAR distributions have the strongest values in the tissue regions closest to the radiating patch with slight variations elsewhere at different frequencies. For both cases, the peak 1g averaged SAR values as a function of frequency are shown in Figure 2.20c, which are lower than 0.24 W/kg throughout the entire targeted passband; well below the 1.6 W/kg specification provided by the FCC.

2.5 CONCLUSION

In this chapter, two new types of recently developed wearable antennas were presented, including lightweight metasurface-enabled antennas as well as planar narrowband and wideband CP filtering antennas. The antenna designs were confirmed by both numerical simulations and experimental measurements. The in-depth on-body performance evaluations of these antennas indicate that they represent excellent candidates for future wearable applications.

REFERENCES

[1] P. S. Hall and Y. Hao, *Antennas and Propagation for Body-Centric Wireless Communications*. Artech House, Norwood, MA, 2012.

[2] G. Z. Yang, *Body Sensor Networks*. Springer, Berlin, Germany, 2006.

[3] C. S. Nikita, *Handbook of Biomedical Telemetry*. Wiley-IEEE Press, Hoboken, NJ, 2014.

[4] E. Sazonov and M. Neuman, *Wearable Sensors: Fundamentals, Implementation and Applications*. Academic Press, Waltham, MA, 2014.

[5] Federal Communications Commission (FCC): The Medical Implant Communication Services (MICS). Available at http://wireless.fcc.gov/services/personal/medicalimplant/

[6] IEEE 802.11 Wireless LAN Working Group. Available at http://www.ieee802.org/11/

[7] FCC, "Federal Communications Commission Revision of Part 15 of the Commission's Rules Regarding Ultra-Wideband Transmission Systems," FCC First Report and Order FCC, April 2002.

[8] *Low-Rate Wireless Personal Area Networks (LR-WPANs) Amendment 4*, IEEE Standard 802.15.4j-2013.

[9] P. S. Hall, Y. Hao, Y. I. Nechayev, A. Alomainy, C. C. Constantinou, C. Parini, M. R. Kamarudin, T. Z. Salim, D. T. M. Hee, R. Dubrovka, A. S. Owadally, W. Song, A. Serra, P.

Nepa, M. Gallo, and M. Bozzetti, "Antennas and propagation for on-body communication systems," *IEEE Antennas Propag. Mag.*, vol. 49, no. 3, pp. 41–58, March 2007.

[10] *IEEE Recommended Practice for Measurements and Computations of Radio Frequency Electromagnetic Fields with Respect to Human Exposure to Such Fields, 100 kHz–300 GHz*, IEEE Standard C95.3-2002, 2002.

[11] Y. I. Nechayev, P. S. Hall, and Z. H. Hu, "Characterization of narrowband communication channels on the human body at 2.45 GHz," *IET Microw. Antennas Propag.*, vol. 4, no. 6, pp. 722–732, June 2010.

[12] R. K. Shaw, B. R. Long, D. H. Werner, and A. Gavrin, "The characterization of conductive textile materials intended for radio frequency applications," *IEEE Antennas Propag. Mag.*, vol. 49, no. 3, pp. 28–40, June 2007.

[13] M. N. Suma, P. C. Bybi, and P. Mohanan, "A wideband printed monopole antenna for 2.45 GHz WLAN applications," *Microw. Opt. Technol. Lett.*, vol. 48, no. 5, pp. 871–873, May 2006.

[14] H. D. Chen, J. S. Chen, and Y. T. Chen, "Modified inverted-L monopole antenna for 2.4/5 GHz dual-band operations," *Electron. Lett.*, vol. 39, no. 22, pp. 1567–1568, November 2003.

[15] A. Alomainy, Y. Hao, and F. Pasveer, "Numerical and experimental evaluation of a compact sensor antenna for healthcare devices," *IEEE Trans. Biomed. Circuits Syst.*, vol. 1, no. 4, pp. 242–249, April 2007.

[16] Z. Wang, L. Z. Lee, D. Psychoudakis, and J. L. Volakis, "Embroidered multiband body-worn antenna for GSM/PCS/WLAN communications," *IEEE Trans. Antennas Propag.*, vol. 62, no. 6, pp. 3321–3329, June 2014.

[17] P. Salonen, L. Sydänheimo, M. Keskilammi, and M. Kivikoski, "A Small Planar Inverted-F Antenna for Wearable Applications," in The Third International Symposium on Wearable Computers, 1999, San Francisco, CA, October 18–19, 1999, pp. 95–100.

[18] P. J. Soh, G. A. E. Vandenbosch, S. L. Ooi, and N. H. M. Rais, "Design of a broadband all-textile slotted PIFA," *IEEE Trans. Antennas Propag.*, vol. 60, no. 1, pp. 379–384, January 2012.

[19] Q. Bai and R. Langley, "Crumpling of PIFA textile antenna," *IEEE Trans. Antennas Propag.*, vol. 60, no. 1, pp. 63–70, January 2012.

[20] A. Alomainy, Y. Hao, A. Owadally, C. G. Parnini, Y. Nechayev, C. C. Constantinou, and P. S. Hall, "Statistical analysis and performance evaluation for on-body radio propagation with microstrip patch antennas," *IEEE Trans. Antennas Propag.*, vol. 55, no. 1, pp. 245–248, January 2007.

[21] L. Akhoondzadeh-Asl, Y. Nechayev, P. S. Hall, and C. C. Constantinou, "Parasitic array antenna with enhanced surface wave launching for on-body communications," *IEEE Trans. Antennas Propag.*, vol. 61, no. 4, pp. 1976–1985, April 2013.

[22] N. Haga, K. Saito, M. Takahashi, and K. Ito, "Characteristics of cavity slot antenna for body-area networks," *IEEE Trans. Antennas Propag.*, vol. 57, no. 4, pp. 837–843, April 2009.

[23] S. Agneessens and H. Rogier, "Compact half diamond dual-band textile HMSIW on-body antenna," *IEEE Trans. Antennas Propag.*, vol. 62, no. 5, pp. 2374–2381, May 2014.

[24] W. Rowe and R. Waterhouse, "Reduction of backward radiation for CPW fed aperture stacked patch antennas on small ground planes," *IEEE Trans. Antennas Propag.*, vol. 51, no. 6, pp. 1411–1413, June 2003.

[25] S. Zhu and R. Langley, "Dual-band wearable textile antenna on an EBG substrate," *IEEE Trans. Antennas Propag.*, vol. 57, no. 4, pp. 926–935, April 2009.

[26] H. R. Raad, A. I. Abbosh, H. M. Al-Rizzo, and D. G. Rucker, "Flexible and compact AMC based antenna for telemedicine applications," *IEEE Trans. Antennas Propag.*, vol. 61, no. 2, pp. 524–531, February 2013.

[27] S. Yan, P. J. Soh, and G. A. E. Vandenbosch, "Low-profile dual-band textile antenna with artificial magnetic conductor plane," *IEEE Trans. Antennas Propag.*, vol. 62, no. 12, pp. 6487–6490, December 2014.

[28] J. Kim and Y. Rahmat-Samii, "Implanted antennas inside a human body: simulations, designs, and characterizations," *IEEE Trans. Microw. Theory Tech.*, vol. 52, no. 8, pp. 1934–1943, June 2004.

[29] Z. N. Chen, G. C. Liu, and T. S. P. See, "Transmission of RF signals between MICS loop antennas in free space and implanted in the human head," *IEEE Trans. Antennas Propag.*, vol. 57, no. 6, pp. 1850–1854, June 2009.

[30] A. Kiourti and K. S. Nikita, "Miniature scalp-implantable antennas for telemetry in the MICS and ISM bands: design, safety considerations and link budget analysis," *IEEE Trans. Antennas Propag.*, vol. 60, no. 8, pp. 3568–3575, August 2012.

[31] C. Liu, Y.-X. Guo, and S. Xiao, "A hybrid patch/slot implantable antenna for biotelemetry devices," *IEEE Antennas Wireless Propag. Lett.*, vol. 11, no. 11, pp. 1646–1649, November. 2012.

[32] A. Alomainy, Y. Hao, C. G. Parnini, and P. S. Hall, "Comparison between two different antennas for UWB on-body propagation measurements," *IEEE Antennas Wireless Propag. Lett.*, vol. 4, no. 6, pp. 31–34, June 2005.

[33] Z. N. Chen, A. Cai, T. S. P. See, X. Qing, and M. Y. W. Chia, "Small planar UWB antennas in proximity of the human head," *IEEE Trans. Antennas Propag.*, vol. 54, no. 4, pp. 1846–1857, April 2006.

[34] M. Klemm, I. Z. Kovacs, F. G. Pedersen, and G. Troster, "Novel small-size directional antenna for UWB WBAN/WPAN applications," *IEEE Trans. Antennas Propag.*, vol. 53, no. 12, pp. 3884–3896, December 2005.

[35] T. S. P. See and Z. N. Chen, "Experimental characterization of UWB antennas for on-body communications," *IEEE Trans. Antennas Propag.*, vol. 57, no. 4, pp. 866–874, April 2009.

[36] M. Contaldo, B. Banerjee, D. Ruffieux, J. Chabloz, E. L. Roux, and C. C. Enz, "A 2.4-GHz BAW-based transceiver for wireless body area networks," *IEEE Trans. Biomed. Circuits Syst.*, vol. 4, no. 6, pp. 391–399, December 2010.

[37] J. Tan, W.-S. Liew, C.-H. Heng, and Y. Lian, "A 2.4 GHz ULP reconfigurable asymmetric transceiver for single-chip wireless neural recording IC," *IEEE Trans. Biomed. Circuits Syst.*, vol. 8, no. 4, pp. 497–509, August 2014.

[38] R. J. Cameron, C. M. Kudsia, and R. R. Mansour, *Microwave Filters for Communication Systems: Fundamentals, Design, and Applications*. John Wiley & Sons, Hoboken, NJ, 2007.

[39] J.-S. Hong, *Microstrip Filter for RF/Microwave Applications*, 2nd ed. John Wiley & Sons, Hoboken, NJ, 2011.

[40] M. Troubat, S. Bila, M. Thévenot, D. Baillargeat, T. Monédière, S. Verdeyme, and B. Jecko, "Characterization of the interactions between a 60-GHz antenna and the human body in an off-body scenario," *IEEE Trans. Microw. Theory Tech.*, vol. 55, no. 6, pp. 1182–1189, June 2007.

[41] H. Zhou, S. Qu, B. Lin, J. Wang, H. Ma, Z. Xin, W. Peng, and P. Bai, "Filter-antenna consisting of conical FSS radome and monopole antenna," *IEEE Trans. Antennas Propag.*, vol. 60, no. 6, pp. 3040–3045, June 2012.

[42] G. Q. Luo, W. Hong, H. J. Tang, J. X. Chen, X. X. Yin, Z. Q. Kuai, and K. Wu, "Filtenna consisting of horn antenna and substrate integrated waveguide cavity FSS," *IEEE Trans. Antennas Propag.*, vol. 55, no. 1, pp. 92–98, January 2007.

[43] M. Barbuto, F. Trotta, F. Bilotti, and A. Toscano, "Filtenna consisting of horn antenna and substrate integrated waveguide cavity FSS," *IEEE Antennas Wireless Propag. Lett.*, vol. 12, pp. 1065–1068, September. 2013.

[44] O. A. Nova, J. C. Bohóquez, N. M. Pena, G. E. Bridges, L. Shafai, and C. Shafai, "Filter-antenna module using substrate-integrated waveguide cavities," *IEEE Antennas Wireless Propag. Lett.*, vol. 10, pp. 59–62, March 2011.

[45] Y. Yusuf and X. Gong, "Compact low-loss integration of high-Q 3-D filters with highly efficient antennas," *IEEE Trans. Microw. Theory Tech.*, vol. 59, no. 4, pp. 857–865, April 2011.

[46] Y. Yusuf, H. Cheng, and X. Gong, "A seamless integration of 3-D vertical filters with highly efficient slot antennas," *IEEE Trans. Antennas Propag.*, vol. 59, no. 11, pp. 4016–4022, November 2011.

[47] Y. Chen, W. Hong, Z. Kuai, and H. Wang, "Ku-band linearly polarized omnidirectional planar filtenna," *IEEE Antennas Wireless Propag. Lett.*, vol. 11, pp. 310–313, April 2012.

[48] J. Zuo, X. Chen, G. Han, L. Li, and W. Zhang, "An integrated approach to RF antenna-filter co-design," *IEEE Antennas Wireless Propag. Lett.*, vol. 8, pp. 141–144, April 2009.

[49] C.-T. Chuang and S.-J. Chung, "Synthesis and design of a new printed filtering antenna," *IEEE Trans. Antennas Propag.*, vol. 59, no. 3, pp. 1036–1042, March 2011.

[50] Z. H. Jiang, M. D. Gregory, and D. H. Werner, "A broadband monopole antenna enabled by an ultrathin anisotropic metamaterial coating," *IEEE Antennas Wireless Propag. Lett.*, vol. 10, pp. 1543–1546, 2011.

[51] H. Mosallaei and K. Sarabandi, "Antenna miniaturization and bandwidth enhancement using a reactive impedance substrate," *IEEE Trans. Antennas Propag.*, vol. 52, no. 9, pp. 2403–2414, September 2004.

[52] F. Yang and Y. Rahmat-Samii, "Reflection phase characterization of the EBG ground plane for low profile wire antenna applications," *IEEE Trans. Antennas Propag.*, vol. 51, no. 10, pp. 2691–2703, October 2003.

[53] A. Vallecchi, J. R. De Luis, F. Capolino, and D. De Flaviis, "Low profile fully planar folded dipole antenna on a high impedance surface," *IEEE Trans. Antennas Propag.*, vol. 60, no. 1, pp. 51–60, January 2012.

[54] C. A. Balanis, *Antenna Theory: Analysis and Design*. Wiley-IEEE Press, Hoboken, NJ, 2005.

[55] R. G. Kouyoumjian and P. H. Pathak, "A uniform geometrical theory of diffraction for an edge in a perfectly conducting surface," *Proc. IEEE*, vol. 62, no. 11, pp. 1448–1461, November 1974.

[56] M. Troubat, S. Bila, M. Thévenot, D. Baillargeat, T. Monédière, S. Verdeyme, and B. Jecko, "Mutual synthesis of combined microwave circuits applied to the design of a filter-antenna subsystem," *IEEE Trans. Microw. Theory Tech.*, vol. 55, no. 6, pp. 1182–1189, June 2007.

[57] J. S. Hong and M. J. Lancaster, "Coupling of microstrip square open-loop resonator for cross-coupled planar microwave filters," *IEEE Trans. Microw. Theory Tech.*, vol. 44, no. 11, pp. 2099–2109, November 1996.

[58] C. Hertleer, H. Rogier, L. Vallozzi, and L. Van Langenhove, "A textile antenna for off-body communication integrated into protective clothing for firefighters," *IEEE Trans. Antennas Propag.*, vol. 57, no. 4, pp. 919–925, April 2009.

[59] E. K. Kaivanto, M. Berg, E. Salonen, and P. de Maagt, "Wearable circularly polarized antenna for personal satellite communication and navigation," *IEEE Trans. Antennas Propag.*, vol. 59, no. 12, pp. 4490–4496, December 2011.

[60] Z. H. Jiang, M. D. Gregory, and D. H. Werner, "A compact, low-profile circularly polarized wearable filtering antenna for off-body communications," *IEEE Trans. Biomed. Circuits Syst.*, vol. 10, no. 2, pp. 328–338, April 2016.

[61] D. M. Pozar, *Microwave Engineering*, 4th ed. John Wiley & Sons, Hoboken, NJ, 2011.

[62] N. Hansen and A. Ostermeier, "Completely derandomized self-adaptation in evolutionary strategies," *Evol. Comput.*, vol. 9, no. 2, pp. 159–195, June 2001.

[63] M. D. Gregory, Z. Bayraktar, and D. H. Werner, "Fast optimization of electromagnetic design problems using the covariance matrix adaptation evolutionary strategy," *IEEE Trans. Antennas Propag.*, vol. 59, no. 4, pp. 1275–1285, April 2011.

[64] Z. H. Jiang and D. H. Werner, "A compact, wideband circularly polarized filtering antenna and its application for wearable devices," *IEEE Trans. Antennas Propag.*, vol. 63, no. 9, pp. 3808–3818, September 2015.

[65] H. F. Pues and A. R. Van de Capelle, "An impedance-matching technique for increasing the bandwidth of microstrip antenna," *IEEE Trans. Antennas Propag.*, vol. 37, no. 11, pp. 1345–1354, November 1989.

[66] M. A. Stuchly and S. S. Stuchly, "Dielectric properties of biological substances-tabulated," *J. Microw. Power*, vol. 15, no. 1, pp. 19–26, January 1980.

[67] K.-M. Luk, K.-F. Lee, and J. S. Dahele, "Analysis of the cylindrical-rectangular patch antenna," *IEEE Trans. Antennas Propag.*, vol. 37, no. 2, pp. 143–147, February 1989.

[68] Z. H. Jiang, D. E. Brocker, P. E. Sieber, and D. H. Werner, "A compact, low-profile metasurface-enabled antenna for wearable medical body-area network devices," *IEEE Trans. Antennas Propag.*, vol. 62, no. 8, pp. 4021–4030, August 2014.

[69] S. N. Makarov, G. M. Noetscher, and A. Nazarian, *Low-Frequency Electromagnetic Modeling for Electrical and Biological Systems Using MATLAB*, Wiley, Hoboken, NJ, 2015. Available: http://www.wiley.com/legacy/wileychi/makarov/index.html?type=Home

[70] C. Furse, D. A. Christensen, and C. H. Durney, *Basic Introduction to Bioelectromagnetics*, 2nd ed. CRC Press, Boca Raton, FL, 2009.

[71] Computer Simulation Technology (CST), Framingham, MA, USA, "HUGO human body model," 2014 [Online]. Available: https://www.cst.com/Applications/Article/HUGO+Human+Body+Model

3

AMC-BACKED FLEXIBLE NEAR-ENDFIRE WEARABLE ANTENNAS FOR ON-BODY COMMUNICATIONS

KUSH AGARWAL AND YONG-XIN GUO

Department of Electrical and Computer Engineering, National University of Singapore, Singapore

3.1 INTRODUCTION

With an increasing number of healthcare monitoring systems using wireless body area network (WBAN) technology [1, 2] for continuous monitoring of patients' physiological data, research has piqued on the development of suitable medical measurement systems and interest has burgeoned in wearable electronics. A WBAN or a body sensor network (BSN) is typically a localized network of electronic devices that communicate within, on, and in close proximity of a human body. These sensor devices should be compact in size, consume little power, and wirelessly transmit and receive vital medical signals for monitoring or controlling basic human body functions, such as heart beat rate, temperature, blood flow, pressure, and location sensing for the elderly. These signals are routed from the human body for diagnostic and healthcare support purposes, to an external remote system using a wearable communication module, as shown in Figure 3.1. The integration of such an electronic system into body-worn textiles offers a large design space and body proximity. The design considerations focus on maintaining a high degree of wearability and a highly reliable communication link. As wearable electronics have to conformally integrate

Electromagnetics of Body-Area Networks: Antennas, Propagation, and RF Systems, First Edition.
Edited by Douglas H. Werner and Zhi Hao Jiang.

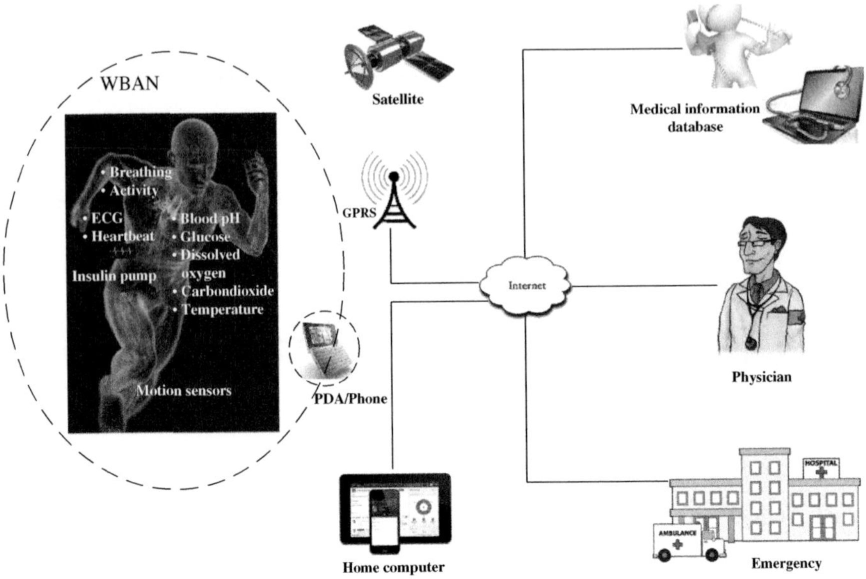

FIGURE 3.1 Wireless body area network for continuous healthcare monitoring and support.

into clothing for better flexibility and comfort, the performance of active circuits and antennas is affected. Furthermore, the design needs to comply with Federal Communications Commission (FCC) requirements concerning the human safety of wearable electronics.

One major research area of wearable antenna technology focuses on ease of integration and device flexibility. This targets conforming antennas with human clothing to provide a comfortable fit with the shape of the body. These antennas are usually composed of fabric-based textile substrates, which offer high flexibility, but introduce additional losses. Moreover, as wearable antennas are placed very close to lossy mediums like the human body, strong radiation pattern distortions and resonance frequency shifting are observed. Thus, the interfering effects of the body need to be accounted for in the design of wearable antennas for body-worn applications. Furthermore, wearable antennas are the only radiating structures in a WBAN communication system and placing them on the human body requires extensive safety analyses of excessive body-absorbed radiation and thermal heating. The FCC limits the permissible specific absorption rate (SAR) values to 2 W/kg averaged over 10 g of tissue and 1.6 W/kg averaged over 1 g of tissue for partial-body exposure to such radiation. Many researchers have reported the performance analysis of numerous body-worn antennas under various adverse conditions when placed on the human body [3–10]. These antenna designs strictly comply with FCC requirements and are safe to operate close to the human body in wearable communication modules.

Metasurfaces are periodically repeated structures in two-dimensional (2D) space that interact with incident radiation in a manner not possible in conventional reflection surfaces. These artificially designed metasurfaces, when combined with antenna technologies, enable compact [11] and low-profile [12] structures, which exhibit considerable antenna performance, despite size reductions. In the case of circularly polarized antennas, properties such as impedance bandwidth, front-to-back ratio, and axial ratio bandwidth are a few parameters that are significantly improved by these metasurfaces. Metasurface-enabled antennas demonstrate stable performance when placed close to the human body in body-worn communication devices. These metasurfaces also tend to reduce the body-absorbed back radiation from the antenna, thus reducing the SAR values and conforming to safety regulations. In [13–17], the authors reported the effects of integrating periodic metamaterial artificial magnetic conductor (AMC) surfaces [18, 19] with boresight radiation antennas to reduce back radiation and make the antennas tolerant to positioning on the human body for single- and multi-band wearable applications.

In [13], the authors presented the design, analysis, and performance of a bicep-mounted flexible EBG inverted-L (FEBGIL) antenna with a non-uniform high impedance surface (NU-HIS) for 1.85 GHz. Improvement in antenna performance has been investigated in [14] for a printed circuit board (PCB)-based slot antenna using two different-sized AMC ground planes for 2.8 GHz wearable applications. In [15], a dual-band CPW-fed wearable antenna is proposed over an AMC surface for 2.45 GHz and 5.8 GHz wireless ISM-band wearable applications. Antenna deformation studies, including bending and crumpling conditions for the same dual-band antenna, are presented in [16]. The authors, in [17], present a monopole antenna that is inkjet printed on paper and is backed by a 3×3 array AMC metasurface. A near-endfire flexible antenna is proposed by the authors in [20] on a symmetric AMC surface for on-body wearable WBAN applications. The antenna and AMC layers are fabricated by screen printing the silver ink on a latex substrate.

In this chapter, we present a multi-layered AMC-backed endfire planar Yagi antenna on a flexible latex substrate, such that the bidirectional-endfire radiation is altered to an off-axis beam tilt in the near-endfire direction. This is accomplished with an AMC of reduced unit cell size that is designed to have 0° reflection phase at 2.44 GHz and is used as reflector ground plane. Three different reflector surfaces—a PEC, a single-layered AMC, and a double-layered AMC—are used and compared as back reflectors for changing the orientation of the radiation pattern. The analysis shows that with the PEC-backed Yagi antenna ($YAGI_{PEC}$), it is difficult to achieve an impedance match for low-profile antenna geometries. An AMC-backed Yagi antenna is able to maintain the impedance matching, while changing the radiation orientation. The double-layered AMC-backed Yagi antenna ($YAGI_{D\text{-}AMC}$) improves the tolerance of the antenna to performance degradations due to the human body when compared to a single-layered Yagi antenna ($YAGI_{S\text{-}AMC}$), thus improving antenna efficiency and reducing peak SAR values. The low-profile near-endfire antenna loaded with AMC surfaces is designed and simulated using CST Microwave Studio (MWS), fabricated, and then tested on both PCB and latex materials to verify the proposed concept. The on-body analysis has been carried out using CST's human tissue-equivalent voxel model. Owing to the fact that the antenna radiation is endfire, only bending along the *H*-plane is analyzed for the distortion study of latex-based antenna performance for on-body voxel conditions.

3.2 AMC-BACKED NEAR-ENDFIRE ANTENNA FOR ON-BODY COMMUNICATIONS

The AMC-backed Yagi antennas are designed to operate in the 2.4 GHz industrial, scientific, and medical (ISM) radio band with near-endfire off-axis radiation targeting on-body WBAN communications. Commercially available electromagnetic (EM) solver CST MWS [21], based on finite integration techniques (FIT), is used for designing and optimizing the antennas with AMC metasurfaces. The cross-sectional and top views of the planar Yagi antenna placed over a double-layered AMC reflector are shown in Figure 3.2. The feed uses a 50 Ω coaxial probe connected to the driver

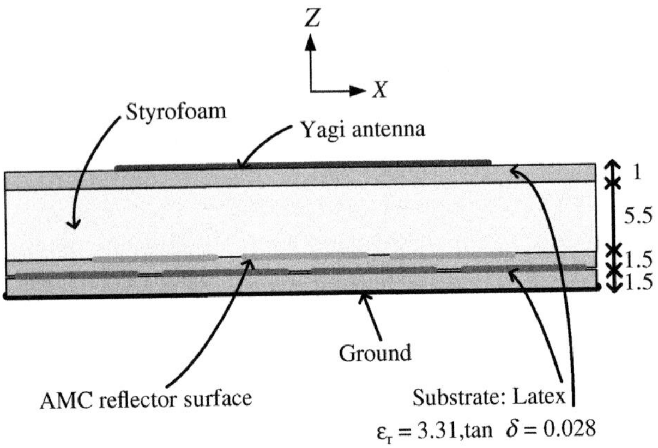

(a)

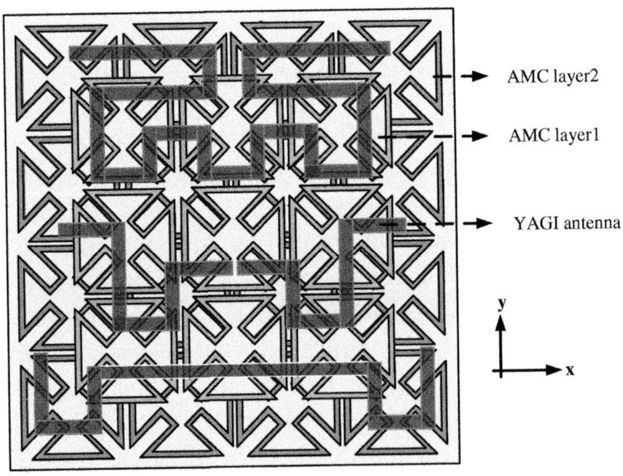

(b)

FIGURE 3.2 Yagi antenna over a double-layered AMC surface: $YAGI_{D\text{-}AMC}$ (a) cross-sectional view and (b) top view of Yagi-antenna over ground-backed AMC reflector (unit: mm). *Source:* Agarwal et al. 2016 [22]. Reproduced with permission of IEEE.

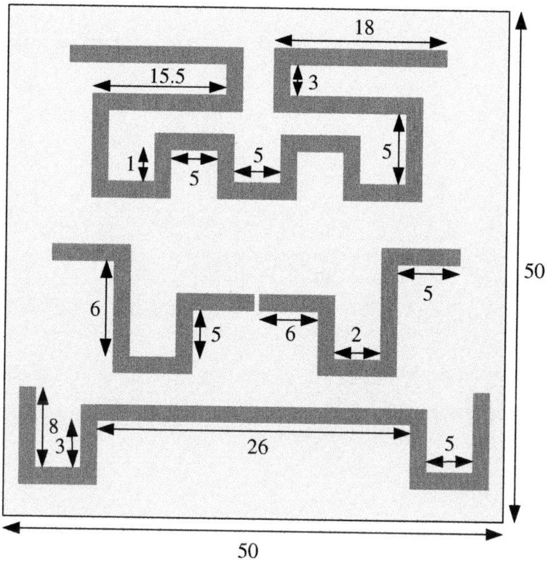

FIGURE 3.3 Planar endfire Yagi-Uda antenna (with optimized dimensions for the director, driver, and reflector elements). (Unit: mm). *Source:* Agarwal et al. 2016 [22]. Reproduced with permission of IEEE.

element of the Yagi antenna balanced by an intermediary transition using a SMD balun. The bidirectional-endfire Yagi, the unidirectional near-endfire $YAGI_{S\text{-}AMC}$, and the $YAGI_{D\text{-}AMC}$ antenna designs are discussed in the subsections 3.2.1, 3.2.2 and 3.2.3.

3.2.1 Bidirectional Yagi Antenna for Endfire Radiation

A conventional planar Yagi antenna, consisting of one director element and one reflector element, is used for endfire radiation at 2.4 GHz. The lateral dimensions of the antenna are fixed at 50 mm × 50 mm. The optimized dimensions for suitable impedance matching at 2.4 GHz for each element are shown in Figure 3.3.

3.2.2 Near-Endfire Yagi Antenna Backed by Single-Layered AMC

The bidirectional-endfire radiation of the YAGI antenna (discussed in the previous subsection 3.2.1) is changed to an off-axis near-endfire radiation pattern using a 0° reflection phase AMC surface. As is well known, PEC and PMC surfaces introduce 180° and 0° reflection phases, respectively [19]. Therefore, the unit cell of the AMC surface is designed to show 0° reflection phase at 2.44 GHz under the floquet boundary conditions. This symmetrical AMC geometry (shown in the inset of Figure 3.4) is designed to reduce the unit cell size and improve the stability of reflection phase response with respect to the direction of incident wave.

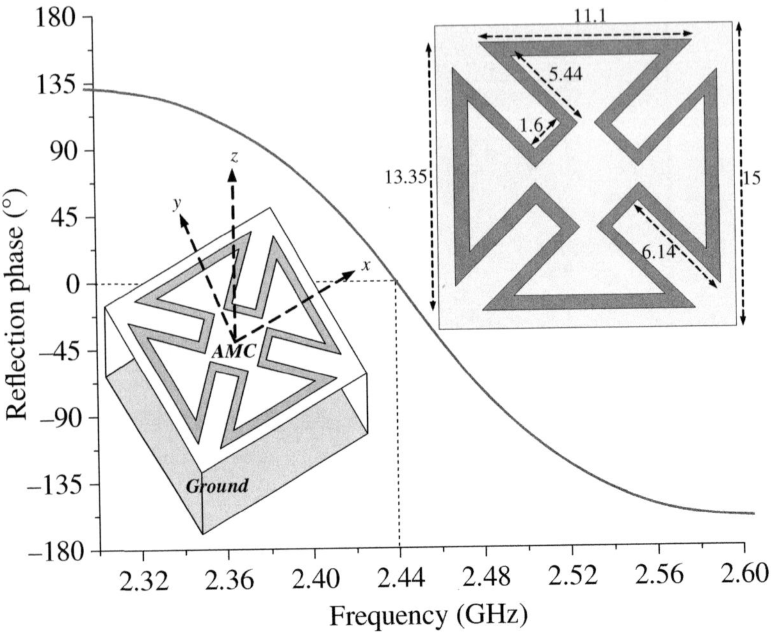

FIGURE 3.4 Reflection phase plot of AMC unit cell with AMC unit cell structure (shown in inset) enclosed by the boundary conditions: (x_{min}, x_{max}, y_{min}, and y_{max} = unit cell), (z_{min} = electric (E_t = 0)), and (z_{max} = open) under the floquet port settings. Also shown are the final optimized dimensions of AMC unit cell. (Unit: mm). *Source:* Agarwal et al. 2016 [22]. Reproduced with permission of IEEE.

The AMC surface is realized using a 3 × 3 array of AMC units cells along the lateral *x*- and *y*- axes. The entire AMC surface is optimized again (AMC unit cell parameters) to take into account the finite nature of the realized substrate and the finite number of AMC units. The reflection phase response of the unit cell is shown in Figure 3.4 for 0° reflection phase at 2.44 GHz. The final optimized dimensions are also shown in Figure 3.4.

The separation distance between the upper surface of the AMC reflector and the Yagi radiator is varied to observe the effect of the separation on the direction of the main lobe. Parameter sweeps and optimizations show that the main lobe shifts from the endfire ($\theta = 90°$) to the boresight ($\theta = 0°$) direction when the separation is reduced to 0 mm. The main lobe direction is near-endfire ($\theta \sim 74°$) when the separation between the AMC metasurface and Yagi radiator is greater than 5 mm and thus a commercially available flexible Styrofoam of 5.5 mm thickness is chosen to integrate the Yagi antenna with the AMC reflector. Figure 3.5a shows the three-dimensional (3D) view of the radiation pattern of the Yagi antenna with and without the AMC surface. It can be seen that the bidirectional radiation of the Yagi antenna is changed to off-axis radiation of $\theta \sim 74°$ when backed by a single-layered AMC surface with

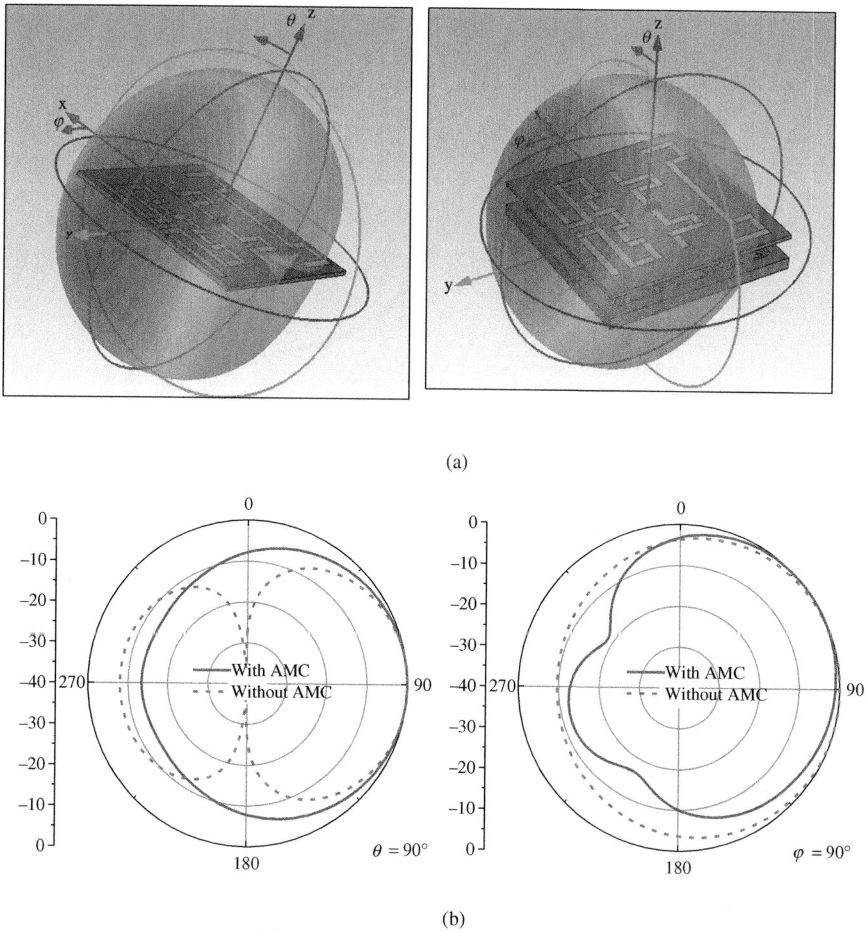

FIGURE 3.5 (a) CST MWS three-dimensional simulation radiation patterns of Yagi antenna without and with AMC configurations at 2.45 GHz, and (b) the normalized endfire radiation polar plots for $\theta = 90°$ (*xy* plane) and $\varphi = 90°$ (*yz* plane). *Source:* Agarwal et al. 2013 [20]. Reproduced with permission of IEEE.

a separation of 5.5 mm. The *xy* plane and *yz* plane normalized radiation polar plots are also shown in Figure 3.5b.

3.2.3 Near-Endfire Yagi Antenna Backed by Double-Layered AMC

This subsection discusses the design of a double-layered AMC surface for improving the antenna front-to-back ratio and reducing the body-absorbed downward antenna radiation. The unit cells for the double-layered AMC surface are optimized together with the upper and lower layers under the finite boundary conditions. It is observed that the 0° reflection phase frequency is achieved for a smaller unit cell size than

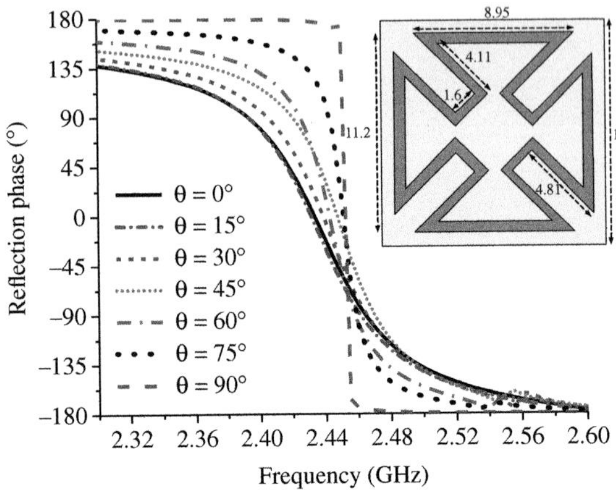

FIGURE 3.6 Reflection phase graph of double-layered AMC surface for incident angle variation. Also shown are the final optimized dimensions of AMC unit. (Unit: mm). *Source:* Agarwal et al. 2016 [22]. Reproduced with permission of IEEE.

the single-layered AMC unit cell for 2.4 GHz as discussed in [12]. Thus, the lower AMC surface layer is made up of an array of 4 × 4 units while the upper AMC layer contains an array of 3 × 3 units as shown in Figure 3.2b under the Yagi radiator for an overall lateral dimension of 50 mm × 50 mm. As the S-AMC and D-AMC reflector surfaces are being used for endfire radiation, the effect of the angle of incidence of the radiation striking the AMC surface is examined for a better understanding of using AMC surfaces for endfire radiation antennas. As seen in Figure 3.6, different incidence angles have differing slopes around the 0° reflection phase frequency. However there is an insignificant shift in the AMC frequency range (−45° to +45°). Also, as the AMC surface is being separated from the Yagi antenna by a separation distance of 5.5 mm, the waves striking the surface will have maximum incidence angle of around 60° or less. It should be noted that the incident wave will be perfectly parallel to the AMC surface at an incident angle of 90°, so the reflection phase plot will be continuous with respect to frequency. The final optimized AMC unit dimensions for a double-layered AMC surface having 0° reflection phase at 2.4 GHz are shown in the inset of Figure 3.6.

3.3 FABRICATING THE ANTENNA CONFIGURATIONS ON FLEXIBLE LATEX SUBSTRATE

Latex is a flexible and conformal translucent material that has physical properties similar to that of rubber. This material can be used as a substrate for wearable applications, as it is water resistant, and unlike textile-based fabric antennas, antennas printed on latex are conducive to mass production with repeatable results. Also, textile-based fabric materials are inherently thinner and require layer stacking to achieve the

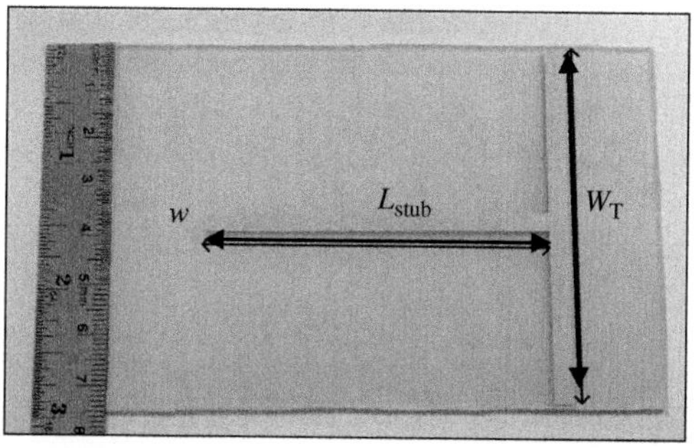

FIGURE 3.7 T-resonator prototype on latex material (with dimensions listed in Table 3.1 for both samples in mm). *Source:* Agarwal et al. 2016 [22]. Reproduced with permission of IEEE.

desired thickness for periodic metasurfaces. Considering these facts, latex is a good choice for designing low-profile AMC-backed wearable antennas from the class of non-woven wearable materials. The electromagnetic properties of the material must be known at the operating frequency band to attain the desired functionality of the wearable antenna. The material characteristics of latex, such as permittivity and loss tangent, were ascertained using the T-resonator method as in [23] and [24]. Two sets of T-resonator samples, as shown in Figure 3.7, were designed as specified in Table 3.1, and measured. The material features a permittivity of 3.31 and a loss tangent of 0.028.

The planar Yagi antenna and AMC reflector surfaces are fabricated by screen printing silver ink on latex substrates, followed by heat curing the printed ink at a temperature of 130°C. This method, when compared to the fabrication of textile material antennas, provides repeatable results with less manufacturing time. Moreover, this is a more precise method for fabricating multi-layered periodic metasurfaces. The Yagi antenna layer is printed on a 1 mm thick latex substrate, while the single- and double-layered AMCs are fabricated using 3 mm and 1.5 mm grounded latex substrates. Commercially available flexible Styrofoam ($\varepsilon_r \approx 1$) of thickness 5.5 mm is used for separating the AMC surface from the Yagi radiator for both $\text{YAGI}_{\text{S-AMC}}$ and $\text{YAGI}_{\text{D-AMC}}$ antenna prototypes.

3.4 INVESTIGATION OF ANTENNA PERFORMANCES IN FREE SPACE

To understand the effect of a lossy medium, such as human tissue, on antenna properties, it is essential to first evaluate the performance of the antenna in free space. The

TABLE 3.1 Dimensions of the T-resonators designed (mm).

Sample	Thickness	w	L_{stub}	W_T
T1	1	2.5	72.2	72.5
T2	1.5	3.7	72.2	72.5

free-space performance of all the antenna configurations, in the terms of return loss, endfire gain, and radiation polar plots, are studied and compared in this section. For comparison, a Yagi antenna is also placed on a PEC reflector ($YAGI_{PEC}$) with a similar Yagi-to-reflector separation distance of 5.5 mm, as in the case of $YAGI_{S\text{-}AMC}$ and $YAGI_{D\text{-}AMC}$ antennas. This will provide insight into the importance of the 0° reflection phase AMC surface when the antennas are placed at a close separation distance to the reflector surface.

3.4.1 Return Loss

The simulated return loss graph for the antennas is shown in Figure 3.8. Note that it is difficult to achieve an impedance match for the $YAGI_{PEC}$ antenna when the Yagi radiator is placed over the PEC surface for a small separation distance of 5.5 mm ($0.044\lambda_o$ at 2.4 GHz). It is observed that the AMC-based antennas have lower return loss bandwidth than the Yagi antenna due to the frequency band gap constraints of periodic structures. The $YAGI_{D\text{-}AMC}$ antenna shows slightly better 10 dB impedance matching than the $YAGI_{S\text{-}AMC}$ antenna around the operating frequency of 2.45 GHz.

3.4.2 Gain

Figure 3.9 shows the endfire gain ($\theta = 90°$, $\varphi = 90°$) of the Yagi antennas, with and without AMC surfaces. It is noted that the endfire gain of AMC configurations is smaller than that of a bidirectional-endfire Yagi antenna due to shift in the direction of main lobe, as discussed in the next subsection 3.4.3. For a proper comparison of gain reduction, the gain in the main lobe is also included for $YAGI_{S\text{-}AMC}$ and $YAGI_{D\text{-}AMC}$ antennas in the figure below and is listed in Table 3.2. The metasurface-based antennas are inherently narrow band and lossy, and hence gain reduction is observed.

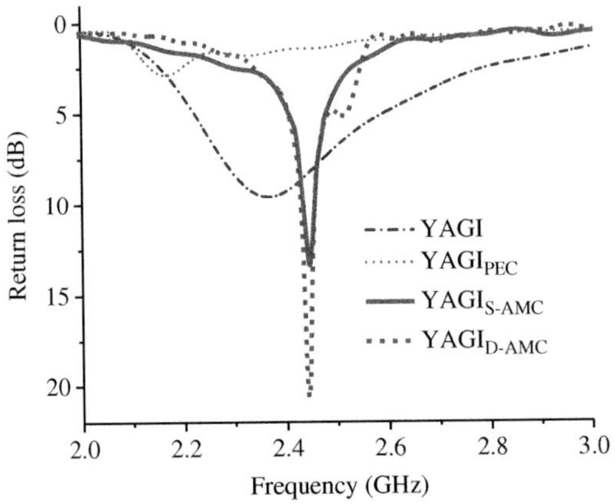

FIGURE 3.8 Return loss graph of latex-based Yagi antenna configurations over PEC, single-layered AMC and double-layered AMC. *Source:* Agarwal et al. 2016 [22]. Reproduced with permission of IEEE.

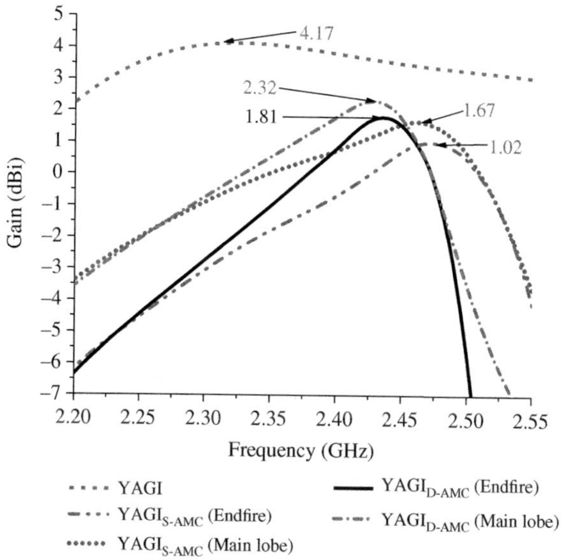

FIGURE 3.9 Gain graph of latex-based Yagi-antenna without and with single- and double-layered AMC configurations in the direction of endfire radiation and main lobe. *Source:* Agarwal et al. 2016 [22]. Reproduced with permission of IEEE.

3.4.3 Radiation Patterns

Simulated normalized radiation polar plots are shown in Figures 3.10a and 3.10b for the xy ($\theta = 90°$) and the yz ($\varphi = 90°$) planes. It is observed that the endfire-bidirectional radiation of the Yagi antenna is changed to an off-axis beam-tilt radiation of ~74° toward the endfire direction using AMC surfaces in YAGI$_{S\text{-}AMC}$ and YAGI$_{D\text{-}AMC}$ antennas (Figure 3.10b). This shift toward the main lobe is highly sensitive and is affected by the separation between the Yagi antenna and the upper AMC surface. As the separation is reduced below 5 mm, the direction of the main lobe starts to shift toward the boresight ($\theta = 0°$) direction. It is also noted that the YAGI$_{D\text{-}AMC}$ antenna improves the back radiation by almost 3 dB compared to the YAGI$_{S\text{-}AMC}$ antenna, and improves by 6 dB compared to the Yagi antenna without the AMC (Figure 3.10a).

TABLE 3.2 Gain of latex-based antenna configurations.

Antenna design	Gain in endfire direction ($\theta = 90°$, $\varphi = 90°$) (dBi)	Gain in main lobe direction (dBi)
YAGI	4.17	4.17
YAGI$_{S\text{-}AMC}$	1.02	1.67
YAGI$_{D\text{-}AMC}$	1.81	2.32

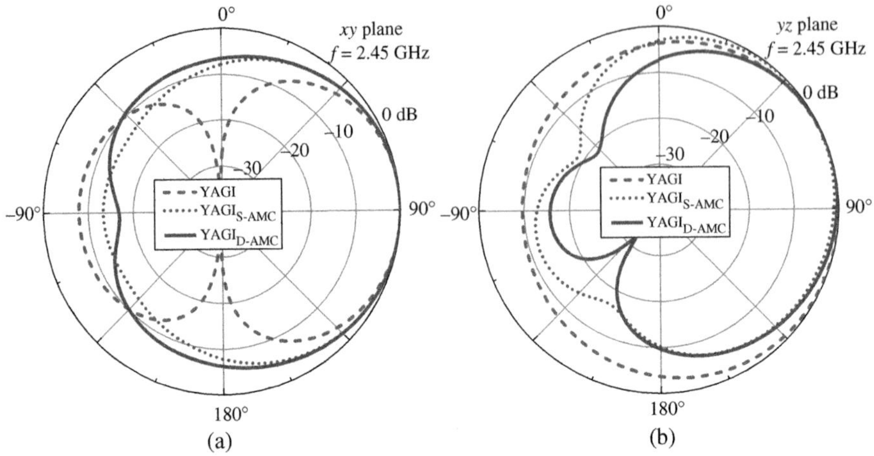

FIGURE 3.10 Normalized endfire radiation polar plots of latex-based Yagi-antenna without and with single- and double-layered AMC configurations at 2.45 GHz: (a) for $\theta = 90°$ (*xy* plane) and (b) for $\varphi = 90°$ (*yz* plane). *Source:* Agarwal et al. 2016 [22]. Reproduced with permission of IEEE.

3.5 INVESTIGATION OF ANTENNA PERFORMANCES ON VOXEL MODEL

It is crucial to study the effects of the human body on the performance stability of antennas and analyze body-absorbed radiation when designing systems that can operate safely within close proximity to the body. Thus, the on-body antenna performance of all antenna configurations (YAGI, $\text{YAGI}_{\text{S-AMC}}$, and $\text{YAGI}_{\text{D-AMC}}$) are simulated using the human tissue-equivalent CST voxel male model "Gustav" for a separation distance of 3 mm between the integrated antenna structure and the surface of the skin. Note that for fair comparison, the separation between the YAGI and the skin surface is kept the same as in the case of the latex Yagi radiator in $\text{YAGI}_{\text{S-AMC}}$ and $\text{YAGI}_{\text{D-AMC}}$ antennas (i.e., 11.5 mm). Three body parts—adult chest, arm and leg (thigh) as shown in Figure 3.11—are used for antenna placement locations for the analysis of frequency detuning, peak SAR values, and antenna radiation efficiency This analysis observes the effect of single- and double-layered AMC surfaces on the antenna properties and is summarized in the subsections 3.5.1, 3.5.2 and 3.5.3.

3.5.1 Frequency Detuning

The return loss graph for the Yagi antenna configurations is plotted in Figure 3.12 for free space and on-chest location conditions. It can be observed that the resonance deteriorates for the antenna configuration without the AMC when it is placed closed to the body model. Note that the YAGI antenna is placed at a larger separation distance of 11.5 mm from the surface of skin to keep the separation distances identical for all the antenna radiators. If we place the YAGI antenna at a close separation

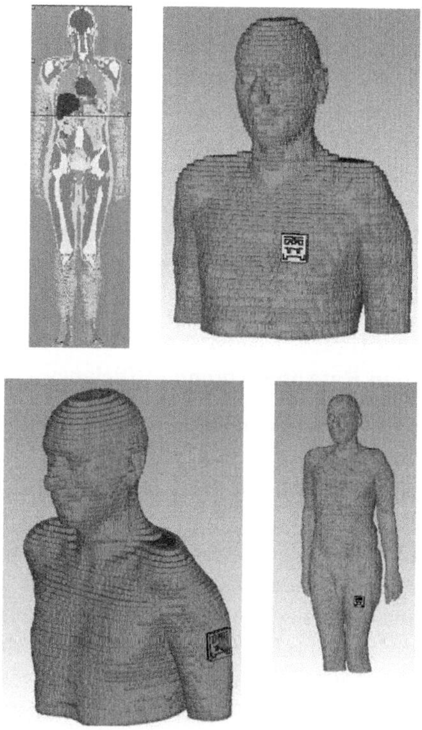

FIGURE 3.11 Three-dimensional CST voxel male model "Gustav" used for on-body antenna performance analysis at body locations of chest, arm, and leg (thigh) for a separation distance of 3 mm between the skin surface and antenna. *Source:* Agarwal et al. 2016 [22]. Reproduced with permission of IEEE.

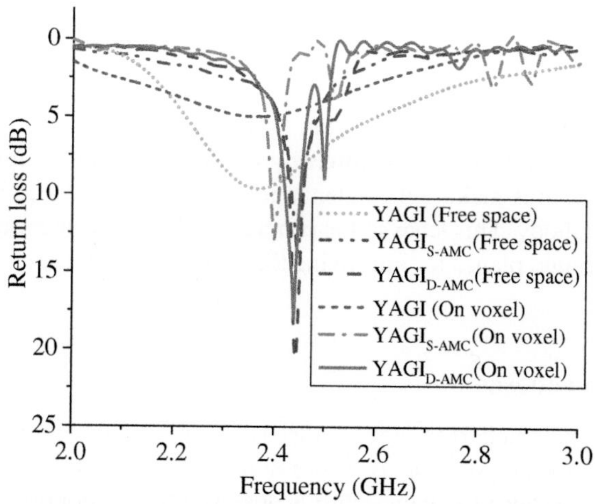

FIGURE 3.12 Return loss graph for the antenna configurations without and with AMC for free space and on-voxel (chest) surrounding conditions. *Source:* Agarwal et al. 2016 [22]. Reproduced with permission of IEEE.

of 3 mm from the voxel surface, a large backward shift in the resonance frequency is observed. On the other hand, the AMC-backed antennas retain their impedance matching when placed close to the voxel model. A small shift to lower frequency of 2.4 GHz is observed for $YAGI_{S\text{-}AMC}$ when placed on voxel as compared to its resonance in free space at 2.45 GHz. $YAGI_{D\text{-}AMC}$ shows almost zero frequency shift on human tissue voxel making the antenna nearly 100% tolerant to the positioning on lossy material like human tissue. This demonstrates how the multi-layered metasurfaces can be used to screen most of the antenna radiation from penetrating the human body and thus stabilize the performance of the wearable electronics placed in close vicinity of the body.

3.5.2 Specific Absorption Rate and Antenna Efficiency

Since the wearable antenna needs to operate in close proximity to the human body, SAR is a crucial factor to evaluate. The FCC regulates the limits of body-absorbed radiation and thermal heating of the tissue in accordance with IEEE C95.1-2005 [25] by restricting the permissible SAR values from partial-body exposure to electromagnetic fields originating from wearable electronic devices [26]. Hence, all the proposed antenna designs are evaluated in accordance with IEEE C95.1-2005 standards for human safety. The antenna efficiency drops from 86% (in the free space) to 29.39% when the YAGI antenna is placed on the voxel chest without an AMC metasurface. The peak recorded SAR value is 4.2 W/kg for 10 g of tissue (where the accepted level is 2 W/kg averaged over 10 g of tissue). The antenna efficiency improves to 74.82% with the SAR level reducing to 1.24 W/kg for 10 g tissue for the $YAGI_{S\text{-}AMC}$ antenna configuration. The $YAGI_{D\text{-}AMC}$ antenna shows further improvement in SAR level to 0.714 W/kg (Figure 3.13) for 10 g of tissue and an antenna efficiency of 78.97%. Various recorded values for peak SAR level for the input power of 1 W on all the body locations are summarized in Table 3.3 for all three antenna configurations. It is observed that the performances of the AMC-backed antennas on voxel arm and leg models are similar to the chest location. Note that if the YAGI antenna is placed at a close separation of 3 mm from the voxel surface, peak SAR values are very high (also included in Table 3.3). The band gap AMC structure delivers reduced radiation from the antenna into the body, lowering the SAR and improving the antenna radiation efficiency, thus making the antenna tolerant to being placed in close proximity to the human body. This leads to the conclusion that AMC metasurfaces can be used as antenna ground planes for improving the performance of the endfire radiation antennas.

3.5.3 Radiation Patterns

The $YAGI_{D\text{-}AMC}$ antenna performance is discussed in this subsection for placement on the voxel's chest right position, as shown in the 3D view in Figure 3.14a. As discussed before, almost zero frequency shift is observed for the double-layered AMC-backed antenna when it is placed on the voxel model as compared to its resonance frequency in free space. The normalized radiation polar plot is shown in Figure 3.14b

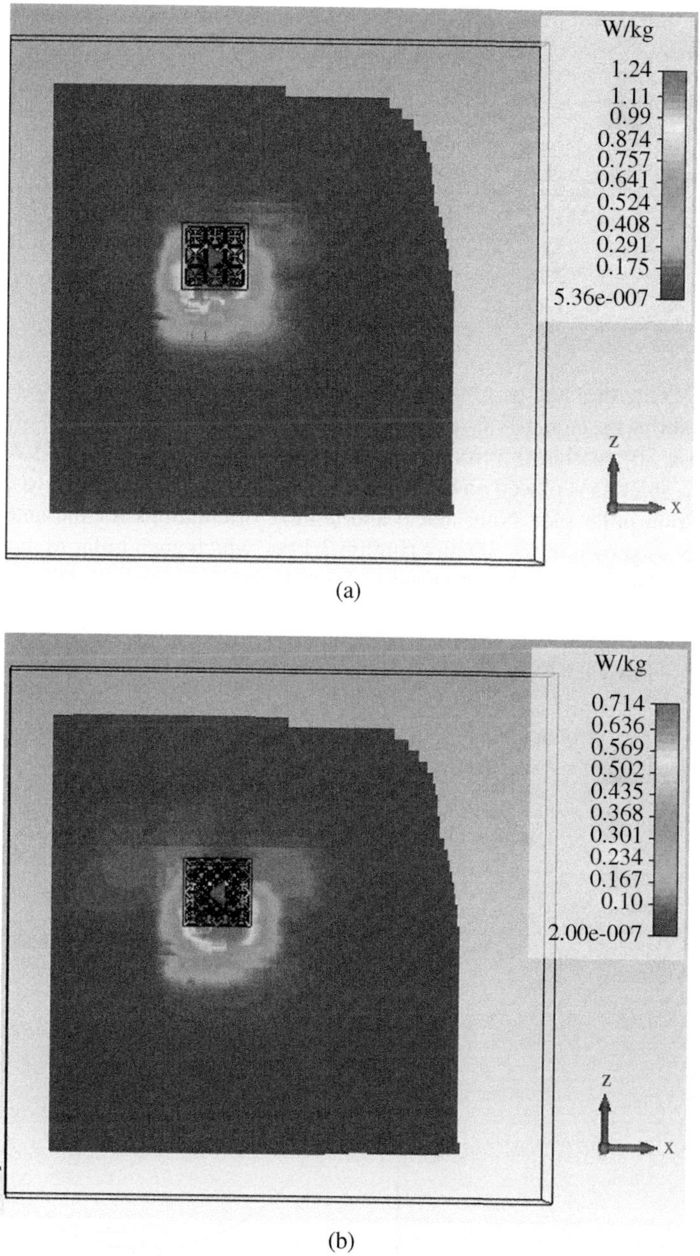

FIGURE 3.13 Peak SAR values over the human voxel model (chest) for integrated antenna configurations: (a) $YAGI_{S\text{-}AMC}$ and (b) $YAGI_{D\text{-}AMC}$ for input power of 1 W. *Source:* Agarwal et al. 2016 [22]. Reproduced with permission of IEEE.

TABLE 3.3 Simulated peak SAR values for different antenna configurations (input power = 1 W).

Body locations→	Chest (W/kg)		Arm (W/kg)		Leg (W/kg)	
Antenna configurations ↓	1 g tissue	10 g tissue	1 g tissue	10 g tissue	1 g tissue	10 g tissue
YAGI	11.4	4.2	18.3	6.76	3.04	1.75
YAGI (3 mm)	23.6	11.1	34.3	14.7	24.7	9.47
$YAGI_{S\text{-}AMC}$	2.06	1.24	1.95	1.31	0.271	0.134
$YAGI_{D\text{-}AMC}$	1.47	0.714	1.81	1.16	0.211	0.111

for $\theta = 90°$ (xy plane) and $\varphi = 90°$ (yz plane). It is observed that the direction of the main lobe shifts by $\theta = 10°$ from the y-axis (earlier 74°) toward the z-axis (now 64°) and $\varphi = 30°$ (earlier 0°) from the y-axis toward the x-axis (now 30°) when the $YAGI_{D\text{-}AMC}$ antenna is placed on the tissue-equivalent voxel as compared to the free-space radiation polar plot. Note that θ and φ have orientations for the antenna on voxel model as shown in the 3D plot (Figure 3.14a), which are similar to those used in the free-space antenna configurations.

3.6 ANTENNA PERFORMANCE UNDER BENDING DEFORMATION

After antenna performance is investigated for a conventional flat Yagi antenna and after the integration of the AMC configurations in free space and on the CST

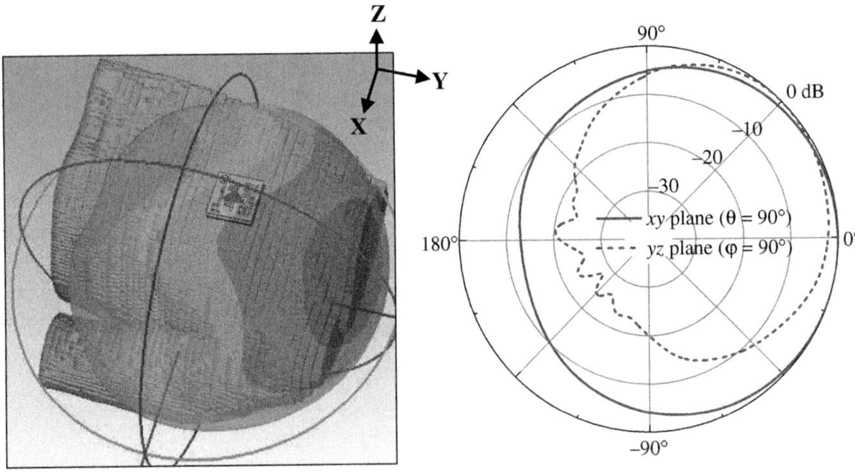

(a) (b)

FIGURE 3.14 (a) Three-dimensional simulated radiation pattern of $YAGI_{D\text{-}AMC}$ antenna configuration at 2.45 GHz for on-voxel chest right location, and (b) normalized radiation polar plot for $\theta = 90°$ (xy plane) and $\varphi = 90°$ (yz plane). *Source:* Agarwal et al. 2016 [22]. Reproduced with permission of IEEE. For a color version of this figure, see the color plate section.

FIGURE 3.15 Photograph of the fabricated double-layered AMC structure on the latex substrate.

MWS human tissue model, performance and SAR values are again evaluated in the presence of antenna deformation. On-body textile/latex wearable antennas usually take the shape of the body, so it is difficult to keep the antennas in flat configurations when placed close to the human body environment (photograph of the fabricated screen-printed double layered AMC structure on the latex substrate demonstrating the level of flexibility is shown in Figure 3.15). As in [16], the bending radius of 40 mm is chosen to approximate a human adult's upper arm for our study, as shown in Figure 3.16 for $YAGI_{D\text{-}AMC}$ antenna along the *H*-plane bending. The bending radius of 40 mm (adult arm) is the most stringent bending deformation test that can be done on an antenna being placed on the three body locations (chest, adult arm, and leg). Since all the proposed antenna configurations discussed in this chapter have endfire radiation, only bending deformation along the *H*-plane is studied. The return loss and radiation patterns are analyzed to evaluate the performances of YAGI, $YAGI_{S\text{-}AMC}$, and $YAGI_{D\text{-}AMC}$ antennas for bending along the *H*-plane of the voxel model.

Figure 3.17 shows the simulated return loss graph for the $YAGI_{S\text{-}AMC}$ and $YAGI_{D\text{-}AMC}$ antennas when bending over the voxel arm of 40 mm radius. It is noted that the bending deformation has almost no effect on the YAGI antenna performance without the AMC, whereas the antenna configurations with the AMC shows slight performance degradation. The $YAGI_{D\text{-}AMC}$ antenna shows a minor shift in resonance frequency. However, the $YAGI_{S\text{-}AMC}$ antenna still has resonance centered at f_c = 2.4 GHz. Also, the variation in 10 dB return loss bandwidth is observed as compared to the flat antenna configurations on voxel. Both the $YAGI_{S\text{-}AMC}$ and $YAGI_{D\text{-}AMC}$ antennas demonstrate operational bandwidth reduction on bending as seen from the Figure 3.17.

The normalized radiation polar plots are shown in Figure 3.18 for the $YAGI_{D\text{-}AMC}$ antenna in flat and bending conditions on the voxel model for $\varphi = 90°$ (*yz* plane) and $\theta = 90°$ (*xy* plane), respectively. Note that the θ and φ have same orientations for

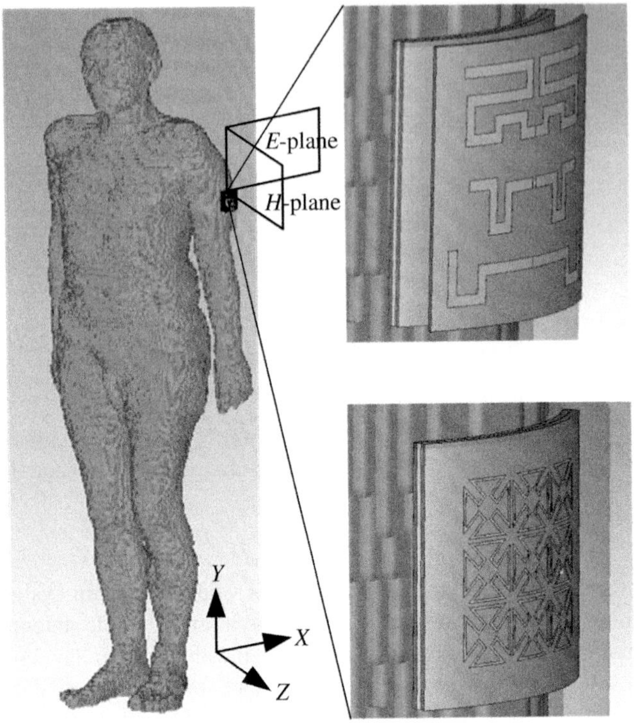

FIGURE 3.16 Three-dimensional voxel model used for the $YAGI_{D\text{-}AMC}$ antenna bending for human upper arm radius of 40 mm along the *H*-plane. *Source:* Agarwal et al. 2016 [22]. Reproduced with permission of IEEE.

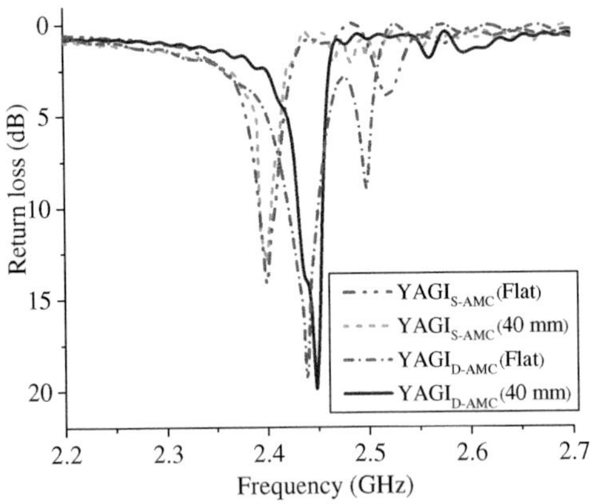

FIGURE 3.17 Return loss graph for the $YAGI_{S\text{-}AMC}$ and $YAGI_{D\text{-}AMC}$ antenna configurations on-voxel (arm) in flat and bending (40 mm radius) conditions. *Source:* Agarwal et al. 2016 [22]. Reproduced with permission of IEEE.

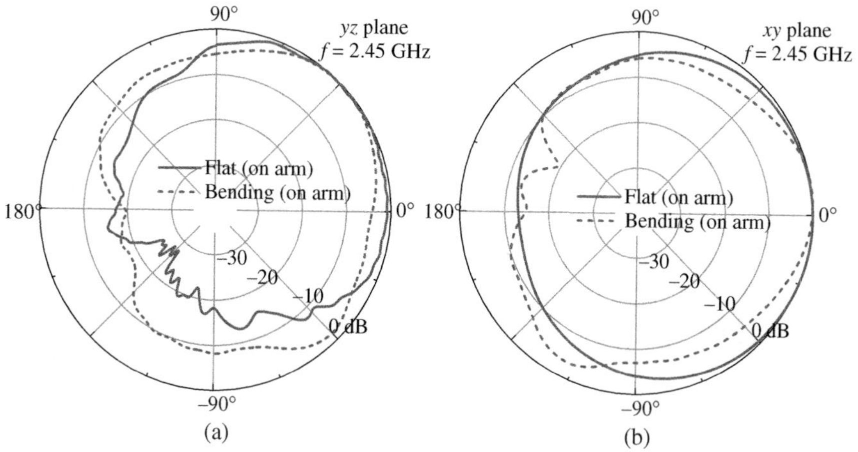

FIGURE 3.18 Normalized radiation polar plots for the $\text{YAGI}_{\text{D-AMC}}$ antenna in flat and bending conditions (arm) for $\varphi = 90°$ (*yz* plane) and $\theta = 90°$ (*xy* plane). *Source:* Agarwal et al. 2016 [22]. Reproduced with permission of IEEE.

voxel as discussed earlier in Figure 3.14a. It can be observed from Figure 3.18a that on bending the antenna with the AMC to a stringent condition of an upper human arm radius of 40 mm, severe distortions in the radiation pattern will result and the endfire gain is reduced by several dB. Also as expected, the $\text{YAGI}_{\text{D-AMC}}$ antenna still preserves its front-to-back ratio under the bending condition similar to that of the flat antenna configuration, as seen in Figure 3.18b. This can be attributed to the fact that the bending deformation along the *H*-plane for an endfire radiation antenna will not affect the radiation pattern in the *xy* plane significantly.

3.7 MEASUREMENT RESULTS

The AMC-backed Yagi antenna prototypes are first designed and optimized on PCB substrate to validate the proposed concept. The planar Yagi radiator is fabricated on Rogers RO4003 substrate, while the single- and double-layered AMCs are fabricated on grounded FR4 substrates, as shown in Figures 3.19 a–3.19c. Commercially available flexible Styrofoam with 5.5 mm thickness is used to separate the Yagi radiator from the AMC surfaces. The flexible antenna prototypes are fabricated by screen printing the silver ink on the latex material for on-body wearable applications, as shown in Figures 3.19 d–3.19f. The latex antenna prototypes are then sliced for overall lateral dimensions of 50 mm × 50 mm, as used in the simulations. The various antenna characteristics in terms of return loss, radiation patterns, and gain measurements are discussed for the PCB- and latex-based antenna prototypes in the subsections 3.7.1, 3.7.2 and 3.7.3.

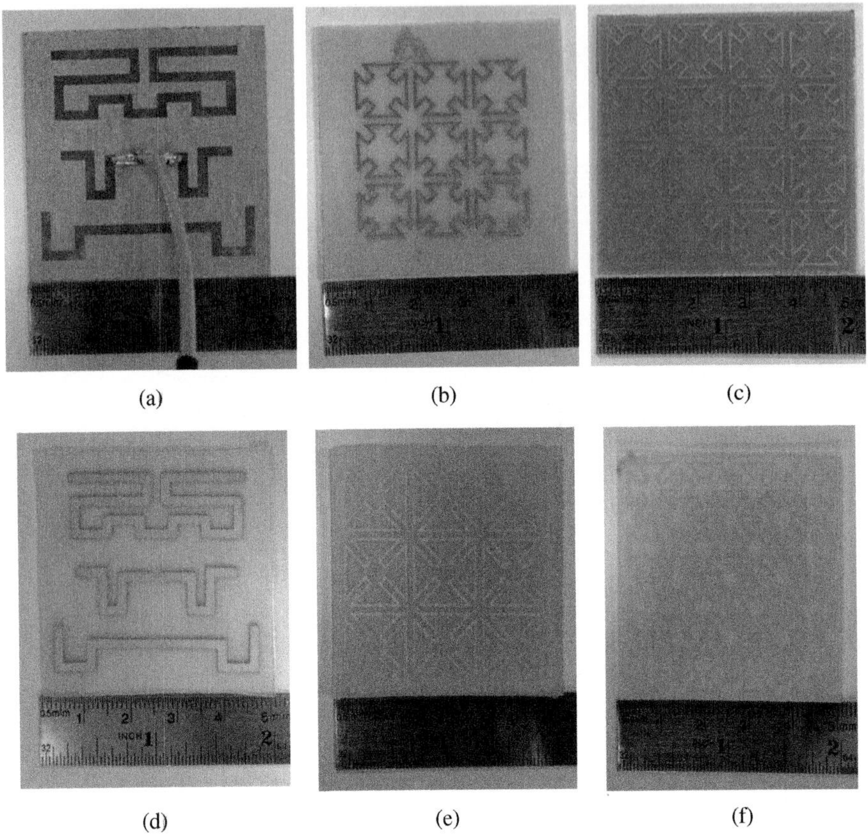

FIGURE 3.19 Antenna prototypes with AMC layers: on FR4 PCB (a) Yagi radiator; (b) double-layered top AMC layer of 3 × 3 units; (c) double-layered bottom AMC layer of 4 × 4 units. Also shown are latex-based AMC layers: (d) Yagi radiator; (e) single layer; and (f) double layer. *Source:* Agarwal et al. 2016 [22]. Reproduced with permission of IEEE. For a color version of this figure, see the color plate section.

3.7.1 Return Loss Measurements

The free-space return loss measurements are acquired using a vector network analyzer R&S ZVL13 for $YAGI_{S\text{-}AMC}$ and $YAGI_{D\text{-}AMC}$ antenna prototypes on PCB, as shown in Figure 3.20. The measured results are in good agreement with the simulations. The recorded value for 10 dB return loss bandwidth for $YAGI_{S\text{-}AMC}$ prototype is 65 MHz (2.43–2.495 GHz), and for $YAGI_{D\text{-}AMC}$ prototype is 60 MHz (2.43–2.49 GHz), respectively.

The return loss measurement graph for the latex-based $YAGI_{S\text{-}AMC}$ and $YAGI_{D\text{-}AMC}$ antenna prototypes in free space are shown in Figure 3.21 with the simulations. A slight mismatch is observed between the simulated results and the actual latex antenna prototype measurements. The recorded value for 10 dB return loss

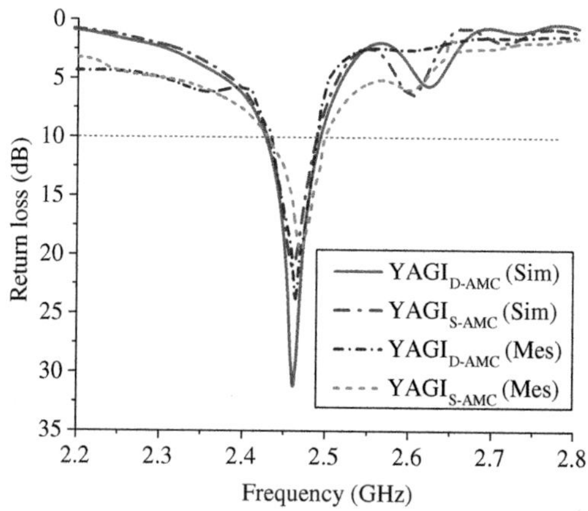

FIGURE 3.20 Measured and simulated return loss graph of YAGI$_{S\text{-}AMC}$ and YAGI$_{D\text{-}AMC}$ antenna prototypes on FR4 PCB. *Source:* Agarwal et al. 2016 [22]. Reproduced with permission of IEEE.

bandwidth for the YAGI$_{S\text{-}AMC}$ antenna prototype is 40 MHz (2.43–2.47 GHz) and for the YAGI$_{D\text{-}AMC}$ antenna prototype is 45 MHz (2.425–2.47 GHz), respectively.

3.7.2 Radiation Pattern Measurements

The far-field radiation patterns are measured in free space using a rotating linearly polarized transmitting horn antenna in an anechoic chamber. The measured and simulated normalized radiation polar plots are shown in Figure 3.22 for PCB-based YAGI$_{S\text{-}AMC}$ and YAGI$_{D\text{-}AMC}$ antenna prototypes at 2.45 GHz. It can be seen that both the *xy* ($\theta = 90°$) and the *yz* ($\varphi = 90°$) plane radiation measured plots have

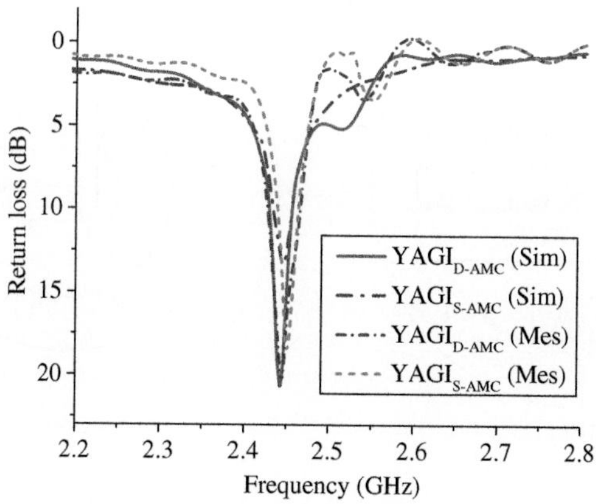

FIGURE 3.21 Measured and simulated return loss graph of YAGI$_{S\text{-}AMC}$ and YAGI$_{D\text{-}AMC}$ antenna prototypes on latex substrate. *Source:* Agarwal et al. 2016 [22]. Reproduced with permission of IEEE.

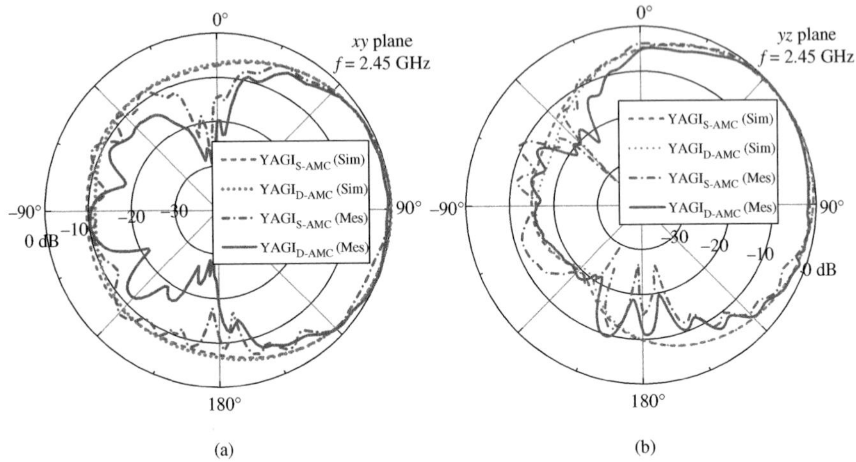

FIGURE 3.22 Measured and simulated normalized radiation patterns of YAGI$_{S\text{-}AMC}$ and YAGI$_{D\text{-}AMC}$ antenna prototypes on FR4 PCB at 2.4 GHz: (a) *xy* plane ($\theta = 90°$) and (b) *yz* plane ($\varphi = 90°$). *Source:* Agarwal et al. 2016 [22]. Reproduced with permission of IEEE.

decent endfire radiation with front-to-back ratio better than 10 dB and beam-tilt radiation of ~74° toward endfire, as observed from the simulations.

The normalized radiation patterns for latex-based antennas are also plotted for the *xy* ($\theta = 90°$) and the *yz* ($\varphi = 90°$) planes in Figure 3.23 for the operating frequency of 2.45 GHz. As seen in Figure 3.23a, both the YAGI$_{S\text{-}AMC}$ and YAGI$_{D\text{-}AMC}$ antenna prototypes have endfire radiation with front-to-back ratio better

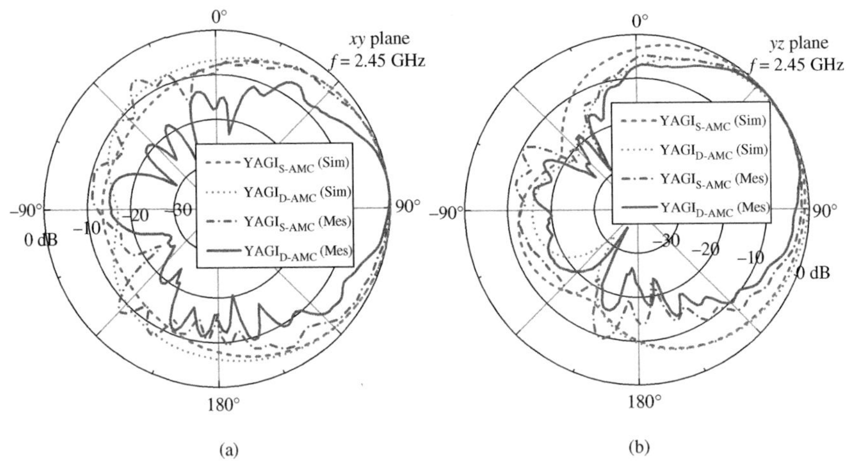

FIGURE 3.23 Measured and simulated normalized radiation patterns of YAGI$_{S\text{-}AMC}$ and YAGI$_{D\text{-}AMC}$ antenna prototypes on latex substrate at 2.4 GHz: (a) *xy* plane ($\theta = 90°$) and (b) *yz* plane ($\varphi = 90°$). *Source:* Agarwal et al. 2016 [22]. Reproduced with permission of IEEE.

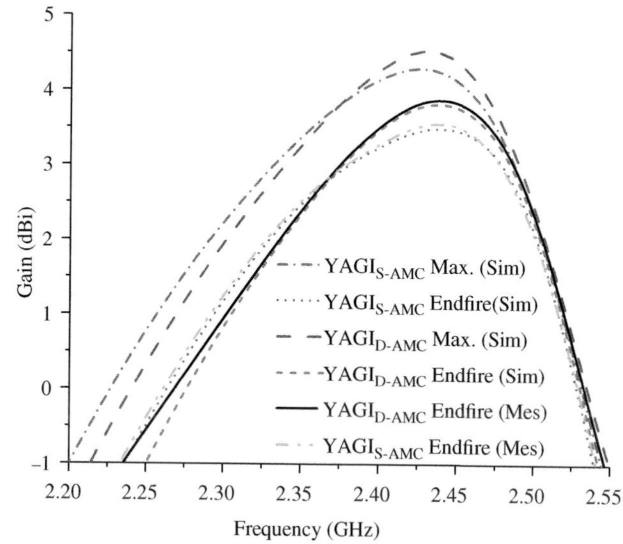

FIGURE 3.24 Simulated (maximum and endfire) and measured (endfire) gain graphs of $YAGI_{S\text{-}AMC}$ and $YAGI_{D\text{-}AMC}$ antenna prototypes on FR4 PCB. *Source:* Agarwal et al. 2016 [22]. Reproduced with permission of IEEE.

than 12 dB. An upward shift in the direction of the main lobe is observed for measured antenna prototype radiation plots compared to simulations, as seen in Figure 3.23b. This shift in the main lobe direction (simulated main lobe: $\theta \sim 74°$) will result in reduced endfire gain. The off-axis beam-tilt radiation toward the endfire direction is very sensitive to the separation between the Yagi antenna and the top surface of the AMC reflector. This upward shift can be attributed to variations in separation due to manual integration of the flexible Yagi radiator, Styrofoam, and AMC surfaces for in-house antenna fabrication.

3.7.3 Gain Measurements

The endfire gain measurements are performed for the $YAGI_{S\text{-}AMC}$ and $YAGI_{D\text{-}AMC}$ antenna configurations in free space in an anechoic chamber using the standard gain comparison method. As shown in Figure 3.24, the measured results for PCB-based antenna prototypes are in perfect agreement with the simulations. The maximum measured gain for the $YAGI_{S\text{-}AMC}$ antenna is 3.8 dBi and for the $YAGI_{D\text{-}AMC}$ antenna is 4.12 dBi at $f = 2.45$ GHz, respectively. Also, maximum simulated gains of 4.44 dBi and 4.78 dBi are observed for $YAGI_{S\text{-}AMC}$ and $YAGI_{D\text{-}AMC}$ PCB antennas in the direction of main lobe (i.e., ~74° toward endfire direction) at 2.45 GHz.

The simulated and measured endfire gain graph is plotted in Figure 3.25 for the latex-based antenna prototypes. A large disagreement is observed between the simulated and the measured results. The maximum measured endfire gain for $YAGI_{S\text{-}AMC}$ antenna is −0.2 dBi in contrast to the simulated gain of 1.02 dBi at the

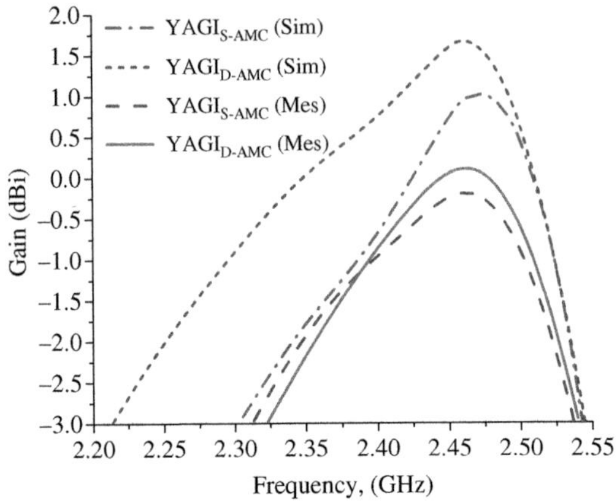

FIGURE 3.25 Measured and simulated endfire gain graphs of $YAGI_{S\text{-}AMC}$ and $YAGI_{D\text{-}AMC}$ antenna prototypes on latex substrate. *Source:* Agarwal et al. 2016 [22]. Reproduced with permission of IEEE.

operating center frequency of 2.45 GHz. Similarly, the $YAGI_{D\text{-}AMC}$ antenna has a maximum measured endfire gain of 0.12 dBi as compared to the simulated gain of 1.81 dBi at 2.45 GHz. This discrepancy in the simulated and measured results can be attributed to the observed shift in direction of the main lobe caused by the imperfect in-house latex antenna fabrication, as discussed in Section 3.7.2.

3.8 CONCLUSION

In this chapter, a novel technique of combining endfire radiation wearable planar antennas with AMCs was discussed for improving the antenna tolerance when positioned on the human body. A conventional planar Yagi antenna was combined with single- and double-layered AMC surfaces to change the direction of radiation from bidirectional-endfire to near-endfire off-axis beam-tilt radiation. The free-space and on-body analyses were performed using the CST MWS voxel model for common antenna properties, such as return loss, gain, frequency shifting, radiation efficiency, and peak SAR level considering body-worn applications. Material property characterization, such as permittivity and loss tangent, was performed for the flexible latex substrate. Both Yagi antennas and AMC surfaces were fabricated using PCB fabrication to verify the concept and on all-flexible latex substrate for body-worn applications. On-voxel analysis shows that the double-layered AMC improves the front-to-back ratio, frequency detuning, radiation efficiency, and peak SAR level compared to the Yagi antenna backed by a single-layered AMC and without AMC. The fabricated double-layered AMC-backed Yagi antenna prototype on latex substrate demonstrates acceptable return loss bandwidth and endfire gain to be used for WBAN applications in 2.4 GHz ISM band.

REFERENCES

[1] E. Jovanov, A. Milenkovic, C. Otto, and P. C. de Groen, "A wireless body area network of intelligent motion sensors for computer assisted physical rehabilitation," *J. Neuroeng. Rehab.*, vol. 2, no. 11, pp. 6, March 2005.

[2] C. C. C. Y. Poon, Y.-T. Zhang, and S.-D. Bao, "A novel biometrics method to secure wireless body area sensor networks for telemedicine and m-Health," *IEEE Commun. Mag.*, vol. 44, no. 4, pp. 73–81, April 2006.

[3] M. Klemm and G. Troester, "Textile UWB antennas for wireless body area networks," *IEEE Trans. Antennas Propag.*, vol. 54, no. 11, pp. 3192–3197, November 2006.

[4] P. S. Hall, Y. Hao, Y. I. Nechayev, A. Alomalny, C. C. Constantinou, C. Parini, M. R. Kamarudin, T. Z. Salim, D. T. M. Hee, R. Dubrovka, A. S. Owadally, W. Song, A. Serra, P. Nepa, M. Gallo, and M. Bozzetti, "Antennas and propagation for on-body communication systems," *IEEE Antennas Propag. Mag.*, vol. 49, no. 3, pp. 41–58, June 2007.

[5] A. Alomainy, A. Sani, A. Rahman, J. G. Santas, and Y. Hao, "Transient characteristics of wearable antennas and radio propagation channels for ultra wideband body-centric wireless communications," *IEEE Trans. Antennas Propag.*, vol. 57, no. 4, pp. 875–884, April 2009.

[6] T. Kennedy, P. Fink, A. Chu, N. Champagne, G. Lin, and M. Khayat "Body-worn E-Textile antennas: the good, the low-mass, and the conformal," *IEEE Trans. Antennas Propag.*, vol. 57, no. 4, pp. 910–918, April 2009.

[7] L. Vallozzi, P. Van Torre, C. Hertleer, H. Rogier, M. Moeneclaey, and J. Verhaevert, "Wireless communication for firefighters using dual-polarized textile antennas integrated in their garment," *IEEE Trans. Antennas Propag.*, vol. 58, no. 4, pp. 1357–1368, April 2010.

[8] E. K. Kaivanto, M. Berg, E. Salonen, and P. de Maagt, "Wearable circularly polarized antenna for personal satellite communication and navigation," *IEEE Trans. Antennas Propag.*, vol. 59, no. 12, pp. 4490–4496, August 2011.

[9] A. A. Serra, P. Nepa, and G. Manara "A wearable two-antenna system on a life jacket for Cospas-Sarsat personal locator beacons," *IEEE Trans. Antennas Propag.*, vol. 60, no. 2, pp. 1035–1042, October 2012.

[10] P. S. Hall and Y. Hao, *Antennas and Propagation for Body-Centric Wireless Communications*. Artech House, Norwood, MA 2012.

[11] K. Agarwal, Nasimuddin, and A. Alphones, "RIS-based compact circularly polarized microstrip antennas," *IEEE Trans. Antennas Propag.*, vol. 61, no. 2, pp. 547–554, February 2013.

[12] K. Agarwal, Nasimuddin, and A. Alphones, "Wideband circularly polarized AMC reflector backed aperture antenna," *IEEE Trans. Antennas Propag.*, vol. 61, no. 3, pp. 1456–1461, March 2013.

[13] S. Bashir, M. Hosseini, R.M. Edwards, M. I. Khattak, and L. Ma, "Bicep Mounted Low Profile Wearable Antenna Based on a Non-uniform EBG Ground Plane-Flexible EBG Inverted-L (FEBGIL) Antenna," in Proceedings of the Loughborough Antennas and Propagation Conference, Loughborough, March 17–18, 2008, pp. 333–336.

[14] Z. Duan, D. Linton, W. Scanlon, and G. Conway, "Improving Wearable Slot Antenna Performance with EBG Structures," in Proceedings of the Loughborough Antennas and Propagation Conference, Loughborough, March 17–18, 2008, pp. 173–176.

[15] Z. Shaozhen and R. Langley, "Dual-band wearable textile antenna on an EBG substrate," *IEEE Trans. Antennas Propag.*, vol. 57, no. 4, pp. 926–935, April 2009.

[16] Q. Bai and R. Langley, "Wearable EBG Antenna Bending and Crumpling," in Proceedings of the Loughborough Antennas and Propagation Conference, Loughborough, November 1–17, 2009, pp. 201–204.

[17] S. Kim, Y.-J. Ren, H. Lee, A. Rida, S. Nikolaou, and M. M. Tentzeris, "Monopole antenna with inkjet-printed EBG array on paper substrate for wearable applications," *IEEE Antennas Wireless Propag. Lett.*, vol. 11, pp. 663–666, June 2012.

[18] A. Foroozesh and L. Shafai, "Investigation into the application of artificial magnetic conductors to bandwidth broadening, gain enhancement and beam shaping of low profile and conventional monopole antennas," *IEEE Trans. Antennas Propag.*, vol. 59, no. 1, pp. 4–20, January 2011.

[19] K. Agarwal, Nasimuddin, and A. Alphones, "Unidirectional wideband circularly polarized aperture antennas backed with artificial magnetic conductor reflectors," *IET Microwaves Antennas Propag.*, vol. 7, no. 5, pp. 338–346, April 2013.

[20] K. Agarwal, Y. X. Guo, B. Salam, and L. C. W. Albert, "Latex Based Near-Endfire Wearable Antenna Backed by AMC surface," in Microwave Workshop Series on RF and Wireless Technologies for Biomedical and Healthcare Applications (IMWS-BIO), 2013 IEEE MTT-S International, Singapore, December 9–11, 2013, pp. 1–3.

[21] CST Microwave Suite Ver. 2012.08, CST Voxel Human Body Model. Available at http://www.cst.com

[22] K. Agarwal, Y. X. Guo, and B. Salam, "Wearable AMC backed near-endfire antenna for on-body communications on latex substrate," *IEEE Transactions on Components, Packaging and Manufacturing Technology*, vol. 6, no. 3, pp. 346–358, March 2016.

[23] D. I. Amey and J. P. Curilla, "Microwave Properties of Ceramic Materials," in Proceedings of the 41st Electronic Components and Technology Conference (ECTC 91), Atlanta, GA, May 11–16, 1991, pp. 267–272.

[24] K.-P. Latti, J.-M. Heinola, M. Kettunen, J.-P. Strom, and P. Silventoinen, "A Review of Microstrip T-Resonator Method in Determination of Dielectric Properties of Printed Circuit Board Materials," in Proceedings of the IEEE Instrumentation and Measurement Technology Conference, IMTC 2005,, Ottawa, May 16–19, 2005, pp. 62–66.

[25] "IEEE Standard for Safety Levels With Respect to Human Exposure to Radio Frequency Electromagnetic Fields, 3 kHz to 300 GHz," IEEE Std C95.1-2005 (Revision of IEEE Std C95.1-1991), 2006, pp. 1–238.

[26] K. Agarwal and Y. X. Guo, "Interaction of Electromagnetic Waves with Humans in Wearable and Biomedical Implant Antennas," in 2015 IEEE Asia-Pacific International Symposium on Electromagnetic Compatibility (APEMC), Taipei, May 25–29, 2015, pp. 154–157.

4

NOVEL ANTENNA DESIGNS AND CHARACTERIZATION METHODOLOGIES FOR MEDICAL DIAGNOSTICS AND SENSING

Harish Rajagopalan and Yahya Rahmat-Samii
Department of Electrical Engineering, University of California, Los Angeles, CA, USA

4.1 INTRODUCTION

About quarter of a century ago, the technology and research community was advertising the concept of "*global connectivity with anyone, at any time, in any place and with any amount of data.*" This dream is now a reality due to the technological advancement in electromagnetics, antenna design, wireless communications, and microwave theory leading to the design of various innovative portable wireless devices (cell phones, laptops, tablets). The next big concept is "*bio-telemetry connecting patients to their doctors and hospitals at anytime, any location and with any amount of monitoring and diagnostics data.*" We are in the midst of this new paradigm becoming a reality. Figure 4.1 shows a summary of some of the human–antenna interaction works (by the author) over the past two decades including EM interaction between handset antennas and human [1–3], wearable antennas for on-body communications [4, 5], implantable antennas for medical devices [6–8], ingestible antennas for medical capsule imaging systems [9, 10], and brain interface systems.

Electromagnetic applications in the field of medicine and healthcare can be classified as diagnostics—magnetic resonance imaging (MRI) and tumor detection [11, 12] and therapeutics—diathermy, hyperthermia for cancer therapy, and microwave

Electromagnetics of Body-Area Networks: Antennas, Propagation, and RF Systems, First Edition.
Edited by Douglas H. Werner and Zhi Hao Jiang.

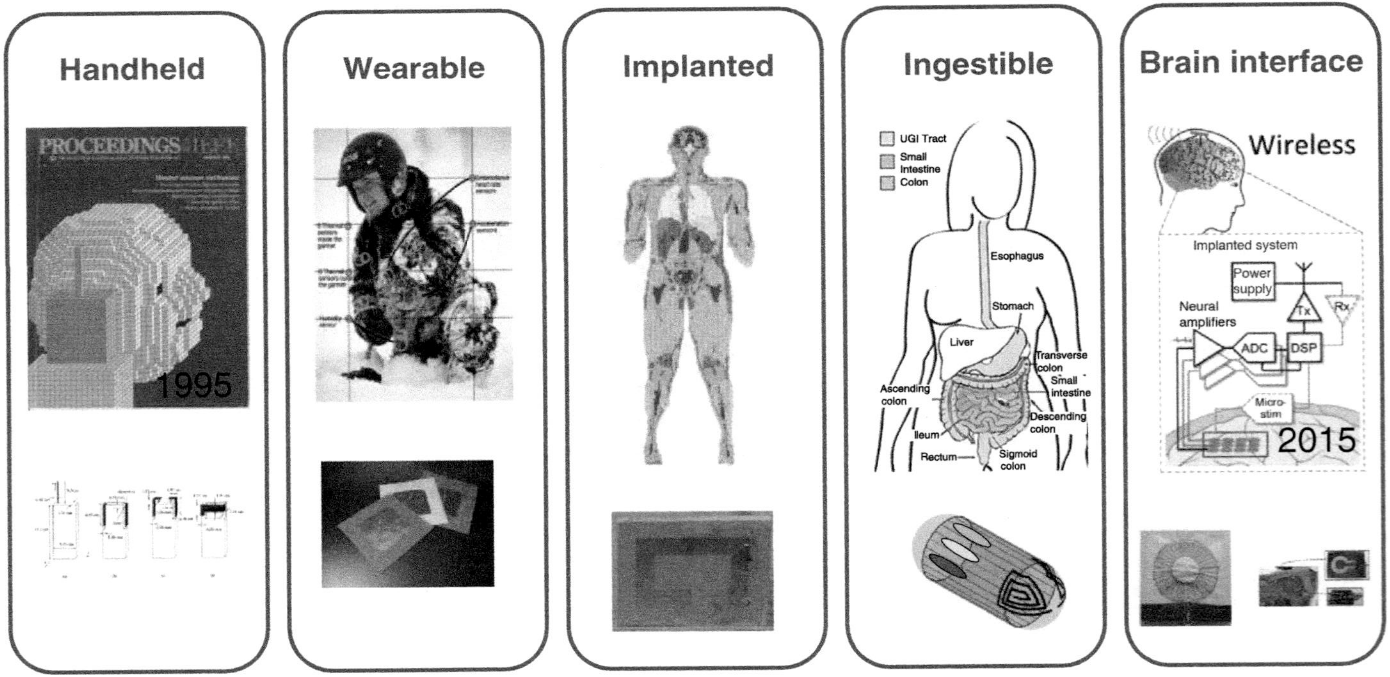

FIGURE 4.1 Personal communications and human–antenna interaction research activities at UCLA antenna research analysis and measurement (ARAM) laboratory, under the directorship of Prof. Yahya Rahmat-Samii covering the era of 1990–2015 and beyond.

angioplasty [13, 14]. In the recent years, innovative use of *electromagnetics* and *wireless technology* has significantly aided the development and successful implementation of state-of-the-art medical devices like pacemakers, imaging pills, and on-body sensors in medicine and healthcare. These devices can reliably perform diagnostic, sensing, and in some cases therapeutic treatments. More importantly these systems help enhance patient lifestyle and comfort to a far greater level.

Electromagnetics and *wireless technology* are also playing an increasingly important role in healthcare processes like patient registration and identification (radio frequency identification (RFID) tags [15, 16]), vital signs monitoring (on-body sensors), telemedicine (reliable online medical data transfer), and others. Devices like smartphones/tablets are being used to monitor and track medical data thus allowing patients to be outside the confines of the hospitals. Some of the technological advancements in healthcare that have emerged due to the successful integration of electromagnetic theory, antenna design, and wireless technology are:

- *Telemedicine:* A physician can remotely monitor a patient while the patient can successfully continue to perform his day-to-day activities. It provides a platform for patient–physician interaction online. The accuracy and reliability of this system relies on accurate medical data monitoring, collection, storage, and transfer.
- *Mobile Health:* Enables patient/physician to access all medical data at any time, place via smartphones and tablets which relies on successful wireless data transfer.
- *On-Body/In-body Sensors:* Ranging from simple fitness trackers to complex pacemakers and imaging capsules, these sensors monitor the patients' vital stats and transfer that information to a nearby data collection center (base station/mobile device). All the data transfer can be done wirelessly.
- *Location Services:* Help track, maintain and store hospital equipment and also manage, track and identify patients in the hospital.

Remote monitoring of the patient is possible which opens a new paradigm for health diagnosis. Figure 4.2 shows some types of home healthcare bio-telemetry systems. In one system as shown in Fig. 4.2a, an implantable/ingestible device communicates and transfers medical data to cell phone. The cell phone can instantaneously transfer the medical information to the hospital where the patient's vital signs (medical data) can be continuously monitored and appropriate action can be taken. This system is helpful when monitoring patients' vital signs during real-time monitoring. Patient mobility is no longer restricted. In the other system as shown in Fig. 4.2b, a home base station which has the ability to store relevant medical data communicates periodically with the implant. The implant contains the implantable/ingestible antenna along with the electronics that allow data and signal processing. This medical data is stored in the home base station and periodically updated in the data collection server which can be remotely accessed by the hospital. This system is good for patient monitoring on a regular basis to check patient

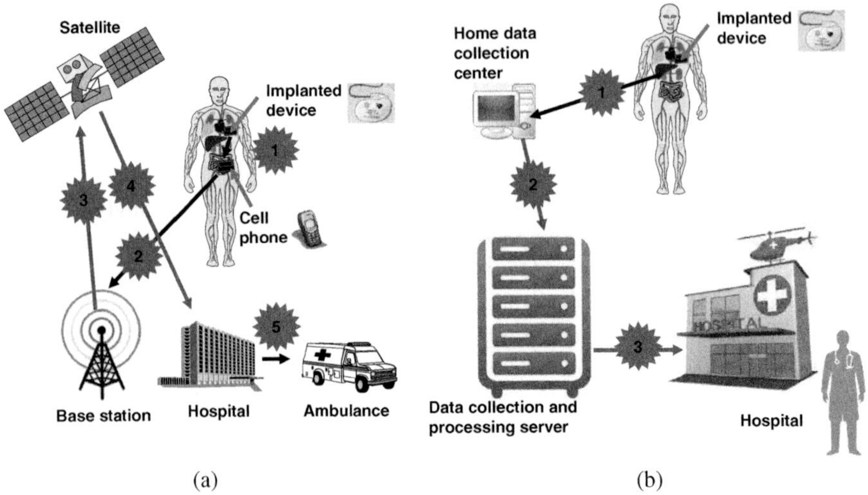

FIGURE 4.2 Sketch of a home healthcare bio-telemetry system. (a) Instantaneous medical data can be transmitted to the hospital via the cell phone link. (b) Medical data can be recorded and stored and reviewed over a span of few hours/days.

progress and recovery. Some other pertinent aspects that also affect these medical system workings are data transfer path/channel between the base station and the on-/in-body sensor, and human body due to its dispersive nature and complex modeling. In order to establish secure and stable wireless communication links for these medical devices, antennas that are placed inside and outside the human body need to be electromagnetically characterized in complete detail.

Typically, while designing antenna solutions for these implantable/ingestible devices, physical constraints and electromagnetic nature of the problem have to be given serious thought. The antenna along with electronics, batteries, and bio-sensors has to fit in the volume of the implantable device. The design of a miniaturized antenna itself can pose severe challenges due to the space constraints and the frequency of operation. The initial conception of the antenna design and the miniaturization process is a critical component as the electrically small antenna can suffer from poor radiation efficiency if not designed with caution. The antenna has to radiate in the presence a very complex environment with high inhomogeneity, frequency-dependent electrical properties and different lossy materials. The interaction of the antenna with other components of the device and the ability to provide robust performance against the battery and other electronics will also need to be taken into account during the antenna design. Finally, the EM radiation has to comply with SAR requirements and emitted power regulations (effective isotropic radiated power (EIRP)). This shows that the design of an implantable/ingestible antenna is a multifold problem where various issues have to be tackled simultaneously for successful implementation. The challenge is to design a miniaturized implantable/ingestible antenna with effective performance to operate in the human body environment for such complex medical systems [17, 18].

4.1.1 Antennas for Implantable Devices: Operating Frequency-Based Classification

In this chapter, based on the frequency of operation, antenna design for *wireless medical implantable devices* is classified into the following categories [18–20]:

Low-Frequency Band: Low-frequency (low kHz) inductive links/implants have long been used for bio-telemetry for implantable medical devices. Heart pacemakers and cochlea implants are two of the early examples of successful implantable devices that integrated electromagnetics successfully into medicine [21–23]. A pacemaker is used to help control abnormal heart rhythms and a cochlea implant can help to provide a sense of sound to a person who is severely hard-of-hearing. Traditionally, communication with the implants was performed over an inductive link. Typically, information from the implanted inductive implant device is accessed by means of an external antenna (device) placed on a patient to establish the inductive communication link. The inductive coupling between the two coils is used to transfer data to and from the implant. The placement of the external coil is sensitive as it determines the performance and speed of the electromagnetic link as shown in Figure 4.3. These links typically suffer from low data rates (1–30 kbps) and extremely short range of communication (10–20 cm).

Medical Implant Communication Service (MICS): FCC allocated MICS band for use by body-worn monitoring devices. These devices operate in the 402–405 MHz band. MICS band provides high-data-rate transmission, communication to about 10 feet, and also lends to miniaturized antenna designs and acceptable human body losses. This band is suited for tissue penetration at relatively low power thus extending the life of the medical implant. Due to higher data rate compared to inductive links medical data can be recorded by data collection center at home without maintaining contact with the patient at all times.

Typically, single layer spiral and meander planar antennas are the most common topologies used for antenna designs in MICS band. The natural geometry of these structures facilitates miniaturization of the antenna structures [24–27]. Figure 4.4a shows a meandered PIFA (planar inverted F-antenna) antenna with a grounding pin at the end of the radiator working at 402 MHz [7]. The printed radiator is located

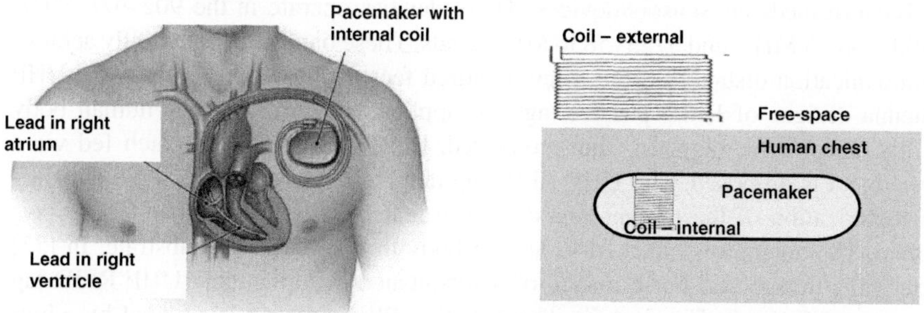

FIGURE 4.3 Typical inductive link for a pacemaker (internal and external coils).

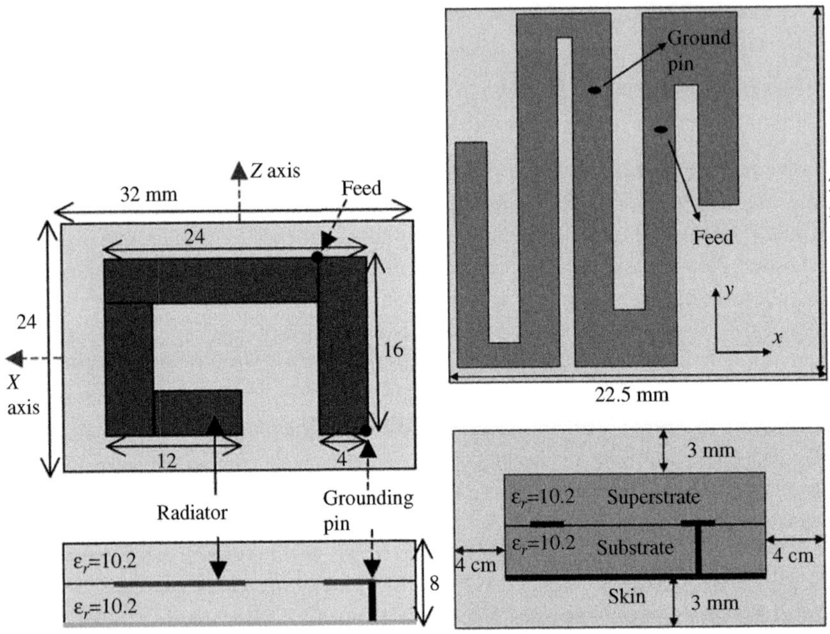

FIGURE 4.4 MICS band antenna design. (a) Meandered PIFA design at 402 MHz. *Source:* Kim and Rahmat-Samii 2004 [6]. Reproduced with permission of IEEE. (b) PSO-optimized serpentine dual-band antenna configuration at 402 MHz and 2.4 GHz. *Source:* Karacolak et al. 2008 [28]. Reproduced with permission of IEEE.

between the substrate and superstrate (biocompatible) dielectric layers. Figure 4.4b shows a dual-band serpentine configuration to optimize antenna surface area working at 402 MHz and 2.4 GHz [28]. A shorting pin is also used for miniaturization purposes. The antenna is optimized using global optimization technique (particle swarm optimization). Some of the miniaturized antenna designs for implantable/ingestible applications for MICS band can be found in [24–32].

ISM and Wi-Fi: ISM and Wi-Fi can also be potentially used for implanted and/or body-worn medicals sensors/devices. These devices operate in the 902–928 MHz, 2400–2483.5 MHz, and 5725–5850 MHz bands. These bands can potentially achieve communication distances up to a few hundred feet. In [33–35] present some UHF antenna designs of RFID passive tags for applications involving the human body as the object to be tagged or bio-monitored. In [34], a suspended patch fed via a nested slot is proposed. The UHF RFID tag design in shown in Fig. 4.5a. Detailed characterization of the tag antenna design to match RFID chip impedance is presented. The antenna is attached to human body through silicone substrate. In [35] details the design and performance of a subcutaneous implantable UHF RFID tag antenna as shown in Figure 4.5b. The radiating PIFA antenna is covered by a biocompatible superstrate of dielectric constant 10.2 and thickness of 1.25 mm and

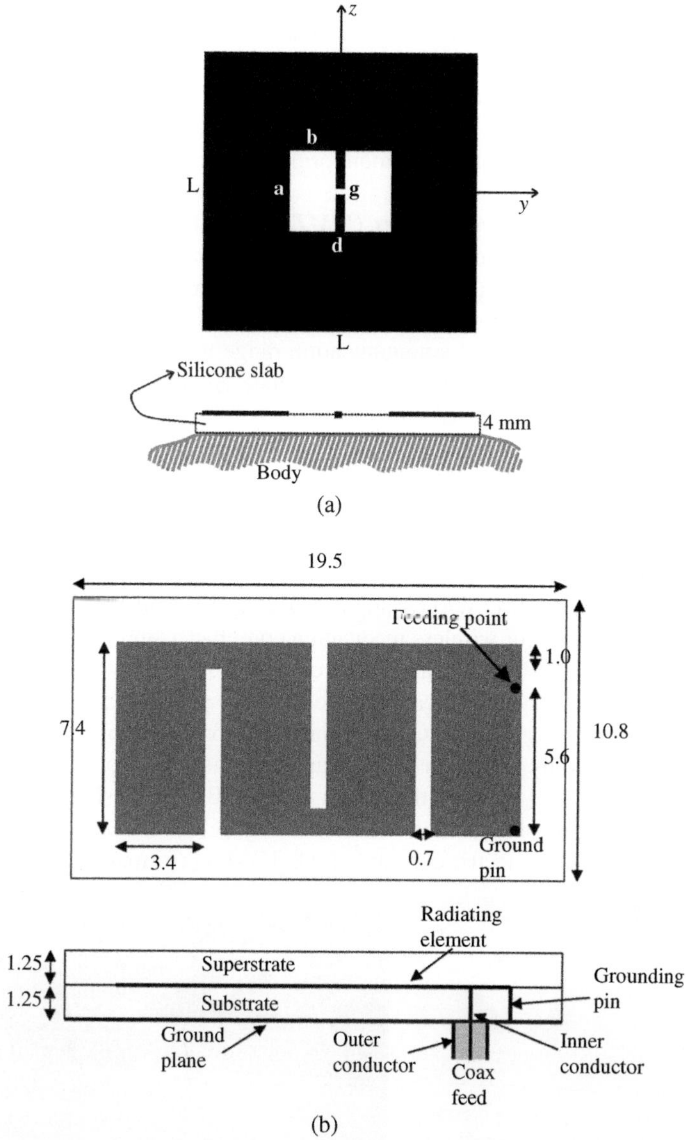

FIGURE 4.5 UHF RFID antenna design. (a) A suspended patch on silicone substrate fed via nested slot. *Source:* Marrocco 2007 [34]. Reproduced with permission of IEEE. (b) Subcutaneous meandered PIFA design with impedance match to the RFID chip. *Source:* Sani et al. 2010 [35]. Reproduced with permission of IEEE.

the antenna is matched to the highly capacitive impedance of the microchip. 2.4 GHz and 5.7 GHz antenna designs and channel propagation characteristics for implantable devices can be found in [36–41]. Some of the designs include cavity slot antenna with a dumbbell shape [37], three-dimensional (3D) folded dipole antennas for implantable retinal prosthesis [38], implantable planar inverted-F antennas designed for intracranial pressure (ICP) monitoring [39], and cardiovascular stents as antennas [40].

Wireless Medical Telemetry Services (WMTS): WMTS uses unlicensed spectrum to communicate data from body sensors to remote monitoring locations. The FCC provided the WMTS a total of 15 MHz of spectrum in three bands, 608–614 MHz, 1395–1400 MHz, and 1427–1429.5 MHz and the devices operating in these bands can have a potential communication range up to several hundred feet. Some of the antenna designs for these bands include applications in wireless capsule endoscopy, implantable retinal prosthesis, and medical compliance [42–46]. In [43] details the design of a miniaturized meandered dipole antenna, as shown in Figure 4.6a, used for retinal prosthesis data telemetry where the antenna design considers the vector alignment concept and space filling techniques. Figure 4.6b shows an electrically small zigzag dipole antenna patterned directly onto the outer surface of a standard-sized capsule [45]. The antenna structure is patterned using screen printed silver colloidal inks for ISM, WMTS applications. The capsule antenna design is experimentally evaluated for wireless medication compliance applications.

Ultra-Wideband (UWB): Medical devices operating at very low power in almost any region of the UWB spectrum can operate at distances up to a few feet. UWB devices can be used for very high data rate communications (>1 Gbps), which makes the technology potentially useful for medical applications that involve large medical database transfers. In [47] presents the design of a UWB capsule-shaped dielectric-loaded implantable wide-slot antenna as shown in Figure 4.7. The antenna is printed on a substrate that has a dielectric constant of 9.8. UWB performance is achieved by

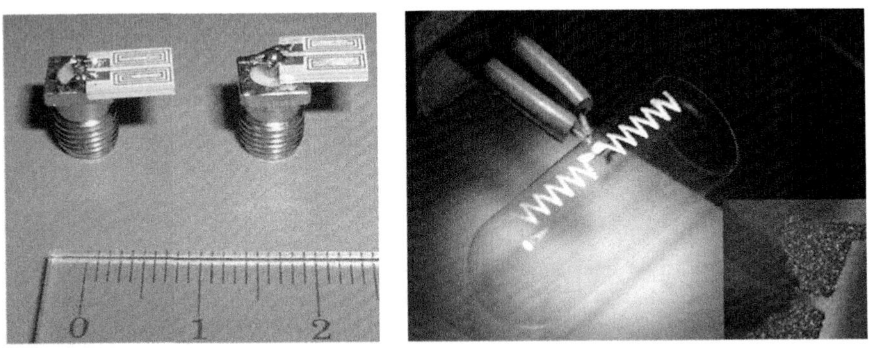

(a) (b)

FIGURE 4.6 WMTS antenna designs. (a) A miniaturized meandered dipole antenna for implantable retinal prosthesis. *Source:* Gosalia et al. 2005 [43]. Reproduced with permission of IEEE. (b) An electrically small zigzag dipole using screen printed silver inks for medical compliance. *Source:* Bashirullah and Euliano 2009 [45]. Reproduced with permission of IEEE.

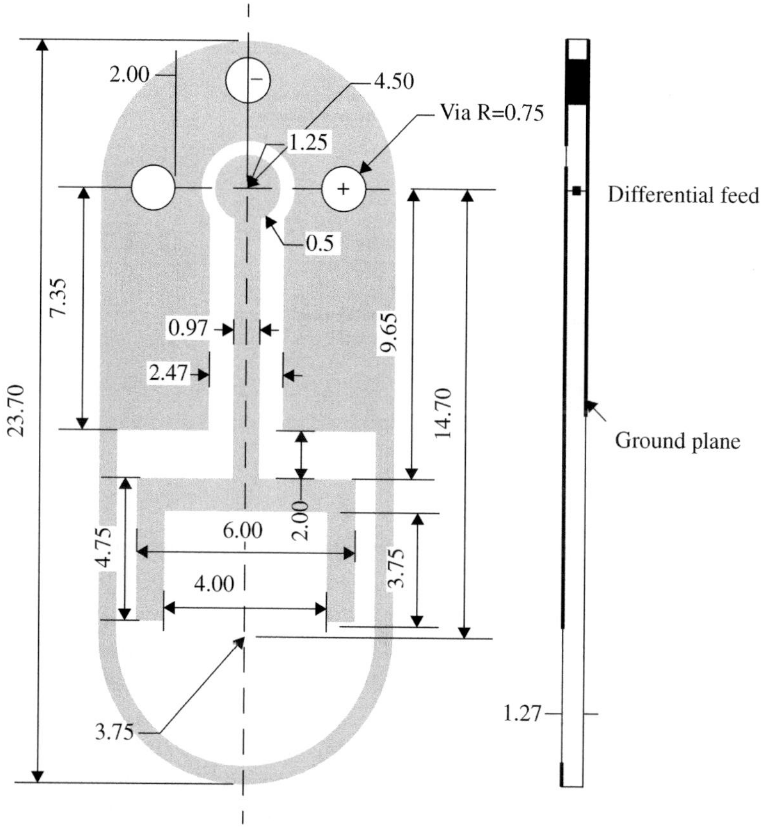

FIGURE 4.7 UWB capsule-shaped dielectric-loaded implantable wide-slot antenna. *Source:* Dissanayake et al. 2009 [47]. Reproduced with permission of IEEE.

loading the antenna with suitable insulators which can deliver broadband matching across a range of dissipative medium properties.

4.1.2 Methodologies for Numerical Simulations and Characterizations

Figure 4.8 shows the different methodologies for the design of implantable/ingestible antennas using numerical simulations and characterizations [48, 49]. In numerical simulations and characterizations, implantable/ingestible antennas are analyzed inside lossy media that simulate biological tissues with different permittivity and conductivity and also vary with frequency. For the implantable/ingestible antenna design, one needs to pay particular attention to RF absorptive properties of the tissues, different interfaces and layers of tissues through which the RF energy travels and corresponding reflections among them, and lastly, the detuning effects on the antenna impedance in these complex environments. A comprehensive database of dielectric properties of body tissues at RF and microwave frequencies is provided in [50–52].

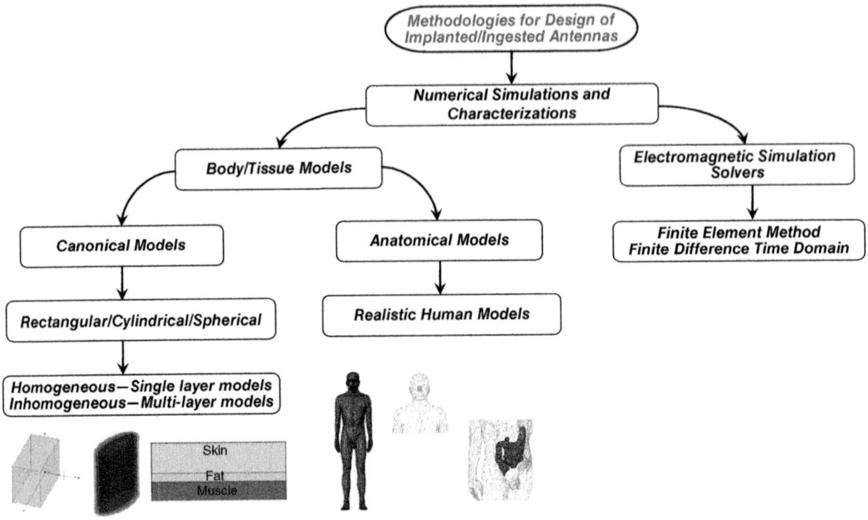

FIGURE 4.8 Schematic diagram showing the methodologies for the antenna design of implantable/ingestible antennas using numerical simulations and characterizations.

The different tissue/body models used for numerical characterizations can be broadly broken into canonical models and anatomical models. Canonical tissue models are typically based on rectangular, spherical, or cylindrical geometries. They can either be used to model single homogenous tissue (muscle/skin) or be used to model multi-layer tissues like skin, fat, and muscle combination. Analytical methods can be applied for analyzing the inside canonical tissue models [53]. Single homogeneous tissue models speed up computation process but ignore reflection among different tissue layers. For implantable medical devices (pacemakers) which are placed deep enough in the muscle where there are minimal fat and skin layers, single tissue models are accurate enough as absorption phenomenon mostly dominates [9]. Multi-layer canonical tissue models increase computation time but can more accurately predict inter-tissue reflection and overall absorption. These models should be used when the medical implant is shallow and RF energy has to go through multiple layers of equal thickness (muscle, fat, skin) [6, 27]. For most realistic and accurate results, anatomical tissue models can be utilized. These human/tissue models are comprised of 300+ objects including bones, organs, and muscles. They have millimeter scale accuracy and frequency-dependent electrical parameters including loss tangent, conductivity, and permittivity of different tissues. These models can be computationally expensive but accurate [9, 40].

Recently, electromagnetic full-wave solvers are being utilized for numerical simulations. Among modern-day full-wave electromagnetic simulators, there are two main types for human body modeling: finite element method (FEM), and finite-difference time domain (FDTD). These solvers achieve greater accuracy for this application because they utilize volume meshing. Ansoft HFSS, CST Microwave Studio, and Remcom XFDTD are some of the commercial solvers available [49].

Based on the different methodologies for numerical simulations and characterizations, a path can be developed for optimized implantable/ingestible antenna designs. As a first pass, a simple homogeneous tissue model such as a box/cylinder with the appropriate dielectric properties can be used for initial antenna design/tuning. Depending on the location and type of medical implant, multi-layer canonical tissue models can be explored for antenna tuning and optimization. The next step can be integrating some other organs in the vicinity of the medical device for greater accuracy of the simulation modeling. Once the initial antenna design is developed, a comprehensive anatomical model can be used for the final design and characterization of the antenna in the wireless medical device.

Figure 4.9 shows the detailed outline of the chapter. The novel antenna designs for medicine diagnostics and sensing are divided into two broad categories: ingestible antenna designs and on-body antenna designs. The ingestible antenna design solutions proposed here can be used for applications such as wireless capsule endoscopy diagnostics and medicine compliance sensing. For the ingestible antenna design for wireless capsule endoscopy, miniaturized conformal space filling antenna design at 1.4 GHz (WMTS band) is discussed in detail. Bio-telemetric link budget calculations and SAR analysis are performed in detail. A reconfigurable antenna system concept is also discussed where the ingestible antenna system can achieve frequency reconfigurablity using wireless power transfer. A comparison is made with other antenna designs for such capsule systems. For the ingestible antenna design for medical compliance sensing, miniaturized conformal space filling antenna design 915 MHz (ISM band) is discussed in detail. Tuning parameters and matching the antenna to chip impedance is discussed. With the aid of full-wave 3D electromagnetic software, high-frequency structure simulator (HFSS), the power reflection coefficient, radiation performance, input impedance are analyzed in order to achieve a final design that meets the specific requirements. Potential batch implementation of such antennas is also discussed. Finally, on-body antennas including UHF RFID tag antennas for patient monitoring is discussed in detail. The tag antenna design is described followed by a simplified circuit model to understand the working of the tag. On-body impedance characterizations are performed and read range measurements are conducted and the results are discussed.

4.2 INGESTIBLE ANTENNA DESIGN AT WMTS BAND: WIRELESS CAPSULE ENDOSCOPY DIAGNOSTICS

Wireless capsule endoscopy has provided a unique platform to map the complete digestive track in real time [54–56]. This non-invasive process in which the patient swallows a vitamin-pill-sized capsule is fast replacing traditional endoscopy as a very practical diagnostic tool (Figure 4.10). Traditional "Scope Tests" would only achieve imaging for 10 out of the 30 feet of the digestive track (the upper 4 feet of the stomach, the colon, and the rectum) leaving the remaining 20 feet of small intestines. With wireless capsule endoscopes, real-time imaging of the small intestines is possible and gastrointestinal tract diseases like intestinal bleeding and ulceration can

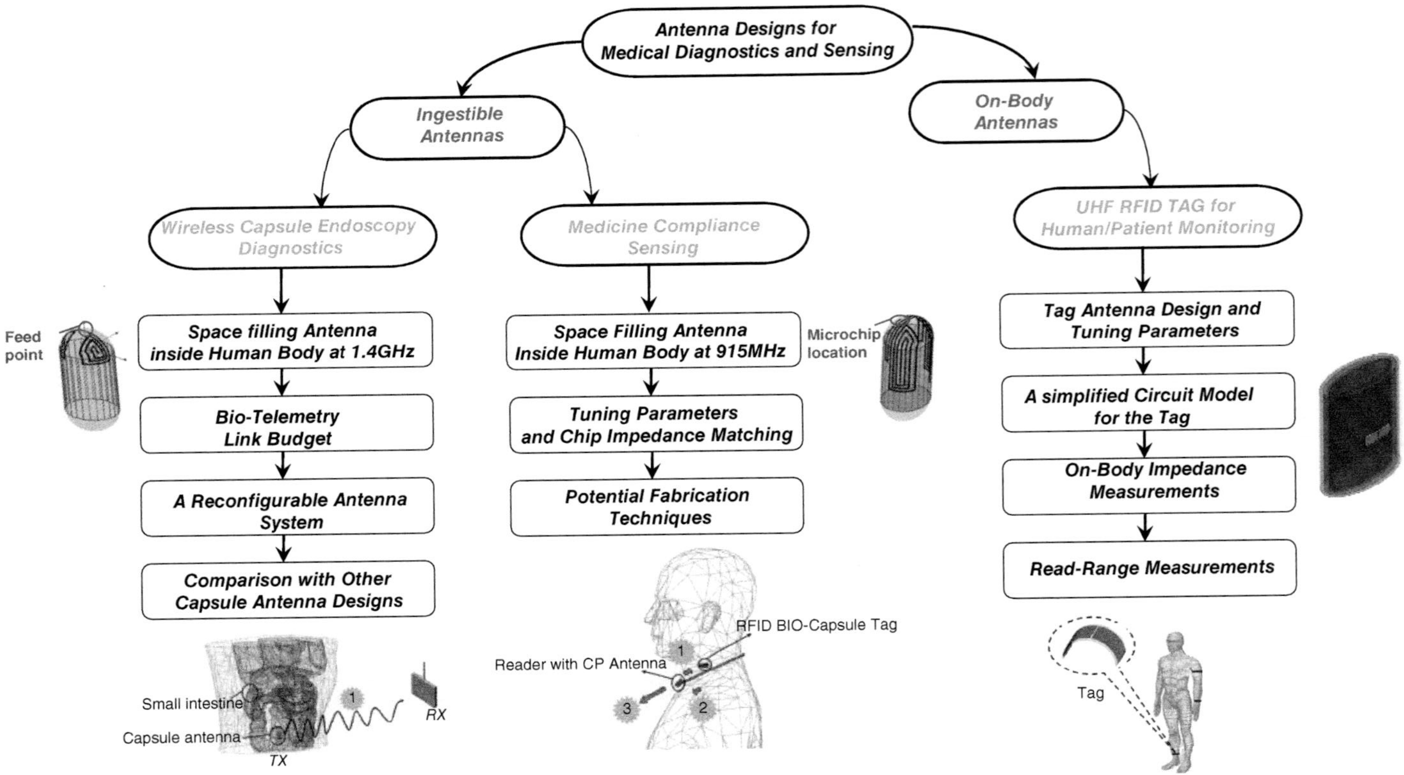

FIGURE 4.9 Schematic diagram showing the methodologies used for the design of ingestible antennas and on-body antenna in this chapter for medical diagnostics and sensing.

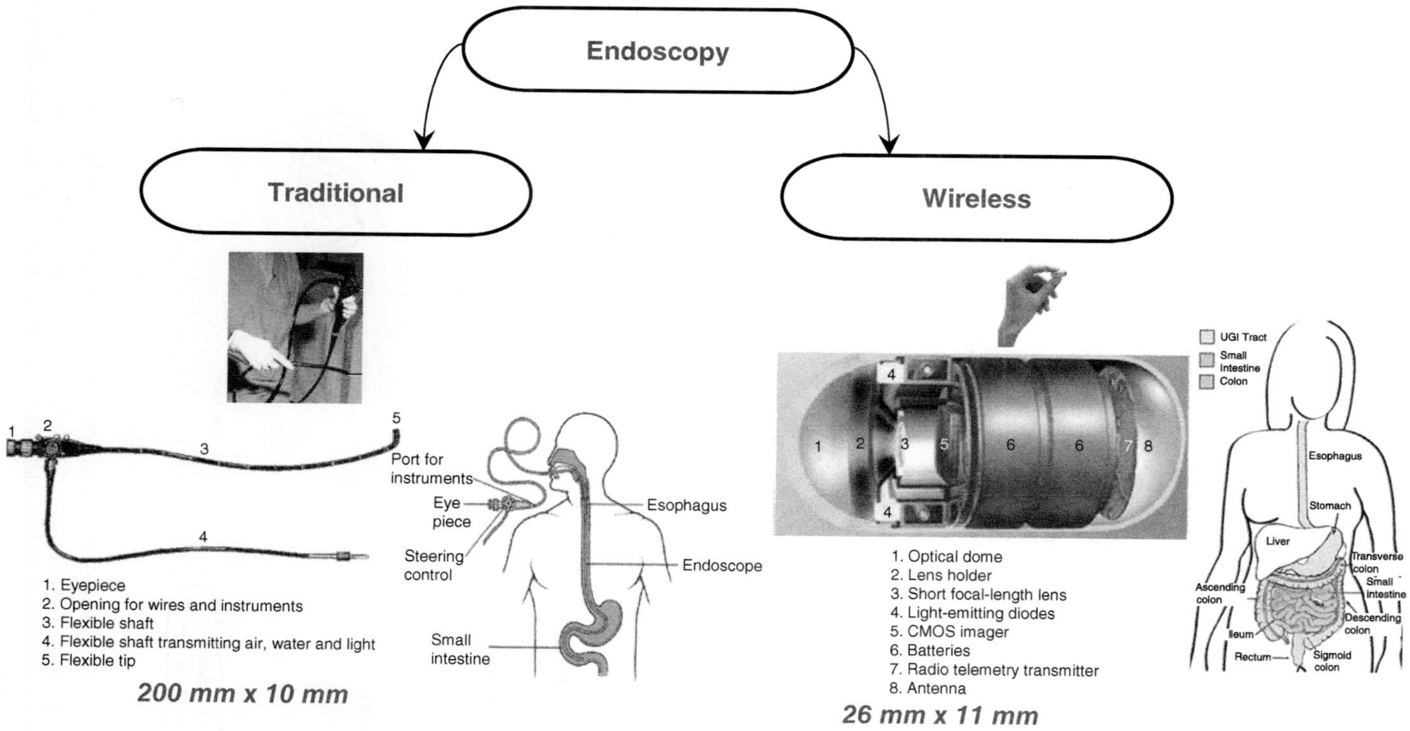

FIGURE 4.10 Traditional endoscopy versus wireless capsule endoscopy.

be diagnosed. The wireless capsule endoscope system typically consists of camera, LEDs, CCD, batteries, transmitter, and an antenna which are integrated inside the capsule (vitamin size pill) [54]. Once the capsule is swallowed, it travels through the GI tract passively and has the ability to transmit video images out of the patient's body through radio telemetry. The antenna design for these capsule systems can be very challenging. Some of the issues that need to be addressed for successful antenna design are: physical size constraints due to size of the capsule (26 mm × 11 mm) leading to antenna miniaturization, the location and orientation of the capsule vary with passage through the digestive track so diverse antenna polarization is necessary, the capsule has many electrical components which may affect the antenna performance so antenna interaction with these components is critical and most importantly, human body effects must be taken into consideration.

4.2.1 Ingestible Antenna Design: Conformal Meandered Offset Dipole Antenna

Figure 4.11a shows the evolution of the conformal design for capsule imager at WMTS frequencies [9, 10, 42]. The offset-fed half-wave dipole has good matching characteristics due to utilization of second resonance but it has linear polarization and

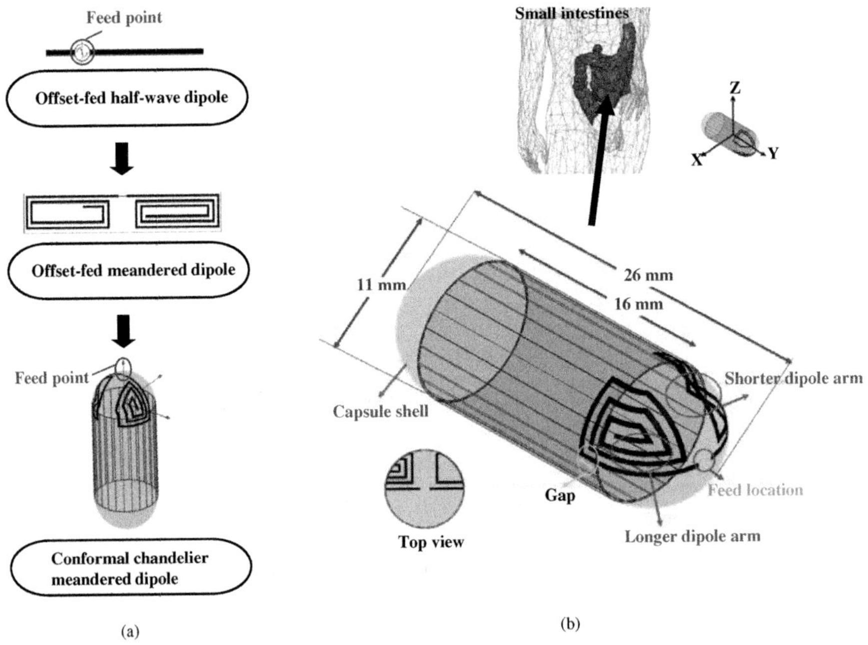

FIGURE 4.11 (a) Evolution of the conformal antenna design for WMTS capsule imager. (b) Antenna design with critical parameters labeled and the capsule antenna placed in the small intestines, inside the human body model. *Source:* Izdebski et al. 2009 [42]. Reproduced with permission of IEEE.

non-effective space utilization. The offset-fed meandered dipole has a miniaturized design but still suffers from space constraints. The conformal chandelier meandered dipole uses space filling technique effectively and conforms to the inner side of the capsule shell thus effectively utilizing the space and also has increased polarization properties due to conformal and offset nature. The antenna is conformed on the inner surface of the capsule shell (dielectric constant of 2.2—polyethylene) away from the camera. As the main function of the capsule imaging system is providing real-time images of the small intestines, it was placed in the location of the small intestines (Ansoft human body model) as shown in Figure 4.11b. The dielectric constant varied from 58.9 to 55.9 and the conductivity varied from 2.2 S/m to 2.7 S/m for frequency range from 1 GHz to 1.8 GHz (electrical parameters for the small intestines).

By varying the length of the offset-fed dipole arms, required matching was obtained at 1.4 GHz. The antenna was matched to 50 Ω within the human body model. Figure 4.12a shows the impedance variation of the capsule antenna inside the human body model. It can be seen that when the gap between the turns is changed from 1.6 mm to 2.4 mm, the overall length of the antenna grows then the resonance shifts lower. The trace width is 0.3 mm. Figure 4.12b shows that there is very good matching at the frequency of interest (1.4 GHz).

The antenna is placed about 5 cm from the surface of the human body model, roughly in the location of the small intestines. Figure 4.13a shows the 3D radiation pattern and two-dimensional (2D) pattern cuts in the *xz* and *yz* planes. The peak gain at 1.4 GHz is about −28.4 dBi. The average gain value in the *yz* plane (from 0° to 180°) is about −36 dBi (used later for link budget). 3D pattern shows that most of the energy is coming out of the human body. Figure 4.13b shows the electric field distribution at the surface of the human body model. The location of the antenna can be easily identified from the hot spot at the surface. Due to the conformality and offset-feed, we achieve some degree of polarization diversity as seen in Figure 4.13c. At boresight, the antenna shows some polarization diversity at 1.4 GHz. This is a very useful property as it achieves robustness in link budget analysis (described later).

4.2.2 Link Budget and SAR Comparison for Different Ingestible Antenna Designs

In the past, the use of helical antenna and spiral antenna has been proposed for these capsule devices (MICS band), but due to the volumetric nature of these antennas, it is possibly ill-suited for miniaturized capsule devices [31,32]. This section will compare the antenna performances of two types of antenna design solutions for capsule endoscopes in a very exhaustive way in terms of link budget and SAR. The inverted conical helical antenna (ICHA), a more volumetric design, and the conformal meandered offset dipole antenna (CMODA—used as a reference design) are compared and characterized for their performance in capsule system [57].

For the ICHA, the gap between the two plates and the radius of the upper metallic plate is used for tuning the frequency and controlling the impedance. The total length of the helical arm is 33.11 mm and the antenna operates in normal mode. For the CMODA, the dipole arm lengths are tuned to match the antenna to the desired

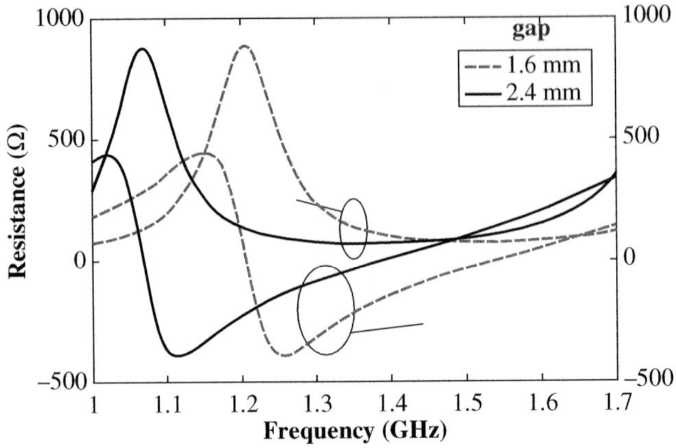

(a)

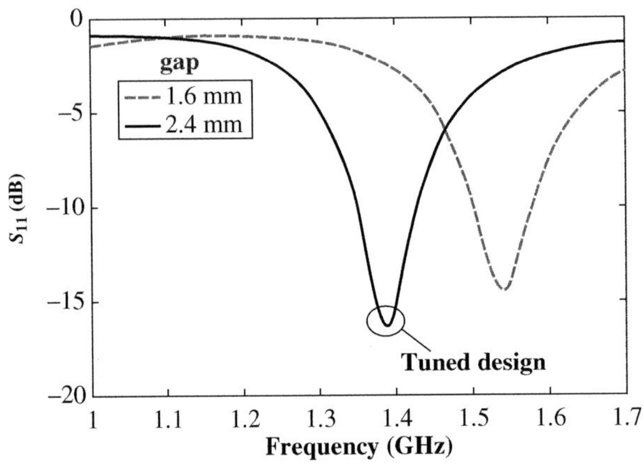

(b)

FIGURE 4.12 (a) Impedance variation of the capsule antenna based on the gap variation. (b) Return loss variation based on the gap variation. *Source:* Izdebski et al. 2009 [10]. Reproduced with permission of IEEE.

impedance. The total length of the antenna is about 70 mm with 45 mm for longer arm and 25 mm for shorter arm. Both the antennas are tuned to 1.4 GHz inside the small intestines. The geometry of the inverted conical antenna can be seen in Figure 4.14. The radiation efficiencies of both these antennas are less than 0.1%. The antenna inside the capsule along with the body acts as a transmitter and the total gain includes the human body loss. For the CDOMA, the average gain value was computed to be approximately −36 dBi (1 W input power). For the ICHA, the average gain value was computed to be approximately −42.33 dBi (1 W input power).

Computation of SAR is important to get an accurate estimate on the safe amount of power that can be delivered to the capsule antenna. In order to calculate the possible maximum acceptable power delivered by the designed ingestible antenna

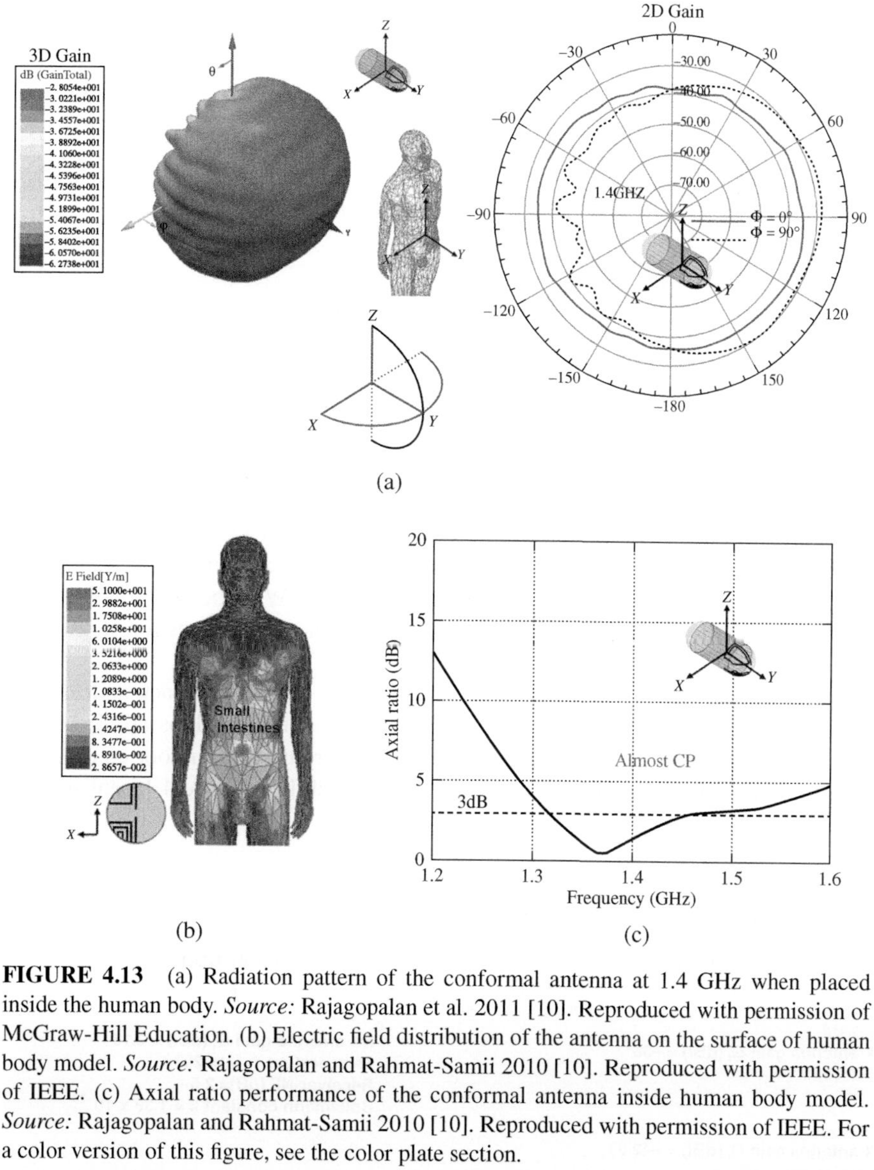

FIGURE 4.13 (a) Radiation pattern of the conformal antenna at 1.4 GHz when placed inside the human body. *Source:* Rajagopalan et al. 2011 [10]. Reproduced with permission of McGraw-Hill Education. (b) Electric field distribution of the antenna on the surface of human body model. *Source:* Rajagopalan and Rahmat-Samii 2010 [10]. Reproduced with permission of IEEE. (c) Axial ratio performance of the conformal antenna inside human body model. *Source:* Rajagopalan and Rahmat-Samii 2010 [10]. Reproduced with permission of IEEE. For a color version of this figure, see the color plate section.

in the human body model, the peak 1 g SAR limitation (1.6 mW/g = 1.6 W/kg) of ANSI was applied. The SAR values were computed from the outer end of the capsule antenna normalized input power of 1 W. For the CDOMA, a maximum SAR of 330 W/Kg (immediately outside the capsule shell) is computed. Away from the antenna, the SAR values drop off rapidly. For the ICHA, a maximum SAR of 250 W/Kg (immediately outside the capsule shell) is computed. In order to satisfy the peak 1 g SAR limitation (1.6 W/kg) of ANSI, the CDOMA and ICHA can deliver

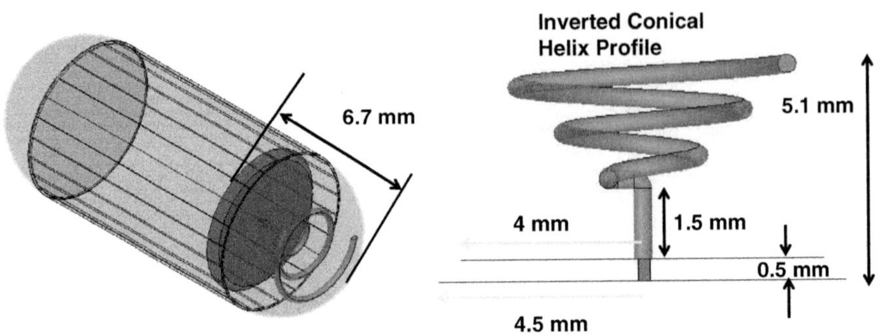

FIGURE 4.14 Design and critical parameters of inverted conical helical antenna. *Source:* Rajagopalan and Rahmat-Samii 2012 [57]. Reproduced with permission of IEEE.

a maximum power of 4.8 mW and 6.4 mW, respectively. The radiation efficiencies of both the antennas inside the human body model are less than 0.1%, thus restricting the maximum possible radiated power to 4.8 μW and 6.4 μW (<25 μW ERC limit). 1 W power is used for normalization; in reality the power delivered to the antenna will be lower than 1 W and typically in the range of 100 μW (depending on the battery lifetime requirement, the transmitter circuit, and other regulations for ingestible devices).

For these capsule systems, medical data (images of the GI tract and small intestines) needs to be transferred. Typically, data rates of 1–2 Mbps are required to transmit clear images at a fast rate. These systems are used for indoor communication (2–5 m) where the patients can have limited mobility. Figure 4.15 enlists all

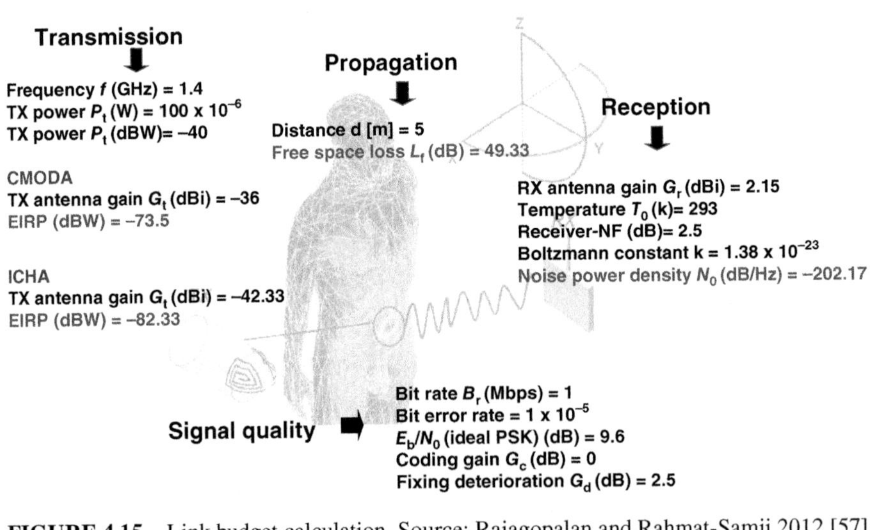

FIGURE 4.15 Link budget calculation. Source: Rajagopalan and Rahmat-Samii 2012 [57]. Reproduced with permission of IEEE.

TABLE 4.1 Summary of the link budget results.

Transmission	
Required C_R/N_0 (dB/Hz)	72.10
Link CDOMA C_C/N_0 (dB/Hz)	78.99
Margin CDOMA (dB)	6.89
Link ICHA C_H/N_0 (dB/Hz)	72.66
Margin ICHA (dB)	0.56

the necessary parameters for the link budget analysis and follows the link calculation listed in [57]. The transmitting antenna gain is calculated for the configuration where the capsule is along y-axis. The receiving antenna is a linearly polarized dipole resonating at 1.4 GHz. The transmit power is limited to 100 μW.

$$C_R/N_0 = {E_b}/{N_0} + 10\log_{10} B_r - G_c + G_d \,(\text{dB/Hz}) \tag{4.1}$$

$$C/N_0 = P_t + G_t + L_f + G_r - N_0 \,(\text{dB/Hz}) \tag{4.2}$$

Using communication theory, a required C_R/N_0 is calculated using equation (4.1). Equation (4.2) shows the calculation of the actual link C/N_0 using Friis formula. If the link C/N_0 exceeds the required C_R/N_0 then it shows that wireless communication and medical data transfer is possible. Table 4.1 shows a summary of the link budgets for both the antenna designs. It shows that in both cases communication is possible but for CDOMA there is a bigger margin for successful communication.

As the capsule system moves inside the human body, there is no way to predict the random movement as the movement of this system is governed by peristalsis. It is important to quantify the link for different orientations of the capsule system. Figure 4.16a shows the different orientations that were simulated. The initial position of the capsule axis before rotation is oriented along x-, y- and z-axis for the xz, yz and xz planes, respectively. The head of the dotted arrow represents the starting position of the capsule (0°). The solid arrow represents the angle in which the capsule system was swept. It is observed that the variation is <5 dB for all the rotations for CDOMA as shown in Figure 4.16b. Figure 4.16c shows that the link variation is about 10 dB for ICHA especially for the yz plane. This potentially means that the margin between the link C/N_0 and the required C_R/N_0 needs to be higher to accommodate this deviation thus restricting the bandwidth and/or distance.

4.2.3 An Adaptive Antenna System with Frequency Reconfigurability Using Wireless Power Transfer Technology

Once swallowed, the capsule travels through different parts of the human body. As the human body is a rich and diverse medium, different parts of the human body have different dielectric and conductive properties. So, the antenna will be detuned when the capsule is in different parts of the body. Adaptive frequency tuning will help retune the antenna to provide best data link performance for its present

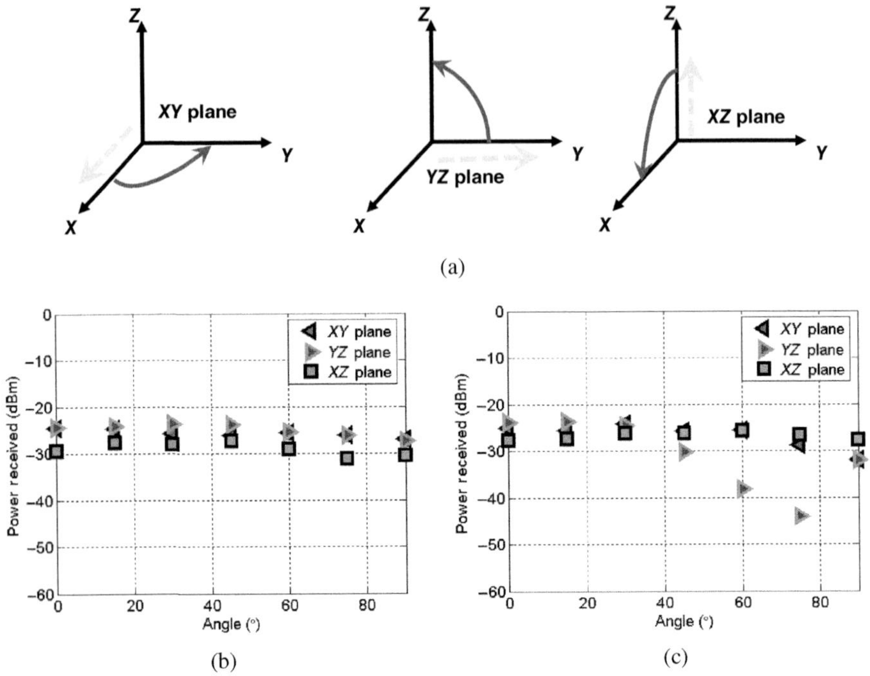

FIGURE 4.16 Link budget robustness based on different orientations. (a) Different orientations. (b) Conformal antenna. (c) Helical antenna. *Source:* Rajagopalan and Rahmat-Samii 2012 [57]. Reproduced with permission of IEEE.

location. A varactor can be used to tune the antenna dynamically. Figure 4.17a shows the different parts that are critical for this adaptive capsule system. The capsule system consists of the following links. Link 1 corresponds to the low-frequency link between the outside transmitting coil (TX_2) and the receiving coil (RX_2) within the capsule and it operates at few MHz. This is the link used for wireless power transfer. Link 2 corresponds to the high-frequency link between the ingestible antenna (TX_1) and the outside receiver (RX_1). This link is used to transfer actual medical data (patient's vitals, biostatistics, video images, etc.) and it operates at 1.4 GHz. Link 3 corresponds to the feedback link between the outside receiver (RX_1) and the outside transmitting coil (TX_2). This link is used to constantly monitor the stability of Link 2. For example, if the received power level in Link 2 drops below a specified threshold, Link 3 will command TX_2 to change the power coupled into the capsule (Link 1) appropriately. This power coupling will drive the varactor voltage accordingly thus tuning the ingestible antenna to restore the link performance. Thus, a smart diagnostic capsule system can be potentially developed with novel miniaturized antenna and wireless power transfer technology.

Two circular coils forming the transmitter (TX_2) and receiver (RX_2) for the low MHz inductive link have radii r_1 and r_2, respectively, as shown in Figure 4.17b. It is assumed that coils are centered on a single axis facing each other in their normal

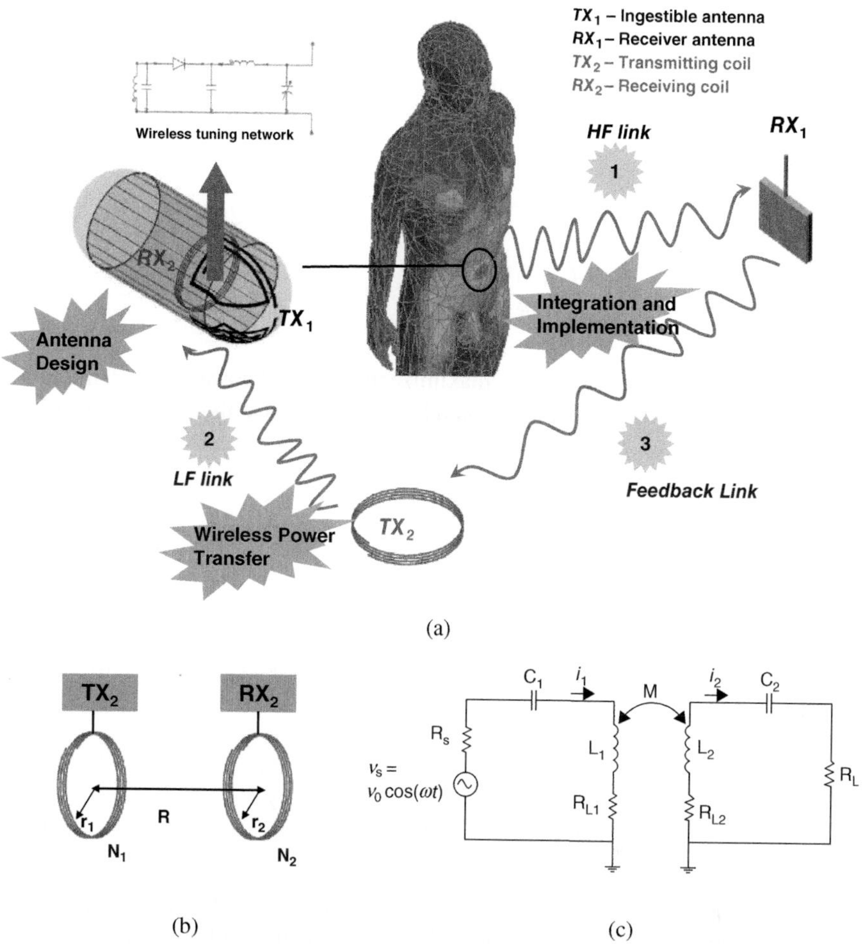

FIGURE 4.17 (a) A frequency reconfigurable adaptive system utilizing conformal antenna and wireless power transfer methodology. (b) Inductively coupled LF link. (c) Equivalent circuit model. *Source:* Azad et al. 2012 [58]. Reproduced with permission of IEEE.

direction. The transmitter and the receiver coil are separated by a distance R. The coils consist of N_1 and N_2 turns and carry i_1 and i_2 currents, respectively. R_{L1} and R_{L2} are the resistances of the coils at the operating frequency (including ohmic loss resistance and radiation resistance), L_1 and L_2 are the self-inductances of the coils and C_1 and C_2 are the capacitors to resonate the coils. R_s and R_L are the source and load impedances, respectively. Q_1 and Q_2 are the loaded quality factors of the transmitter and the receiver; and $Q_{1,int}$ and $Q_{2,int}$ are the intrinsic quality factors of the transmitting and receiving coils. By selecting different turn ratio of the voltage transformer, one can obtain the desired source and load terminations. Figure 4.17c shows the equivalent circuit model of the inductive link.

As the distance $R \gg r_1$ *and* r_2, we use the weak coupling condition, where the effect of the mutual coupling between the two coils on the impedance seen from the transmitter side can be ignored. The power coupled in an *inductively coupled near-field link* shown in Figure 4.17b is given by [58]

$$P_r(\omega = \omega_0) = 4P_t k^2 Q_1 Q_2 \left(1 - \frac{Q_1}{Q_{1,\text{int}}}\right)\left(1 - \frac{Q_2}{Q_{2,\text{int}}}\right)$$
$$k = \left(\frac{\sqrt{r_1 r_2}}{R}\right)^3 \qquad R \gg r_1, r_2 \tag{4.3}$$

where k is the coupling coefficient. To maximize the received power through the coupling, the critical coupling condition should be selected both at the transmitter and at the receiver, and the received power under this condition is given by

$$P_r(\omega = \omega_0) = P_t k^2 Q_1 Q_2 = \frac{P_t k^2 Q_{1,\text{int}} Q_{2,\text{int}}}{4}$$
$$Q_1 = \frac{Q_{1,\text{int}}}{2}, \quad Q_2 = \frac{Q_{2,\text{int}}}{2} \tag{4.4}$$

The inductance and the resistance of the coils can be calculated from [59],

$$L = \frac{\mu_0 N^2 \pi r}{2} \tag{4.5}$$

$$R = R_{\text{ac}} + R_r = \rho \frac{rN}{r_i \delta} + \left(\frac{177NS}{\lambda^2}\right)^2 \tag{4.6}$$

where N is the number of turns, r is the coil radius, R_{ac} is the AC resistance of the coil, R_r is the radiation resistance of the coil, ρ is the resistivity of the coil conductor, r_i is the radius of the wire conductor, δ is the skin depth, and S is the surface area of the coil.

Table 4.2 shows the different parameters of both coils for R = 600 mm, f = 100 MHz, and P_t = 0.3 W. The calculation of the parameters of the coils and the link is a first pass attempt as it assumes that the coils are aligned. The effects of coil misalignment are ignored. The receiver coil inside the capsule system can have any orientation so one way to potentially mitigate coil–coil misalignment is to have a three-axis coil on the TX_2. The TX_2 coil can be big to be able to provide enough inductive energy. The limitation on the size of this coil is based on magnetic field exposure to human body by this coil. The RX_2 coil is constrained by space and size. Based on the IEEE Standard for safety levels for human exposure to EM fields, the maximum permissible exposure to magnetic field in a controlled environment at 100 MHz is 0.2 μT [60]. For the TX_2 coil configuration detailed above, based on P_t = 0.3 W and R = 600 mm, the magnetic field at the body surface (B_{body}) is 0.18 μT. Assuming

TABLE 4.2 TX_2 **and** RX_2 **coil parameters.**

Parameter	Value
N_1, N_2	5, 20
r_1, r_2	50 mm, 5 mm
r_{i1}, r_{i2}	0.5 mm, 0.05 mm
L_1, L_2	2.46 μH, 3.94 μH,
R_{L1}, R_{L2}	1.89 Ω, 5.17 Ω,
$Q_{1,int}$, $Q_{2,int}$	820, 479
R_s, R_L	50 Ω, 50 Ω
Q_1, Q_2	29.9, 45
C_1, C_2	1 pF, 0.6 pF
k	1.8×10^{-5}
P_r	9.88 μW (−20.1 dBm)

that the capsule system is 5 cm (R') from the skin surface, the magnetic field at the capsule system (B_{cap}) is 0.008 μT and it is given by

$$B_{cap} = \frac{\mu_o N_1 I_1 r_1^2}{2\left((R+R')^2 + r_1^2\right)^{3/2}} e^{-\alpha R'} \tag{4.7}$$

Figure 4.18a shows the adaptive tuning network design. In order to produce a steady DC voltage from a half-wave rectified AC source, a smoothing circuit is needed. A capacitor is placed across the DC output of the rectifier. There will still remain an amount of AC ripple voltage where the voltage is not completely smoothed. The amplitude of the remaining ripple depends on how much the load discharges the capacitor between the peaks of the waveform. In order to further reduce this ripple, an LC π-filter (pi-filter) can be used. This complements the reservoir capacitor, with a series inductor, and a second filter capacitor, which in this case can be the varactor. This filter combination makes sure that a steadier DC output can be obtained across the terminals of the varactor. The series inductor presents high impedance at the ripple current frequency.

The voltage for the tuning network is tapped across the capacitor, C_2. Based on the power received in the RX_2 coil, the voltage across the capacitor is a product of the voltage across R_{L2} and $Q_{2,int}$. The maximum voltage across the varactor is 4.1 V (after subtracting the 0.7 V drop across diode). Typical varactor (0.15 cm × 0.15 cm) offers a capacitance tuning range from 2.4 pF to 1 pF over a DC supply voltage from 0.5 V to 3 V. The varactor is placed across the feed terminals. It can be seen that for 3 V variation, frequency of the antenna can be tuned from 1.25 GHz to 1.4 GHz as seen in Figure 4.18b. This methodology can potentially be used to drive microchips and sensors present inside the capsule system thus making batteryless capsule systems a reality.

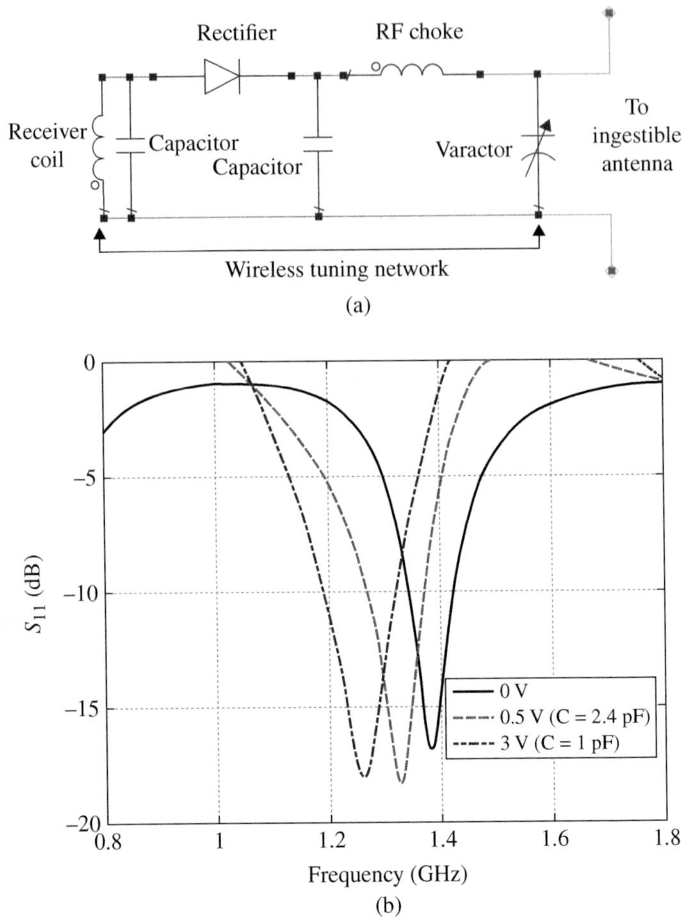

FIGURE 4.18 (a) Adaptive tuning network. (b) The frequency tuning range of the conformal antenna for different varactor capacitances.

4.3 INGESTIBLE ANTENNA DESIGN AT ISM BAND: MEDICAL COMPLIANCE SENSING

Electromagnetics, in particular RFID, can become a major player in medicine and thus, in health care. Due to the non-line-of-sight operation, larger read range, and ability to store more information, RFID tags serve as a possible replacement to bar codes. RFID wrist bands have been successfully used to identify patients in hospitals. Implantable RFID chips [61] have been reported for storing information and tracking location. RFID technology can drastically improve the processing times associated with medical record retrieval, dispensing of medications, personnel management, continuous medical storage of patient statistics, patient monitoring and tracking, drug delivery control and thus improve the efficiency of the overall healthcare system.

Medical non-compliance has become a significant problem in healthcare. A patient's failure to comply with prescription medication and instructions has become a major obstacle in treating illnesses. Medical non-compliance causes approximately 125,000 deaths annually in the United States. Along with this it leads to 10–25% of hospital admissions and costs billions in emergency-room visits and other avoidable medical expenses in the United States [62]. Non-compliance has a varied spectrum of causes ranging from patient forgetfulness to willful failure to take medication due to side effects or other factors. It is therefore of paramount importance to enable monitoring of the ingestion of medication by the physician and any patient-consented support network.

In the past, several medicine monitoring schemes have been proposed including tagging of medicine containers. However, all the proposed monitoring schemes cannot confirm whether the patient has actually ingested the medicine. Recently, Eastman Kodak™ Company has proposed a system to monitor medicine ingestion by tagging each pill with an RFID Tag [63]. Figure 4.19 shows a representation of a possible medical monitoring system with an RFID pill tag for medicine compliance sensing [42, 64].

The medical compliance sensing system can potentially consist of RFID pill tag, a small reader in the form of a necklace/belt/patch, and a main receiver for data collection and storage as shown in Figure 4.19. The proposed medical compliance sensing system will work in the following way: Every prescribed pill has an RFID tag. The patient swallows the capsule pill with the RFID tag and the microchip embedded

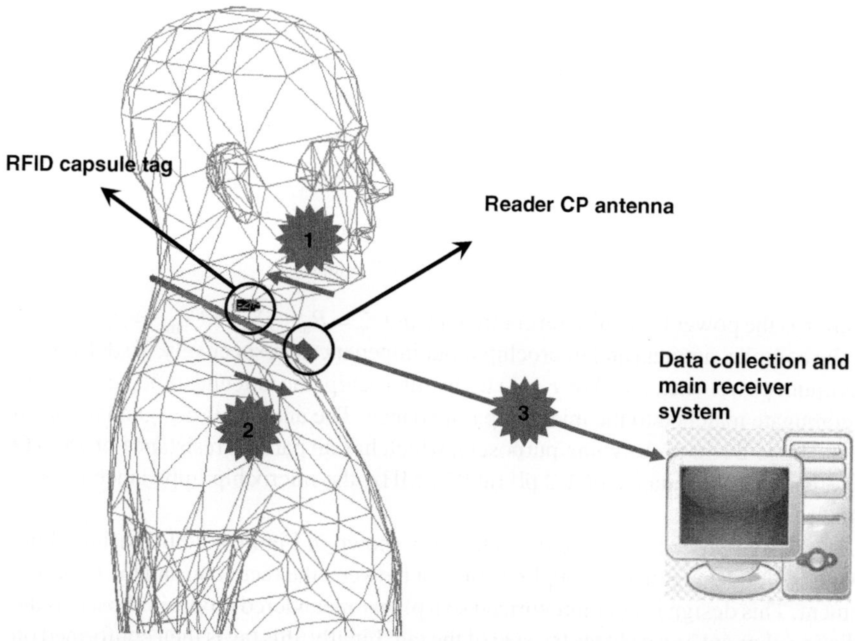

FIGURE 4.19 A representation of a possible medical monitoring system.

inside the capsule. Once the pill is inside the human body, the RFID tag is complex conjugate matched to the chip providing perfect matching inside human body environment. The body-worn reader interrogates the RFID tag. The tag sends backscattered modulated information that consists of the medication details (amount of dosage, name of the medication, etc.). Once the body-worn reader successfully reads the tag, it sends the information to the main receiver/cell phone for data storing and cataloguing, thus, enabling the patient to keep track of his medicine intake. The RFID tag is designed such that it is mismatched to the microchip in free space and outside human body, and only matched to the chip when the pill is inside the human body environment. The antenna and the chip will be designed from biologically inactive materials (noble metals like silver). Once the capsule pill is ingested and the stomach acids open the pill delivering the medication, the antenna and the chip will be harmlessly excreted in a short time. As soon as the body-worn reader reads the RFID tag, it will send a signal to the patient's cell phone. This information will be immediately uploaded to the internet. Thus, through the use of social networking, the patient's personal network of family and friends will have access to the patient prescription program and thus effectively communicate and monitor the patient medication compliance remotely. In order to protect patient medical privacy, the patient will be able to select the network of people the information is shared with. This medical data will also be integrated into the patient's electronic medical record using a secure data collection algorithm, enabling clinicians to remotely monitor patient compliance.

One of the key parameters for the success of monitoring process is the design of the tag. The tag antenna design has the following specific requirements: miniaturization due to size of the capsule (26 mm × 11 mm), easy integration with the capsule pill, understanding human body effects on the antenna performance, bandwidth to cover UFH RFID band (860–960 MHz) and easy tuning ability to tune to any available microchip. The tag input impedance has to be matched to the microchip impedance to ensure adequate power transfer to the microchip. The power reflection coefficient between the antenna and the microchip is given by

$$\rho = 1 - \tau = 1 - \frac{4R_a R_{mc}}{|Z_a + Z_{mc}|^2} \tag{4.8}$$

where τ is the power transmission coefficient and $Z_a = R_a + jX_a$ and $Z_{mc} = R_{mc} + jX_{mc}$ are the complex antenna and microchip input impedances, respectively. To deliver the maximum power (i.e., $\rho = 0$ or $\tau = 1$) to the microchip, the antenna impedance has to be conjugate matched to the microchip impedance. The antenna is designed to match Alien Higgs-2 (for prototyping purposes), which has an parallel resistance of 1500 Ω and a parallel capacitance of 1.2 pF (at 915 MHz, the microchip impedance is $14 - j145$ Ω).

The design procedure of the tag antenna is shown in Figure 4.20a. The initial tag design is based on inductive coupling concept between the feed loop and the radiating element. This design is not miniaturized so a planar meandered dipole is chosen as the radiating element as it reduces the size of the tag. Finally, this tag is then conformed on the inside of the capsule shell. The design is based on a space filling conformal planar

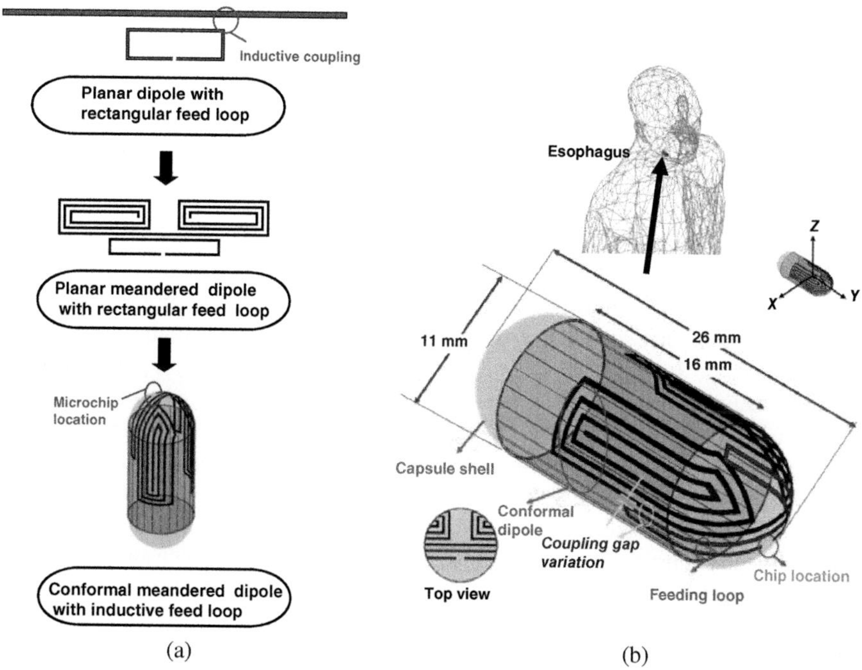

FIGURE 4.20 (a) Evolution of the conformal RFID capsule antenna design for ISM band. (b) Antenna design with critical parameters labeled and the capsule antenna placed in the esophagus region. *Source:* Rajagopalan and Rahmat-Samii 2010 [42]. Reproduced with permission of IEEE.

balanced meandered dipole concept. The inductive loop feeding scheme was selected as it is a simple way to match the tag impedance to any arbitrary chip impedance. The input impedance of the tag antenna is given by [65]

$$Z_a = R_a + jX_a = Z_{loop} + \frac{(2\pi fM)^2}{Z_d} \tag{4.9}$$

where Z_a is the antenna impedance, Z_d and Z_{loop} are the impedances of the planar meandered dipole and rectangular feed loop, respectively, and M is the mutual inductance between the dipole and the loop. The radiating element length is chosen such that the tag resonates close to 900 MHz.

There is independent control of both the resistance and reactance of the tag. The resistance of the tag depends on the mutual inductance between the dipole and the feed loop, and the reactance of the tag depends on the self-inductance of the loop. This makes the tag very promising to match to different parts of the body and also to different chip impedances. Figure 4.20b shows the tag conformed inside the capsule shell (dielectric constant—2.2) with all the critical parameters labeled. These parameters including the coupling gap and feeding loop length are varied to understand the effects on the impedance of the tag. The RFID bio-capsule tag is placed about 4 cm from the surface of the neck roughly in the esophagus region. The variation of dielectric constant and conductivity with frequency was incorporated into the Ansoft human body model. The dielectric constant varied from 55.6 to 54.4 and the conductivity varied from 0.88 S/m to 1.05 S/m for frequency range from 0.7 GHz to 1.2 GHz.

4.3.1 Tuning Parameters and Impedance Control

Figure 4.21 shows the input impedance curves for the tag for different gap coupling. As the gap coupling between the conformal dipole and feed loop becomes larger, the mutual inductance decreases thus decreasing the input resistance. The input reactance changes because as the gap coupling is increased, it decreases the overall feed loop length (tag footprint is kept constant). Figure 4.22 shows the input impedance curves for the tag for different feeding loop length. As the feeding loop length becomes larger, the mutual inductance between the dipole and the feed loop increases thus increasing the input resistance. The input reactance also increases as the feeding loop increases due to additional inductance. Figure 4.23 shows the resistance and reactance for the capsule RFID tag design optimized for the esophagus region. It also shows the

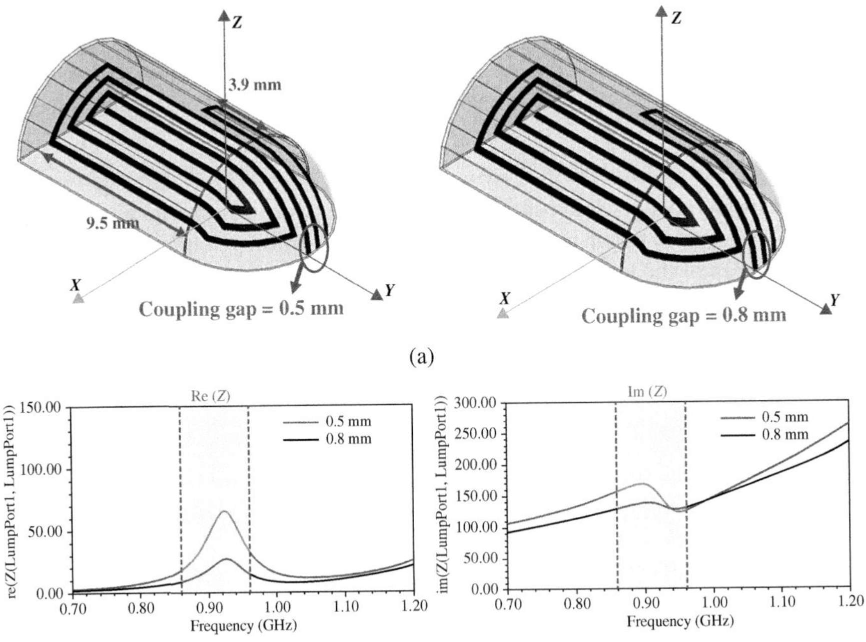

FIGURE 4.21 (a) Variation of parameter—coupling gap (b) Resistance variation based on the coupling gap change. (c) Reactance variation based on the coupling gap change.

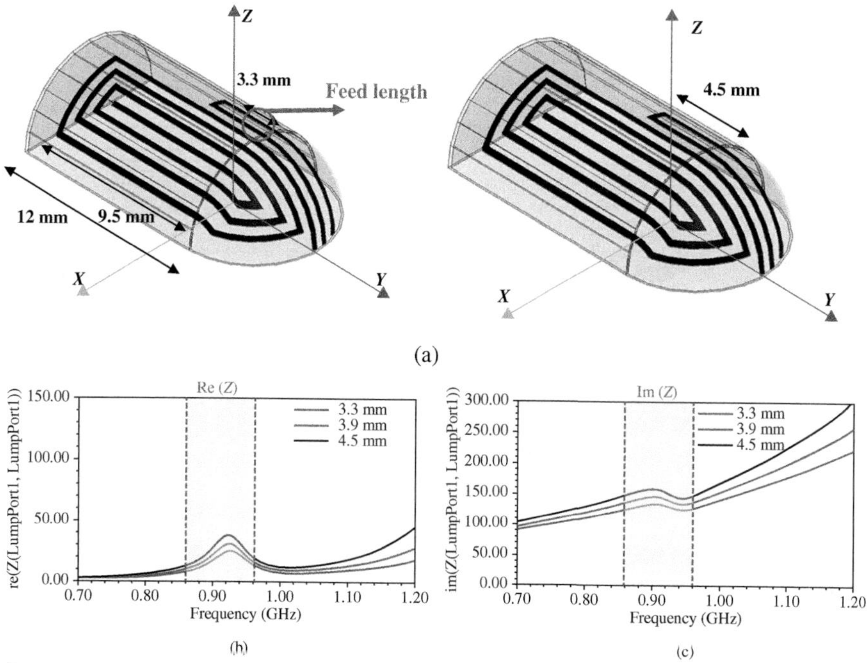

FIGURE 4.22 (a) Variation of parameter—feeding loop length. (b) Resistance variation based on the feeding loop length change. (c) Reactance variation based on the feeding loop length change.

power reflection coefficient, and it can be seen that the tag is matched well in ISM band inside the human body and not matched outside in free space.

Figure 4.24a shows the electric field distribution at the surface of the human body model, when the tag is placed in the esophagus location of the human body. The location of the tag can be clearly identified from the hot spot. The small necklace reader (with the antenna) will identify the tag when it passes through this region thereby making a medical record. Figure 4.24b shows the radiation pattern cuts for the two different planes. Approximate gain value is −40 dBi (including human body effect) on average.

4.3.2 Potential Batch Fabrication Process

A batch fabrication process is developed where these capsule tags can be potentially mass produced. The process consists of three steps: depositing metal thin films (1 μm aluminum) on flexible biocompatible polymer substrates (50 μm Polyimide—kapton), patterning the photoresist onto the substrate into suitable antenna configurations, and finally etching the aluminum and stripping the photoresist. Two representative designs (planar dipole and meandered dipole) of antennas on

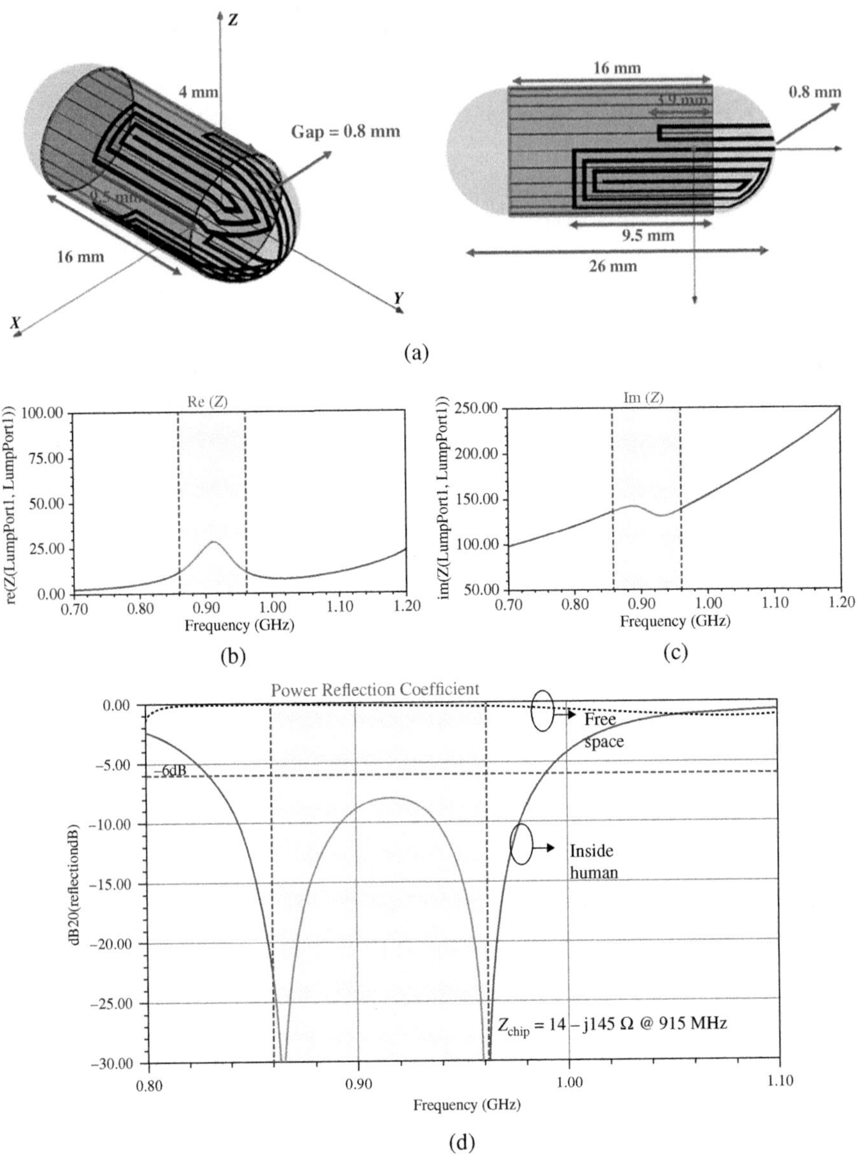

FIGURE 4.23 (a) Optimized design. (b) Resistance curve. (c) Reactance curve. (d) Power reflection coefficient (free space vs. inside human body).

flexible polymers are shown in Fig. 4.25. Due to the flexible nature of the polymer, the patterned antenna can be conformed to the inside of the capsule pill. Eventually, the antenna and the chip have to be designed from biologically inactive materials so that when the tag opens up after coming in contact with stomach acid, it does not cause any potential issues.

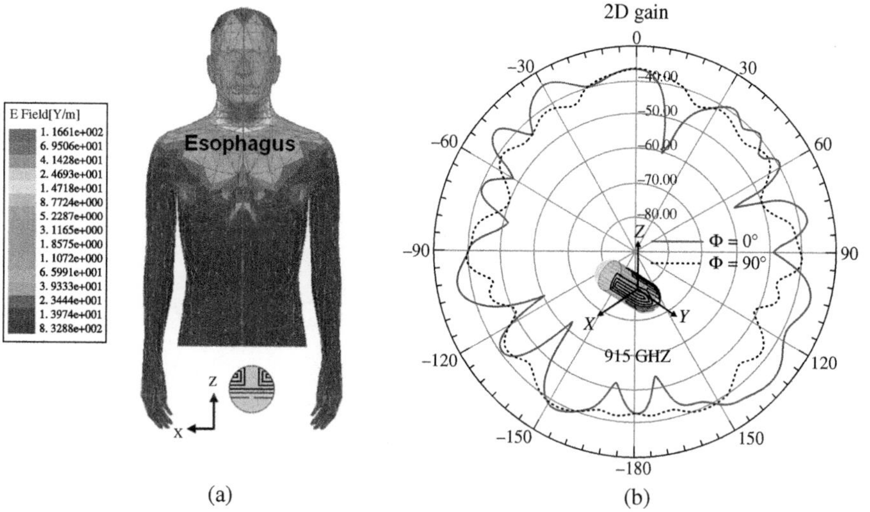

FIGURE 4.24 (a) Electric field distribution of the capsule tag antenna. (b) 2D radiation pattern of the tag. For a color version of this figure, see the color plate section.

4.4 ON-BODY ANTENNA AT UHF BAND: RFID TAG FOR PATIENT MONITORING

Far-field (UHF-915 MHz) RFID tags work on the principle of traveling waves between the reader and the tag [66, 67]. The UHF (860–960 MHz) band is of particular interest due to the long read range possibility [68, 69]. At the UHF frequency, high-dielectric (human) objects degrade the performance of the tag affecting the tag impedance, its ability to couple power to the microchip and the read range of the tag. These RFID tag problems are well documented in the literature [70, 71]. Body-matched tags (suspended patch type) have been proposed for real-time biomonitoring of patients [16]. Tooth-attached RFID tags [72] have been reported for storing patient information and tracking location. There is a need to develop novel RFID tag antenna designs which successfully deal with all the challenges presented

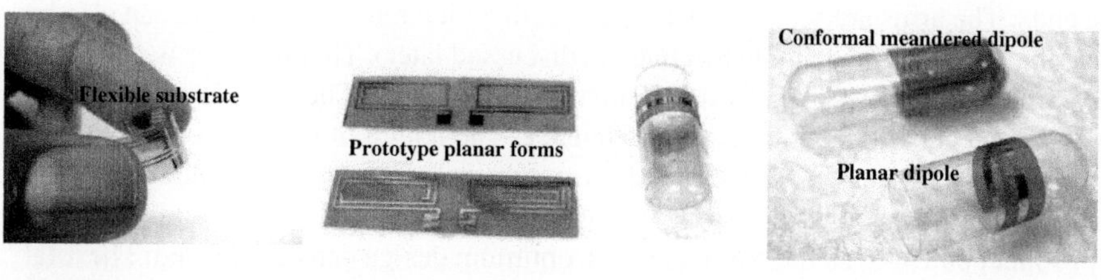

FIGURE 4.25 Fabricating the planar capsule tags on flexible substrate using metal thin film deposition.

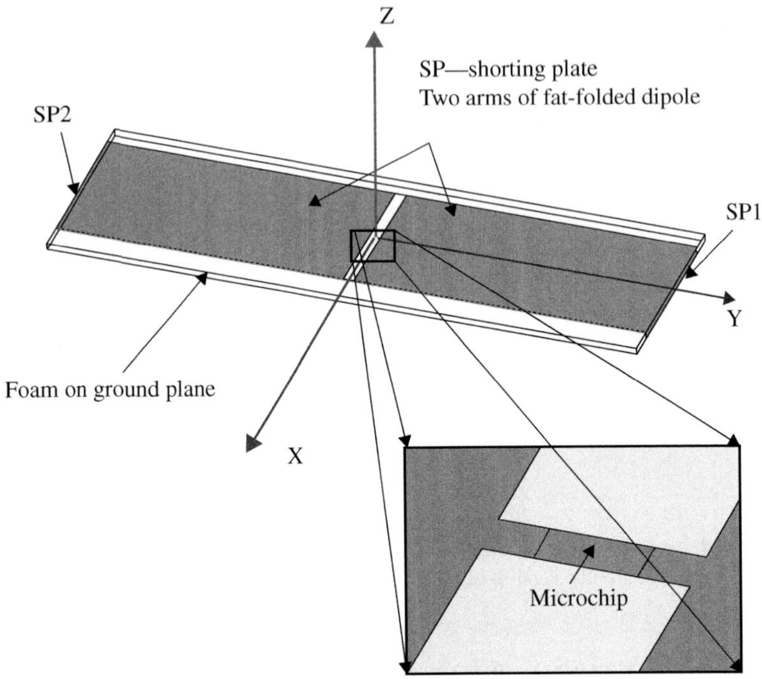

FIGURE 4.26 Schematic of the proposed RFID bio-tag. *Source:* Rajagopalan and Rahmat-Samii 2010 [15]. Reproduced with permission of IEEE.

including low-profile structure, low-cost fabrication, platform-tolerant behavior, and nonplanar surfaces.

The work in [15] proposes the design of a conformal RFID tag for remote human monitoring. The rectangular tag is designed such that it can be mounted onto common clothing like belts, collars, and waist bands for patient localization and monitoring. The design is based on a simple microstrip-type structure, which is low profile and inherently platform tolerant. Figure 4.26 shows the schematic of the tag design. The antenna operates in the US UHF band (902–928 MHz). The design is conjugate matched to Alien Higgs-2 microchip with impedance of $14 - j145\ \Omega$ at 915 MHz.

The RFID bio-tag consists of two quarter-wave arms shorted to the ground plane at the ends. The arms are separated by a gap width which radiates (structure can also be described by fat-folded dipole mechanism discussed later). This structure is placed on a 0.4 mm FR4 layer backed by a 1.4 mm thick foam layer. The foam layer is present to improve bandwidth and radiation efficiency. The length of the patches determines the frequency of interest. The width of the patches controls the impedance of the tag. The gap width can be used for fine tuning the frequency and for bandwidth control. By proper control of all the parameters, an optimum design can be achieved. The total dimensions of the antenna are optimized to 116 mm × 40 mm × 1.8 mm. The gap width is tuned to 2 mm and the single arm size is 57 mm × 30 mm.

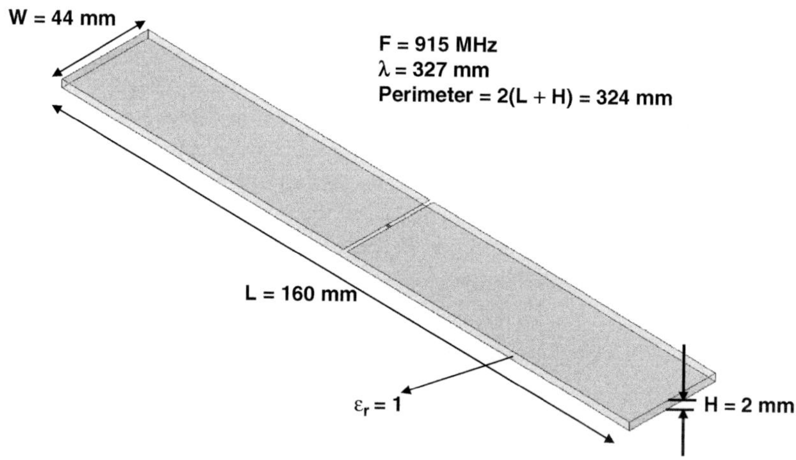

FIGURE 4.27 Simplified version of the bio-tag for circuit modeling.

4.4.1 Simplified Circuit Model for Planar Fat-Type Folded Dipole Bio-tag

The RFID bio-tag can also be described using a fat-folded dipole mechanism. A simple circuit model is used to describe the design and working of the tag. For simplicity, air substrate is considered. Figure 4.27 shows the fat-dipole bio-tag (W/L >0.1) used for circuit model analysis.

Figure 4.28a shows the updated transmission line model for fat-type planar folded dipole structure. The antenna mode (radiation mode), transmission line mode I (two-plate line mode), and transmission line mode II (slot line mode) form the complete transmission line model. The regular (thin-type planar where W/L <0.1) folded dipole operates as an unbalanced transmission line. The current on the regular folded dipole can be decomposed into two distinct modes [60]: an antenna mode where the currents

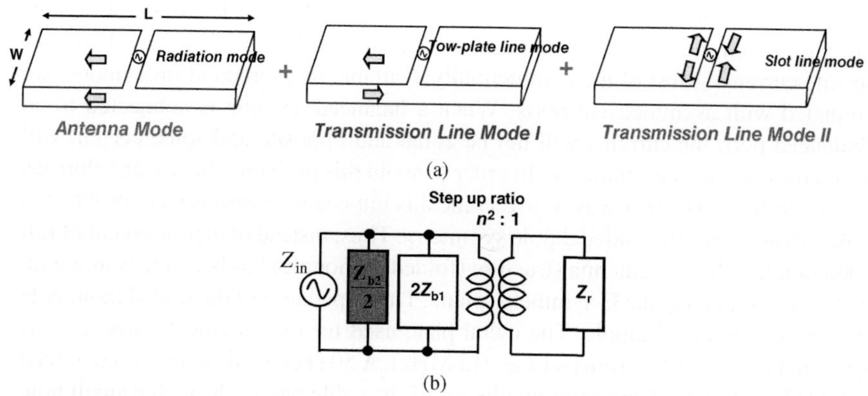

FIGURE 4.28 Transmission line model for this bio-tag. (b) Updated circuit model for fat-type planar folded dipole.

flow in the same direction on the top and bottom generating significant radiation and a transmission line mode where the currents flow in opposite directions yielding little radiation. For the fat-type planar folded dipole structure, there is presence of a third mode, the slot line mode where currents are shown in Figure 4.28a.

Figure 4.28b shows the updated circuit model for this fat-type planar folded dipole. Equation (4.10) describes the short-circuit transmission line corresponding to transmission line mode I and equation (4.11) describes the open-circuit transmission line corresponding to transmission line mode II. Equation (4.12) shows the total input impedance of the fat-type folded dipole.

$$Z_{b1} = jZ_{01} \tan \beta\ell \ (\ell = L/2) \tag{4.10}$$

$$Z_{b2} = -jZ_{02} \cot \beta\frac{W}{2} \tag{4.11}$$

$$\frac{1}{Z_{in}} = \frac{1}{n^2 Z_y} + \frac{1}{2Z_{b1}} + \frac{2}{Z_{b2}} \tag{4.12}$$

Z_r is the equivalent resistance of the dipole for the antenna mode, Z_{01} is the characteristic impedance of microstrip line for transmission line mode I, Z_{02} is the characteristic impedance of slot line for transmission line mode II and Z_{in} is the input impedance. As both the top and bottom conductor have same width, step up ratio is 4 ($n = 2$). Equation (4.11) is not present in the simple thin-type folded dipole model [59].

Figure 4.29 shows the currents distribution on the top plate of a thin-type planar folded dipole (regular) and a fat-type planar folded dipole. For the thin-type planar folded dipole, the current distribution for top and bottom (not shown here) plates are in the same direction and the currents are dominated by radiation currents. For the fat-type planar folded dipole, currents on top and bottom (not shown here) plates are in same direction but due to presence of transmission line currents, current nulls are observed. The slot line mode currents are clearly visible in the center feed region.

For measurement purposes, two different prototypes of the folded dipole structures were built. Both the arms of a balanced folded dipole antenna should be driven by opposite currents. Most of the commercially available measurement instruments are terminated with asymmetrical ports. When a balanced antenna is connected to an unbalanced port, the currents will not be equal and opposite and some current will flow on the outside of a conductor. In order to avoid this problem, the use of balun has been suggested. One other way to overcome this impedance measurement problem is to take advantage of the folded dipole symmetry. Thus, instead of measurement of full folded dipole, a half of antenna structure (folded monopole) has been measured with a metal plate replacing the E-symmetry plane. The impedance of the folded monopole is half that of a folded dipole. The metal plate used for measurement purposes was approximately 30 cm × 30 cm (~1λ at 915 MHz). A 50 Ω coaxial cable is used to feed the folded monopole. The center conductor of the cable passes through a small hole in the metal plate. The outer conductor of the cable is soldered to the metal plate. It is made sure that the gap between the top strip of the folded monopole and the metal plate is maintained throughout the experiment. The calibration is performed without

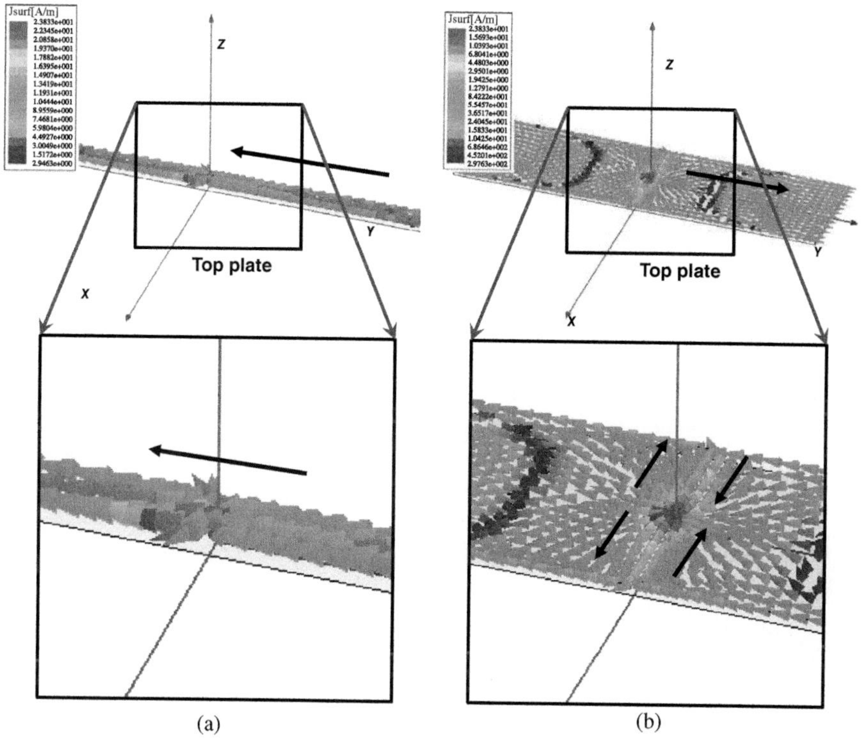

FIGURE 4.29 (a) Currents on a thin-type planar folded dipole. (b) Currents on a fat-type planar folded dipole.

the coaxial cable so the reference plane is below the metal plate. Port extension option of the network analyzer is utilized to make sure that the calibration plane is right at the metal plate.

For the thin-type planar folded dipole shown in Fig. 4.30a, the new model is much closer to the old model as the slot line mode currents are minimal. The differences between the circuit model and HFSS simulations are primarily due to feed capacitive effects not taken into consideration in analytical equations. For the fat-type planar folded dipole shown in Fig. 4.30b, the new model is closer to numerical simulation and measurements as the slot line mode becomes more dominant. The difference between the new model and simulation is attributed to dipole radiation impedance. Better agreement is expected using calculation of dipole impedance using numerical techniques.

4.4.2 Tag Characterizations on Human Body in Terms of Impedance and Radiation Pattern

A four-layer stratified elliptical cylinder is used to represent the human body. This model (skin plus fat, muscle, bone and organs) represents a more realistic human

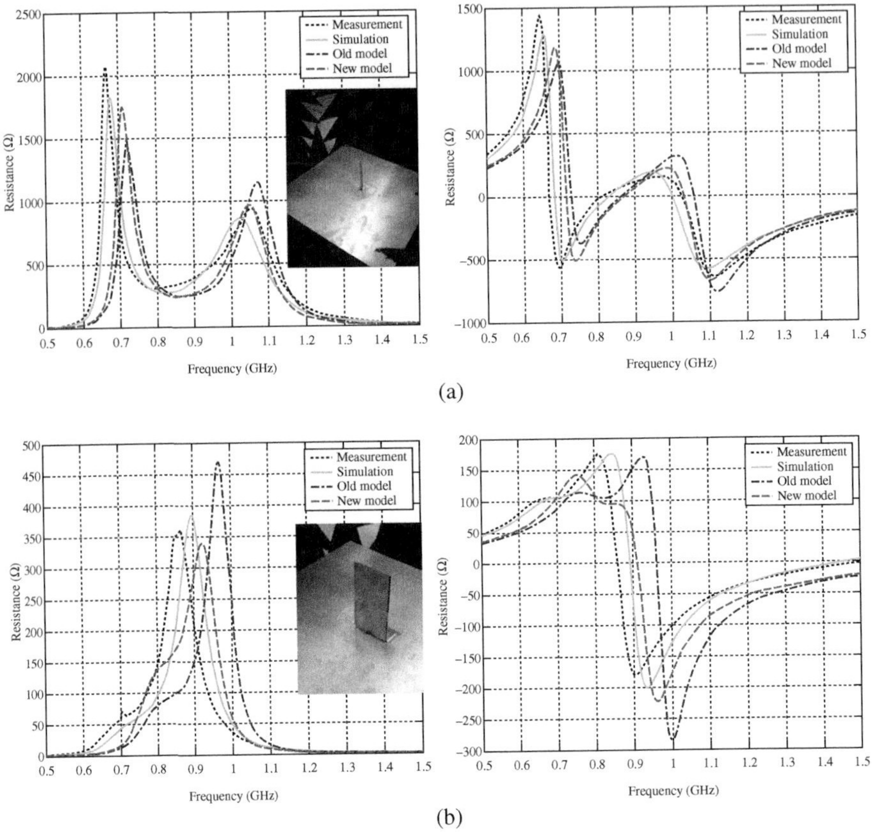

FIGURE 4.30 Measurements using half the structure (folded monopole). (a) Thin-type planar folded monopole (W = 4 mm). (b) Fat-type planar folded monopole (W = 44 mm).

body torso, and four-layer parameters are mentioned in Figure 4.31. The dielectric constant and conductivity for all the layers are obtained at 915 MHz. The variation of these properties with frequency has been incorporated into the simulation model. The dimensions of all the layers are also tabulated [16]. The tag is placed on the human body model in the horizontal position in the possible scenario that it can be mounted onto common clothing like belts and waist bands.

The fabricated tag was placed on the human body in the horizontal as shown in Figure 4.32a. A quarter-wave balun was used for input impedance measurements. A systematic calibration scheme was used to de-embed the balun from the actual measurements. The balun was soldered to the feed point of the antenna (no chip present). Monopole approach was not used for impedance as the tag had to be placed against human.

Figure 4.32b shows the comparison of the input impedance of the tag between the simulated model and measured results when the tag is oriented horizontally. There is very good agreement especially in the band of interest. Figure 4.32c shows the

915 MHz

Layer	ε_r	σ (S/m)	Thin man (cm)
Skin + fat	14.42	0.24	33.5 × 16.8
Muscle	54.99	0.95	31.0 × 14.2
Bone	20.76	0.34	28.4 × 10.5
Organs	52	0.93	27.2 × 8.40

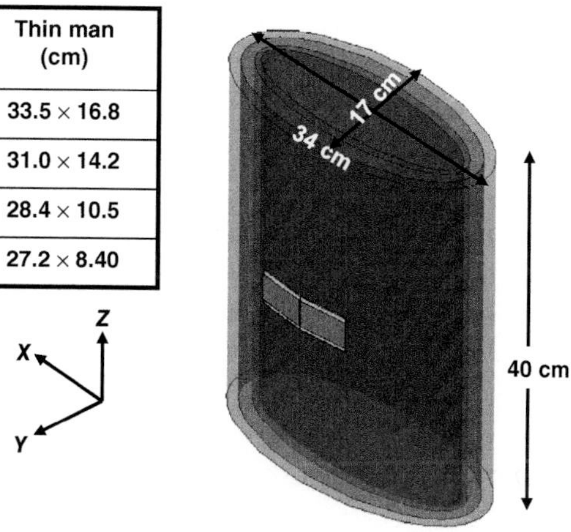

FIGURE 4.31 The parameters of the complex four-layer elliptical cylinder used for representing the human body. *Source:* Rajagopalan and Rahmat-Samii 2010 [15]. Reproduced with permission of IEEE.

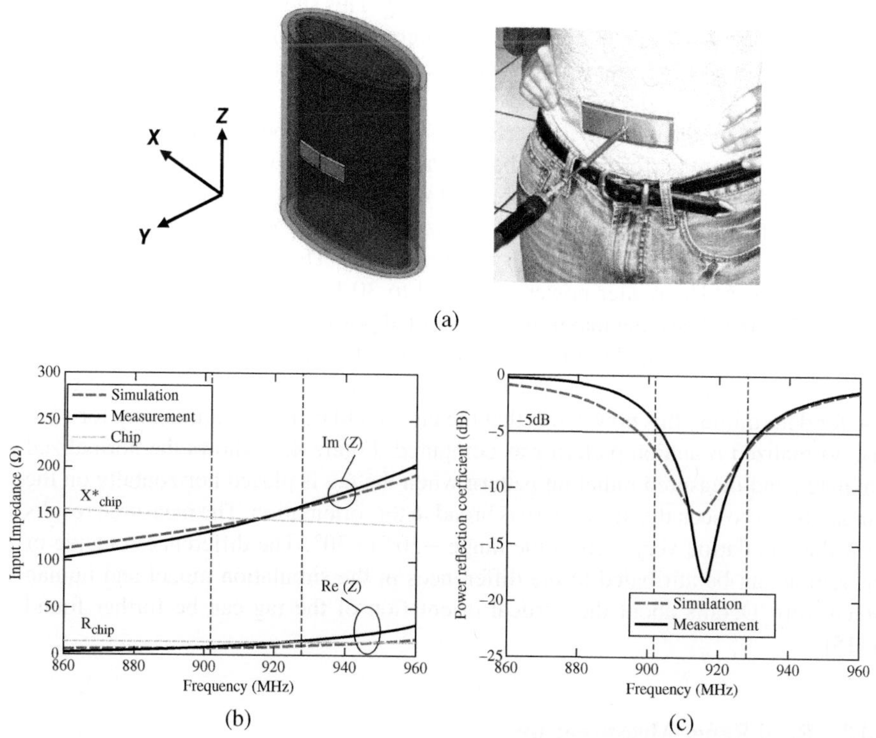

FIGURE 4.32 (a) Horizontal alignment of the tag on human. (b) Impedance curves for the horizontal tag orientation. (c) Power reflection coefficient curve for the horizontal tag orientation. *Source:* Rajagopalan and Rahmat-Samii 2010 [15]. Reproduced with permission of IEEE.

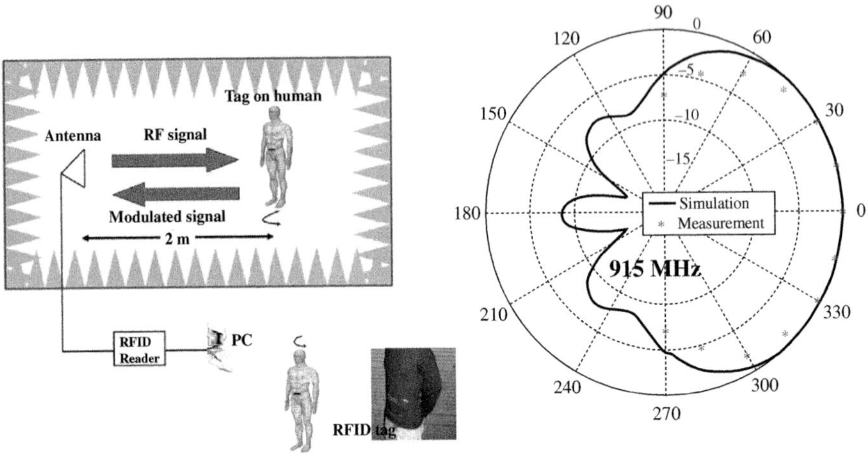

FIGURE 4.33 Radiation pattern measurements for the horizontal tag orientation. *Source:* Rajagopalan and Rahmat-Samii 2010 [15]. Reproduced with permission of IEEE.

comparison of the power reflection coefficient of the tag between the simulation and measurement when the tag is oriented horizontally. This shows how efficiently power is transferred between the antenna and the microchip. It can be clearly seen that in the band of interest there is at least –5 dB matching (necessary condition is minimum −3 dB match).

In order to gain insight into the performance of the on-body tag, radiation pattern measurements were performed. The balun was removed and the microchip was soldered to the feed location. The measurement was performed in the far-field chamber. The tag was placed on the human in the horizontal orientation. Mercury4 reader with M/A-COM dual antenna was used for reading the tag. The reader was controlled by a host computer. The reader power was fixed to 30.1 dBm and the reader antenna gain was 5.9 dBi. Thus, the maximum transmitted power was fixed at 4 W EIRP. The pattern was plotted using the following procedure: The minimum transmitted power required for reading the tag was determined at each measurement angle. The criterion for determining this power was that the tag should be read continuously for 45 s. The normalized radiation pattern was computed. Figure 4.33 shows the normalized simulated and measured radiation pattern when the tag is placed horizontally on the human. It can be seen that the pattern is broad in this orientation. The measured results track the simulation very well in the range −30° to 30°. The differences outside of this region can be attributed to the differences in the simulation model and human dimensions. Details about the vertical orientation of the tag can be further found in [15].

4.4.3 Read Range Measurements

When the RFID bio-tag is placed on the human body, absorption of power due to the body causes reduction in gain and distortion in pattern. As the body has high dielectric constant and certain conductivity, the antenna has to be designed to match the body. Typically, when the tag is placed on human body, the realized gain is in the range of 0 dB to −5 dB depending on the design of the tag. The read range for an RFID tag can be computed using the equation given by [55]

$$d_{\max}(\theta,\varphi) = \frac{\lambda}{4\pi}\sqrt{\frac{\text{EIRP}}{P_{\text{chip}}}\tau G_{\text{tag}}(\theta,\varphi)} \tag{4.13}$$

where EIRP is the effective isotropic radiated power, P_{chip} is the chip sensitivity, G_{tag} is the gain of the tag antenna, d is the distance between the tag and the reader, f is the frequency of operation, and c is the speed of the light. The maximum radiated power by the reader is constrained to 4 W EIRP within 902–928 MHz band (USA). Figure 4.34a shows the read range versus realized tag gain for different chip sensitivities. For tags with realized gain ranging from 0 to −5 dB and chip sensitivity of about 40 μW, a read range of 4 to 8 m is estimated.

Read range measurements were performed in open environment in order to avoid reflection from the surroundings. The maximum read range was defined as the maximum line-of-sight distance from the reader antenna where the reader detects the tag continuously without interruption. The output power of the reader is fixed at values similar to radiation pattern studies (30.1 dBm). Figure 4.34b shows the read range measurement results. The plot shows the maximum boresight read range distance at which the tag can be successfully read. The reader power was varied from 0.2 W (23 dBm) to 1 W (30 dBm). Initial investigations have shown that when the tag is placed on human, approximately 3.5 m maximum boresight read range is obtained.

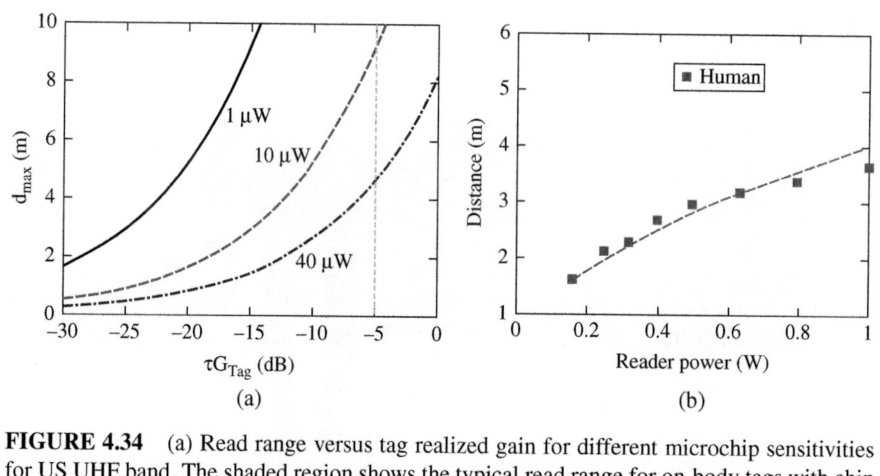

FIGURE 4.34 (a) Read range versus tag realized gain for different microchip sensitivities for US UHF band. The shaded region shows the typical read range for on-body tags with chip sensitivity of 40 μW (−13 dBm). (b) Actual read range measurements with bio-tag on human. *Source:* Rajagopalan and Rahmat-Samii 2010 [15]. Reproduced with permission of IEEE.

4.5 FUTURE OUTLOOK

Since the days of the first pacemakers and electronic digestible pill systems, implantable and ingestible medical devices have undergone a major transformation. The advancement of different technologies including electronics, battery, packaging, and encapsulation has made these devices multifunctional, long term, and reliable. These devices have contributed significantly to the quality of patient life. The technologies that support remote monitoring of patients will continue to be a major area of interest in the field of medicine. Therefore, advanced body-centric wireless communication system is the future of electromagnetics in medicine/healthcare. Figure 4.35 shows a wireless body-sensing system which incorporates portable data receiver, on-body sensors/antennas, and miniature implants/ingestible devices thus forming off-body, on-body, and in-body communication links. The off-body link corresponds to the link between the on-body or in-body device with external reception unit, the on-body link corresponds to the link between different on-body devices, and finally the in-body link corresponds to the link between the in-body devices with the on-body sensor.

By increasing the functionality of existing medical devices and by adding more in-body and on-body sensors/devices, body-centric wireless communication can become quite exhaustive and challenging. As these systems become more and more complex, one has to make sure that they are compatible with existing technologies and always ensure electromagnetic compatibility and other FCC regulations. Due to

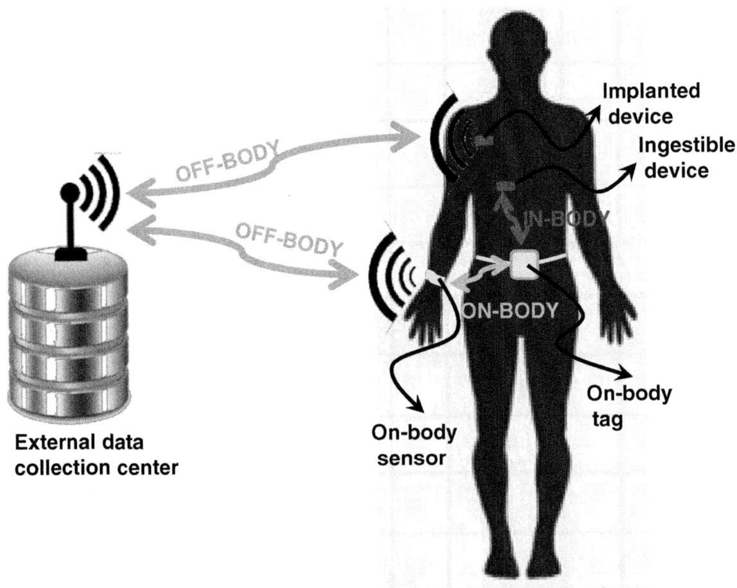

FIGURE 4.35 Future outlook and challenges for wireless communication and EM propagation for in-body and on-body devices.

the sensitivity of the information and patient confidentiality concerns, secure wireless protocols for relevant transfer and understanding of the medical data need to be developed. Further miniaturization of these devices without compromising normal functioning leads to more sophisticated antenna designs for reliable and robust wireless links. Novel antenna design strategies need to be developed for different in-body, on-body sensors/devices [73]. Wireless channel modeling, EM propagation, and link budget analysis need to be addressed in more detail for this complex environment. Battery life can pose a challenge for implantable/ingestible devices, thus calling for novel solutions to utilize wireless power transfer. In order to extend the lifetime of an implantable device, wireless data transmission has to be coupled with wireless power transfer. This scenario calls for antenna designs to share different working functionalities.

Lastly, we will briefly discuss two applications which can contribute toward the future of medicine in a significant way: Brain–machine interface systems [74] and nano-biosensors-based capsule systems [75].

4.5.1 A Brain–Machine Interface System

A brain–machine interface system can help synchronize and interface the use of prosthetic limbs in the same manner as biological limbs which will greatly benefit patients with a wide range of disabilities. A brain–machine interface system is shown in Figure 4.36. These systems typically require neural sensors/electrodes for brain stimulation as well as an external data center to decode and store neural signal/data and also control the mechanical actuators in the prosthetic limbs. The feedback from the touch and position sensors on the prosthetic limbs is rerouted to the neural implant. One of the main challenges for such systems is design of miniature implantable devices embedding neural sensors/electrodes, integrated electronics, and transmitting antennas for power and data telemetry. The neural implant should be biocompatible, highly

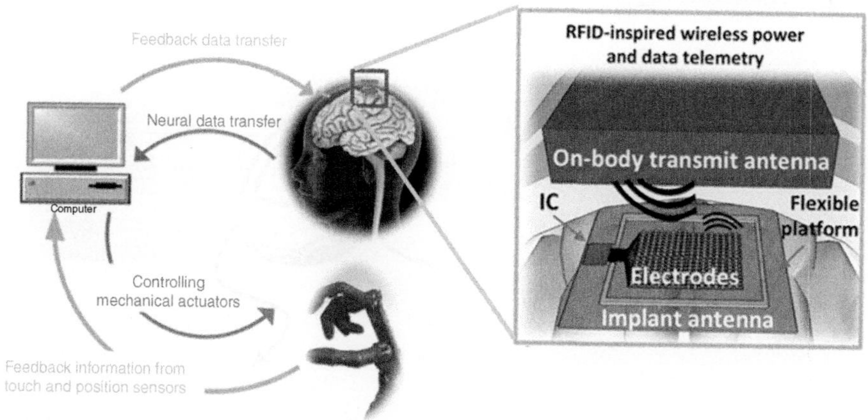

FIGURE 4.36 The wireless brain–machine interface system. *Source:* Moradi et al. 2014 [74]. Reproduced with permission of IEEE.

miniaturized, and wirelessly powered so that it can last a lifetime. The design of such a system is detailed in [74].

The neural implant consists of an implanted loop antenna integrated with an electrode array for neural sensing. Wireless power transfer is used to keep the neural implant alive at all times. On the transmitting-antenna side, a loop antenna is designed with the highest possible magnetic field for strong magnetic coupling to the implanted loop and short-range RFID backscattering technology is utilized for wireless powering and communication with the implant.

4.5.2 A Diagnostic Capsule System with Nano-Biosensors

The existing wireless diagnostic capsules have several limitations. The collected physiological data such as GI tract images, vital signs, etc. lack an overall clinical relevance unless accompanied with a tissue biopsy for *in vitro* analyses, most capsules are typically powered by an internal battery cell so they have a fixed lifetime and there is no guided way to control the orientation of the capsule leading to less effective link budgets and capsule performance. A smart diagnostic capsule system can be potentially developed with nano-biosensors, wireless power transfer capability, and novel miniaturized antenna robust to capsule orientation which can potentially address most of the limitations currently in these capsule imaging systems. One such nano-biosensor-based capsule system is shown in Fig. 4.37.

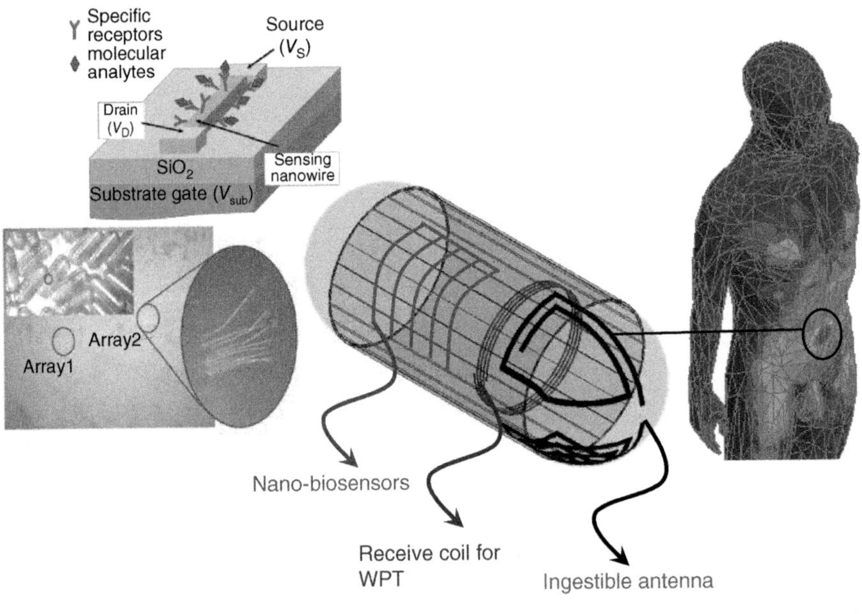

FIGURE 4.37 A diagnostic capsule system with nano-biosensors. *Source:* Rajagopalan et al. 2012 [75]. Reproduced with permission of IEEE. For a color version of this figure, see the color plate section.

For the nano-biosensors, a transformative amplifying nanowire FET sensor concept has been experimentally demonstrated which seamlessly integrates a sensing nanowire and a nanowire FET amplifier [75]. Once the molecular analytes are bound to the immobilized receptors, E-field emanated from those analyte charges will terminate in the channel and modulate its conductivity (i.e., measured signal) thus, detecting the bio-markers. Silicon nanowire arrays can be formed on the curved capsule surface to detect bio-markers in the GI tract (e.g., hemoglobin). Silicon nanowire arrays are formed by electron-beam lithography.

For the wireless power transfer a receiver coil can be placed strategically inside the capsule system. The receiving section potentially comprise of a three receiving coil (RC) set (in order to main robustness of the inductive link irrespective of the orientation of the capsule), a rectifier, and a regulating circuit. The transmit coil produces a strong alternating magnetic field. This magnetic field couples with the receiving coil and induces an alternating current in the receiving coil. The rectifier converts this into direct current, and finally the DC is smoothened by the regulating circuit. The main requirements for wireless power transfer are that it must be able to deliver stable and sufficient power, the level of the electromagnetic field generated from transmit coil should be within safely limits, and the receiving coil and circuit should be small enough so that it can be assembled into a small size capsule [76].

In order to ensure that the wireless link is always maintained irrespective of the capsule random orientation, we need to design robust antenna solutions which are insensitive to capsule rotations. The conformal meandered antenna [10] proposed above has those desirable characteristics. Thus, the novel diagnostic capsule system can potentially enhance the functionality of the existing medical systems.

4.6 CONCLUSION

The concept of "*bio-telemetry connecting patients to their doctors and hospitals at anytime, any location and with any amount of monitoring and diagnostics data*" can become a reality in the next few years due to the advancement of various medicine implantable and ingestible devices. Over the last decade, antenna designs for implantable devices have been proposed for several wireless data telemetry applications such as artificial eyes, cochlear implants, brain and cardiac pacemakers monitoring, heart rate monitoring, and glucose monitoring. Antenna designs for ingestible devices have also been proposed for applications in capsule endoscopy systems for pressure monitoring, GI tract imaging, and others. In this chapter, different antenna designs for implantable/ingestible medical devices were classified with the frequency of operation criterion. A brief summary of the various methodologies for the design and development of these antennas using numerical simulations and characterizations was also provided. A conformal capsule antenna design was discussed in detail for medical diagnostic applications. The antenna performance parameters were discussed in the human body environment, a thorough comparison was conducted between the proposed design and some other existing antenna topologies in terms of link budget estimation, and finally a wireless power transfer enabled frequency reconfigurable system was proposed and initial inductive link budget calculations were provided. Next, a medical monitoring and sensing system was proposed with the capsule antenna system using RFID technology. The antenna performance and tuning parameters were discussed and potential batch fabrication of these biocompatible antennas were detailed. Lastly, an RFID bio-tag was discussed which can be used for patient monitoring, including vitals and location, in hospital environments. A thorough investigation was performed in terms of impedance and read range characteristics. Future outlook for the medical monitoring and diagnostics within the paradigm of the bio-telemetry health care system was also highlighted.

REFERENCES

[1] M. A. Jensen and Y. Rahmat-Samii, "EM interaction of handset antennas and a human in personal communications," *Proc. IEEE*, vol. 83, no. 1, pp. 7–17, January 1995.

[2] Y. Rahmat-Samii, K. W. Kim, M. Jensen, K. Fujimoto, and O. Edvardsson, "Antennas and humans in personal communications," in *Mobile Antenna Systems Handbook*, 2nd ed., eds. K. Fujimoto and J. R. James, pp. 361–460. Artech House Inc., 2001.

[3] Y. Rahmat-Samii and Z. Li, "Handset communication antennas, including human interactions," in *Wireless Networks*, invited chapter, eds. G. Franceschetti and S. Stornelli, pp. 95–159. Academic Press, Elsevier, June 2006.

[4] P. Salonen and Y. Rahmat-Samii, "Wearable antennas: advances in the design, characterization and application," in *Antennas and Propagation for Body-Centric Wireless Communications*, eds. P. Hall and Y. Hao, pp. 151–188. Artech House Inc., 2006.

[5] L. Ukkonen and Y. Rahmat-Samii, "Embroidered textile antennas for body-centric wireless identification and sensing devices," in *Handbook of Antenna Technologies*. Springer, 2015.

[6] J. Kim and Y. Rahmat-Samii, "Implanted antennas inside a human body: simulations, designs, and characterizations," *IEEE Trans. Microw. Theory Tech.*, vol. 52, pp. 1934–1943, August 2004.

[7] Y. Rahmat-Samii and J. Kim, *Implanted Antennas in Medical Wireless Communications*, Synthesis Lectures on Antennas. Morgan & Claypool Publishers, June 2006.

[8] A. J. Johnasson, A. Karlsson, W. G. Scanlon, N. E. Evans, and Y. Rahmat-Samii, "Medical implant communication antennas," in *Antennas and Propagation for Body-Centric Wireless Communications*, invited chapter, eds. P. Hall and Y. Hao, pp. 241–270. Artech House Inc., MA, 2006.

[9] P. Izdebski, H. Rajagopalan, and Y. Rahmat-Samii, "Conformal ingestible capsule antenna: anovel chandelier meandered design," *IEEE Trans. Antennas Propag. (Special Issue on Body-Centric Antennas)*, vol. 57, no. 4, pp. 900–909, April 2009.

[10] H. Rajagopalan, P. Izdebski, and Y. Rahmat-Samii, "Antennas in medicine: ingestible capsule antennas," in *Frontiers in Antennas, Next Generation Design and Engineering*, invited chapter, ed. F. B. Gross, pp. 305–338. McGraw-Hill, 2011.

[11] E. C. Fear, S. C. Hagness, P. M. Meaney, M. Okoniewski, and M. A. Stuchly, "Enhancing breast tumor detection with near-field imaging," *IEEE Microw. Mag.*, vol. 3, no. 1, pp. 48–56, March 2002.

[12] A. Rosen, M. A. Stuchly, and A. V. Vorst, "Applications of RF/microwaves in medicine," *IEEE Trans. Microw. Theory Tech.*, vol. 50, no. 3, pp. 963–974, March 2002.

[13] M. J. Hagmann, "Optimization of helical coil applicators for hyperthermia," *IEEE Trans. Microw. Theory Tech.*, vol. 36, pp. 148–150, 1988.

[14] F. Sterzer, "Microwave medical devices," *IEEE Microw. Mag.*, vol. 3, no. 1, pp. 65–70, March 2002.

[15] H. Rajagopalan and Y. Rahmat-Samii, "Conformal RFID Antenna Design Suitable for Human Monitoring and Metallic Platforms," in Proceedings of the 4th European Conference on Antennas and Propagation (EuCAP '10), Barcelona, Spain, April 2010, pp. 1–5.

[16] G. Marrocco, "Body-Matched Antennas for Wireless Biometry," in Proceedings of the European Conference on Antennas and Propagation: EuCap 2006, Nice, France, November 2006.

[17] P. S. Hall and Y. Hao, *Antennas and Propagation for Body-Centric Wireless Communication*, 2nd ed. Artech House, 2012.

[18] F. Merli, "Implantable Antennas for Biomedical Applications", Ph.D Dissertation, Ecole Polytechnique Feédeérale de Lausanne (EPFL), Switzerland, 2011.

[19] Wireless Medical Technologies. Available at www.fr.com/files/Uploads/attachments/FinalRegulatoryWhitePaperWirelessMedicalTechnologies.pdf

[20] A. Kiourti, K. A. Psathas, and K. S. Nikita, "Implantable and ingestible medical devices with wireless telemetry functionalities: a review of current status and challenges," *Bioelectromagnetics*, vol. 35, no. 1, pp. 1–15, January 2014

[21] J. G. Webster, *Design of Cardiac Pacemakers*. IEEE Press, New York, 1995.

[22] J. Cavuoto, "Neural engineering's image problem," *IEEE Spectrum*, vol. 41, no. 4, pp. 32–37, April 2004.

[23] A. J. Johansson, "Wireless communication with medical implants: antennas and propagation," *Thesis*, Lund University, 2004.

[24] V. Shirvante, F. Todeschini, X. Cheng, and Y.-K. Yoon, "Compact Spiral Antennas for MICS Band Wireless Endoscope Toward Pediatric Applications," in 2010 IEEE Antennas and Propagation Society International Symposium(APSURSI), Toronto, ON, July 11–17, 2010, pp. 1–4.

[25] A. Kiourti and K. S. Nikita, "Miniature scalp-implantable antennas for telemetry in the MICS and ISM bands: design, safety considerations and link budget analysis," *IEEE Trans. Antennas Propag.*, vol. 60, no. 8, pp. 3568–3575, August 2012.

[26] J. Kim and Y. Rahmat-Samii, "Planar inverted-F antennas on implantable medical devices: meandered type versus spiral type," *Microw. Opt. Tech. Lett.*, vol. 48, no. 3, pp. 567–572, 2006.

[27] P. Soontornpipit, C. Furse, and Y. C. Chung, "Design of implantable microstrip antenna for communication with medical implants," *IEEE Trans. Microw. Theory Tech.*, vol. 52, no. 8, pp. 1944–1951, August 2004.

[28] T. Karacolak, A. Z. Hood, and E. Topsakal, "Design of a dual-band implantable antenna and development of skin mimicking gels for continuous glucose monitoring," *IEEE Trans. Microw. Theory Tech.*, vol. 56, no. 4, pp. 1001–1008, April. 2008.

[29] M. Norris, J.-D. Richerd, and D. Raynes, "Sub minature Antenna Design for Wireless Implants," in Proceedings of the IET Seminar on Antennas and Propagation for Body-Centric Wireless Communications, London, 2007, pp. 57–62.

[30] F. Merli, L. Bolomey, J.-F Zurcher, G. Corradini, E. Meurville, and A. K. Skrivervik, "Design, realization and measurements of a miniature antenna for implantable wireless communication systems," *IEEE Trans. Antennas Propag.*, vol. 59, no. 10, pp. 3544–3555, October 2011.

[31] S. I. Kwak, K. Chang, and Y. J. Yoon, "The helical antenna for the capsule endoscope," in *Proc. IEEE Antennas and Propag. Soc. Int Symp.*, vol. 2B, pp. 804–807, July 2005.

[32] S. H. Lee and Y. J. Yoon, "Fat arm spiral antenna for wideband capsule endoscope systems," in *Proc. IEEE Radio and Wireless Symp.*, 2010, pp. 579–582.

[33] D. Valderas, C. Schmidt, and X. Chen, "Broadband Implanted UHF RFID Antenna," in 2010 IEEE Antennas and Propagation Society International Symposium(APSURSI), Toronto, ON, July 11–17, 2010, pp. 1–4.

[34] G. Marrocco, "RFID antennas for the UHF remote monitoring of human subjects," *IEEE Trans. Antennas Propag.*, vol. 55, no. 6, pt. 2, pp. 1862–1870, June 2007.

[35] A. Sani, M. Rajab, R. Foster, and Y. Hao, "Antennas and propagation of implanted RFIDs for pervasive healthcare applications," *Proc. IEEE*, vol. 98, no. 9, pp. 1648–1655, 2010.

[36] G. Collin, A. Chami, C. Luxey, P. Le Thuc, and R. Staraj, "Small electrical antenna for saw sensor biotelemetry," *Microw. Opt. Tech. Lett.*, vol. 51, no. 10, pp. 2286–2293, 2009.

[37] H. Usui, M. Takahashi, and K. Ito, "Radiation Characteristics of an Implanted Cavity Slot Antenna into the Human Body," in IEEE Antennas and Propagation Society International Symposium 2006, July 9–14, 2006, pp. 1095–1098.

[38] S. Soora, K. Gosalia, M. S. Humayun, and G. Lazzi, "A comparison of two and three dimensional dipole antennas for an implantable retinal prosthesis," *IEEE Trans. Antennas Propag.*, vol. 56, no. 3, pp. 622–629, March 2008.

[39] R. Warty, M. R. Tofighi, U. Kawoos, and A. Rosen, "Characterization of implantable antennas for intracranial pressure monitoring: reflection by and transmission through a scalp phantom," *IEEE Trans. Microw. Theory Tech.*, vol. 56, no. 10, pp. 2366–2376, October 2008.

[40] E. Y. Chow, O. Yuehui, B. Beier, W. J. Chappell, and P. P. Irazoqui, "Evaluation of cardiovascular stents as antennas for implantable wireless applications," *IEEE Trans. Microw. Theory Tech.*, vol. 57, no. 10, pp. 2523–2532, 2009.

[41] S. Radiom, M. Baghaei-Nejad, G. Vandenbosch, L.-R. Zheng, and G. Gielen, "Far-field RF Powering System for RFID and Implantable Devices with Monolithically Integrated On-Chip Antenna," in 2010 IEEE Radio Frequency Integrated Circuits Symposium (RFIC), Anaheim, CA, May 23–25, 2010.

[42] H. Rajagopalan and Y. Rahmat-Samii, "Novel Ingestible Capsule Antenna Designs for Medical Monitoring and Diagnostics," in 2010 Proceedings of the Fourth European Conference on Antennas and Propagation (EuCAP 2010), Barcelona, Spain, April 12–16, 2010.

[43] K. Gosalia, M. S. Humayun, and G. Lazzi, "Impedance matching and implementation of planar space-filling dipoles as intraocular implanted antennas in a retinal prosthesis," *IEEE Trans. Antennas Propag.*, vol. 53, no. 8, pp. 2365–2373, August 2005.

[44] K. Gosalia, G. Lazzi, and M. Humayun, "Investigation of a microwave data telemetry link for a retinal prosthesis," *IEEE Trans. Microw. Theory Tech.*, vol. 52, no. 8, pp. 1925–1933, August 2004.

[45] R. Bashirullah and N. Euliano, "Capsule Antennas for Medication Compliance Monitoring," in IEEE Radio and Wireless Symposium, 2009, San Diego, CA, January 18–22, 2009, pp. 123–126.

[46] H. Yu, C.-M. Tang, and R. Bashirullah, "An Asymmetric RF Tagging IC for Ingestible Medication Compliance Capsules," in IEEE Radio Frequency Integrated Circuits Symposium, 2009, Boston, MA, June 7–9, 2009, pp. 101–104.

[47] T. Dissanayake, K. P. Esselle, and M. R. Yuce, "Dielectric loaded impedance matching for wideband implanted antennas," *IEEE Trans. Microw. Theory Tech.*, vol. 57, no. 10, pp. 2480–2487, October 2009.

[48] E. Y. Chow, M. M. Morris, and P. P. Irazoqui, "Implantable RF medical devices," *IEEE Microw. Mag.*, vol. 14, no. 4, pp. 64–73, June 2013.

[49] A. Kiourti and K. S. Nikita, "A review on implantable patch antennas for biomedical telemetry: challenges and solutions," *IEEE Antennas Propag. Mag.*, vol. 54, no. 3, pp. 210–228, June 2012.

[50] C. Gabriel, S. Gabriel, and E. Corthout, "The dielectric properties of biological tissues: I. Literature survey," *Phys. Med. Biol.*, vol. 41, no. 11, pp. 2231–2249, 1996.

[51] S. Gabriel, R. W. Lau, and C. Gabriel, "The dielectric properties of biological tissues: II. Measurements in the frequency range 10 Hz to 20 GHz," *Phys. Med. Biol.*, vol. 41, no. 11, pp. 2251–2269, 1996.

[52] S. Gabriel, R. W. Lau, and C. Gabriel, "The dielectric properties of biological tissues: III. Parametric models for the dielectric spectrum of tissues," *Phys. Med. Biol.*, vol. 41, no. 11, pp. 2271–2293, 1996.

[53] J. Kim and Y. Rahmat-Samii, "On the applicability of simplified spherical human heads for implanted antennas in biomedical communications", *Electromagnetics*, vol. 25, pp. 491–511, 2005.

[54] W. A. Qureshi, "Current and future applications of the capsule camera," *Nature Rev. Drug Disc.*, vol. 3, pp. 447–450, May 2004.

[55] G. Iddan, G. Meron, A. Glukhovsky, and P. Swain, "Wireless capsule endoscopy," *Nature*, vol. 405, p. 417, 2000.

[56] S. N. Adler and Y. C. Metzger, "PillCam COLON capsule endoscopy: recent advances and new insights," *Therapeutic Advances in Gastroenterology*, vol. 4, no. 4, pp. 265–268, 2011.

[57] H. Rajagopalan and Y. Rahmat-Samii, "Wireless medical telemetry characterization for ingestible capsule antenna designs", *IEEE Antennas Wireless Propag. Lett.*, vol. 11, pp. 1679–1682, 2012.

[58] U. Azad, H. C. Jing, and Y. E. Wang, "Link budget and capacity performance of inductively coupled resonant loops," *IEEE Trans. Antennas Propag.*, vol. 60, no. 5, pp. 2453–2461, May 2012.

[59] C. A. Balanis, *Antenna Theory*, 3rd ed. John Wiley & Sons, New York, 2005.

[60] *IEEE Standard for Safety Levels With Respect to Human Exposure to Radio Frequency Electromagnetic Fields, 3 kHz to 300 GHz*, IEEE C95.1-1991, 1992.

[61] K. R. Foster and J. Jaeger, "RFID inside: the murky ethics of implanted chips," *IEEE Spectrum*, pp. 24–29, March 2007.

[62] Available at http://www.epill.com/statistics.html (e-pill Medication Reminders Home Page).

[63] A. Solanas and J. Castellà-Roca, "RFID technology for the health care sector," *Recent Patents Electrical Eng.* 2008, vol. 1, no. 1, pp. 22–31, 2008.

[64] H. Rajagopalan and Y. Rahmat-Samii, "Ingestible RFID Bio-capsule Tag Design for Medical Monitoring," in 2010 IEEE Antennas and Propagation Society International Symposium (APSURSI), Toronto, ON, July 11–17, 2010, pp. 1–4.

[65] H. W. Son and C. S. Pyo, "Design of RFID tag antennas using an inductively coupled feed," *Electron. Lett.*, vol. 41, no. 18, pp. 994–996, 2005.

[66] G. Marrocco, "Gain-optimized self-resonant meander line antennas for RFID applications," *IEEE Antennas Wireless Propag. Lett.*, vol. 2, no. 21, pp. 302–305, 2003.

[67] Q. Xianming and Y. Ning, "A folded dipole antenna for RFID," *Proc. IEEE Antennas and Propag. Soc. Int. Symp.*, pp. 97–100, June 2004.

[68] G. Marrocco, "The art of UHF RFID antenna design: impedance matching and size-reduction techniques," *IEEE Antennas Propag. Mag.*, vol. 50, no. 1, pp. 66–79, February 2008.

[69] K. V. S. Rao, P. V. Nikitin, and S. F. Lam, Antenna design for UHF RFID tags: a review and a practical application," *IEEE Trans. Antennas Propag.*, vol. 53, no. 12, pp. 3870–3876, December 2005.

[70] P. R. Foster and R. A. Burberry, "Antenna problems in RFID systems," *Proc. Inst. Elect. Eng. Colloquium RFID Tech.*, pp. 3/1–3/5, October 1999.

[71] D. M. Dobkin and S. M. Weigand, "Environmental Effects on RFID Tag Antennas," in Proceedings of the IEEE MTT-S International Microwave Symposium, Sunnyvale, CA, June 2005, pp. 135–138.

[72] H. Ishihata, T. Tomoe, K. Takei, T. Hirano, K. Yoshida, S. Shoji, H. Shimauchi, and H. Horiuchi, "A radio frequency identification implanted in a tooth can communicate with the outside world," *IEEE Trans. Info. Tech. Biomed.*, vol. 11, no. 6, pp. 683–685, November 2007.

[73] A. K. Skrivervik and F. Merli, "Design Strategies for Implantable Antennas," in Proceedings of the 4th European Conference on Antennas and Propagation, Loughborough, UK, November 2011.

[74] E. Moradi, K. Koski, T. Björninen, L. Sydänheimo, J. M. Rabaey, J. M. Carmena, Y. Rahmat-Samii, and L. Ukkonen, "Antenna applications corner: miniature implantable and wearable on-body antennas: towards the new era of wireless body-centric systems," *IEEE Antennas Propag. Mag.*, vol. 56, no. 1, pp. 271–291, 2014.

[75] H. Rajagopalan, K.-S. Shin, J. Kina, C. O. Chui, and Y. Rahmat-Samii, "A Smart Diagnostic Capsule with a Novel Antenna and Nano-biosensors," in Proceedings of the 6th European Conference on Antennas and Propagation (EuCAP), Prague, March 26–30, 2012, pp. 190–193.

[76] M. R. Basar, M. Y. Ahmad, J. Cho, and F. Ibrahim, "Application of wireless power transmission systems in wireless capsule endoscopy: an overview," *Sensors*, vol. 14, pp. 10929–10951, 2014.

5

BASIC PERFORMANCE CHARACTERISTICS OF WEARABLE ANTENNAS OVER A WIDE FREQUENCY RANGE

Koichi Ito
Center for Frontier Medical Engineering, Chiba University, Chiba-shi, Japan

5.1 INTRODUCTION

In recent years, body-centric wireless communications [1,2] or human body communications (sometimes referred as intrabody communications) [3] have become a very active area of research because of their numerous attractive applications, such as personal healthcare systems, personal entertainment, smart or intelligent homes, identification systems. Body-centric wireless communications were first devised and demonstrated by Zimmerman [4]. His prototype device operated at a very low frequency of 330 kHz. Since then, many researchers have followed his pioneering work and some extended it to higher frequencies, such as 10 MHz [5–10]. Meanwhile, ultra-high frequency (UHF) bands such as 403 MHz, 868 MHz, and 2.45 GHz have also become emerging trends for body-centric wireless communications [1,2, 11–13]. Hence, the prospective frequencies are in an extremely wide range, and an objective idea on how to select the proper frequency band for individual applications is required [14].

Today, research on body-centric wireless communications is diversified in terms of

- Domains: on-body, in-body, and off-body [1]
- Frequencies: MHz–GHz (ISM, MICS, etc.)

Electromagnetics of Body-Area Networks: Antennas, Propagation, and RF Systems, First Edition.
Edited by Douglas H. Werner and Zhi Hao Jiang.

- Transmission mechanism:
  - Electric current
  - Quasi-electrostatic/magnetic coupling
  - space/creeping/surface wave propagation

In this Chapter, the basic performance characteristics of wearable antennas for body-centric wireless communications will be described over a wide frequency range.

Section 5.2 overviews frequency dependence of the communication channels between 3 MHz and 3 GHz for wearable antennas mounted on the human body [14]. Since it is difficult to measure the received signal for such a wide frequency range, all the results are obtained by using numerical analyses based on the finite-difference time-domain (FDTD) method [15]. A static human body model which has a posture of standing in free space is employed in order to characterize various frequency channels. Specifically, electric field distributions around the human body, received open voltage, and power transmission efficiency with ideal matching circuits are discussed.

Section 5.3 first describes the influence of a surrounding environment (namely a metal floor and a wall close to the human body) on the electric field distribution and received open voltages around the human body [16]. Then, the effects of body movement (walking) on received open voltages is presented [14]. Here, a dynamic human body model which simulates walking motion is employed to assess the channel variations.

In Section 5.4, two different scenarios are demonstrated to apply the basic results described above to practical situations: a commercialized wall-mounted identification (ID) tag system [17] and a human vital data acquisition system using a dual-mode antenna [18].

5.2 FREQUENCY DEPENDENCE OF COMMUNICATION CHANNELS BETWEEN WEARABLE ANTENNAS MOUNTED ON THE HUMAN BODY

5.2.1 Calculation Models

It is necessary to choose an appropriate antenna to cover a wide frequency range from 3 MHz to 3 GHz, because frequency characteristics of a communication channel largely depend on the performance of the antenna to be used. In order to properly assess the frequency characteristics over such an extremely wide range, an antenna (or "electrode" for much lower frequencies) should be simple and retain its basic behavior over the range. To address this requirement, a typical top-loaded monopole antenna (shown in Figure 5.1) is introduced both for transmitting and receiving [14]. Its first resonant frequency is approximately 4 GHz, and its radiation pattern is omnidirectional (tangential to the body surface) over the frequency range of interest. The

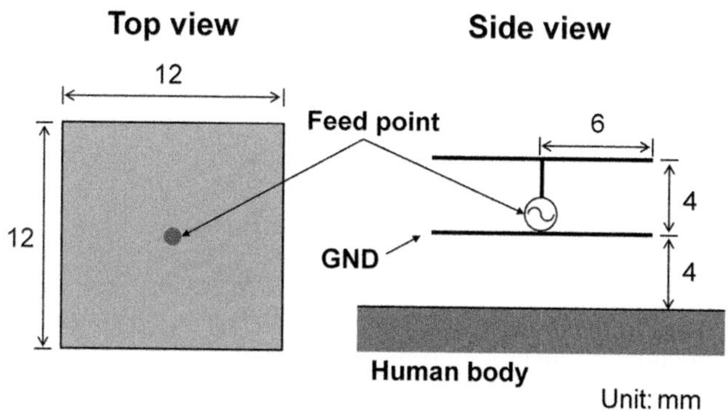

FIGURE 5.1 Geometry of the antenna. *Source:* Ito et al. 2012 [14]. Reproduced with permission of IEEE.

transmitting antenna is excited by use of a current source with an infinite internal impedance. The terminals of the receiving antennas are open-ended in order to observe open voltages.

Figure 5.2 shows the human body model used in the simulations, and depicts the locations of a transmitter (TX) and five receivers (RX) [14]. The human body model was created by employing Poser®, which is a commercial 3D graphics software package [19]. The human body model measures 1.72 m in height and is composed of uniform tissue which is electrically equivalent to muscle. A transmitting antenna (TX) is mounted on the abdomen, and receiving antennas (RX) are mounted on the chest,

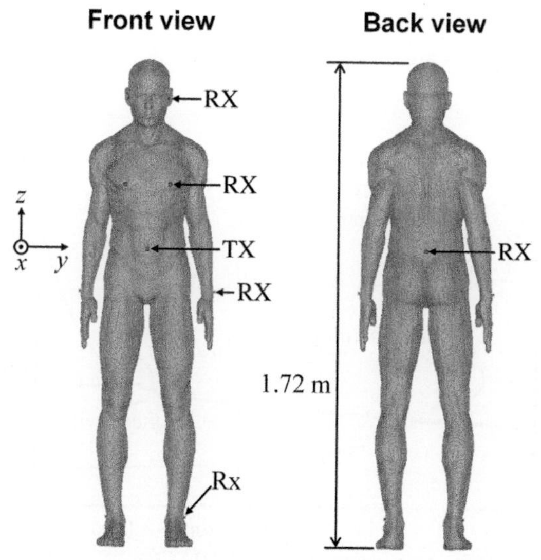

FIGURE 5.2 Human body model used in the simulations (in free space). *Source:* Ito et al. 2012 [14]. Reproduced with permission of IEEE.

wrist, ankle, ear, and back, respectively. For all antennas, a 4 mm separation exists between the antennas and the body surface, as shown in Figure 5.1.

Dielectric properties of human tissue are dependent on frequency. In order to solve the problem in the FDTD method with some degree of accuracy, the complex relative permittivity of the human tissue can be approximated by the Debye equation

$$\varepsilon_r(\omega) = \varepsilon_\infty + \frac{\varepsilon_s - \varepsilon_\infty}{1 + j\omega\tau} + \frac{\sigma_s}{j\omega\varepsilon_0} \tag{5.1}$$

where ω is the angular frequency, ε_0 is the permittivity of free space, ε_s is the static permittivity at zero frequency, ε_∞ is the optical permittivity at infinite frequency, τ is the relaxation time, and σ_s is the static conductivity. As mentioned above, the human body model is composed of muscle-equivalent tissue, however, a Debye-type dispersion does not accurately model the actual muscle over the entire frequency range because the actual muscle has gentler dispersion [20]. For this reason, the FDTD calculation was separated into three ranges of frequencies; 3–30 MHz, 30–300 MHz, and 300 MHz to 3 GHz. The parameters in (5.1) were determined for each range so as to be continuous at boundaries between neighboring ranges as shown in Figure 5.3 [14]. As a result, the discrepancy between literature and approximated values are within 10% over the whole frequency range, where the literature value is fitted by a 4-pole Cole–Cole equation [21]. The specific values of the parameters are summarized in Table 5.1 [14]. Other computational parameters (cell size, dimension of the computational domain, precision of the absorbing boundary condition, and the maximum frequency of exciting pulse) were also determined for each range listed in Table 5.1. The cell size in a region including the antennas and the human body is 2 mm. The "maximum frequency" is defined to have an intensity of −60 dB at the peak frequency. A time step of 3.84×10^{-12} s is used to satisfy the stability condition and 20,000 iterations are completed in the simulation. With these conditions,

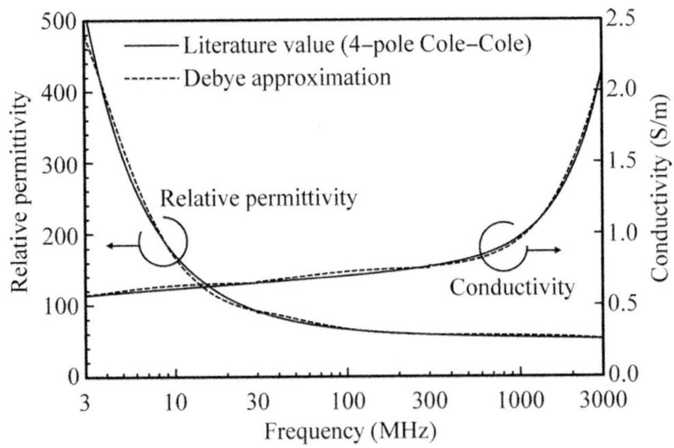

FIGURE 5.3 Frequency dependence of dielectric properties of muscle. *Source:* Ito et al. 2012 [14]. Reproduced with permission of IEEE.

TABLE 5.1 Computational parameters for respective frequency ranges. ***Source:*** **Ito et al. 2012 [14]. Reproduced with permission of IEEE.**

Parameter	3–30 MHz	30–300 MHz	300 MHz–3 GHz
Static permittivity, ε_s	706	105	58
Optical permittivity, ε_∞	81	57	40
Relaxation time, τ	3.98×10^{-8} s	3.18×10^{-9} s	3.54×10^{-10} s
Static conductivity, σ_s	0.52 S/m	0.63 S/m	0.75 S/m
Cell size	2–20 mm		2–5 mm
Computational domain	1.4 m × 1.4 m × 2 m		1 m × 1 m × 2 m
Number of PML layers	20		10
Maximum frequency of pulse	500 MHz		4 GHz

the field of interest converges sufficiently, and undesired reflections from the PML are excluded. All the results were obtained by applying the Fourier transform to the transient fields excited by a pulse, not to those by a sinusoidal wave. It has been confirmed that the results obtained with this procedure agree with those performed by the method of moment [9, 10].

5.2.2 Electric Field Distributions

Figure 5.4 shows the calculated electric field distributions inside and around the human body at 3 MHz, 30 MHz, 300 MHz, and 3 GHz [16]. The observation plane is

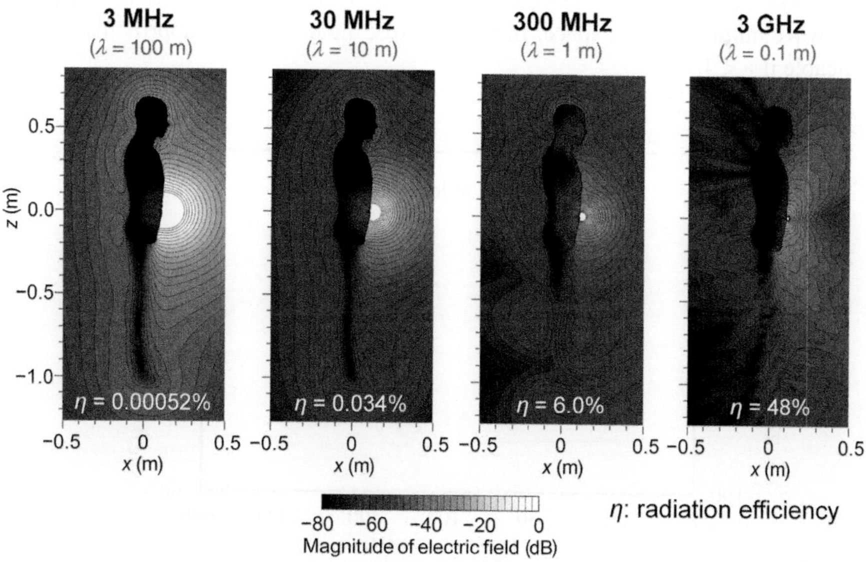

FIGURE 5.4 *E*-field distributions at different frequencies in free space. *Source:* Ito and Haga 2010 [16]. Reproduced with permission of IEEE. For a color version of this figure, see the color plate section.

the *xz* plane, which includes the feeding point. Here, equal input power is assumed to clarify the radiation efficiency for each frequency. By comparing the *E*-field distributions at 3 MHz and 30 MHz, it can be noted that shapes of the contour lines are almost the same, however, the *E*-field intensities at 30 MHz are lower by about 20 dB. Radiation efficiencies at 3 MHz and 30 MHz are 0.00052% and 0.034%, respectively. These results imply that the consumed power is mostly due to absorptions inside the human body; the amount of the absorbed power compared with the stored energy around the body is greater at higher frequencies. This is due to the fact that the impedance of the muscle increases with frequency. At 300 MHz, the *E*-field intensities are similar to those at 30 MHz, but several dips are generated and radiation efficiency increases to about 6%. This is due to the fact that the height of the human body is somewhat longer than the wavelength (1 m). When the frequency is 3 GHz, the *E*-field intensities behind the body are relatively low, but radiation efficiency reaches 48%. Furthermore, many ripples in the *E*-field are generated all over the body.

5.2.3 Received Open Voltage and Ideal Power Transmission Efficiency

In order to indicate transmission "efficiency" for an on-body communication system, two different indexes are described below: received open voltage and ideal power transmission efficiency.

The calculated received open voltage of the receiver located at the chest is shown in Figure 5.5. Here, the received open voltage is normalized by the input voltage of the transmitter. The voltage keeps flat below at about 200 MHz and then increases with frequency. Considering that the distance between the transmitting and receiving antennas is $r = 0.24$ m, the frequency when $kr = 1$ is about 200 MHz, where k is the wave number in free space. In other words, the quasi-static approximation is applicable if $r < 1/k$.

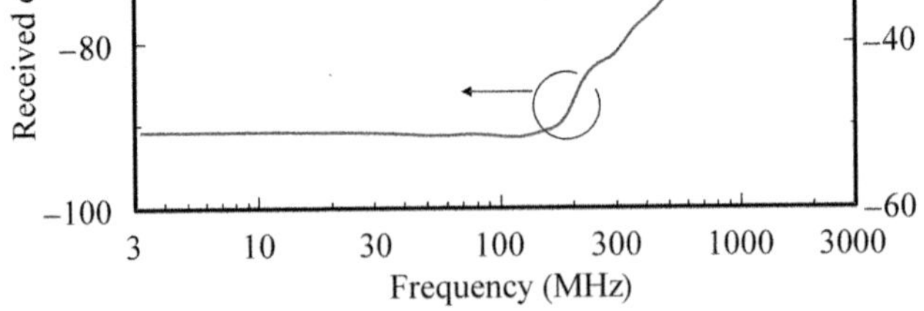

FIGURE 5.5 Received open voltage and ideal power transmission efficiency. (Open voltage is normalized by the input voltage of the transmitter.)

Assessments in terms of the received open voltage described above are appropriate at low frequency because the receiving circuits for low frequency have high input impedance in order to detect the reactive field. However, impedance matching should be considered at higher frequencies. For this reason, the assessments in terms of power transmission efficiency are essential in properly characterizing the system's performance. Power transmission is defined as the ratio of the receiving power to the input power when the impedance matching is perfect for both the transmitter and the receiver. For the maximum power transmission, it is necessary to consider the mutual coupling between the transmitter and the receiver. The matching condition can be found by employing conjugate-image impedance theory [22]. Without lengthy derivations, the ideal power transmission efficiency G_{21} can be expressed by self-impedance and mutual impedance of the transmitter and the receiver as follows [14]:

$$G_{21} = \frac{|Z_{21}|^2}{R_{11}R_{22}} \cdot \frac{1}{(1+\theta_{\mathrm{R}})^2 + \theta_X^2} \tag{5.2}$$

where

$$\theta_{\mathrm{R}} = \sqrt{\left(1 - \frac{R_{21}^2}{R_{11}R_{22}}\right)\left(1 + \frac{X_{21}^2}{R_{11}R_{22}}\right)} \tag{5.3}$$

$$\theta_X = \frac{R_{21}X_{21}}{R_{11}R_{22}} \tag{5.4}$$

and R_{11} and X_{11} are the real and imaginary parts of the self-impedance of transmitting antenna Z_{11}, and the same rule is applied to the self-impedance of receiving antenna Z_{22}, and the mutual impedance Z_{21}. Incidentally, the ideal power transmission efficiency called here is the same as "ultimate gain" defined in [21].

Figure 5.5 also shows calculated ideal power transmission efficiency when the receiver is located at the chest. The efficiency increases with decreasing frequency below about 200 MHz. This is due to the increasing input power with frequency as shown in Figure 5.6 which simulates the input power of the transmitter when a 1 V input voltage is assumed [14]. For the same reason, the efficiency does not increase as much at higher frequencies. A local maximum in efficiency can be observed at around 500 MHz. Although the power transmission efficiency is a better gauge than the open voltage, it requires substantially more computation time because the self-impedance for all the antennas must be obtained.

Additionally, it should be noted that the input power is very low at low frequencies, as shown in Figure 5.6. This is due to the fact that the electric fields around the human body are almost the reactive field which doesn't consume the power unlike the radiating field. This is one of the unique features of low frequency communications.

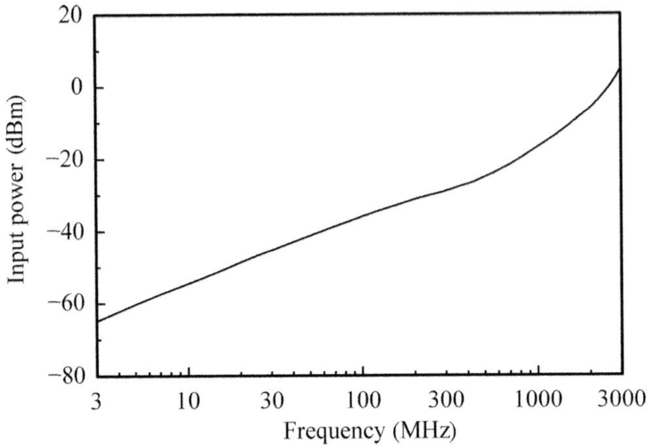

FIGURE 5.6 Input power when 1 V input voltage is assumed. *Source:* Ito et al. 2012 [14]. Reproduced with permission of IEEE.

5.3 INFLUENCES OF SURROUNDING ENVIRONMENT AND BODY MOVEMENT

5.3.1 Influence of a Floor and a Wall

This subsection describes the influence of a surrounding environment (namely a metal floor and a wall close to the human body) on the electric field distribution and received open voltages around the human body [16].

Figure 5.7 shows the perfect electric conductor (PEC) floor which is 20 mm beneath the soles of the human body model and the PEC wall which is 0.5 m from the body center. Both the floor and the wall are infinite in size.

Figure 5.8 illustrates the calculated electric field distributions inside and around the human body at 3 MHz, 30 MHz, 300 MHz, and 3 GHz for the two cases (a) only with floor and (b) with floor and wall. The calculations were done in the same manner as for Figure 5.4 (in free space). Electric field distributions at 3 MHz and 30 MHz for both the cases have almost no change compared with Figure 5.4. However, standing waves are observed at 3 GHz in the region between the transmitter and the floor/wall.

Figures 5.9a–5.9e plots the frequency dependence of the received open voltages for the receivers placed at various locations on the human body model. In each figure, the results of three cases are compared; just in free space, only with the floor, and with the floor and the wall. According to these results, it is obvious that the received open voltages at the chest are the highest and the most stable among the five locations. On the contrary, the voltages at the back are the lowest and rather unstable for frequencies greater than 40 MHz. The voltages at the ankle are naturally affected by the presence of the floor.

Figure 5.10 compares the received open voltages at different locations for the case with the floor [14]. Of course, the voltages are different with different locations, but

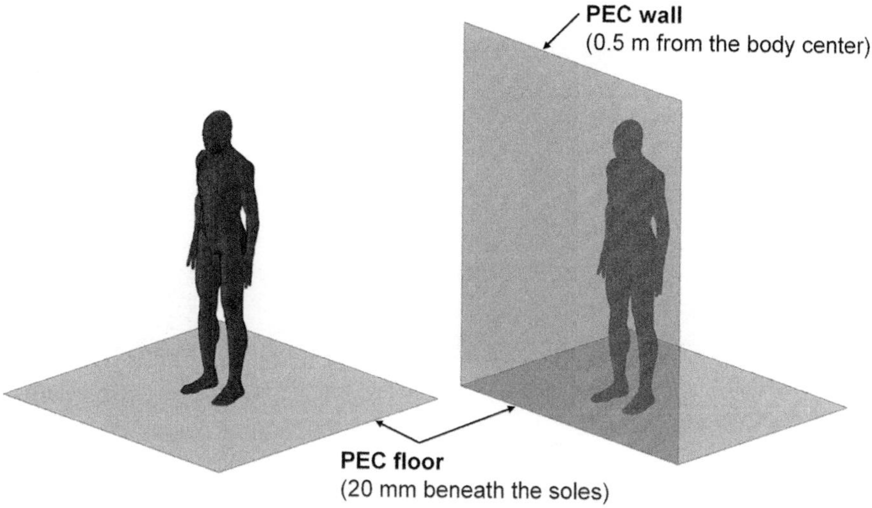

FIGURE 5.7 Effects of surrounding environment.

they are quite stable in a range below 30 MHz regardless of the locations. It is interesting to note that the voltages at the ankle and the ear are almost the same in magnitude at lower frequency, however, the difference between them is large at around 50–300 MHz. This is owing to non-uniformity of the electric field distributions at and around the resonant frequencies of the human body.

Figure 5.11 compares the ideal power transmission efficiencies at different locations for the case with the floor [14]. In contrast to the received open voltages in Figure 5.10, the power transmission efficiencies increase with decreasing frequency in a range below 200 MHz. This is due to the increasing input power with frequency as shown in Figure 5.6.

5.3.2 Influence of Body Movement

This subsection describes the effects of body movement (walking) on received open voltages [14]. For statistic assessments, a dynamic human body model which simulates walking motion is used, as shown in Figure 5.12 [14]. The height of the dynamic model is the same as that of the standing model in Figure 5.2. One period of a walking motion (two steps) is divided into 30 frames. For symmetric right and left sides of the motion, only the first half of the frames shown in Figure 5.12 are employed. A transmitter is mounted on the abdomen, and nine receivers are mounted on the body at the right and left chests, right and left wrists, right and left ankles, right and left ears, and back, respectively. By combining the received voltages observed on the right and left halves of the body, the dynamic characteristics of the

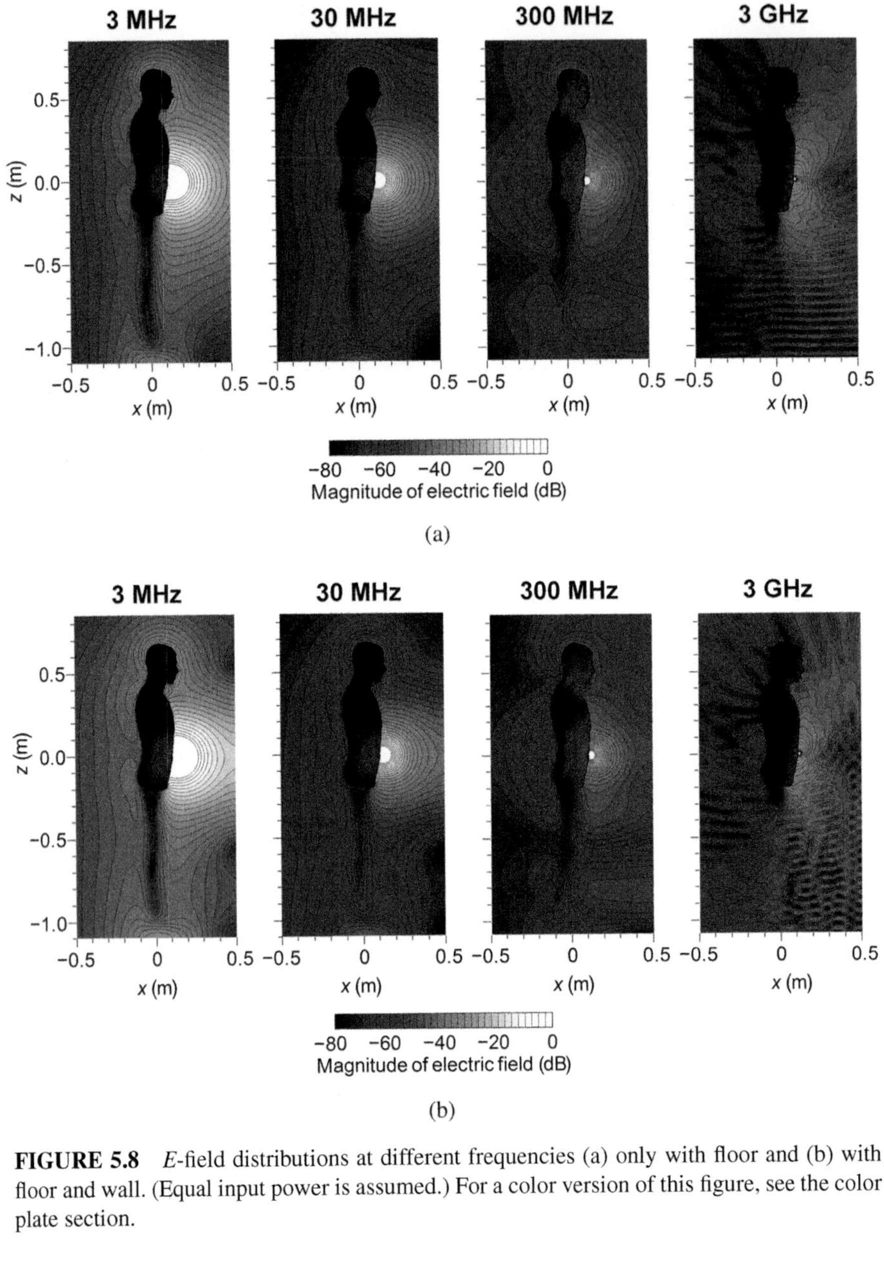

FIGURE 5.8 *E*-field distributions at different frequencies (a) only with floor and (b) with floor and wall. (Equal input power is assumed.) For a color version of this figure, see the color plate section.

communication channel are obtained. The computational conditions are the same as those in Section 5.2 (see Table 5.1).

Because of space limitation, only results observed at the wrist and the ankle, which have large variations, are described. Figure 5.13 shows the dynamic characteristics of the received open voltage observed at the wrist [14]. In addition, Figure 5.14 shows

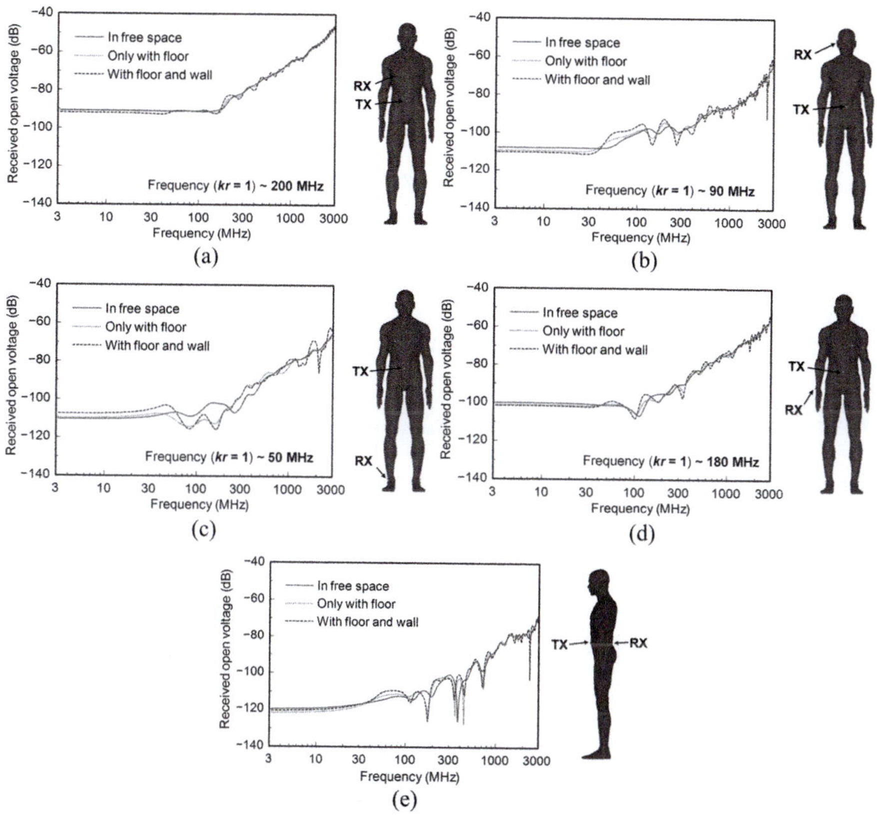

FIGURE 5.9 Received open voltages normalized by the input voltage of transmitter at different locations (a) at chest, (b) at ear, (c) at ankle, (d) at wrist, and (e) at back.

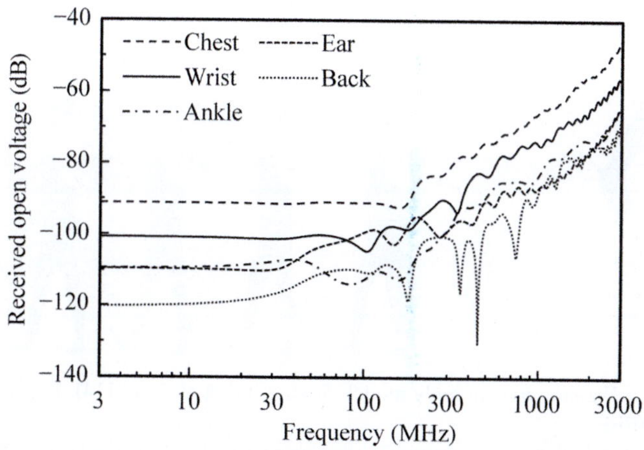

FIGURE 5.10 Received open voltages normalized by the input voltage of transmitter at different locations (with floor). *Source:* Ito et al. 2012 [14]. Reproduced with permission of IEEE.

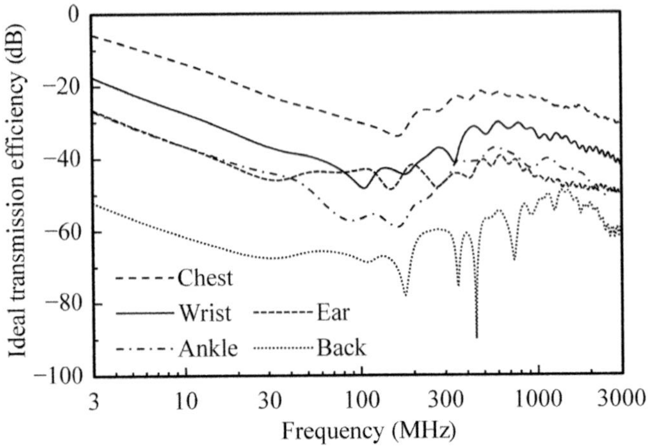

FIGURE 5.11 Ideal power transmission efficiencies at different locations (with floor). *Source:* Ito et al. 2012 [14]. Reproduced with permission of IEEE.

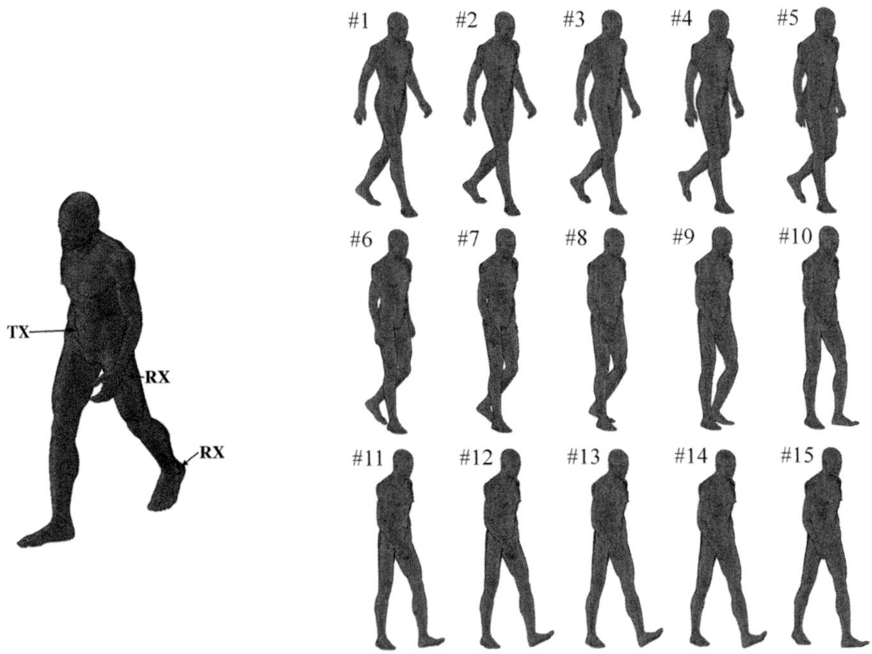

FIGURE 5.12 Walking human body model. *Source:* Ito et al. 2012 [14]. Reproduced with permission of IEEE.

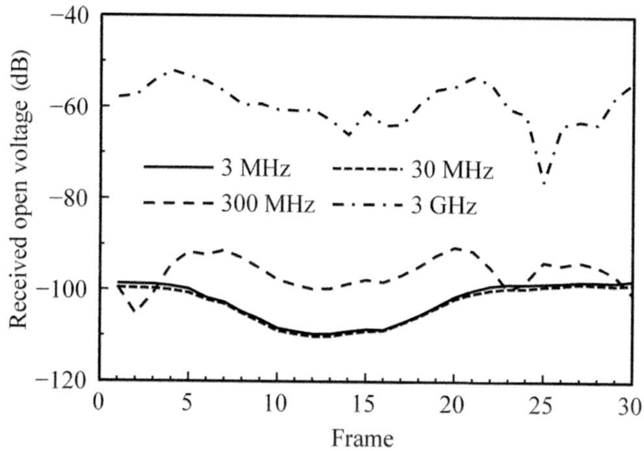

FIGURE 5.13 Dynamic characteristics of received open voltage at wrist. *Source:* Ito et al. 2012 [14]. Reproduced with permission of IEEE.

the frequency dependences of the average and the standard deviation of the received open voltage [14]. All the results are normalized by the input voltage. According to Figure 5.13, the variations of the received voltage at the wrist are large even at 3 MHz and 30 MHz. Specifically, the voltages fluctuate by approximately 12 dB at those frequencies. When the voltages on the wrist are low (at frames #10 to #16), the arm is swung toward the back; therefore, the shadow fading should be considered even at low frequencies. However, the curve is relatively smooth because multipath fading does not appear at these frequencies. By contrast, the voltages at 300 MHz and 3 GHz have several dips during the period of motion. In particular, the result at

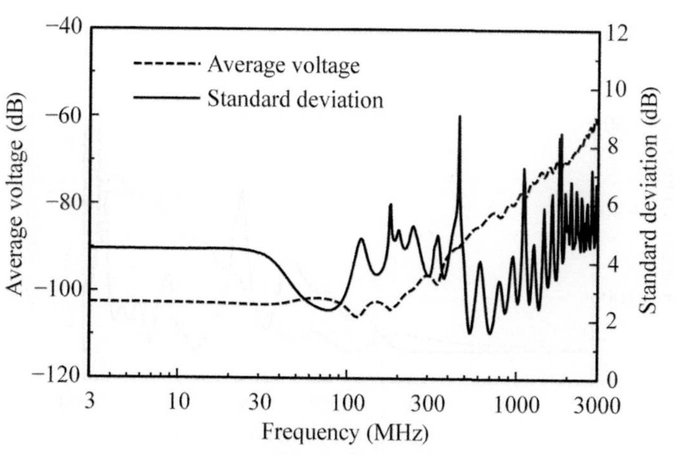

FIGURE 5.14 Average and standard deviation of received open voltage at wrist. *Source:* Ito et al. 2012 [14]. Reproduced with permission of IEEE.

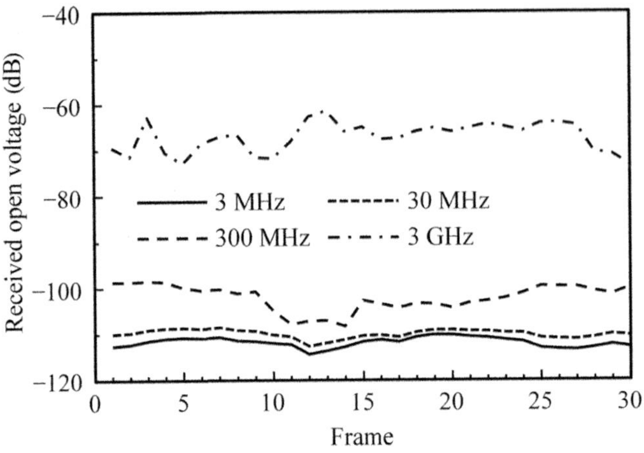

FIGURE 5.15 Dynamic characteristics of received open voltage at ankle. *Source:* Ito et al. 2012 [14]. Reproduced with permission of IEEE.

3 GHz shows a null point at frame #25. In the frequency domain, the standard deviation has many ripples, especially above 300 MHz. These results are due to the fact that standing waves are generated around the arm. The deep null points are observed only at several frequency points, likely due to the finiteness of the frame number. In other words, the fast fading is significant. If it is necessary to accurately assess the channel characteristics considering the fast fading, then we have to enlarge the division number of the postures.

In the same manner, Figure 5.15 shows the dynamic characteristics of the received open voltage observed at the ankle, and Figure 5.16 shows the frequency dependences

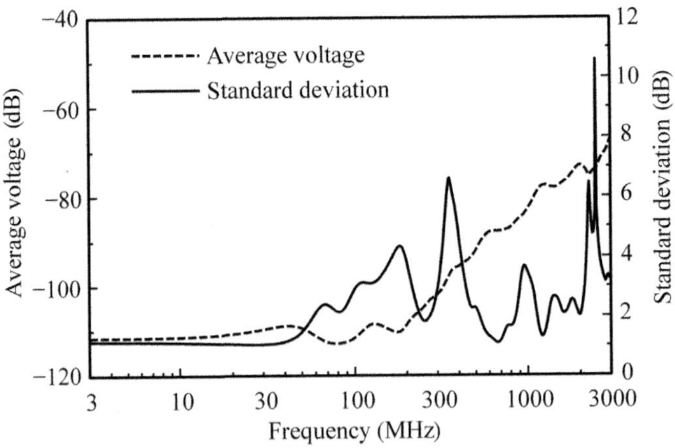

FIGURE 5.16 Average and standard deviation of received open voltage at ankle. *Source:* Ito et al. 2012 [14]. Reproduced with permission of IEEE.

of the average and the standard deviation [14]. According to Figure 5.15, the received voltage at the ankle is relatively low at 3 MHz and 30 MHz. However, several dips during the period of motion are observed at 300 MHz and 3 GHz, similar to those at the wrist. The standard deviation in the frequency domain is high at several periodic frequency points in a range of 70–400 MHz (at most 6.6 dB). This is likely due to the resonance phenomenon of the human body. Also, since it is difficult to implement diversity antennas at those frequencies, the 70–400 MHz frequency band is not appropriate for some applications, such as motion captures.

Overall, the standard deviations are relatively high for frequencies above 100 MHz. For frequencies higher than about 1 GHz, the diversity technique can enhance the communication performance. The diversity technique may be difficult to employ at around several hundreds of megahertz due to the communication wavelength and human body resonances. In addition, the standard deviations of the received voltages, as well as the averages, are approximately flat in a range below 30 MHz. Therefore, the quasi-static approximation would be applicable for this range.

5.4 PRACTICAL APPLICATIONS

In this section, two different scenarios are demonstrated to apply the basic results, described above, to practical situations: a commercialized wall-mounted ID tag system [17], and a human vital data acquisition system using a dual-mode antenna [18].

5.4.1 Wall-Mounted ID Tag System

Figure 5.17 illustrates the setup of the calculation model for a wall-mounted ID tag system with a receiver equipped on a wall. Here, the floor and the wall are concrete slabs with typical dielectric properties. These surfaces are backed by PEC. The floor

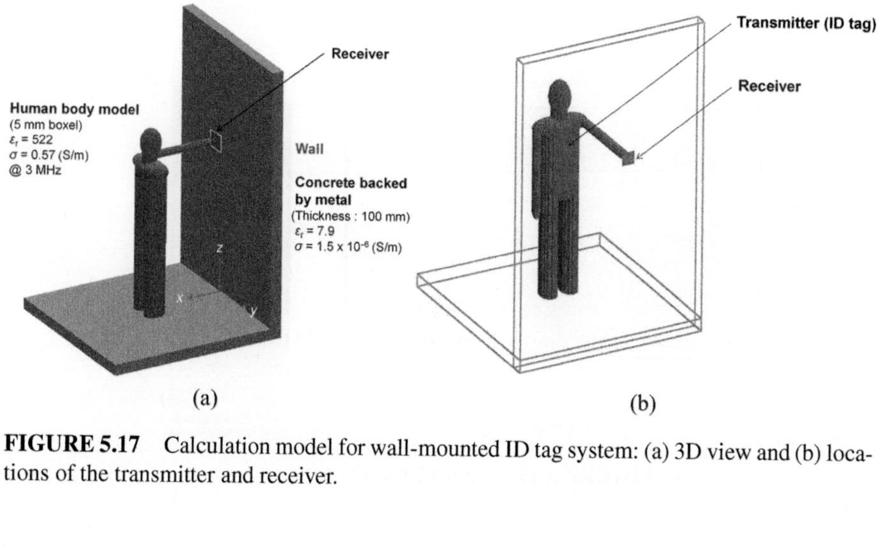

(a) (b)

FIGURE 5.17 Calculation model for wall-mounted ID tag system: (a) 3D view and (b) locations of the transmitter and receiver.

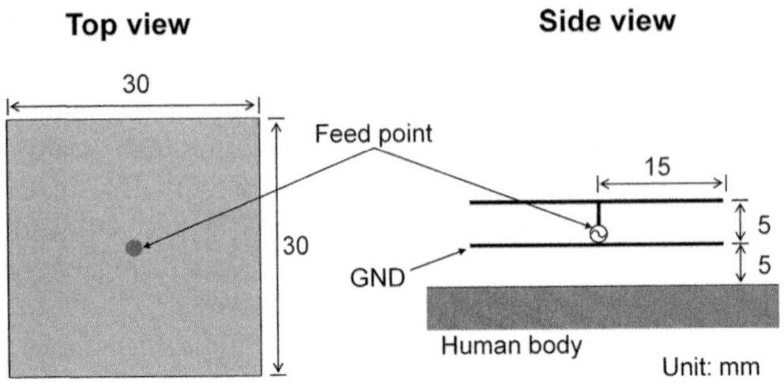

FIGURE 5.18 Geometry of the transmitter.

and the wall are extended into the PML, thus they represent infinite structures. The human body model in this scenario has homogenous dielectric properties equal to muscle at 3 MHz and consists of several elliptical cylinders and spheroids for simplicity. The dimensions are determined by referring to a realistic numerical model of a Japanese adult male with average height and weight. There is a 20 mm air gap between the sole of the human body model and the floor. Figure 5.17b shows the locations of the transmitter and the receiver. Figures 5.18 and 5.19 show the structure of the transmitter and the receiver, respectively, which are similar to Figure 5.1. The transmitter is placed 5 mm away from the body surface and is excited by a 1 V delta gap voltage source with 50 Ω internal resistance. The receiver is attached on the wall and is positioned in front of the left arm.

Calculated electric field distributions are shown in Figure 5.20. The two observation planes studied are (a) a plane along the body center and (b) a plane including the left arm. Here, the electric field levels are normalized with the value at the feeding

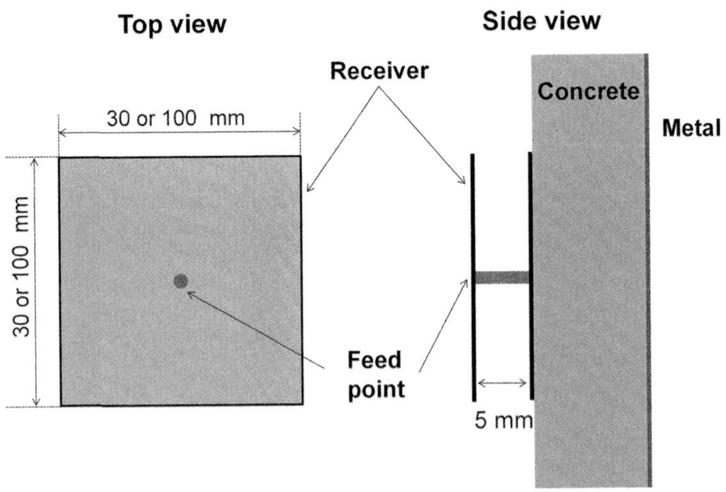

FIGURE 5.19 Geometry of the receiver.

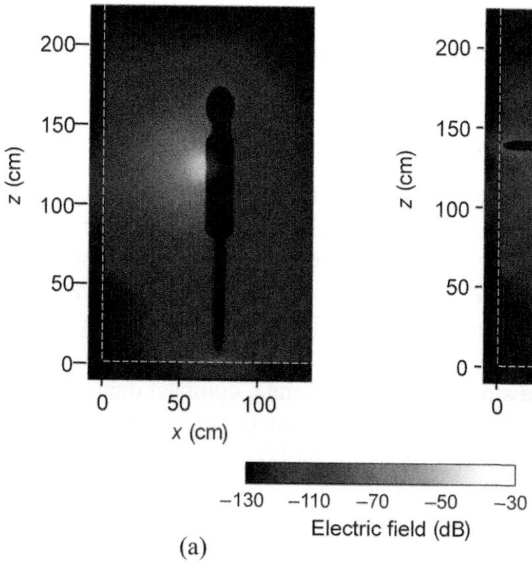

FIGURE 5.20 Electric field distributions at 3 MHz. For a color version of this figure, see the color plate section.

point. White dashed lines indicate the surfaces of the floor and the wall. According to these results, the electric field is smoothly distributed over the body, as well as the left arm.

Figure 5.21 shows the dependence of the received open voltages on the distance between the tip of the left arm and the receiver. There are two curves representing the differences in the receiver size. From the results, it can be confirmed that the received open voltage decreases rapidly as the distance increases. For practical design of a wall-mounted ID tag system, it is important to be able to set a threshold in terms of

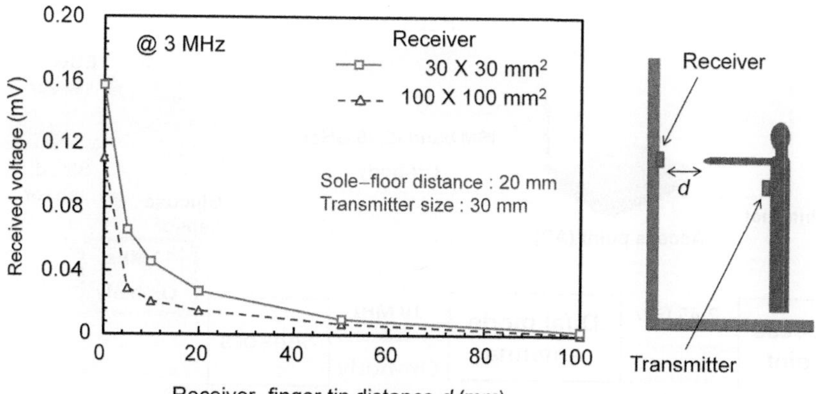

FIGURE 5.21 Received voltage versus receiver–finger tip distance.

received voltage for a particular situation. The results described here will give some guidance for choosing the desired threshold voltage.

5.4.2 Human Vital Data Acquisition System Using a Dual-Mode Antenna

Figure 5.22 illustrates an example of a human vital data acquisition system which can be applied to a personal healthcare system. Various bio-signals from different sensors are sent to a repeater via on-body communications and these data are transmitted to an external device in off-body mode. Relatively low frequency bands (tens of MHz) are more suitable for on-body mode, while the 2.45 GHz ISM band is a good candidate for off-body mode [14].

In order to realize such a single wearable antenna, compact dual-mode antennas were proposed and reported [18]. The dual-mode antenna can operate in two different modes: off-body mode and on-body mode. In on-body mode, the antenna performs as a parallel-plate electrode which generates an electric field distribution at 10 MHz. Therefore, low power consumption can be realized for on-body communications, as described in Section 5.2. In off-body mode, the antenna radiates at the 2.45 GHz ISM band to communicate with an external device.

Figure 5.23 shows the structure of a compact dual-mode antenna which has a feeding pin, a signal electrode, and a folded ground electrode [18]. The size of the antenna was reduced to 30 mm × 33 mm × 4 mm. For practical use, a 2 mm thick plastic spacer ($\varepsilon_r = 3$) is placed between the antenna and the human body. Due to the folded ground

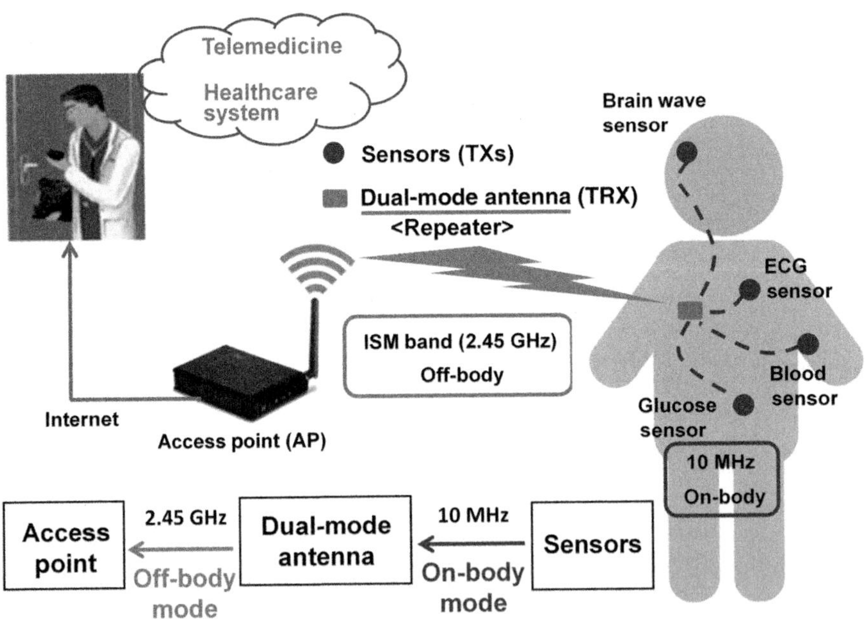

FIGURE 5.22 Human vital data acquisition system.

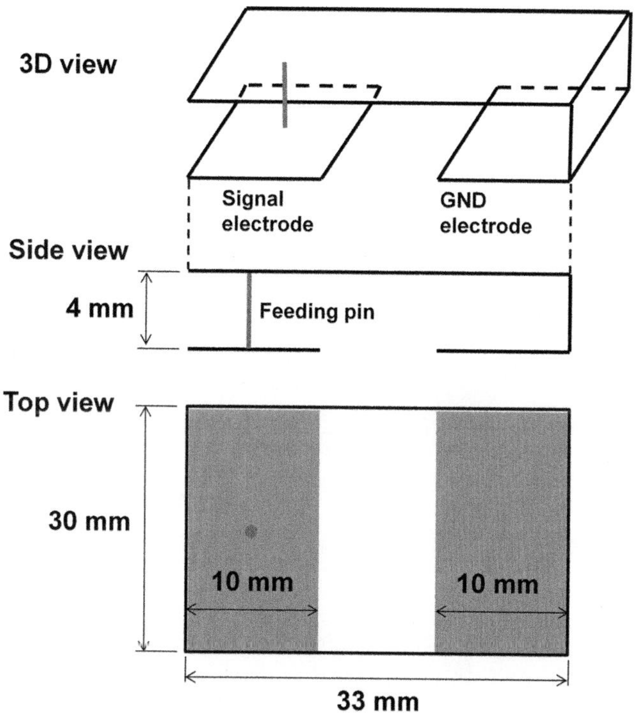

FIGURE 5.23 Structure of a compact dual-mode antenna. *Source*: Lin and Ito 2014 [18]. Reproduced with permission of IEEE.

structure, the antenna resonates at 2.45 GHz. Regarding the polarization generated from an antenna, vertical polarization is suitable for on-body mode and horizontal polarization for off-body mode, as shown in Figure 5.24. The antenna in Figure 5.23 is designed to generate such proper polarizations.

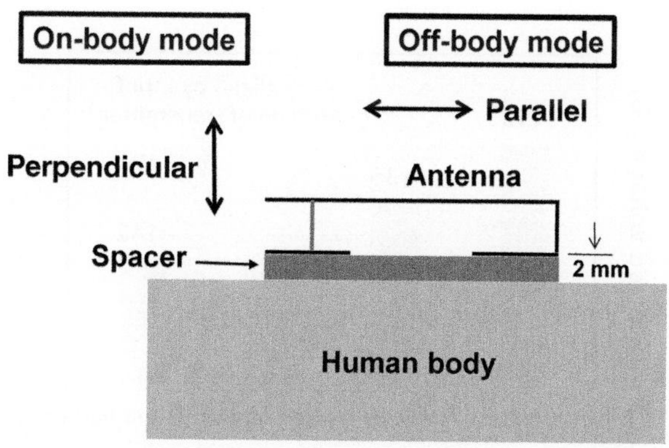

FIGURE 5.24 Proper polarization of dual-mode antenna.

TABLE 5.2 Computational parameters for two modes.

Mode	On-body mode	Off-body mode
Frequency	10 MHz	2.45 GHz
Cell size	1–2 mm	0.5–2 mm
Computation domain	1.5 m × 1.5 m × 2 m	1 m × 1 m × 2 m
Number of PML layers	16	8

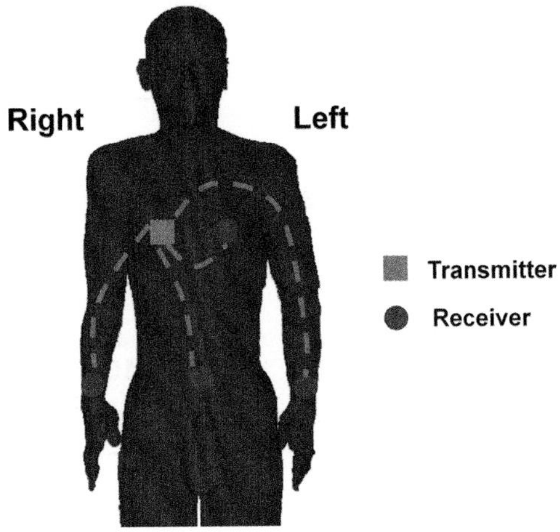

FIGURE 5.25 A transmitter and four receivers on the human model.

For on-body mode, a high-resolution human body model (1750 mm in height) [14] was used for the FDTD simulation. Table 5.2 shows the computational parameters used for the two modes. As shown in Figure 5.25, one transmitter is attached to the right chest while four receivers are attached to the left chest, abdomen, and right and left wrists. Figure 5.26 presents the variation in received voltage for the four different on-body links on the human body. From the results, the relative received voltages are

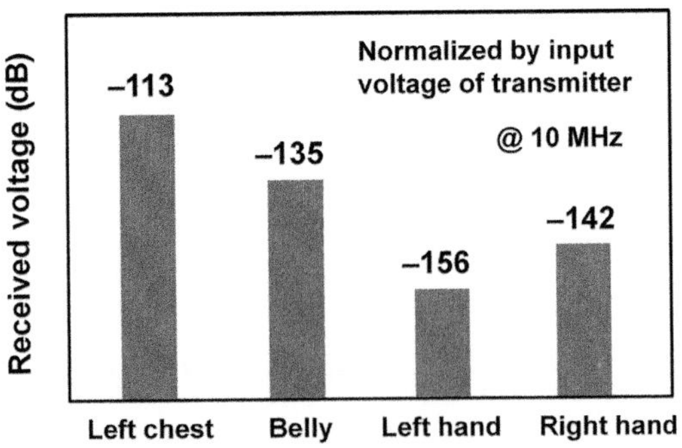

FIGURE 5.26 Simulated received voltages at four receivers.

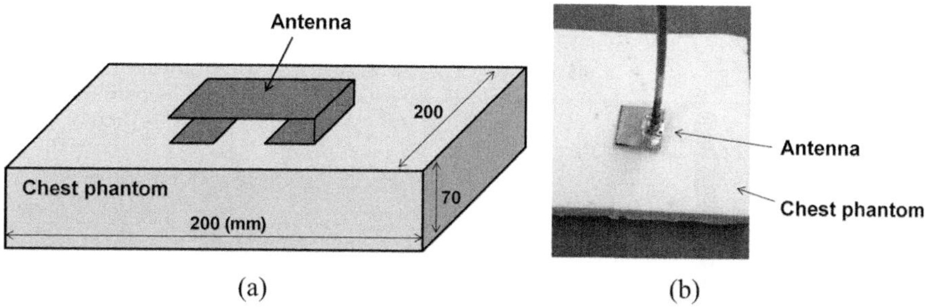

FIGURE 5.27 Evaluation of reflection coefficient (S_{11}) at 2.45 GHz band. (a) Simulation model and (b) measurement with phantom.

around −110 to −160 dB depending on the distance between the transmitter and the receiver.

Figures 5.27a and 5.27b show the simulation model, and the measurement setup with a phantom, respectively, for evaluation of the reflection coefficient at 2.45 GHz band. The chest phantom (dimensions of 200 mm × 200 mm × 70 mm; $\varepsilon_r = 35.2$, and $\sigma = 1.16$ S/m at 2.45 GHz) can be utilized just for evaluation of the reflection coefficient.

Figure 5.28 shows the simulated and measured reflection coefficients of the dual-mode antenna for off-body communication [18]. Here, the antenna is attached to the central part of the chest phantom, as well as to the right chest of the high-resolution human body model. The results confirmed rather good impedance matching in the ISM band and small differences between the chest and the human body phantoms used.

Figure 5.29 shows simulated radiation patterns for the dual-mode antenna attached on the chest phantom (dashed line) as well as for the high-resolution human model (solid line) at 2.45 GHz [18]. It was found that the radiation patterns are relatively

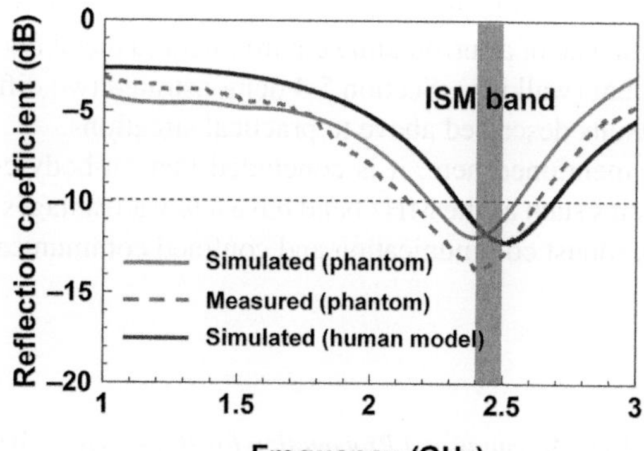

FIGURE 5.28 Simulated and measured reflection coefficients (S_{11}). *Source*: Lin and Ito 2014 [18]. Reproduced with permission of IEEE.

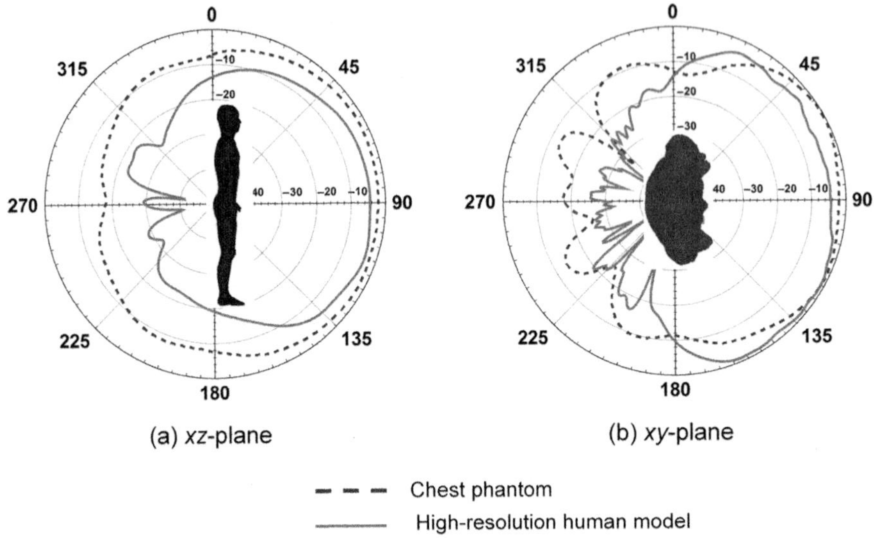

FIGURE 5.29 Simulated radiation patterns at 2.45 GHz. *Source*: Lin and Ito 2014 [18]. Reproduced with permission of IEEE.

omnidirectional in the half-sphere in front of the human body. Unlike the reflection coefficients above, the radiations patterns are somewhat different between the chest phantom and the human body model.

5.5 CONCLUSION

This Chapter has described the basic performance characteristics of wearable antennas for body-centric wireless communications over a wide frequency range. Section 5.2 has overviewed frequency dependence of the communication channels between 3 MHz and 3 GHz for wearable antennas mounted on the human body. Section 5.3 has described the influence of a surrounding environment (a metal floor and a wall) as well as body movement (walking). Section 5.4 demonstrated two different scenarios to apply the basic results described above to practical situations.

From the results mentioned here, it is concluded that on-body communications using lower frequencies such as the MHz band have a few advantages in terms of low power consumption, robust communication and confined communication range.

REFERENCES

[1] P. S. Hall and Y. Hao, *Antennas and Propagation for Body-Centric Wireless Communications*, 2nd ed. Artech House, Norwood, MA, 2012.

[2] Special issue on antennas and propagation for body-centric wireless communications, *IEEE Trans. Antennas Propag.*, vol. 57, no. 4, pp. 833–1016, April 2009.

[3] M. Seyedi, B. Kibret, D. T. H. Lai, and M. Faulkner, "A survey on intrabody communications for body area network applications," *IEEE Trans. Biomed. Eng.*, vol. 60, no. 8, pp. 2067–2079, August 2013.

[4] T. G. Zimmerman, "Personal area networks: near-field intrabody communication," *IBM Syst. J.*, vol. 35, no. 3–4, pp. 609–617, April 1996.

[5] K. Hachisuka, Y. Terauchi, Y. Kishi, K. Sasaki, T. Hirota, H. Hosaka, K. Fujii, M. Takahashi, and K. Ito, "Simplified circuit modeling and fabrication of intrabody communication devices," *Sens. Actuat. A Phys.*, vol. 130–131, pp. 322–330, April 2006.

[6] N. Cho, J. Yoo, S. Song, J. Lee, S. Jeon, and H. Yoo, "The human body characteristics as a signal transmission medium for intrabody communication," *IEEE Trans. Microwave Theory Tech.*, vol. 55, no. 5, pp. 1080–1086, May 2007.

[7] K. Fujii, M. Takahashi, and K. Ito, "Electric field distributions of wearable devices using the human body as a transmission channel," *IEEE Trans. Antennas Propag.*, vol. 55, no. 7, pp. 2080–2087, July 2007.

[8] A. Sasaki and M. Shinagawa, "Principles and demonstration of intrabody communication with a sensitive electrooptic sensor," *IEEE Trans. Instrum. Meas.*, vol. 58, no. 2, pp. 457–466, February 2009.

[9] N. Haga, K. Saito, M. Takahashi, and K. Ito, "Proper derivation of equivalent-circuit expressions of intra-body communication channels using quasi-static field," *IEICE Trans. Commun.*, vol. E95-B, no. 1, pp. 51–59, January 2012.

[10] N. Haga, K. Saito, M. Takahashi, and K. Ito, "Effects of posture and earth ground on intrabody communication channels using quasi-static field," *IEICE Trans. Commun.*, vol. J95-B, no. 2, pp. 257–264, February 2012 (in Japanese).

[11] T. Aoyagi, Iswandi, M. Kim, J. Takada, K, Hamaguchi, and R. Kohno, "Body Motion and Channel Response of Dynamic Body Area Channel," in Proceedings of the 5th European Conference on Antennas and Propagation, Rome, April 11–15, 2011, pp. 3293–3297.

[12] A. R. Chandran, G. A. Conway, and W. G. Scanlon, "Compact Low-Profile Patch Antenna for Medical Body Area Networks at 868 MHz," in Proceedings of the IEEE International Symposium on Antennas and Propagation, San Diego, CA, July 5–11, 2008, pp. 1–4.

[13] G. A. Conway, W. G. Scanlon, and S. L. Cotton, "The Performance of On-Body Wearable Antennas in a Repeatable Multipath Environment," in Proceedings of the IEEE International Symposium on Antennas and Propagation, San Diego, CA, July 5–11, 2008.

[14] K. Ito, N. Haga, M. Takahashi, and K. Saito, "Evaluations of body-centric wireless communication channels in a range from 3 MHz to 3 GHz," *Proc. IEEE*, vol. 100, no. 7, pp. 2356–2363, July 2012.

[15] A. Taflove, *Advances in Computational Electrodynamics: The Finite-Difference Time-Domain Method.* Artech House, Norwood, MA, 1998.

[16] K. Ito and N. Haga, "Basic Characteristics of Wearable Antennas for Body-Centric Wireless Communications," in Proceedings of the 2010 Loughborough Antennas and Propagation Conference, Loughborough, November 8–9, 2010, pp. 42–47.

[17] K. Ito, R. Suga, N. Haga, M. Takahashi, and K. Saito, "Link Characteristics Between Wearable Antenna and Receivers Equipped on External Objects at MHz Band," in Proceedings of the Fourth European Conference on Antennas and Propagation, Barcelona, April 12–16, 2010, C29P2-4.

[18] C-H. Lin and K. Ito, "Compact Wearable Dual-mode Antennas for Body-centric Wireless Communications," in 2014 IEEE International Symposium on Antennas and Propagation, Memphis, TN, July 6–11, 2014, pp. 307–308.

[19] E-frontier Web site. Available at http://www.e-frontier.com/

[20] S. Gabriel, R. W. Lau, and C. Gabriel, "The dielectric properties of biological tissues: II. Measurements in the frequency range 10 Hz to 20 GHz," *Phys. Med. Biol.*, vol. 41, pp. 2251–2269, April 1996.

[21] IFAC Web site. Available at http://niremf.ifac.cnr.it/tissprop/

[22] S. Roberts, "Conjugate-image impedances," *Proc. IRE*, vol. 34, no. 4, pp. 198–204, April 1946.

6

IMPLANTED ANTENNAS AND RF TRANSMISSION IN THROUGH-BODY COMMUNICATIONS

TERENCE SHIE PING SEE[1], ZHI NING CHEN[1,2], AND XIANMING QING[1]

[1] *Institute for Infocomm Research, A*STAR, Singapore*

[2] *Department of Electrical and Computer Engineering, National University of Singapore, Singapore*

6.1 INTRODUCTION

The last decade has witnessed a rapid surge of interest in new sensing, monitoring, and rehabilitation devices for health care and the use of wearable/wireless devices for clinical applications. There has been a strong tendency to use implanted electronic devices for therapeutic and diagnostic purposes. Recently, the Medical Implant Communication Service (MICS) technologies have been investigated with great interest. The MICS, which was allocated by the Federal Communication Committee (FCC) on a shared, secondary basis in 1999, refers to a specification for using a frequency band between 402 MHz and 405 MHz in communication with medical implants [1, 2]. It allows bidirectional radio communication between a pacemaker or other electronic implants and external devices. The maximum transmit power is limited to 25 μW, or −16 dBm, in order to reduce the risk of the interfering with other users within the same band. Other frequency bands considered for implant communication include the Industrial, Scientific, and Medical (ISM) bands (902–928 MHz, 2.40–2.48 GHz, 5.725–5.875 GHz), Wireless Medical Telemetry Services (WMTS) bands (608–614 MHz, 1.395–1.400 GHz), and ultra-wideband (UWB) of 3.1–10.6 GHz.

Electromagnetics of Body-Area Networks: Antennas, Propagation, and RF Systems, First Edition.
Edited by Douglas H. Werner and Zhi Hao Jiang.

The frequency band of 902–928 MHz is commonly abbreviated as the 915 MHz ISM band. In this band, there are no restrictions to the application or the duty cycle. Furthermore, the allowed power output is considerably higher. The FCC Section 15.249 allows 50 mV/m of electrical field strength at a distance of 3 m within the frequency band of 902–928 MHz, which corresponds to an Effective Isotropic Radiated Power (EIRP) of −1.23 dBm. A higher output power up to 30 dBm is permitted if the system employs some form of spread spectrum such as frequency hopping or direct sequence spread spectrum, since they are less likely to interfere with other systems and are also immune to interference from other systems.

The MICS system is able to transmit information from an implanted antenna that is embedded inside the human body to an external device by using a wireless communication link. Using the MICS, a health-care provider can set up a wireless link between an implanted device and a base station, which allows physicians to establish high-speed, ease of use, reliable, and short-range access to patient's health data in real time. The MICS system is able to reduce the frequency of diagnosis by a doctor and to alleviate the physical or mental burden of the patients. Furthermore, it is noninvasive as the communication can be established without wire piercing of skin, which reduces the risk of infection during a medical diagnosis [3].

Traditional endoscopy which requires a long and thin flexible tube through the patient's mouth is generally intrusive and requires the patient to be sedated intravenously. Moreover, only the images of the upper 4 ft of the stomach, colon, and rectum can be captured. The remaining 20 ft of the intestine cannot be diagnosed. With the development of a capsule endoscope system, the images of the entire digestive tract can be transmitted, which aids in the early detection of obscure gastrointestinal bleeding, gastroparesis, Crohn's disease, Celiac disease as well as in the surveillance of small bowel tumors and polyps. Using capsule endoscopy is a noninvasive, safe, and well-tolerated procedure, although the inability to sulfate the bowel due to capsule retention and perform biopsy as well as the lack of therapeutic capabilities are some of the concerns associated with this method [4,5]. Generally a capsule endoscope consists of a wireless integrated circuit (IC) transceiver, light-emitting diode (LED), camera, sensors, and an antenna. Due to the space limitations, the antenna designed for a capsule endoscope is required to be physically small, which leads to an increased challenge for it to be matched to IC transceiver at a specific frequency [6,7].

Paralysis is a debilitating disease of the neuromuscular system identified by partial or complete loss of motor functions, which has affected millions of people worldwide. Among the common causes, stroke, spinal cord injury, traumatic brain injury, and multiple sclerosis are highly cited. Over 5 million people in the United States and nearly 40,000 people in Singapore suffer from the disease in some form or another. Among those, tetraplegia, a type of paralysis that hinders all four limbs, is the most difficult to manage. Most of the care for paralysis is largely palliative. At present, studies targeting nerve repair through stem cell research and tissue engineering are still in their infancy. The damaged nerves regrow but very slowly, and often the damage is too severe to heal. A potential treatment for paralysis is to route control signals from the brain around the injury by artificial connections. Such signals could then control electrical stimulation of muscles, thereby restoring volitional movement to paralyzed limbs. Recent studies have shown that tetraplegic patients can consciously control the activity of nerve cells or neurons in the motor cortex responsible for upper limb movements, even after several years of paralysis [8, 9]. This finding has raised great expectations among the paralyzed patients for a neuromotor prosthetic device, a medical implant that aims to replace or restore lost motor functions in paralyzed humans by routing movement-related signals from the brain, around damaged parts of the nervous system, to external effectors such as robot arm, wheel chair, or even patients' own muscles via functional electrical stimulation. Experiments on monkeys and humans have established the proof of concept by recording from a relatively small number of neurons than initially thought; suggesting that neuromotor prosthesis based upon intracortical neuronal ensemble spiking activity could provide a valuable new neurotechnology to restore independence for humans with paralysis. In these studies, the subjects have successfully controlled a computer cursor or a robot arm through an implant, intracortical neural probe (array), tested into the region of motor cortex responsible for hand movements.

Despite the encouraging results, the neuromotor prosthetic devices are not ready for clinical use as they undermine the patient safety by tethering the subject to a rack of instruments through cable connections for power/data transmission. This approach not only creates inconvenience to the subject but also exposes the subject to a great concern by putting him or her at a risk of brain infection [10, 11]. Moreover, the sophisticated instruments for decoding neural signals and generating action commands demand the presence of a technician to monitor, troubleshoot, and calibrate. Furthermore, the probe-tissue interface, where the inflammatory response to foreign body, triggers a scar tissue formation around the probes encapsulating and isolating them from neurons. This isolation degrades the recording stability and signal quality over time, deteriorating the device performance in chronic implants. Therefore, in order to ensure patient safety and reduce the risk of infection, the use of the cables should be eliminated and data should be sent to the external controller wirelessly [12, 13]. The wireless communication for implantable devices has drawn much attention from both academia and industry as it possesses several advantages, such as minimizing restrictions in daily activities, providing the possibility of remote control and monitoring, and facilitating less invasive surgical procedures [14, 15]. As inductive biotelemetry faces the constraints of limited communication range and sensitivity to inter-coil misalignment [16], the antenna-enabled biotelemetry is preferred. However, as these medical devices are used within or at close proximity to the human body, the associated electromagnetic effects due to the interaction between the RF signals and human tissues have led to various regulatory bodies imposing maximum limits for the specific absorption rate (SAR), which determines the amount of RF energy that is absorbed by the human tissues. The antennas on the head as well as the controller play a critical role in ensuring the quality and reliability of the received signals. Most of the studies have been conducted with the head antennas for various applications such as monitoring, diagnosis, and treatment [17–21].

The antenna design for head implants is extremely challenging as the small form factor is required, even at low operating frequencies. Moreover, due to the lossy, inhomogeneous, and dispersive surrounding tissues, the impedance matching is often affected. Hence, the antenna needs to be co-designed with its surrounding tissues and biocompatible casing in order to achieve optimum efficiency. In order to reduce the SAR and enhance the transmission with the external receiver, an antenna with directional radiation is preferred. This chapter will focus on RF transmission characterization as well as the antenna designs for wireless capsule endoscopy and neural motor prosthetic devices.

6.2 ANTENNAS FOR WIRELESS CAPSULE ENDOSCOPY

6.2.1 RF Transmission Characterization

The characterization of RF transmission through/in body is critical for the link budget calculation of a communication system [22–30]. Full-duplex wireless communication through the human body possesses many engineering challenges. The dielectric properties of the human body tissues are inherently lossy, and the fluid within has a frequency dependent conductivity and permittivity. The communication channel is expected to change as the capsule is transported through the body natural bowel movement. The various physical limitations, such as path loss and noise figure of each component, have to be practically defined before specifying the design parameter. In this section, two methodologies of the human body channel characterization are discussed.

(a) Measurement of the RF transmission using animal tissues such as pig tissue.
(b) Simulation of the RF transmission in human body.

The experimental study was conducted using dipole and loop antennas at 403 MHz and 915 MHz as shown in Figure 6.1.

6.2.1.1 Measurement Methodology In order to characterize the transmission attenuation inside the body, the measurements were conducted by using two different

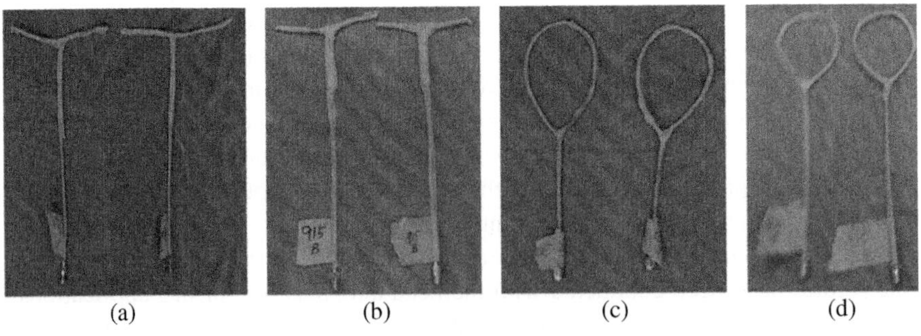

(a) (b) (c) (d)

FIGURE 6.1 Dipole and loop antennas at 403 MHz and 915 MHz: (a) dipoles at 403 MHz, (b) dipoles at 915 MHz, (c) loops at 403 MHz, and (d) loops at 915 MHz.

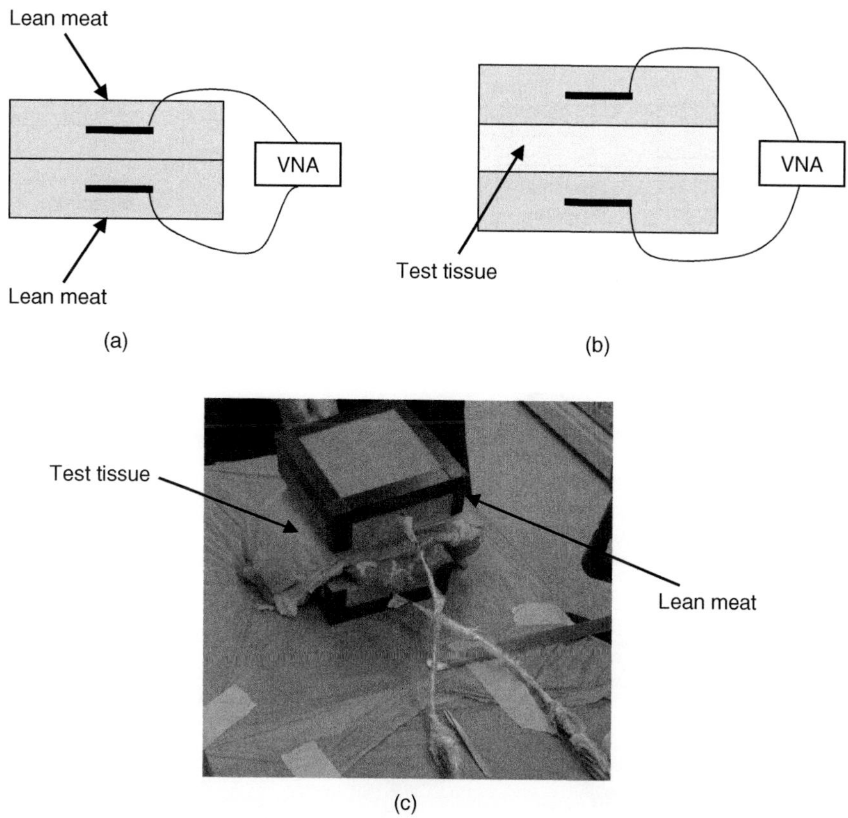

FIGURE 6.2 Measurement setup 1: (a) reference, (b) adding of the measured tissue sample, and (c) photo of the practical measurement setup.

setups; (i) setup 1: vector network analyzer and (ii) setup 2: spectrum analyzer and signal generator. The two setups were chosen to ensure that the attenuation data obtained are reliable and consistent.

In setup 1, the reference measurement was taken by placing a pair of identical antennas inside two slices of lean meat as shown in Figure 6.2. The antennas were connected directly to the ports of a vector network analyzer and the $|S_{21}|_{ref}$ was recorded. Next, a piece of tissue was placed in between the two slices of lean meat. The test tissue can either be lean meat, fat, or lean meat with fat of a certain thickness. The $|S_{21}|_{tissue}$ with the presence of the test tissue was then recorded. The attenuation can be calculated by Δ (in dB) $= |S_{21}|_{ref} - |S_{21}|_{tissue}$. From this, the attenuation per mm can be calculated by taking Δ/t, where t is the thickness of the tested tissue (in mm).

In Setup 2, the reference measurement was taken by placing a pair of identical antennas inside two slices of lean meat as shown in Figure 6.3. One of the antennas was connected to a signal generator while the other antenna was connected to a spectrum analyzer. The received signal level was recorded from the spectrum analyzer.

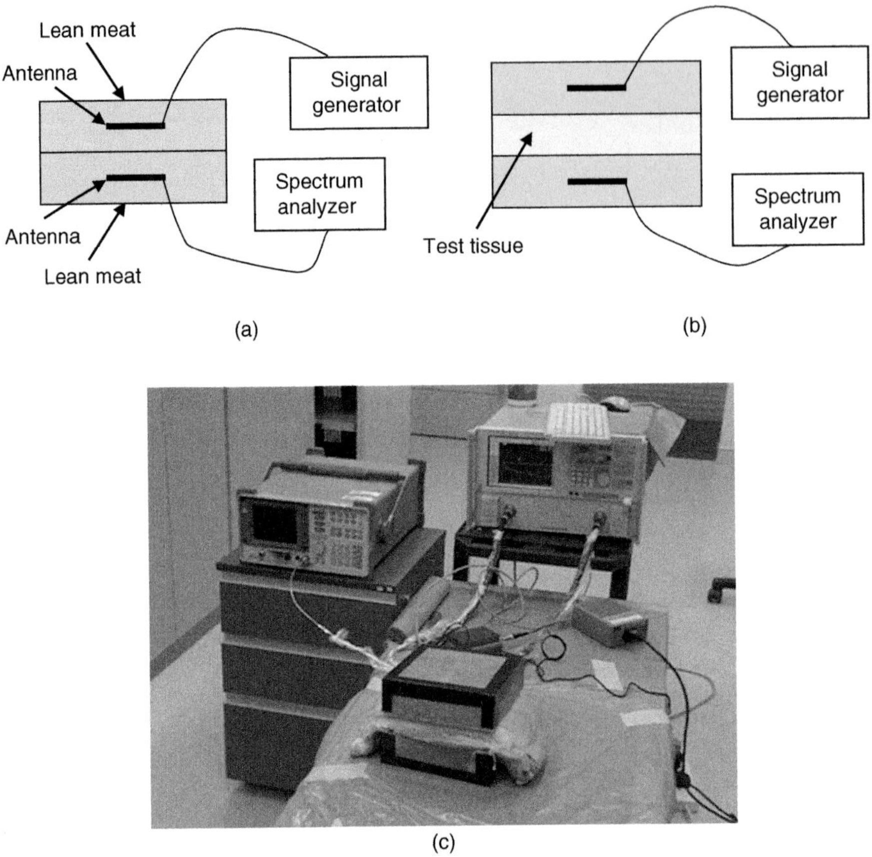

FIGURE 6.3 Measurement setup 2: (a) reference, (b) adding of the measured tissue sample, and (c) photo of the practical measurement setup.

Similar to setup 1, the attenuation can be calculated by inserting a piece of tissue in between the two slices of meat.

In order to characterize the transmission attenuation through the body, the measurements were conducted by using the setup as shown in Figure 6.3. The reference measurement was conducted with a dipole or loop antenna implanted inside the lean meat and a probe on the surface of the lean meat as shown in Figure 6.4. The probe was connected to the spectrum analyzer and the P_{ref} was recorded. Next, a piece of fat tissue with a thickness of 25 mm was placed directly on top of the lean meat (i.e., $d = 0$ mm). The attenuation per centimeter of the fat tissue was then calculated.

6.2.1.2 Results Figure 6.5 shows the measured $|S_{21}|_{ref}$ against frequency using Setup 1. Tables 6.1(a) and 6.1(b) show the measured $|S_{21}|$ and attenuation with the different test tissues using dipole and loop antennas at 403 MHz and 915 MHz, respectively.

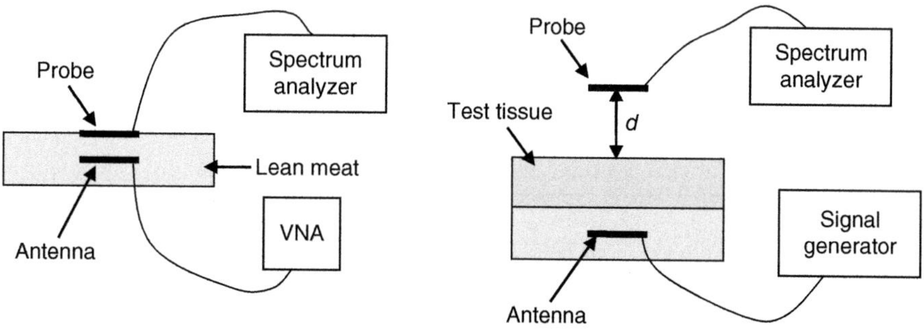

FIGURE 6.4 Measurement setup for through-body transmission measurement.

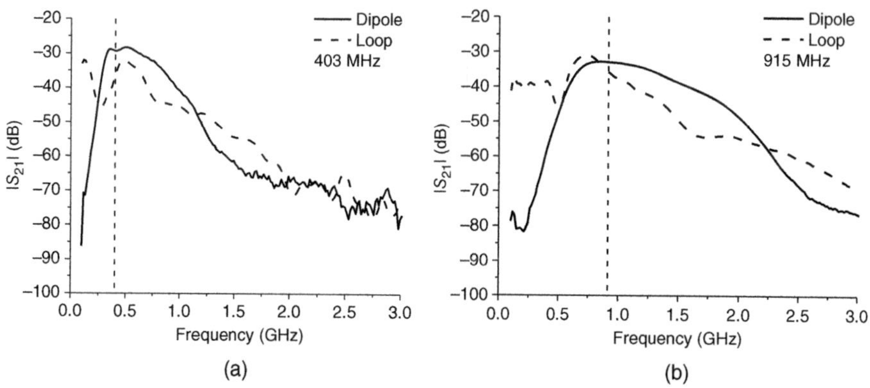

FIGURE 6.5 $|S_{21}|_{ref}$ at (a) 403 MHz and (b) 915 MHz.

TABLE 6.1 (a). Measured $|S_{21}|$ and attenuation using dipole antenna.

	403 MHz			915 MHz		
Dipole antenna	$\|S_{21}\|$ (dB)	Δ (dB)	Δ/mm (dB/mm)	$\|S_{21}\|$ (dB)	Δ (dB)	Δ/mm (dB/mm)
Reference	−29.4			−32.9		
Lean meat, $t = 25$ mm	−35.9	6.5	0.260	−42.4	9.5	0.380
Fat, $t = 25$ mm	−36.1	6.7	0.268	−43.4	10.5	0.420
Lean meat + Fat, $t = 50$ mm	−40.5	11.1	0.222	−51.4	18.5	0.370

TABLE 6.1 (b). Measured $|S_{21}|$ and attenuation using loop antenna.

	403 MHz			915 MHz		
Loop antenna	$\|S_{21}\|$ (dB)	Δ (dB)	Δ/mm (dB/mm)	$\|S_{21}\|$ (dB)	Δ (dB)	Δ/mm (dB/mm)
Reference	−36.3			−35.8		
Lean meat, $t = 25$ mm	−42.7	6.4	0.256	−48.2	12.4	0.496
Fat, $t = 25$ mm	−43.0	6.7	0.268	−46.7	10.9	0.436
Lean meat + Fat, $t = 50$ mm	−48.6	12.3	0.246	−53.4	17.6	0.352

TABLE 6.2 (a). Measured power level and attenuation using dipole antenna.

	403 MHz			915 MHz		
Dipole antenna	Power level (dBm)	Δ (dB)	Δ/mm (dB/mm)	Power level (dBm)	Δ (dB)	Δ/mm (dB/mm)
Reference	−25.6			−28.9		
Lean meat, $t = 25$ mm	−32.7	7.1	0.284	−37.8	8.9	0.356
Fat, $t = 25$ mm	−33.7	8.1	0.324	−38.5	9.6	0.384
Lean meat + Fat, $t = 50$ mm	−37.8	12.2	0.244	−47.5	18.6	0.372

TABLE 6.2 (b). Measured power level and attenuation using loop antenna.

	403 MHz			915 MHz		
Loop antenna	Power level (dBm)	Δ (dB)	Δ/mm (dB/mm)	Power level (dBm)	Δ (dB)	Δ/mm (dB/mm)
Reference	−27.8			−32.1		
Lean meat, $t = 25$ mm	−33.5	5.7	0.228	−39.5	7.4	0.296
Fat, $t = 25$ mm	−35.5	7.7	0.308	−40.5	8.4	0.336
Lean meat + Fat, $t = 50$ mm	−41.1	13.3	0.266	−49.8	17.7	0.354

Tables 6.2(a) and 6.2(b) show the measured receiving power level from the spectrum analyzer with the different test tissues using dipole and loop antennas at 403 MHz and 915 MHz, respectively.

The simulation was also conducted using homogeneous layers of skin, muscle (lean meat), and fat tissues. The dispersive dielectric constant (ε_r) and conductivity (σ) of these tissues are listed in Table 6.3. The attenuation data are shown in Table 6.4.

Generally, the attenuation is higher when the frequency is increased. At 403 MHz, the attenuation with lean meat and fat is approximately 0.2 dB/mm using dipole antenna and 0.3 dB/mm using loop antenna. At 915 MHz, the attenuation increases to about 0.3 dB/mm using dipole antenna and 0.4 dB/mm using loop antenna. The measurements were validated by the simulation results.

Next, the probe was placed at different heights above the test tissue that is a composition of skin, muscle, and fat tissues. With respect to the power level at $d = 0$

TABLE 6.3 Tissue properties.

	Skin		Muscle		Fat	
	ε_r	σ	ε_r	σ	ε_r	σ
403 MHz	46.7	0.69	57.1	0.80	5.58	0.04
915 MHz	41.3	0.41	55.0	0.95	5.46	0.05

TABLE 6.4 (a). Simulated attenuation using dipole antenna.

	403 MHz		915 MHz	
Dipole antenna	Δ (dB)	Δ/mm (dB/mm)	Δ (dB)	Δ/mm (dB/mm)
Muscle, $t = 25$ mm	6.16	0.246	7.40	0.296
Muscle + Fat + Skin, $t = 25$ mm	6.20	0.248	8.28	0.331

TABLE 6.4 (b). Simulated attenuation using loop antenna.

	403 MHz		915 MHz	
Loop antenna	Δ (dB)	Δ/mm (dB/mm)	Δ (dB)	Δ/mm (dB/mm)
Muscle, $t = 25$ mm	4.76	0.190	6.39	0.256
Muscle + Fat + Skin, $t = 25$ mm	5.53	0.221	8.03	0.321

mm, the attenuation in air as the probe moves away from the surface of the test tissue was computed. The measured results are given in Table 6.5 and the attenuation is computed as shown in Table 6.6.

The measured attenuation at $d = 0$ mm is generally higher than the simulations. This could be due to the presence of the air gap between the probe and the surface of the test tissue during measurements, which is caused by the uneven surface of the test tissue. However, the trend is consistent, that is, the attenuation of using dipole antenna is lower than that of the loop antenna at both the 403 MHz and 915 MHz.

The spectrum analyzer was used because the received signal level is very weak, which is caused by the large attenuation as the probe moves away from the test tissue. Generally, the attenuation increases with the d. From the measured results, the rate of attenuation in air for the loop antenna is slower than the dipole antenna with increasing d. The results from this study suggest that the loop antenna is more suitable to be used for microcapsule endoscopy.

TABLE 6.5 (a). Measured power level and attenuation using dipole antenna.

	403 MHz			915 MHz		
Dipole antenna	Power level (dBm)	Δ (dB)	Δ/mm (dB/mm)	Power level (dBm)	Δ (dB)	Δ/mm (dB/mm)
Reference	−59.6			−61.2		
$d = 0$ mm	−71.3	11.7	0.468	−71.9	10.7	0.428
$d = 5$ mm	−73.4	2.1		−75.5	3.6	
$d = 12$ mm	−75.3	4.0		−78.7	6.8	
$d = 24$ mm	−78.2	6.9		−85.0	13.1	
$d = 48$ mm	−84.2	12.9		−89.5	17.6	
$d = 72$ mm	−87.5	16.2		−94.5	22.6	

TABLE 6.5 (b). Measured power level and attenuation using loop antenna.

	403 MHz			915 MHz		
Loop antenna	Power level (dBm)	Δ (dB)	Δ/mm (dB/mm)	Power level (dBm)	Δ (dB)	Δ/mm (dB/mm)
Reference	−57.7			−61.0		
$d = 0$ mm	−74.4	16.7	0.668	−73.2	12.2	0.488
$d = 5$ mm	−77.0	2.6		−78.3	5.1	
$d = 12$ mm	−81.8	7.4		−82.5	9.3	
$d = 24$ mm	−88.5	14.1		−86.9	13.7	
$d = 48$ mm	−90.3	15.9		−89.5	16.3	
$d = 72$ mm	−87.8	13.4		−91.9	18.7	

6.2.2 Channel Characterization

6.2.2.1 Fading Characteristics The simulation is conducted using the CST Microwave Studio software. In this study, the channel includes the antenna effects. Table 6.7 shows the simulation parameters and the medium properties [31]. Figure 6.6 shows the setup diagram for the simulation. The centers of the internal and external

TABLE 6.6 (a). Simulated attenuation using dipole antenna.

	403 MHz		915 MHz	
Dipole antenna	Δ (dB)	Δ/cm (dB/mm)	Δ (dB)	Δ/cm (dB/mm)
$d = 0$ mm	6.71	0.268	8.00	0.320
$d = 12$ mm	7.49		8.67	
$d = 24$ mm	10.2		12.1	
$d = 36$ mm	12.5		14.9	
$d = 48$ mm	14.5		17.2	
$d = 60$ mm	16.3		19.1	
$d = 72$ mm	17.9		20.7	

TABLE 6.6 (b). Simulated attenuation using loop antenna.

	403 MHz		915 MHz	
Loop antenna	Δ (dB)	Δ/cm (dB/mm)	Δ (dB)	Δ/cm (dB/mm)
$d = 0$ mm	8.69	0.348	9.98	0.399
$d = 12$ mm	3.31		13.7	
$d = 24$ mm	8.06		19.7	
$d = 36$ mm	12.6		17.4	
$d = 48$ mm	16.8		17.0	
$d = 60$ mm	20.8		17.3	
$d = 72$ mm	24.3		18.0	

TABLE 6.7 Simulation parameters and medium properties.

Simulation parameters			
Frequency span (MHz)	920–925		
Delay resolution (μs)	0.2 (1/BW)		
Bins, N	801		
Frequency step, Δf (kHz)	6.25 (BW (N-1))		
Maximum excess delay (μs)	160 (1/Δf)		
Medium properties at 922.5 MHz			
	Muscle	Fat	Skin
Relative permittivity ε_r	55	5.56	41.3
Conductivity σ (S/m)	0.95	0.05	0.41
Attenuation constant α (Np/m)	23.8	3.98	12.0
Phase constant β (rad/m)	145.3	45.8	124.8
Penetration depth (mm)	46	276	91.8

antenna are fixed at the center of the 200 mm × 200 mm cross section. The distance d is varied from 15 mm to 60 mm in steps of 5 mm.

Figure 6.7 shows the magnitude responses of the channel for d = 15 mm and 60 mm in the frequency band 920–925 MHz. From the figures, it can be concluded that it is a flat fading channel.

6.2.2.2 Path Loss Figure 6.6 and 6.8 show the plot of the path loss (dB) versus distance (mm) for the case of perfect polarization match. The simulation results can be fitted linearly into the equation

$$\mathrm{PL(dB)} = 0.19 \times d(\mathrm{mm}) + 23, \ d \geq 15\ \mathrm{mm} \tag{6.1}$$

The path loss is about 0.19 dB/mm.

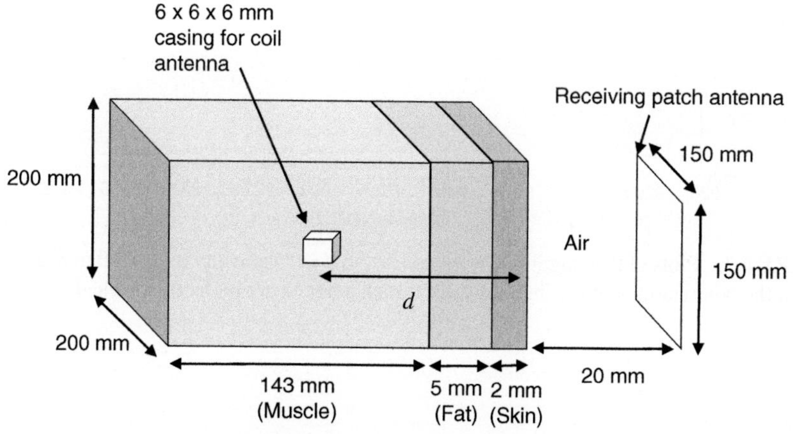

FIGURE 6.6 Simulation setup.

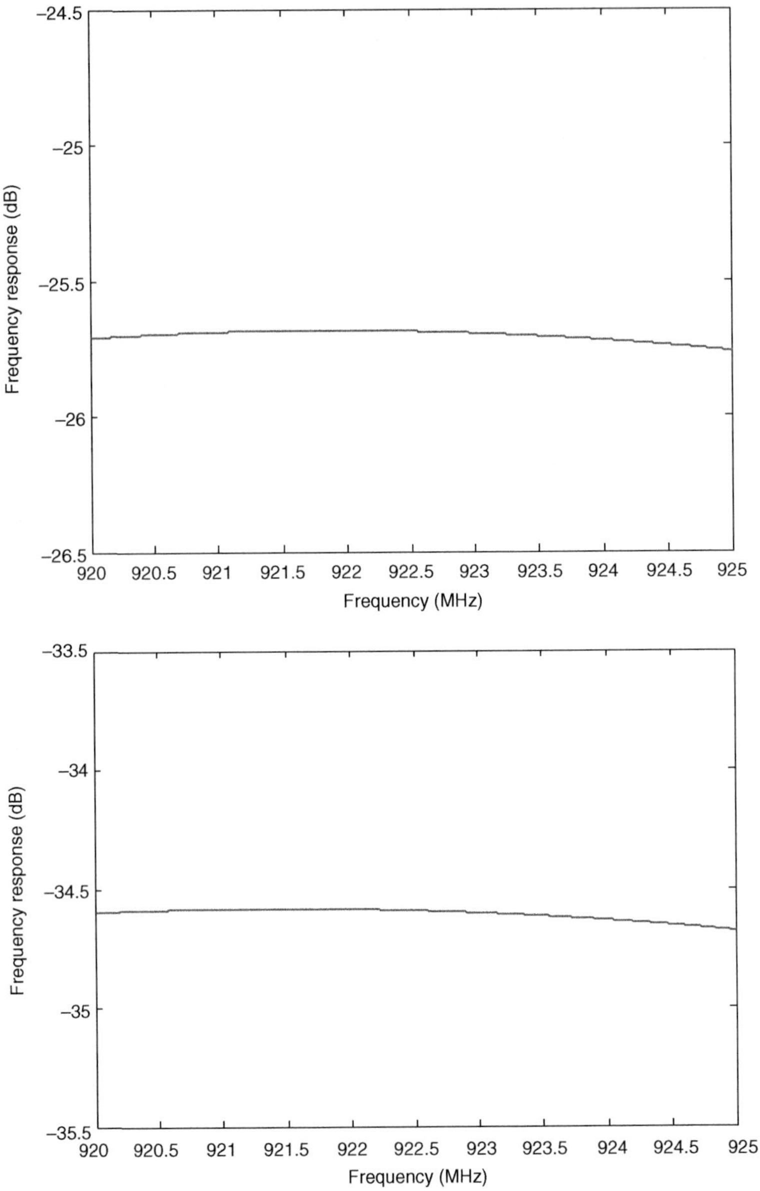

FIGURE 6.7 Plots of the magnitude response for (a) $d = 15$ mm and (b) $d = 60$ mm, assuming that the polarizations of the internal and external antenna are perfectly matched.

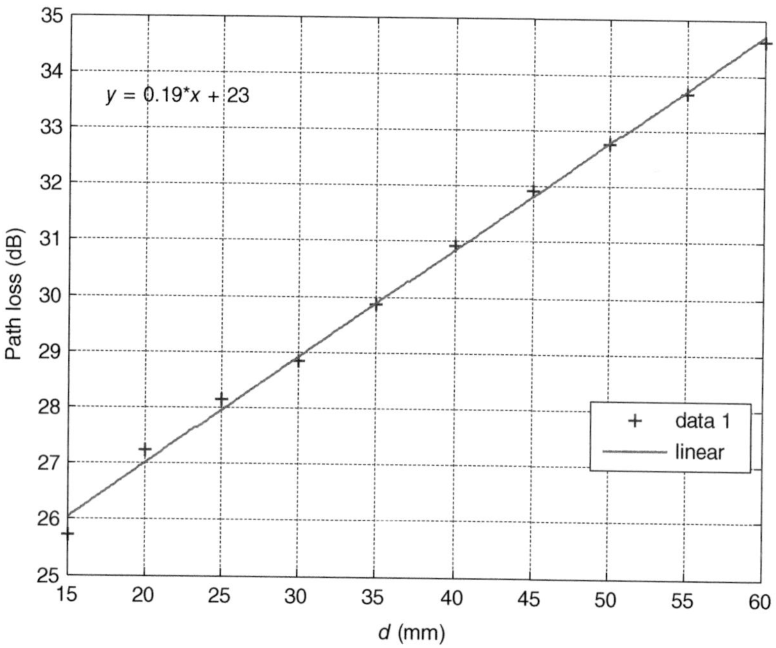

FIGURE 6.8 Plot of the path loss (dB) versus distance *d* (mm) and its linear fit, assuming that the polarizations of the internal and external antenna are perfectly matched.

6.2.2.3 Polarization Mismatch Figure 6.9 shows the example of the internal coil antenna being co-polarized or cross-polarized by having the radiation null pointing toward the external circularly polarized antenna. Due to the external antenna being circularly polarized, the strength of the received signal will not be affected by rotating the internal antenna in *xz*-plane. However, a 90° rotation in the *yz*-plane will result in the significant reduction in the signal strength. An increase of 16.65 dB in the path loss is experienced as shown in Figure 6.10.

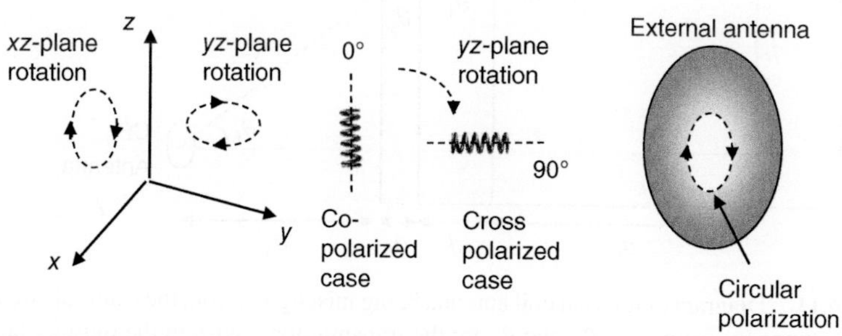

FIGURE 6.9 Diagram of internal coil antenna's orientation with respect to the external antenna.

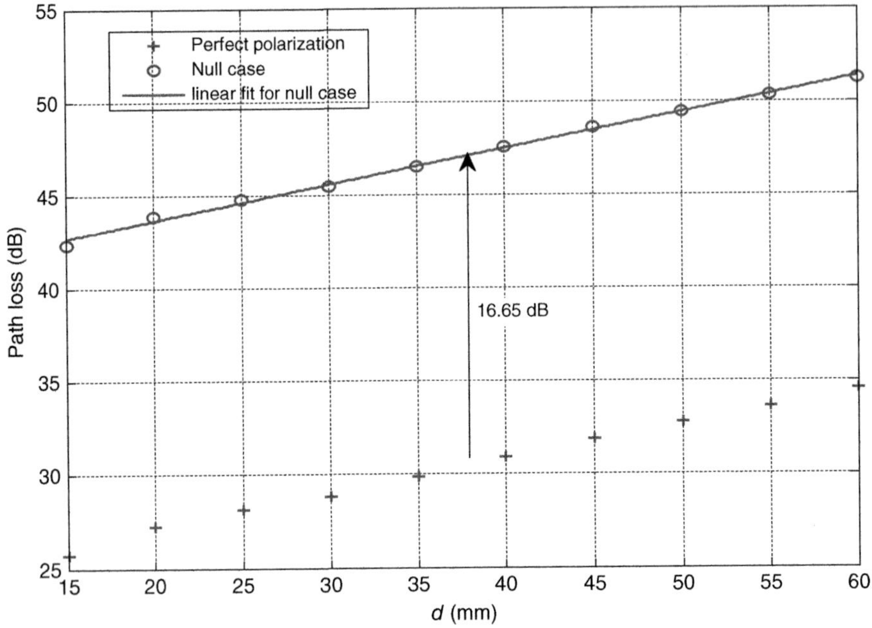

FIGURE 6.10 Plot of the path loss (dB) versus distance d (mm) for the case of co- and cross-polarization.

6.2.2.4 Misalignment Figure 6.11 shows a diagram of the internal coil antenna being misaligned from the external circularly polarized antenna by h mm. There exists a transmission angle as the electromagnetic wave propagates from a medium to another. Assuming that the electromagnetic propagation can be regarded as uniform plane waves, these transmission angles and the incident angle θ_i are calculated for the

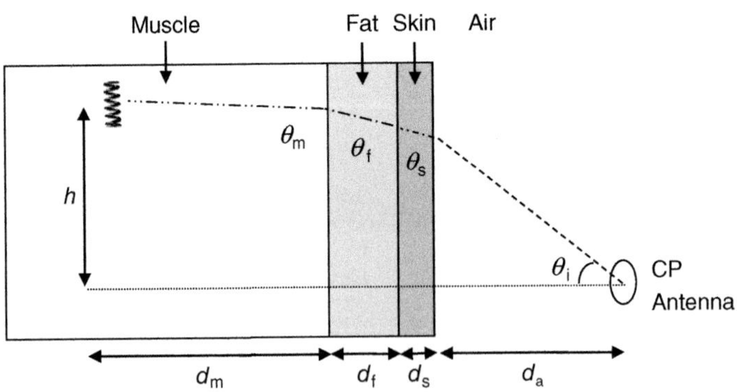

FIGURE 6.11 Diagram of internal coil antenna being misaligned from the external circularly polarized antenna by h mm. θ_m, θ_f, and θ_s are the transmission angles in the muscle, fat, and skin region, respectively. (All angles are w.r.t the normal, i.e., angle of incidence as shown for θ_i).

case of $d_a = 20$ mm, $d_s = 2$ mm, $d_f = 5$ mm, $d_m = 8$ mm, and $h = 90$ mm as $\theta_I = 76.97°$, $\theta_s = 8.75°$, $\theta_f = 25.88°$, and $\theta_m = 7.49°$. It can be seen that the propagation distance in the body does not differ much from that in the case of no misalignment. In fact, it is 15.7 mm as compared to 15 mm. The simulation is conducted to get the total loss for this setup

$$\text{Total loss} = \text{path loss} + \text{misalignment loss} = 41.96\ \text{dB}.$$

Assuming that the loss in the air is negligible (as compared to the loss in the body) and using the path loss model, the path loss is 25.98 dB with a propagation distance of 15.7 mm.

$$\text{Misalignment loss for } 76.97° = \text{total loss} - \text{path loss} = 15.98\ \text{dB}.$$

Suppose the worst case scenario is a misalignment of $h = 90$ mm for a coil antenna 60 mm from the surface of the skin, the propagation distance is 61.03 mm and $\theta_i = 76.05°$. Using misalignment loss of 15.98 dB and path loss model, we can conservatively estimate the total loss to be

$$\begin{aligned}\text{Total loss} &= \text{path loss for } 61.03\ \text{mm} + \text{misalignment loss for } 76.97° \\ &= 34.6 + 15.98\ \text{dB} \\ &= 50.58\ \text{dB}.\end{aligned}$$

The "worst case" scenario can be defined as shown in Figure 6.12, where an estimation of the total loss can be given by

$$\begin{aligned}\text{Total loss} &= \text{path loss} + \text{misalignment loss} + \text{cross polarization loss} \\ &= 34.6 + 15.98 + 16.65 \\ &= 67.23\ \text{dB}.\end{aligned}$$

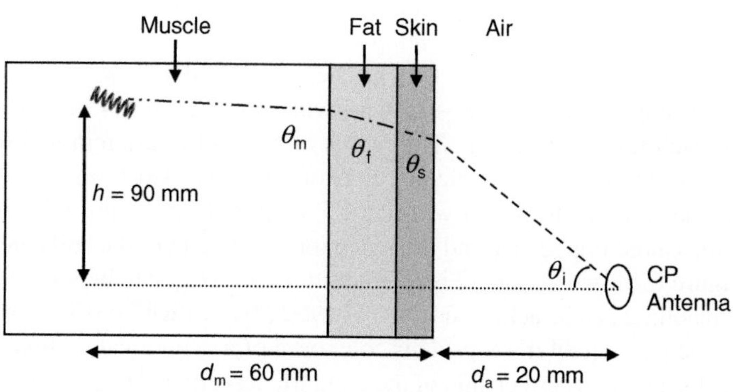

FIGURE 6.12 Diagram of "worst-case" scenario, where the internal coil antenna is positioned $d_m = 60$ mm deep from the surface of the skin and shifted up by $h = 90$ mm and with its radiation null pointing toward the external circularly polarized antenna.

FIGURE 6.13 Coil antenna. *Source*: See et al. 2011 [28]. Reproduced with permission of IEEE.

6.2.3 Antenna Designs

6.2.3.1 Implant Antenna Design In consideration of the variation of the location and orientation of the capsule as it travels inside the digestive tract, the radiation pattern of the capsule antenna should be omnidirectional and has mixed polarization. Moreover, since the radiation characteristics of the implant antenna are affected by the presence of the lossy and dispersive tissues, it is necessary to characterize the radio link from the capsule antenna to the external antenna, which is critical for the link budget calculation of a communication system

A coil antenna as shown in Figure 6.13 has been designed to fit within the capsule in this study. It consists of five turns of wire with diameter of 0.4 mm. The spacing between each turn is 0.52 mm. The radius of the coil is 2.1 mm and the total length is 5 mm. An impedance-matching network is connected to the two ends of the coil antenna in order to match with the RF chip with an output impedance of 200 Ω. The measured impedance and the reflection coefficient of the coil antenna are shown in Figure 6.14, where the $|\Gamma|$ is referenced to 200 Ω. It can be seen that good impedance matching can be achieved from 850–950 MHz with $|\Gamma| < 0.1$. As shown in Figure 6.15, the package of the capsule is composed of a cylindrical section and two hemispherical sections on either side of the cylindrical section. The package is made of polycarbonate with a dielectric constant of around 3.0. The radius of the cylinder is 5.5 mm and the length is 15 mm. The antenna is positioned inside the hemispherical section at one side.

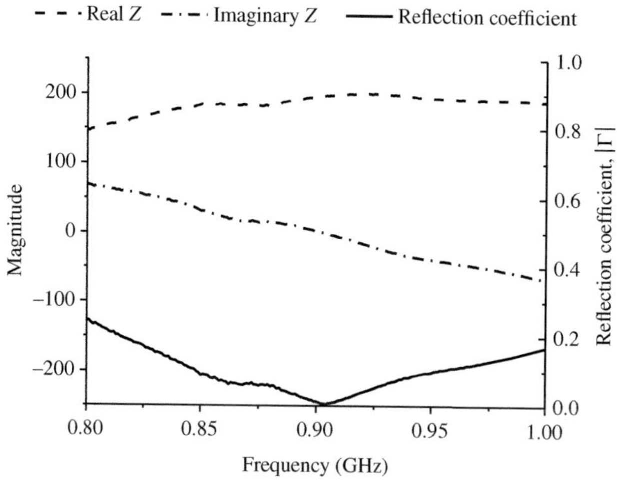

FIGURE 6.14 Measured impedance and reflection coefficient. *Source*: See et al. 2011 [28]. Reproduced with permission of IEEE.

6.2.3.2 External Antenna Design The external antenna should be low-profile and has a directional radiation pattern. Also, due to the random nature of the orientation of the capsule inside the digestive tract, the external antenna is preferred to be circularly polarized. The developed circularly polarized antenna is shown in Figure 6.16. The overall size of the antenna is 100 mm × 100 mm × 4.524 mm. The antenna was simulated using CST Microwave Studio, and good axial ratio performance and impedance matching can be achieved over the frequency range of 920–925 MHz as shown in Figure 6.17.

6.2.3.3 Through-Body RF Transmission A layered homogeneous model that considers the case of with and without visceral fat can replicate the propagation medium by a source in the small intestine. As such, the channel will be set up as a layered homogeneous layer that comprises skin, subcutaneous fat, muscle, and visceral fat. The small intestine is treated as muscle tissue because of the similar electrical properties. Based on this setup, the channel measurement is conducted for

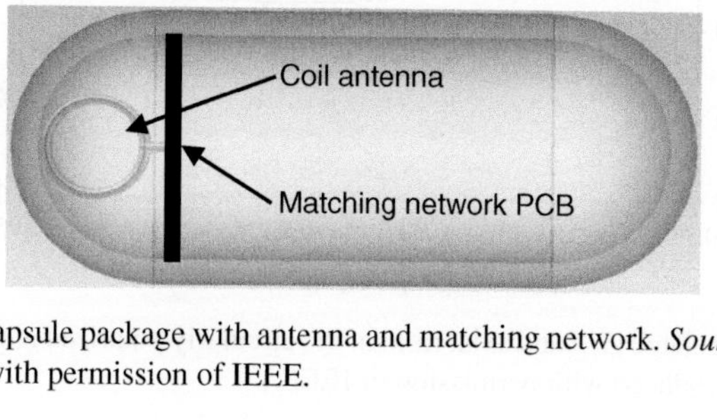

FIGURE 6.15 Capsule package with antenna and matching network. *Source*: See et al. 2011 [28]. Reproduced with permission of IEEE.

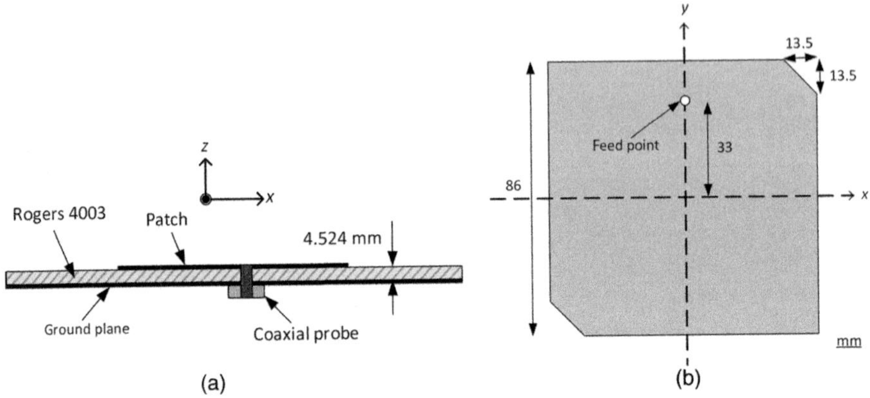

FIGURE 6.16 External circularly polarized antenna: (a) side view and (b) top view. *Source*: See et al. 2011 [28]. Reproduced with permission of IEEE.

the investigations of the following:

i) Capsule antenna polarization
ii) Path loss in muscle medium
iii) Path loss with varying visceral fat thickness
iv) Path loss in air
v) Interference picked up by another external antenna
vi) Estimation of the worst case loss for link budget calculation

The measurement setup is shown in Figure 6.18a. First, the minced meat fills up the acrylic box to a height of 120 mm. The capsule casing which contains the internal

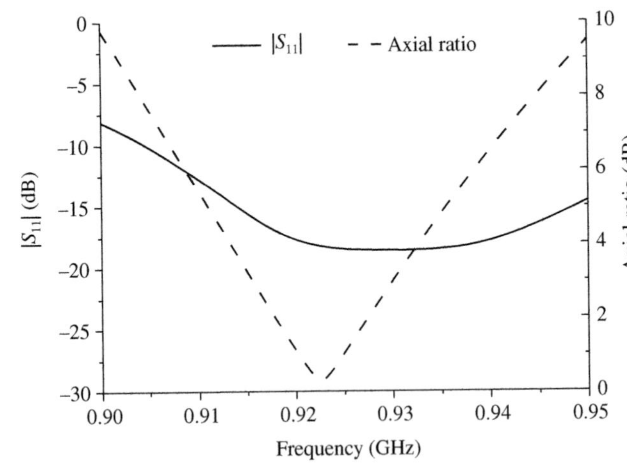

FIGURE 6.17 Simulated $|S_{11}|$ and axial ratio of the circularly polarized antenna. *Source*: See et al. 2011 [28]. Reproduced with permission of IEEE.

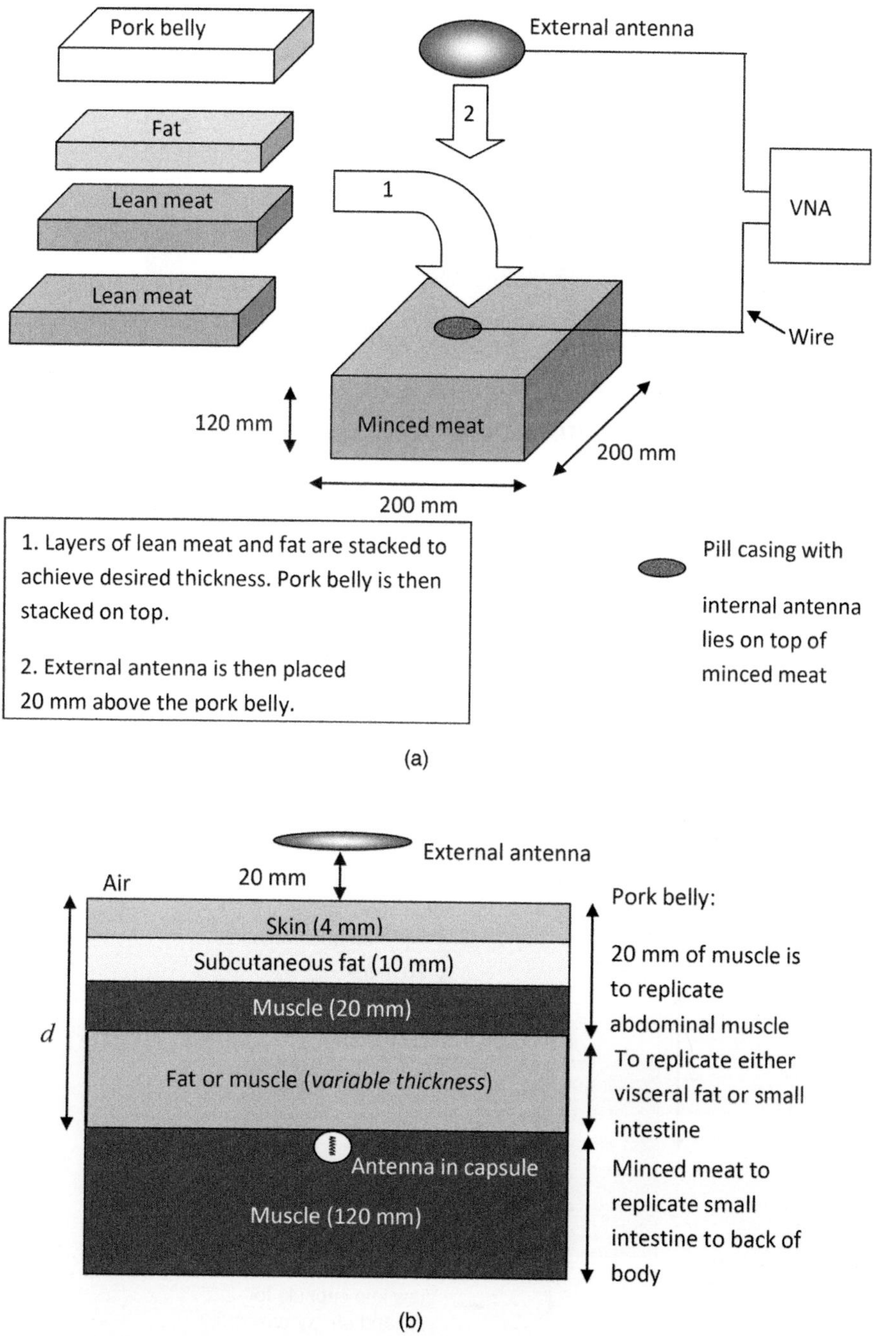

FIGURE 6.18 Measurement setup: (a) 3D view, (b) side view, and (c) photo of the setup.

(c)

FIGURE 6.18 (*Continued*)

antenna is placed on top of the minced meat. To replicate the different depths of the capsule in the small intestine from the belly surface, layers of lean meat and fat are stacked on top of the minced meat until the desired thickness. The pork belly which replicates the skin, subcutaneous fat, and abdominal muscle is then stacked on top. Last, the external circularly polarized antenna is placed 20 mm away from the pork belly surface. Figure 6.18b shows a side view of the measurement setup. For the case with or without visceral fat, fat or lean meat is stacked respectively. Figure 6.18c shows a photo of the measurement setup. The polystyrene foam which has electrical properties close to that of air is used with the external circularly polarized antenna to ensure the 20 mm spacing.

To conduct channel measurements for the various objectives that are listed above, modifications were made to the general measurement setup that is shown in Figure 6.18b. Figure 6.19 shows the setup for the path loss measurement with different

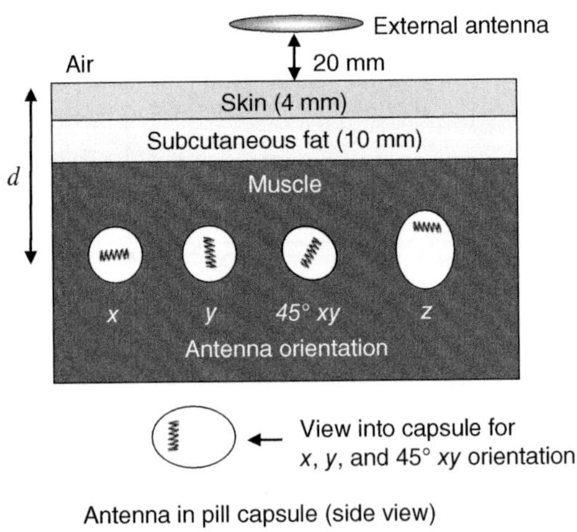

FIGURE 6.19 Measurement setup for the different capsule antenna orientations.

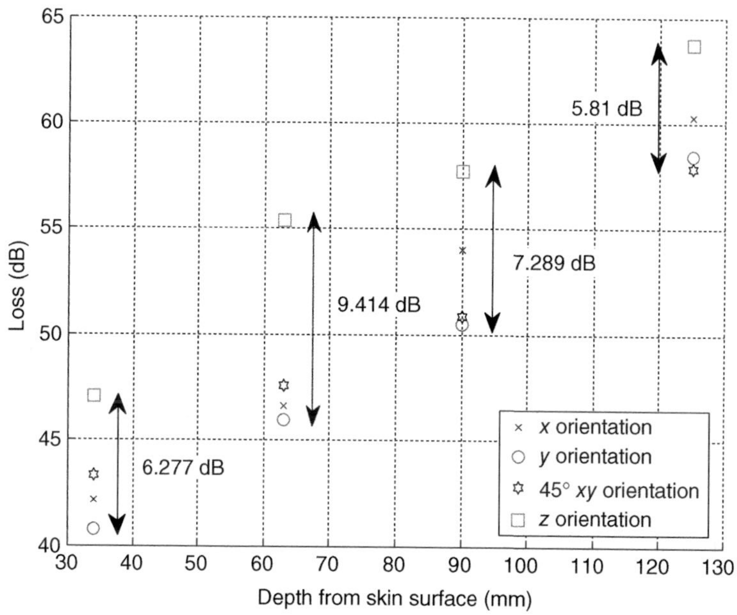

FIGURE 6.20 Measured path loss for the different capsule antenna orientations.

antenna orientations. In this study, the principal axes (x, y, and z) and the orientation of 45° in the xy-plane are considered. For each orientation, the capsule is aligned to the center of the external circularly polarized antenna and measurements are conducted for various depths of the capsule antenna from skin surface, $d = 34$, 63, 90, 125, and 150 mm. The results for the various antenna orientations are plotted in Figure 6.20.

Figure 6.21 shows the setup for path loss measurement in muscle medium. The thickness of the muscle medium is varied by stacking pieces of lean meat. The results

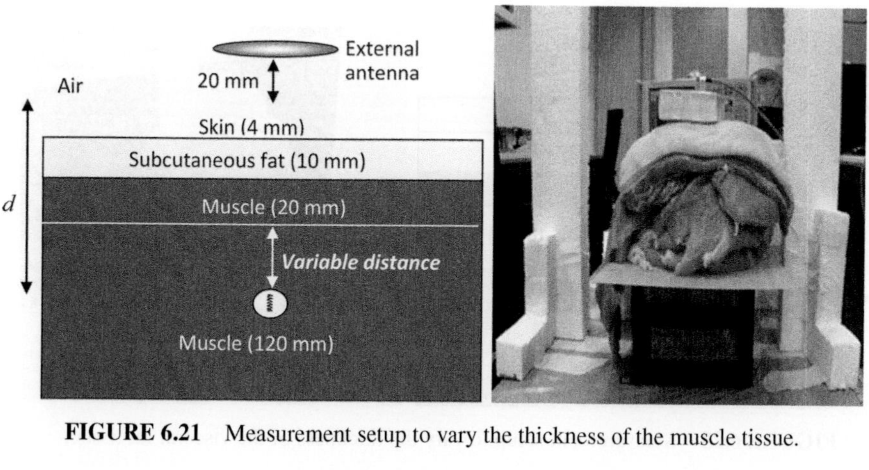

FIGURE 6.21 Measurement setup to vary the thickness of the muscle tissue.

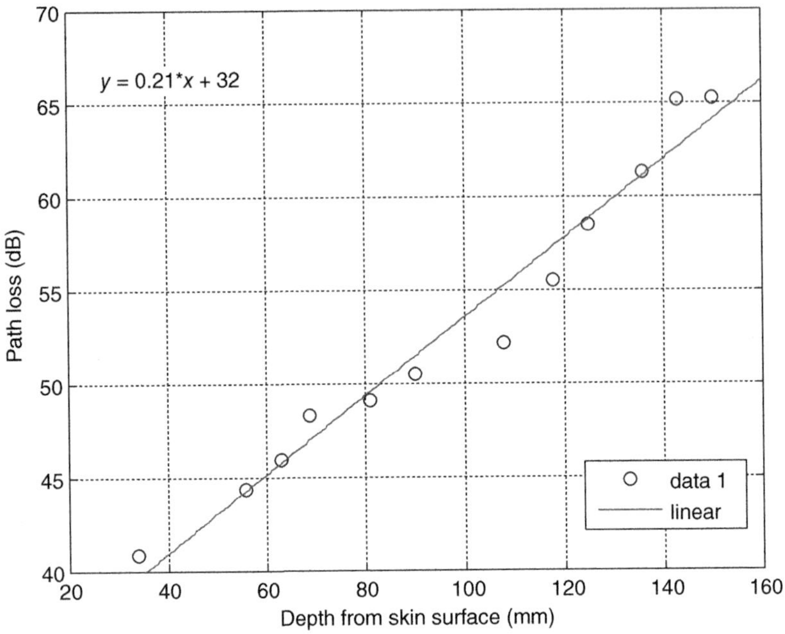

FIGURE 6.22 Measured path loss for the different thicknesses of the muscle tissue.

are shown in Figure 6.22. Figure 6.23 shows the setup for path loss measurement with varying visceral fat thickness. Similarly, this is achieved by stacking pieces of fat. The results are shown in Figure 6.24. Figure 6.25 shows the setup for path loss measurement in air, which is achieved by stacking pieces of polystyrene foam. The depth of the capsule antenna from the skin surface is maintained at 90 mm. The results are shown in Figure 6.26. Figure 6.27 shows the setup for the measurement of

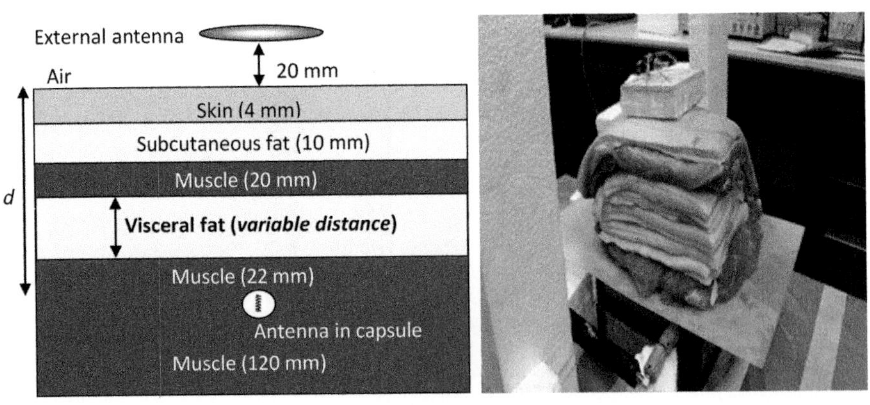

FIGURE 6.23 Measurement setup to vary the thickness of the visceral fat tissue.

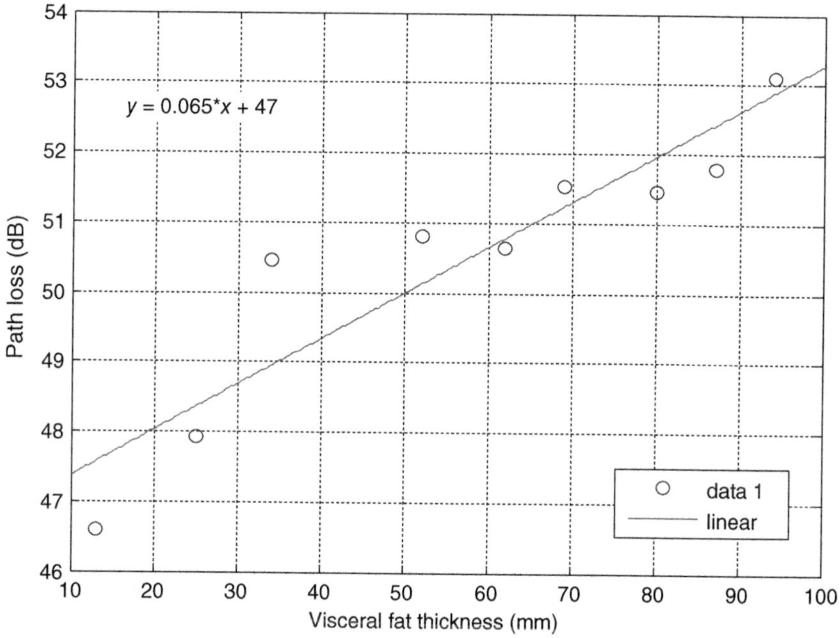

FIGURE 6.24 Measured path loss for the different thicknesses of the visceral fat tissue.

interference picked up by another external circularly polarized antenna. The second circularly polarized antenna picks up the signal from the capsule antenna that is intended for the first circularly polarized antenna. To achieve the same impedance matching, the second circularly polarized antenna is placed 20 mm from another pork belly and connected to the Vector Network Analyzer (VNA) while the first circularly

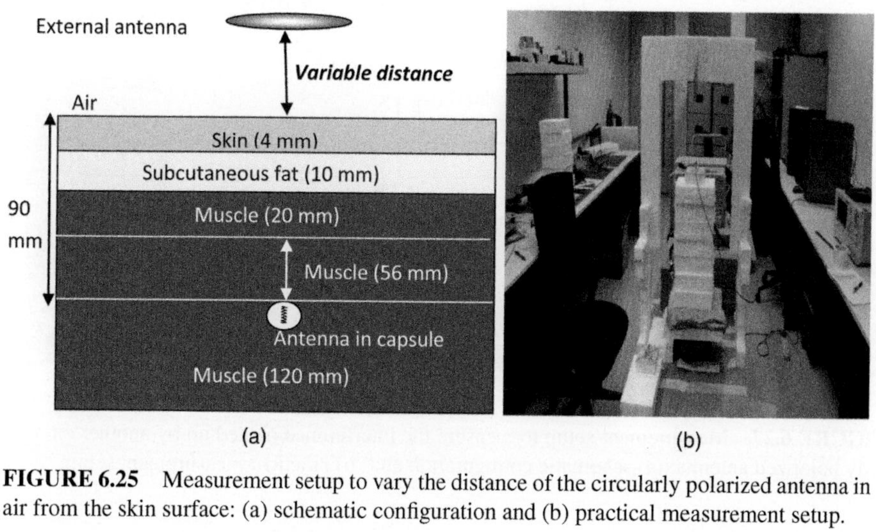

FIGURE 6.25 Measurement setup to vary the distance of the circularly polarized antenna in air from the skin surface: (a) schematic configuration and (b) practical measurement setup.

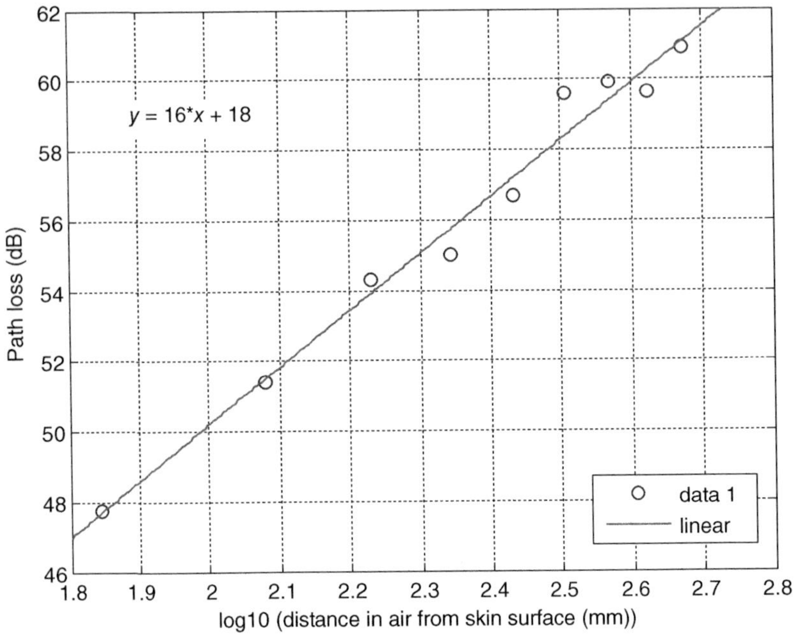

FIGURE 6.26 Measured path loss for the different distances of the circularly polarized antenna in air from the skin surface.

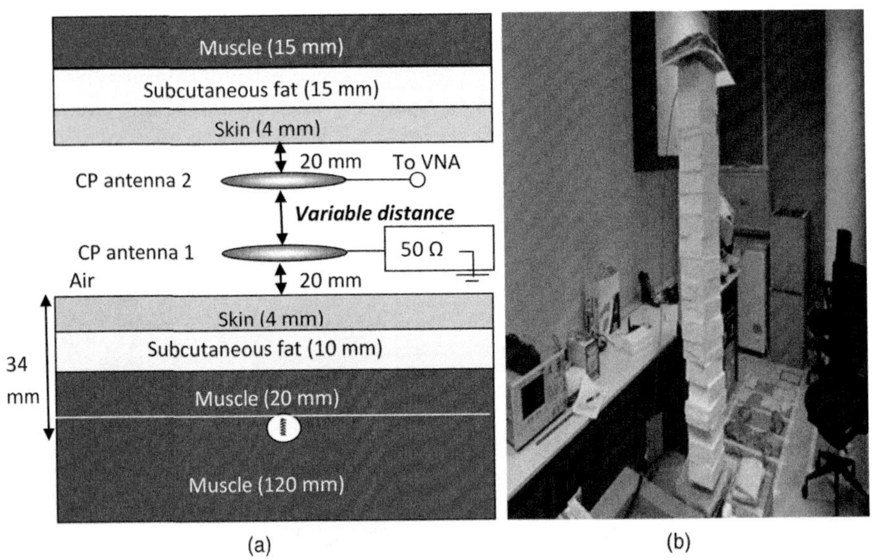

FIGURE 6.27 Measurement setup to measure the interference picked up by another circularly polarized antenna: (a) schematic configuration and (b) practical measurement setup.

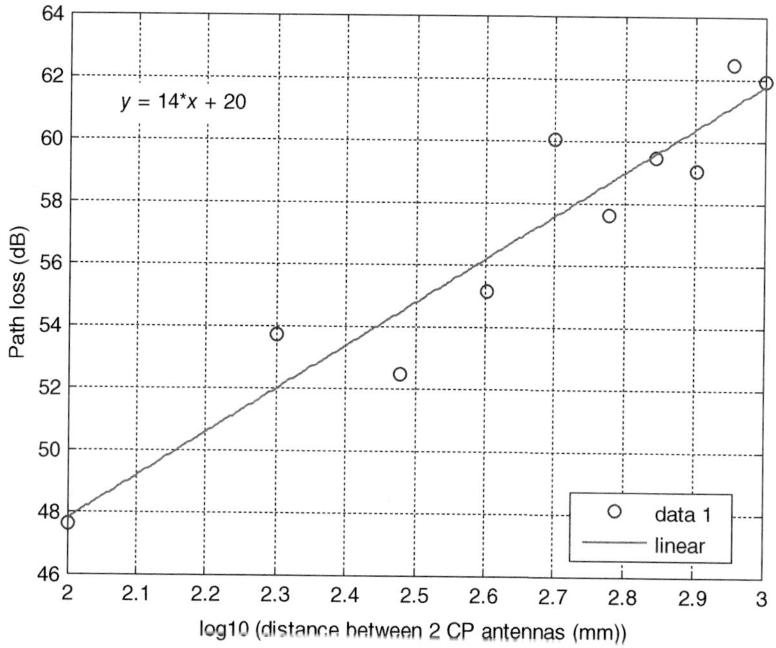

FIGURE 6.28 Measured path loss for the different distances between the circularly polarized antennas in air.

polarized antenna is terminated with a 50 Ω load. The depth of the capsule antenna from the skin surface is maintained at 34 mm. The results are shown in Figure 6.28.

For the estimation of the worst-case loss, the "worst-case" capsule position is assumed to be shifted by 75 mm upward and sideways from the belly center and 150 mm deep from the belly surface, as shown in Figure 6.29. This "worst-case" capsule position is obtained from examining the HUGO model. Both cases of with and without visceral fat are considered. For the case with visceral fat, a piece of 94 mm fat is inserted between the lean meats. For both the cases, measurements are taken for three capsule antenna orientations (*x*, *y*, and *z*) that are shown in Figure 6.19. Generally, the link loss for the "worst-case" is around 65 dB, which should be taken into account in the link budget calculation.

6.2.4 System Integration

6.2.4.1 Link Budget The proposed link budget of the system is shown in Table 6.8.

6.2.4.2 Animal Testing A field trial was conducted at the National University Hospital using a live pig. The capsule antenna was inserted into the intestine of the pig, which was about 105 mm deep from the skin surface. The location of the

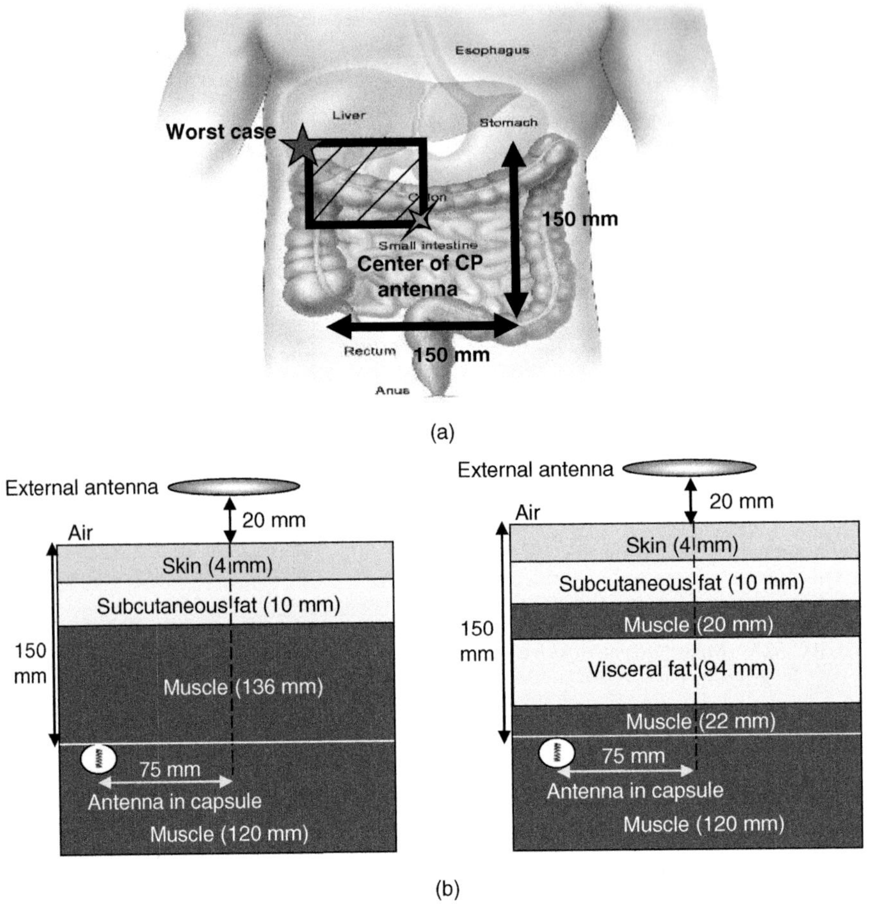

FIGURE 6.29 Experiment to determine the worst-case path loss: (a) front view representation and (b) measurement setups.

circularly polarized antenna was varied across an area of 40 cm × 50 cm as shown in Figure 6.30. Generally, the link loss is lower than 71 dB at Positions 1, 2, 4, 5, 7, 8, 10, and 11, which is acceptable. A higher link loss is recorded at Positions 3, 6, 9, and 12 because of the further distance between the two antennas but is not a cause for concern, since the cross-sectional area of the human body is much smaller as shown in Figure 6.29.

In order to conserve the space inside the capsule, the coil antenna is integrated with the RF module as shown in Figure 6.31. In order to match the antenna to the differential impedance of 200 Ω such that the losses due to impedance mismatch can be minimized, a matching network has been implemented.

A sample of the capsule is shown in Figure 6.32. The field trial was carried out using live pigs. The capsule was inserted into the pig using an endoscope and was able

TABLE 6.8 Link budget.

Parameter	Uplink Value	Downlink Value
Carrier frequency (MHz)	915.00	915.00
Bandwidth (MHz)	2.00	1.00
Coded data rate (Mbps)	2.00	1.00
Code rate	0.875	0.333
Uncoded data rate (Mbps)	1.75	0.33
Theoretical noise floor (dBm)	**−110.97**	**−113.98**
RECEIVER	**13.00**	**13.00**
RF noise figure (dB)	8.00	8.00
Implementation loss (dB)	5.00	5.00
Receiver antenna gain (dBi)	0.00	0.00
Receiver noise floor (dBm)	**−97.97**	**−100.98**
Required SNR at 1E-6 BER (dB)	8.70	10.50
Receiver sensitivity (dBm)	**−89.27**	**−90.48**
TRANSMITTER	**−16.02**	**−16.02**
Transmitter power (dBm)	−16.02	−16.02
Transmitter antenna gain (dBi)	0.00	0.00
Total loss (dB)	67.23	67.23
Available link margin (dB)	**6.02**	**7.23**

to be passed out of the body naturally after 2 days. Figures 6.33a–6.33d show some of the photos taken during the animal trial. From the measurements, good RF transmission between the capsule antenna and the external circularly polarized antenna can be achieved, which is important to reduce the frame error rate so as to obtain clear images. The images inside the gastrointestinal tract of the live pig are shown

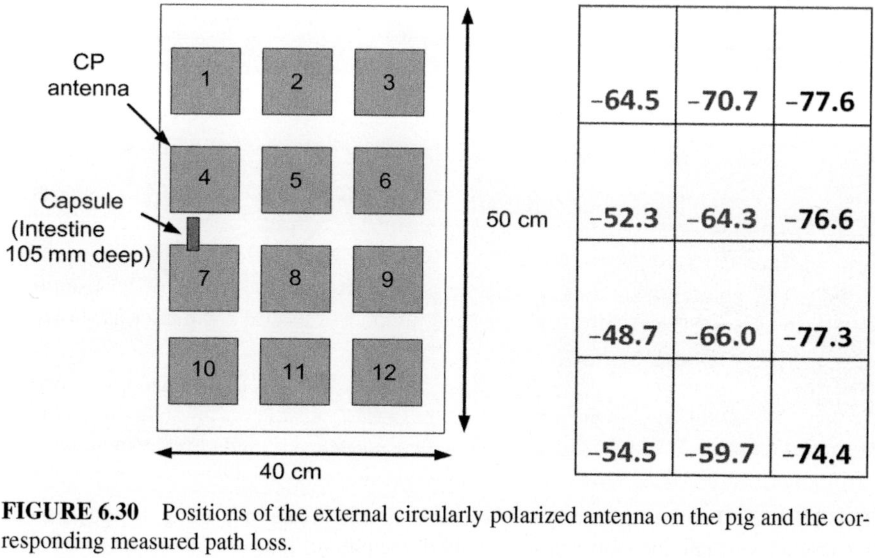

FIGURE 6.30 Positions of the external circularly polarized antenna on the pig and the corresponding measured path loss.

FIGURE 6.31 Photo of capsule antenna integrated with the RF module.

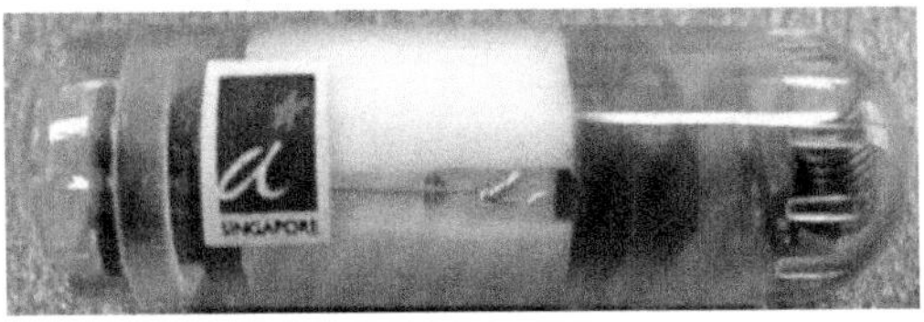

FIGURE 6.32 Photo of wireless capsule.

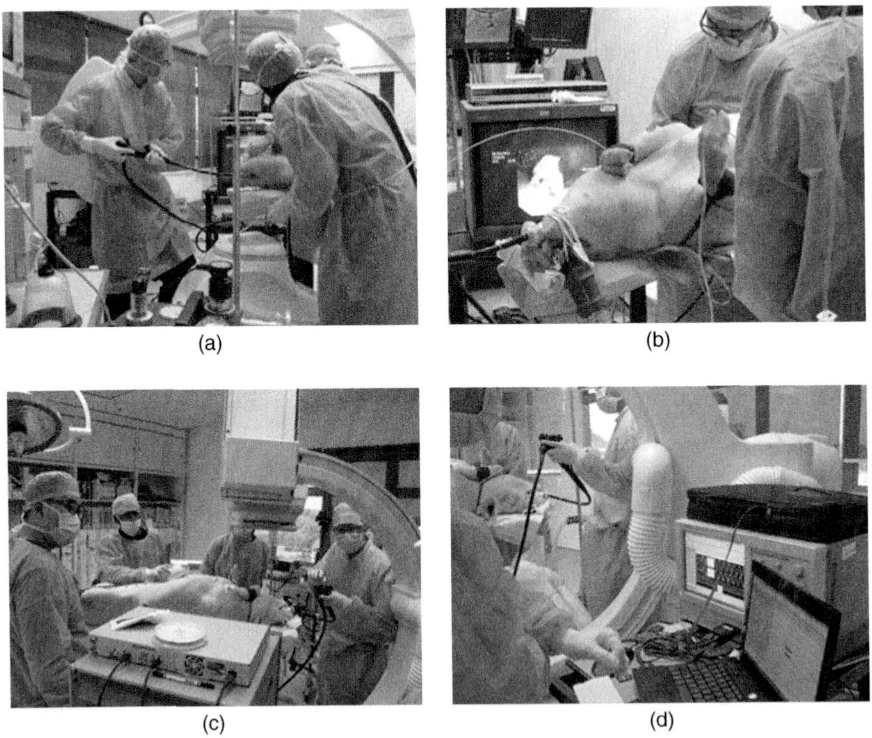

FIGURE 6.33 Photos: (a) insertion of capsule into the sedated pig using an endoscope, (b) capsule tested into the gastrointestinal tract of the pig, (c) images of the gastrointestinal tract with the external antenna directly placed on the skin surface, (d) QPSK waveforms for the I and Q channels from the logic analyzer.

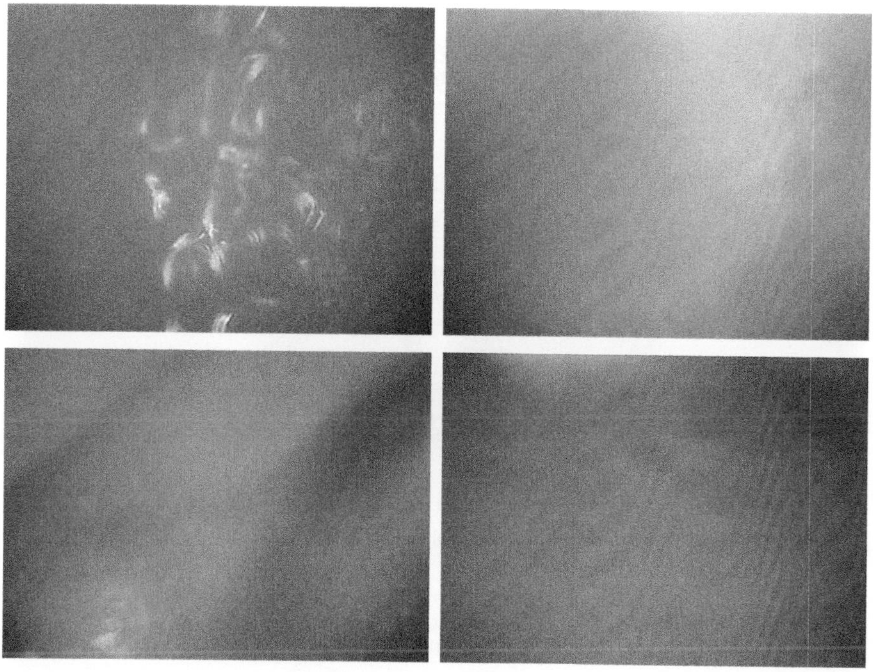

FIGURE 6.34 Photos of the images captured inside the gastrointestinal tract of the live pig. For a color version of this figure, see the color plate section.

in Figure 6.34, which are reasonably clear. With this, the basic functionality of the capsule for wireless endoscopy has been successfully demonstrated.

6.3 ANTENNAS IN WIRELESS IMPLANTABLE NEUROPROBE MICROSYSTEM FOR MOTOR PROSTHESIS

6.3.1 Antenna Configuration on Head

In this study, the head antenna can either be placed close to the top of the head or directly beneath the skin layer at the top of the head as shown in Figure 6.35. The external antenna can be placed nearby and the communication distance will depend on the link quality and reliability.

6.3.2 Antenna Designs at 434 MHz Band

6.3.2.1 Meandered Dipole Head Antenna The head antenna design is shown in Figure 6.36. It is fabricated on a Rogers4003C ($\varepsilon_r = 3.38$, $\tan\delta = 0.003$) substrate. It comprises a series of meandering strips of width 0.25 mm on both the top and bottom sides of the Printed Circuit Board (PCB), which are connected by vias of 0.4 mm diameter. The overall size of the PCB substrate is 25 mm × 10 mm × 1.6 mm. Beneath the PCB, the antenna is co-designed with the casing and the tissues with the properties as shown in Table 6.9 [31]. The impedance

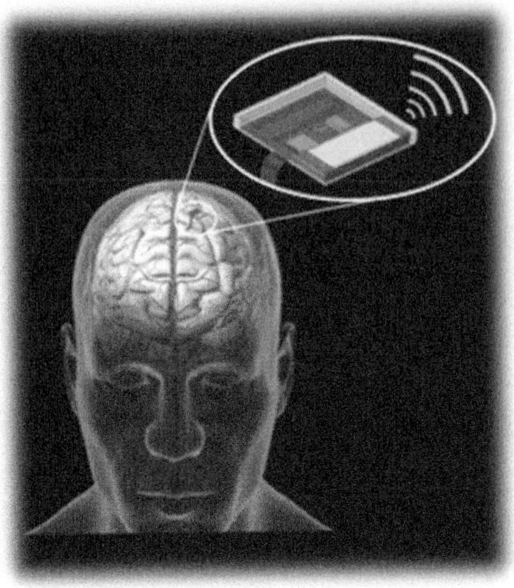

FIGURE 6.35 An illustration of the placement of the antenna on the head.

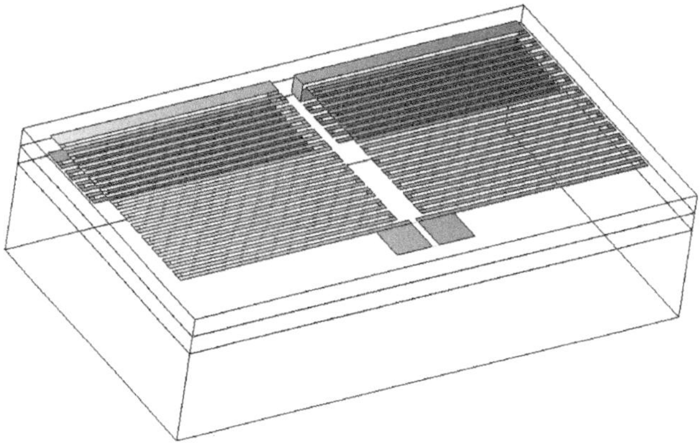

FIGURE 6.36 Meandered dipole antenna.

TABLE 6.9 Tissue parameters at 434 MHz.

	Skin	Fat	Bone
Relative permittivity	46.1	5.57	13.1
Conductivity (S/m)	0.702	0.0417	0.0944
Thickness (mm)	5	2.5	5

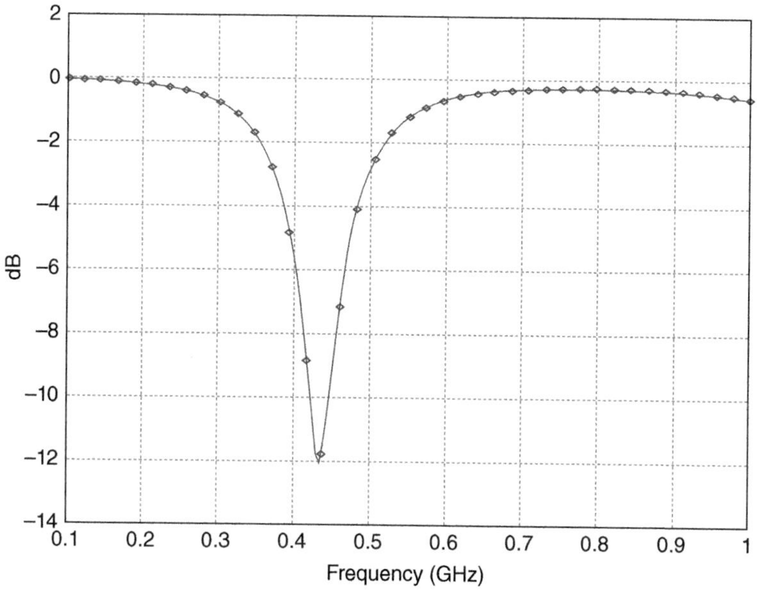

FIGURE 6.37 Simulated reflection coefficient.

matching of the antenna is shown in Figure 6.37. It can be seen that the antenna has a bandwidth of 400–468 MHz for the $|S_{11}| < -6$ dB.

In order to increase the antenna gain, the meandered dipole shown in Figure 6.36 is modified and shown in Figure 6.38(a). It comprises a series of meandering strips of width 0.5 mm on the top side of a Rogers 4003C ($\varepsilon_r = 3.38$, $\tan\delta = 0.003$) substrate with an overall size of 30 mm × 50 mm × 0.8 mm. In order to evaluate the antenna system performance, a pair of identical dipoles is placed face-to-face to each other. From Figure 6.39, it can be seen that the transmission level is increased by about 30 dB. Furthermore, the antenna system bandwidth, which is defined as the bandwidth where the transmission level is 6 dB below the maximum, can be broadened by increasing the width of the strips to 2 mm as shown in Figure 6.38b. The overall size of the PCB substrate is 60 mm × 50 mm × 0.8 mm. From Figure 6.39, it can be seen that the antenna system bandwidth is increased from 37 MHz to 60 MHz when the width of the strips is increased.

6.3.2.2 ***Compact Planar Dipole External Antenna*** The external antenna design is a compact planar dipole loaded with a high permittivity periodic structure (HPPS), which is presented for wide operating bandwidth and high gain. The effective permittivity features the nonresonant response in the bandwidth of interest and yields a sub-wavelength-sized unit cell. The HPPS is able to achieve the high effective permittivity broad bandwidth. This high permittivity phenomenon is attributed to the strong electromagnetic coupling between the opposite crossly positioned overlapping square patches. As shown in Figure 6.40, the HPPS is fabricated on two pieces of Rogers

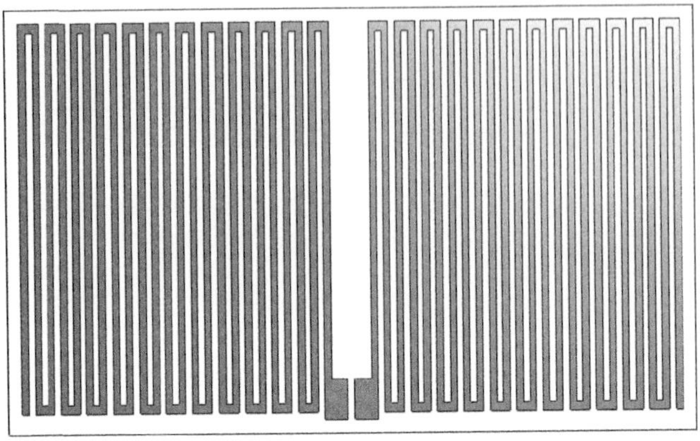

(a)

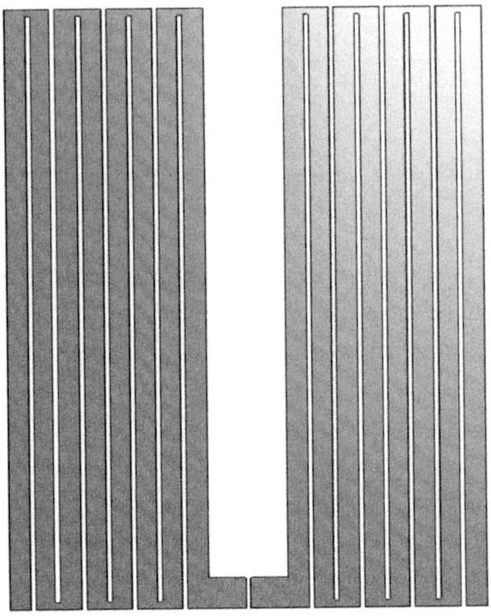

(b)

FIGURE 6.38 Meandered dipole designs: (a) strip width of 0.5 mm and (b) strip width of 2 mm.

4003C substrate, each having a thickness of 32 mils. The dipole is positioned on top of the HPPS and has a length of 66 mm. The air thickness between the HPPS and the ground plane is 50 mm. The size of the ground plane is 250 mm × 250 mm.

The HPPS loaded dipole antenna generates much stronger and more evenly distributed magnetic field and surface current in a larger area of the antenna aperture. Such strongly and uniformly distributed field associated with the anisotropic HPPS greatly enhances the illumination efficiency of the antenna aperture and thus improves the gain of the antenna.

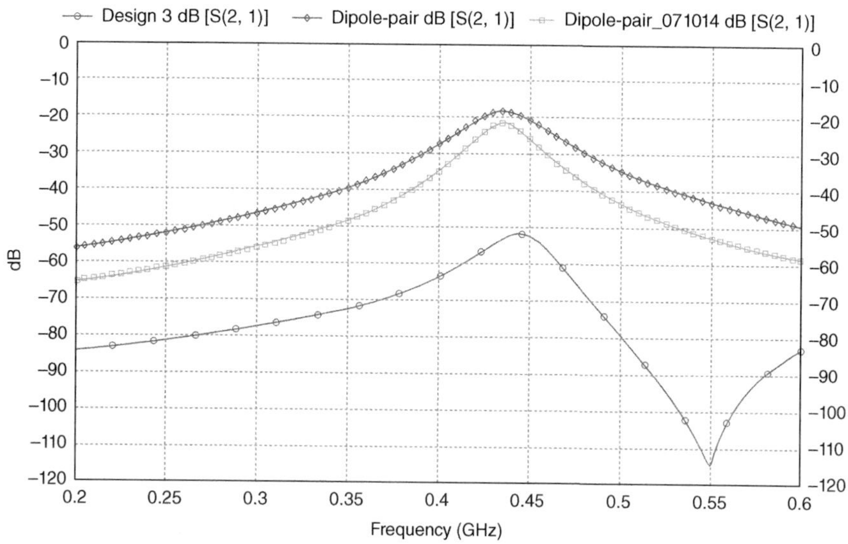

FIGURE 6.39 Antenna system bandwidth comparison.

As shown in Figure 6.41, the antenna features a bandwidth of 423–447 MHz for the $|S_{11}| < -6$ dB. Moreover, the transmission loss between the head antenna in Figure 6.36 and the external antenna at 10 cm separation is 28 dB at 434 MHz, which fulfils the link budget requirements.

6.3.2.3 Wideband Sleeve Monopole External Antenna Figure 6.42 shows the geometry of the sleeve monopole antenna. The stripline structure is used to feed the antenna. The transmission line and planar monopole radiator are sandwiched in between the Roger 4003C substrates. The foam is used to fill the outer sleeve. The

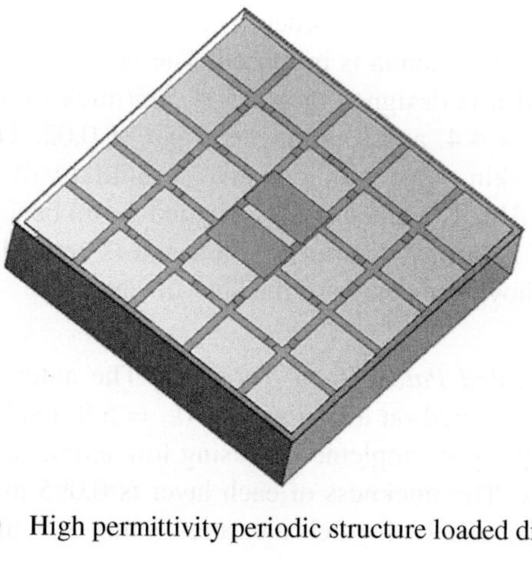

FIGURE 6.40 High permittivity periodic structure loaded dipole antenna.

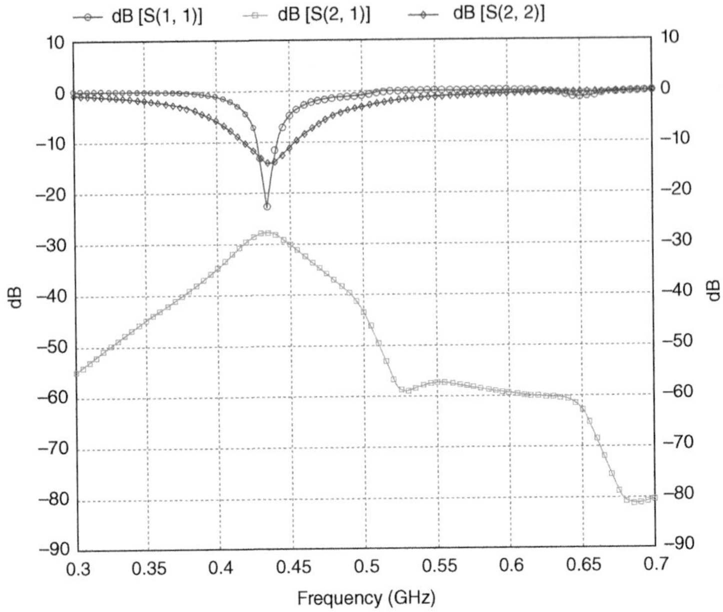

FIGURE 6.41 Simulated antenna reflection coefficient and transmission.

broadband performance can be achieved by adjusting the length and shape of the radiator, the length and width of the stripline, as well as the length of the sleeve. The impedance bandwidth of 225–510 MHz for $|S_{11}| < -10$ dB can be achieved as shown in Figure 6.43. This antenna features the omnidirectional pattern in the *E*-plane and symmetrical pattern in the *H*-plane as shown in Figure 6.44.

6.3.3 Antenna Designs at 4 GHz Band

6.3.3.1 Bow-Tie Head Antenna In order to operate across a wide frequency band of 3.5–4.5 GHz, a bow-tie antenna is proposed. The proposed antenna has the size of 23 mm × 5 mm and it is designed on a 0.508 mm thick FR4 substrate with the dielectric constant, ε_r of 4.4, and loss tangent, tanδ of 0.02. The head antenna is embedded beneath the skin tissue with a relative permittivity of 40.9 and a conductivity of 2.7 S/m at 4 GHz. The antenna is embedded 2 mm beneath the skin. An air box with a volume of 26 mm × 11 mm × 0.708 mm is created to encapsulate the antenna. Figure 6.45 shows the design of the bow-tie antenna.

6.3.3.2 Aperture-Coupled Patch Head Antenna The antenna design is shown in Figure 6.46. It is fabricated on a Ferro A6M ($\varepsilon_r = 5.9$, tan$\delta = 0.002$) substrate that is composed of 10 layers implemented using low temperature co-fired ceramics (LTCC) technology. The thickness of each layer is 0.095 mm. The patches are printed on the top layer and each patch measures 9 mm × 8 mm. The size of the

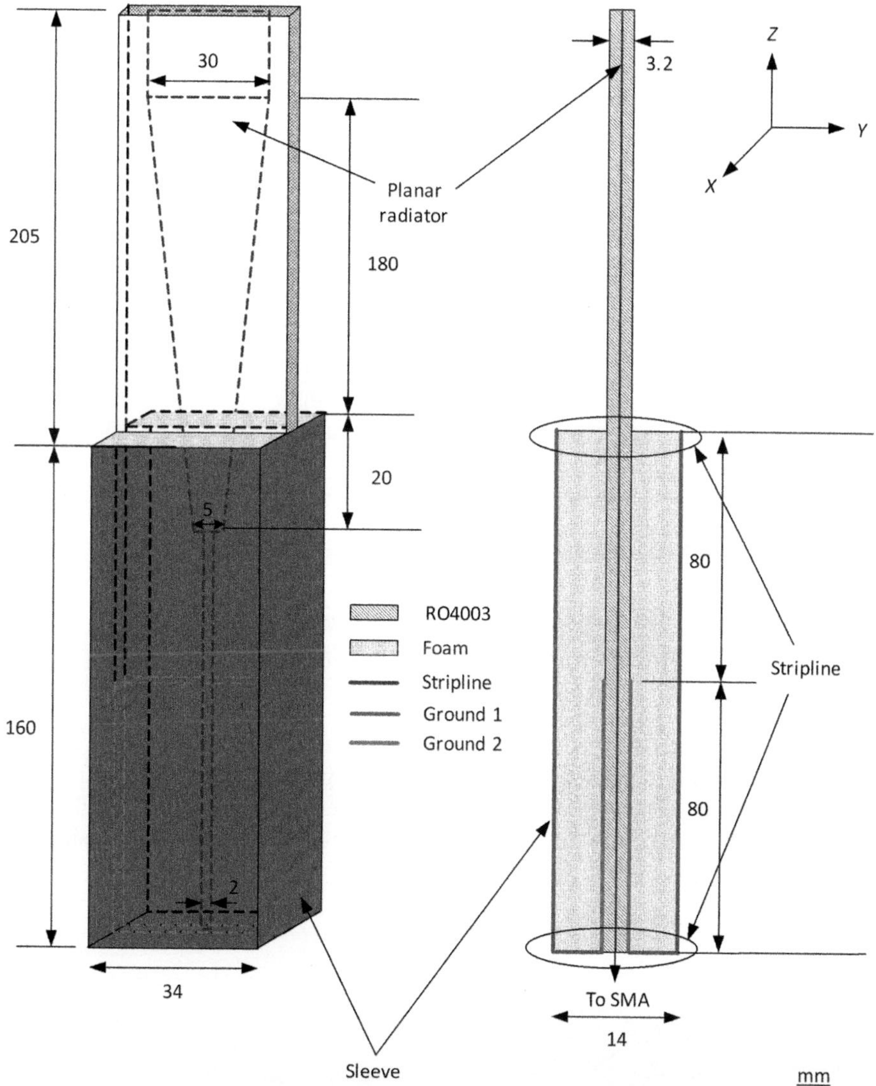

FIGURE 6.42 Sleeve monopole antenna.

substrate and the ground plane is 24 mm × 10 mm. The differential feeding structure is positioned one layer beneath the patches. The width of the differential microstrip lines is 0.77 mm and the gap between them is 0.7 mm, which provide a differential impedance of about 100 Ω. In order to achieve the required impedance matching, the length, width, and separation of the microstrip lines beneath the patches are 6.2 mm, 0.9 mm, and 0.35 mm, respectively. The ends of the differential microstrip lines are shorted to provide the aperture coupling, which broadens the impedance bandwidth. For measurement convenience, a cavity with size of 2.2 mm × 2 mm × 0.095 mm

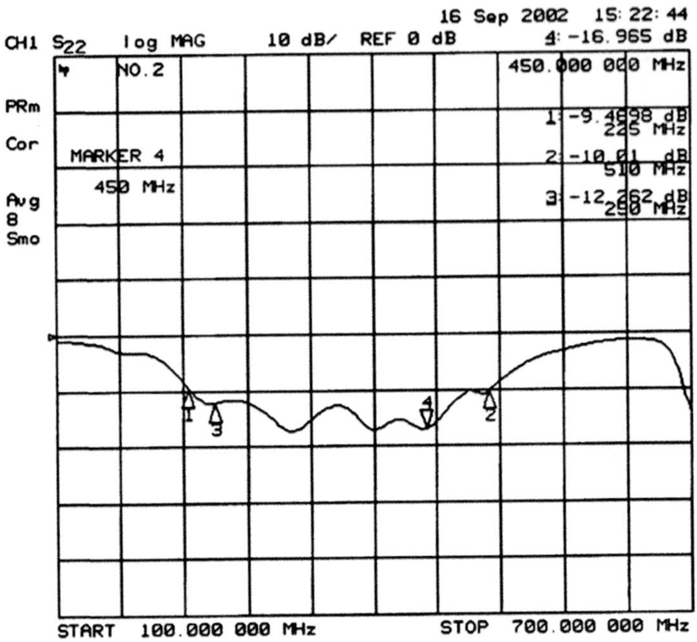

FIGURE 6.43 Measured reflection coefficient.

was constructed so that the feed lines are exposed to be soldered to the coaxial cable. In order to isolate the antenna from the tissue, a casing is required to be constructed to house the antenna. The antenna casing is made of Teflon ($\varepsilon_r = 2.1$, $\tan\delta = 0.0002$), which is biocompatible with the tissues. The thickness of the wall of the casing is 2 mm.

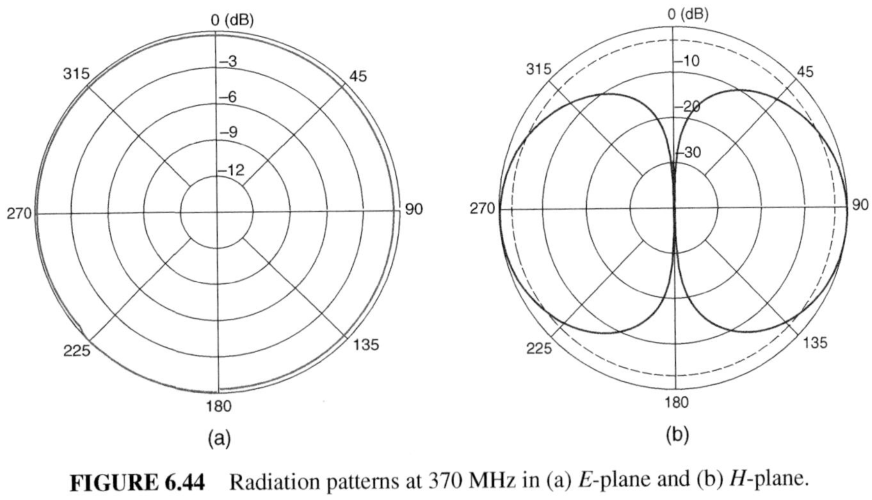

FIGURE 6.44 Radiation patterns at 370 MHz in (a) *E*-plane and (b) *H*-plane.

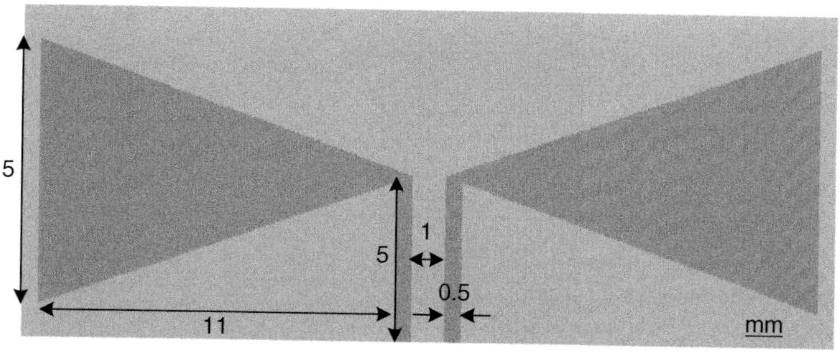

FIGURE 6.45 Bow-tie antenna.

Since the antenna performance is highly dependent on the presence of the lossy and dispersive tissues around it, it is important to take this issue into account and co-design the antenna with the casing and the surrounding tissues. As shown in Figure 6.47, the head is modeled as a three-layered homogeneous structure comprising skin, tendon, and cortical bone. The overall size of the tissues is 125 mm × 87 mm × 12.5 mm, which is based on the size of the top of the head. The three-layered model is sufficient in this study to predict the antenna performance in terms of the impedance, radiation, and peak SAR. The top of the casing is 2 mm below the surface of the skin and

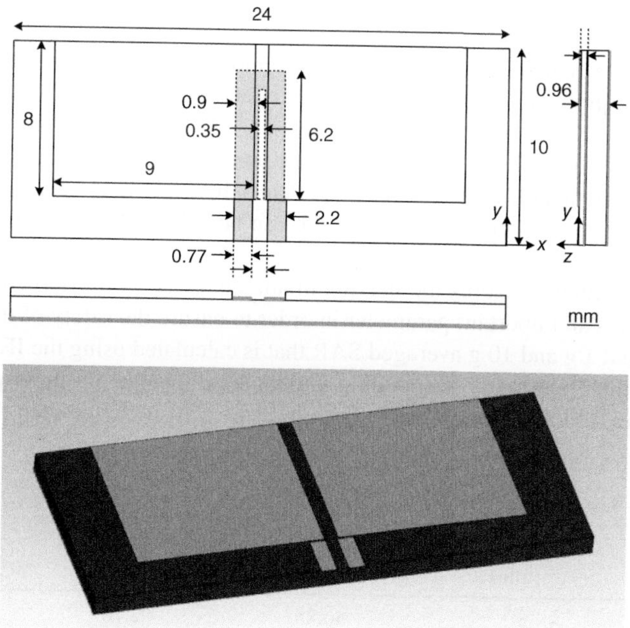

FIGURE 6.46 Aperture-coupled patch antenna.

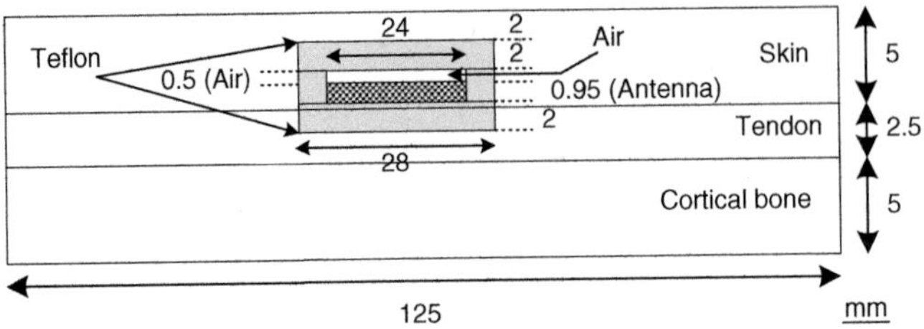

FIGURE 6.47 Modeling of the antenna placement inside the tissues. *Source*: See et al. 2015 [32]. Reproduced with permission of IEEE.

there is an air gap of 0.5 mm above the antenna. Table 6.10 tabulates the thickness and properties (relative permittivity, conductivity, and density) of the tissues at 4 GHz [31], where the properties of the tissues are consistent with those at the top of the head. The antenna was modeled and simulated using the CST Microwave Studio software.

With the differential impedance referenced to 100 Ω, the simulated and measured reflection coefficients are shown in Figure 6.48. The simulated bandwidth of the antenna for −10 dB reflection coefficient is about 3.76–4.22 GHz while the measured bandwidth is about 4.1–4.5 GHz. The slight upward frequency shift is mainly due to the fabrication tolerance. The average thickness of the fabricated samples is about 0.935 mm, which is slightly smaller than the simulated thickness of 0.95 mm. The measured values in the x- and y-directions are generally about 0.7% lower than the design value.

The radiation pattern of the antenna implanted inside the head at 4 GHz is shown in Figure 6.49, which shows that the main radiation is directed outward from the head. This is desirable in order to increase the signal transmission between the head implant and the external receiver.

The directivity and gain of the antenna at the boresight are shown in Figure 6.50. It can be observed that the directivity of more than 5 dBi and a boresight gain of more than −8 dBi across the impedance bandwidth can be achieved when the antenna is implanted inside the head.

The sSAR, which measures the maximum amount of energy absorbed by a unit mass of tissue, is an important parameter in order to ensure the safety of the patient. The peak spatial 1 g and 10 g averaged SAR that is calculated using the IEEE C95.3 averaging method and the corresponding maximum allowable net input power are given in Figure 6.51 for the delivered peak power of 1 W. It can be seen that due to

TABLE 6.10 Tissue parameters at 4 GHz.

Tissue	Thickness (mm)	Relative permittivity	Conductivity (S/m)	Density (kg/m^3)
Skin	5	40.85	2.70	1125
Tendon	2.5	40.24	3.17	1151
Cortical bone	5	10.53	0.728	1850

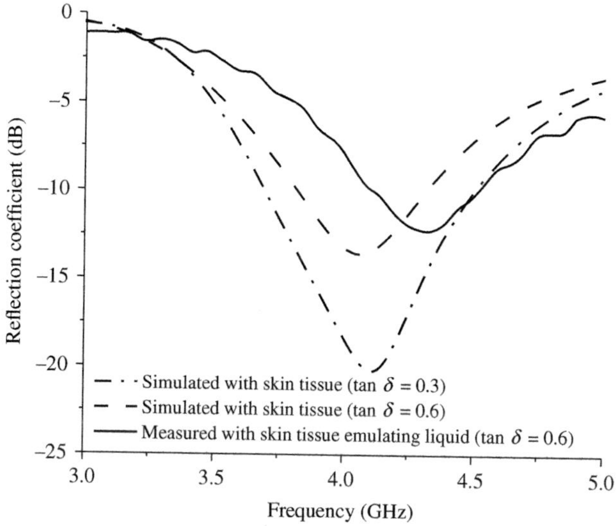

FIGURE 6.48 Reflection coefficient.

the directional radiation of the antenna, the peak spatial 1 g and 10 g averaged SAR is less than 53 W/kg and 17 W/kg across the bandwidth of 4.1–4.5 GHz, respectively. Taking the origin to be at the position shown in Figure 6.46, the peak SAR occurs at the location (−14.5, 5.5, 3.7). In order to meet the ANSI C95.1–1999 regulation of 1.6 W/kg in 1 g of tissue and ANSI C95.1–2005 regulation of 2 W/kg in 10 g of tissue, the power that is delivered to the implant antenna should be reduced to less than

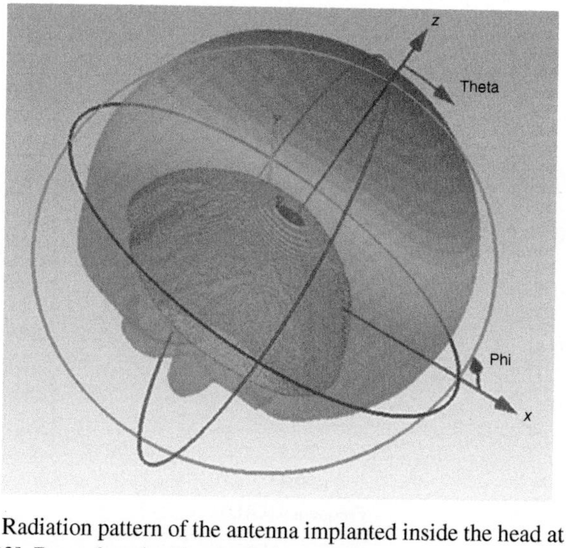

FIGURE 6.49 Radiation pattern of the antenna implanted inside the head at 4 GHz. *Source*: See et al. 2015 [32]. Reproduced with permission of IEEE.

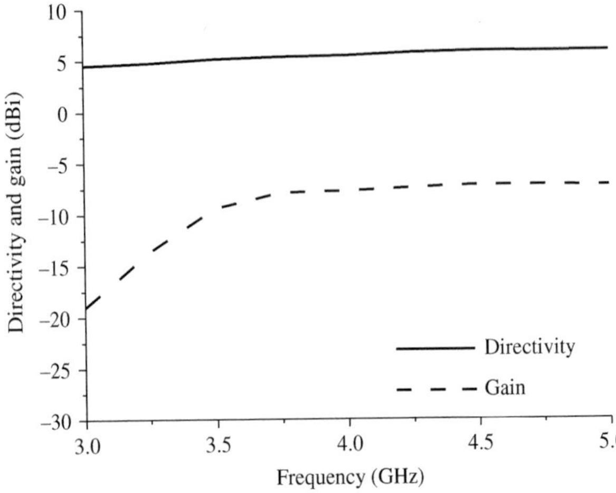

FIGURE 6.50 Directivity and gain at boresight of antenna implanted inside the head. *Source*: See et al. 2015 [32]. Reproduced with permission of IEEE.

28 mW and 118 mW, respectively. Generally, it is important that the peak SAR is kept low. Other than the health considerations, from the system performance perspective in terms of the communication range, the higher SAR will require a greater attenuation of the input power to the antenna in order to meet the regulations, which will lead to a reduction in the transmission range.

6.3.3.3 External Antenna Designs A bow-tie antenna, a linearly polarized patch antenna, and a circularly polarized patch antenna are proposed to operate at 4 GHz.

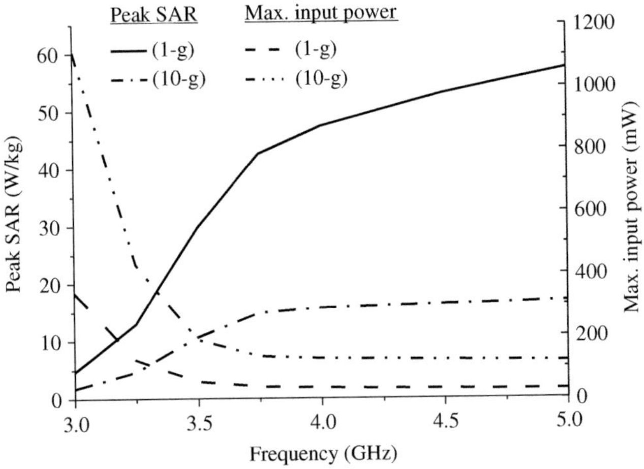

FIGURE 6.51 Peak SAR (1 g and 10 g average) and corresponding maximum input power.

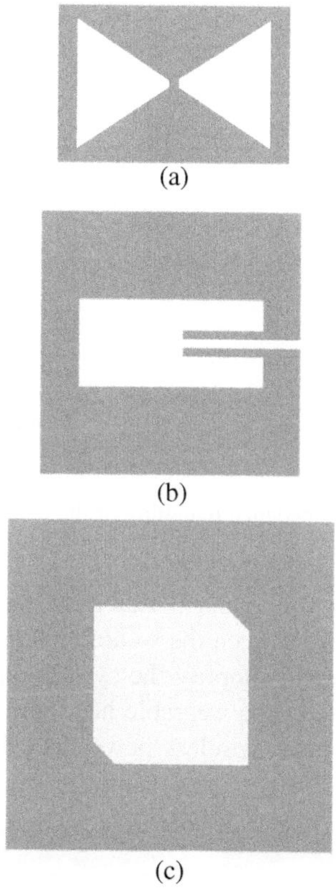

FIGURE 6.52 External antenna designs: (a) bow-tie antenna, (b) linearly polarized patch, and (c) circularly polarized patch.

The bow-tie antenna has a size of 20 mm × 13 mm, and it is designed on a 0.508 mm thick FR4 substrate with the dielectric constant, ε_r of 4.4 and loss tangent tanδ of 0.02. The linearly polarized patch antenna has a patch size of 21 mm × 10 mm and a ground plane size of 30 mm × 30 mm. It is designed on a piece of 0.508 mm thick FR4 PCB substrate. The circularly polarized patch is edge-truncated with a patch size of 19 mm × 19 mm and a ground plane size of 40 mm × 40 mm. It is designed on a piece of 1.524 mm thick FR4 PCB substrate. It has an axial ratio (AR) of 2 dB at 4 GHz. Figure 6.52 shows the designs of the external antenna and the corresponding impedance-matching response is shown in Figure 6.53.

6.3.4 Primate Demonstration

The head antenna is deployed on a 100-channel wireless neural recording system. The system consists of three 32-channel implantable probe units for neural signal acquisition, a wearable head unit for neural signal acquisition, wireless telemetry,

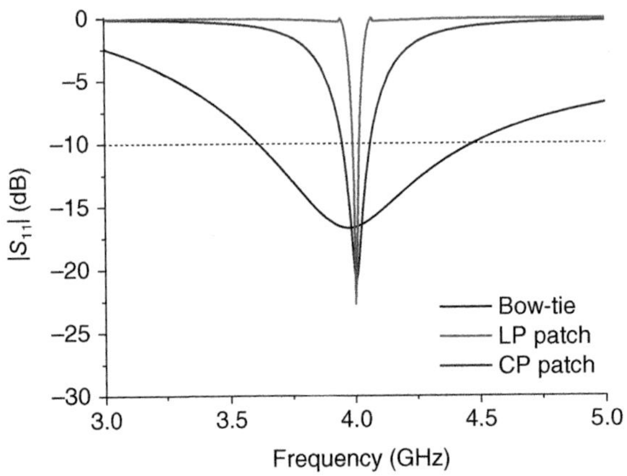

FIGURE 6.53 Impedance matching of the external antennas.

and wireless power transfer as well as an external base station. Wireless data telemetry and wireless power transfer between the wearable unit and the base station help to remove the tethering wires and simplify the system interconnections so that the patient mobility is not restricted. The wearable head unit comprises a 100-channel neural signal acquisition IC and a wireless power and data link IC. The external

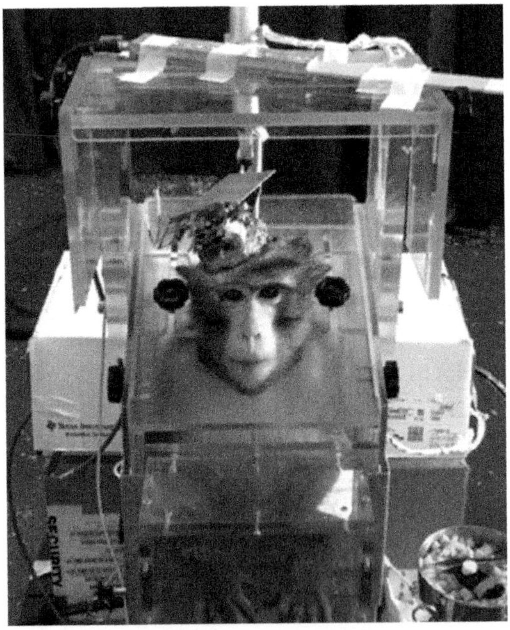

FIGURE 6.54 A wireless neuroprobe system demonstration on the primate. For a color version of this figure, see the color plate section.

antenna is connected to the base station located on the chair of the primate, which includes a wideband data telemetry receiver and a wireless power transmitter. The neural signal acquisition IC together with the wireless power and telemetry IC is mounted over the skull together with a coupling coil and an antenna. Both the ICs are connected with a flexible cable for power, command, and data communication. The photos of the demonstration of the system on the primate are shown in Figure 6.54.

The integrated wireless system permits self-motion, employing a robotic platform. The primate is able to use its motor cortical activity to control a robotic platform in a self-motion task. These results pave the way for the implementation of this technology in human tetraplegic patients, as well as for the future integration of robotic platform and robotic arm control to endow these patients with an unprecedented level of autonomy.

6.4 CONCLUSION

This Chapter has reviewed the key issues of RF or microwave signal propagation through body/head and implanted/external antennas of implanted medical devices. It has clearly indicated that the most challenges of antenna design for body/head implants are caused, in particular, by the small form factor of the antennas. The key performances of the implanted/external antennas such as the impedance matching and antenna efficiency are affected by the lossy, inhomogeneous, and dispersive surrounding tissues. Our experience has shown that the co-design of antenna with its surrounding tissues and biocompatible casing is needed in order to achieve optimum efficiency. The study in this Chapter has also shown that to reduce the SAR and enhance the transmission with the external receiver, an antenna with directional radiation is preferred. In this chapter, we have shown the characterization of RF transmission through tissues as well as the antenna designs for wireless capsule endoscopy and neural motor prosthetic devices.

REFERENCES

[1] "Medical implant communications service (MICS) federal register," *Rules. Reg.*, vol. 64, no. 240, pp. 69926–69934, December 1999.

[2] FCC, Guidelines for Evaluating the Environmental Effects of Radio Frequency Radiation, ET Docket No. 93-62, Washington, DC, 1996.

[3] P. S. Hall and Y. Hao, *Antennas and Propagation for Body-Centric Wireless Communications,* Artech House, MA, 2006.

[4] G. Iddan, G. Meron, A. Glukhovsky, and P. Swain, "Wireless capsule endoscopy," *Nature*, vol. 405, pp. 417, June 2000.

[5] W. A. Qureshi, "Current and future applications of the capsule camera," *Nat. Rev. Drug Discov.*, vol. 3, pp. 447–450, May 2004.

[6] P.M. Izdebski, H. Rajagopalan, and Y. Rahmat-Samii, "Conformal ingestible capsule antenna: A novel chandelier meandered design," *IEEE Trans. Antennas Propag.*, vol. 57, no. 4, pp. 900–909, April 2009.

[7] S. H. Lee and Y. J. Yoon, "A Dual Spiral Antenna for Ultra-Wideband Capsule Endoscope System," in International Workshop on Antenna Technology: Small Antennas and Novel Metamaterials, Chiba, March 4–6, 2008, pp. 227–230.

[8] C. Ethier, E. R. Oby, M. J. Bauman, and L. E. Miller, "Restoration of grasp following paralysis through brain-controlled stimulation of muscles," *Nature*, vol. 485, pp. 368–371, May 2012.

[9] P. H. Peckam, M. W. Keith, K. L. Kilgore, J. H. Grill, K. S. Wuolle, G. B. Thrope, P. Gorman, J. Hobby, M. J. Mulcahey, S. Caroll, V. R. Hentz, and A. Wiegner, "Efficacy of an implanted neuroprosthesis for restoring hand grasp in tetraplegia: a multicenter study," *Arch. Phys. Med. Rehabil.*, vol. 82, pp. 1380–1388, October 2001.

[10] A. Umemura, J. L. Jaggi, H. I. Hurtig, A. D. Siderowf, A. Colcher, M. B. Stern, and G. H. Baltuch, "Deep brain stimulation for movement disorders: morbidity and mortality in 109 patients," *J. Neurosurg.*, vol. 98, pp. 779–784, April 2003.

[11] R. Merkel, *Intervening in the Brain: Changing Psyche and Society*. Springer, Berlin, New York, 2007.

[12] U. M. Jow and M. Ghovanloo, "Optimization of data coils in a multiband wireless link for neuroprosthetic implantable devices," *IEEE Trans. Biomed. Circuits. Syst.*, vol. 4, no. 5, pp. 301–310, October 2010.

[13] B. Gosselin, "Recent advances in neural recording microsystems," *Sensors*, vol. 11, pp. 4572–4597, April 2011.

[14] N. Gopalsami, I. Osorio, S. Kulikov, S. Bukyo, A. Martynov, and A. C. Raptis, "SAW microsensor brain implant for prediction and monitoring of seizures," *IEEE Sens. J.*, vol. 7, no. 7, pp. 977–982, July 2007.

[15] C. Cavallotti, M. Piccigallo, E. Susilo, P. Valdastri, A. Menciassi, and P. Dario, "An integrated vision system with autofocus for wireless capsular endoscopy, "*Sens. Actuators A Phys.*, vol. 156, no. 1, pp. 72–78, November 2009.

[16] G. C. Crumley, N. E. Evans, J. B. Burns, and T. G. Trouton, "On the design and assessment of a 2.45 GHz radio telecommand system for remote patient monitoring," *Med. Eng. Phys.*, vol. 20, pp. 750–755, March 1999.

[17] K. Jaehoon and Y. Rahmat-Samii, "Implanted antennas inside a human body: simulations, designs, and characterizations," *IEEE Trans. Microw. Techn.*, vol. 52, no. 8, pp. 1934–1943, August 2004.

[18] K. M. S. Thotahewa, J. -M. Redoute, and M. R. Yuce, "SAR, SA, and temperature variation in the human head caused by IR-UWB implants operating at 4 GHz," *IEEE Trans. Microw. Techn.*, vol. 61, no. 5, pp. 2161–2169, May 2013.

[19] H. Bahrami, B. Gosselin, and L. A. Rusch, "Design of a Miniaturized UWB Antenna Optimized for Implantable Neural Recording Systems," in Proceedings of the IEEE 10th International Conference on New Circuits and Systems, Montreal, QC, June 17–20, 2012, pp. 309–312.

[20] M. K. Hosain, A. Z. Kouzani, S. Tye, D. Mortazavi, and A. Kaynak, "Compact Stacked Planar Inverted-F Antenna for Passive Deep Brain Stimulation Implants," in Proceedings of the IEEE 34th International Conference on Engineering in Medicine and Biology Society (EMBC), San Diego, CA, August 28–September 1, 2012, pp. 851–854.

[21] A. Kiourti and K. S. Nikita, "Miniature scalp-implantable antennas for telemetry in the MICS and ISM bands: design, safety considerations and link budget analysis," *IEEE Trans. Antennas Propag.*, vol. 60, no. 8, pp. 3568–3575, August 2012.

[22] Z. N. Chen, G. C. Liu, and T. S. P. See, "Transmission of RF signals between MICS loop antennas in free space and implanted in the human head," *IEEE Trans. Antennas Propag.*, vol. 57, no. 6, pp. 1850–1854, June 2009.

[23] T. S. P. See, X. Qing, and Z. N. Chen, "Design and Characterization of On-Head Antenna Systems for Neural Motor Prosthesis," in Proceedings of the European Conference on Antennas and Propagation, Lisbon, May 13–17, 2015, pp. 1–6.

[24] Z. N. Chen, A. Cai, T. S. P. See, X. Qing, and M. Y. W. Chia, "Small planar UWB antennas in proximity of the human head," *IEEE Trans. Microw. Theory Techn.*, vol. 54, no. 4, pp. 1846–1857, June 2006.

[25] T. S. P. See, X. Qing, Z. N. Chen, C. K. Goh, and T. M. Chiam, "RF Transmission in/through the Human Body at 915 MHz," in Proceedings of the IEEE International Symposium on Antennas and Propagation, Toronto, ON, July 11–17, 2010, pp. 1–4.

[26] X. Qing, Z. N. Chen, T. S. P. See, C. K. Goh, and T. M. Chiam, "Characterization of RF Transmission in Human Body," in Proceedings of the IEEE International Symposium on Antennas and Propagation, Toronto, ON, July 11–17, 2010, pp. 1–4.

[27] X. Qing, Z. N. Chen, T. S. P. See, C. K. Goh, and T. M. Chiam, "RF Transmission Characteristics in/through the Human Body," in Proceedings of the IEEE Conference on Cybernetics and Intelligent Systems, Singapore, June 28–30, 2010, pp. 20–23.

[28] T. S. P. See, Z. N. Chen, and X. Qing, "Implanted and External Antennas for 915-MHz Capsule Endoscopy," in Proceedings of the International Workshop on Antenna Technology, Hong Kong, March 7–9, 2011, pp. 29–32.

[29] C. K. Goh, X. Qing, Z. N. Chen, and T. S. P. See, "Effect of Wireless Charging Antennas on Transmission of An Antenna Pair Through Human Body," in Proceedings of the Asia-Pacific Conference on Antennas and Propagation, Singapore, August 27–29, 2012, pp. 57–58.

[30] X. Qing, T. S. P. See, Z. N. Chen, T. M. Chiam, Nasimuddin, and C. K. Goh, "Characterization of in- and near-body RF transmission loss for biomedical implants," *J. Med. Imaging Health Inform.*, vol. 3, no. 1, pp. 112–119, March 2013.

[31] S. Gabriel, R. W. Lau, and C. Gabriel, "The dielectric properties of biological tissues: III. Parametric models for the dielectric spectrum of tissues," *Phys. Med. Biol.*, vol. 41, no. 11, pp. 2271–2293, November 1996.

[32] T. S. P. See, X. Qing, W. Liu, and Z. N. Chen, "A wideband ultra-thin differential loop-fed patch antenna for head implants," *IEEE Trans. Antennas Propag.*, vol. 63, no. 7, pp. 3244–3248, July 2015.

7

ANTENNAS, PHANTOMS, AND BODY-CENTRIC PROPAGATION AT MILLIMETER-WAVES

NACER CHAHAT,[1] ADRIAN TANG,[1] ANDA GURALIUC,[2]
MAXIM ZHADOBOV,[2] RONAN SAULEAU,[2] AND GUIDO VALERIO[3]

[1] *NASA Jet Propulsion Laboratory, California Institute of Technology, Pasadena, CA, USA*
[2] *Institute of Electronics and Telecommunications of Rennes, UMR CNRS 6164, University of Rennes 1, Rennes, France*
[3] *Laboratoire d'Electronique et Electromagnétisme, Sorbonne Universités – Université Pierre et Marie Curie, Paris, France*

7.1 INTRODUCTION

Over the past few years, wearable antennas for body-centric wireless communication systems have been an increasingly important area of research, showing great promise for monitoring and communication in a wide range of applications such as health care, entertainment, military, identification systems, sports, smart homes, and space, for example [1] and [2].

Wearable devices refer to electronic technologies that are incorporated into items of clothing and accessories, which can comfortably be worn on the body. Most recently, many commercial products have emerged in daily life, such as the Google Glass, Smart Watches, Smart Clothes (Figure 7.1a), and so on.

While these devices were originally limited to specific applications (e.g., defense applications to increase overall effectiveness of soldiers on the battlefield), they are now integrated into our garments, watches, smartphones, and glasses. A monitoring capability is also under investigation for vital sign monitoring and indoor

Electromagnetics of Body-Area Networks: Antennas, Propagation, and RF Systems, First Edition.
Edited by Douglas H. Werner and Zhi Hao Jiang.

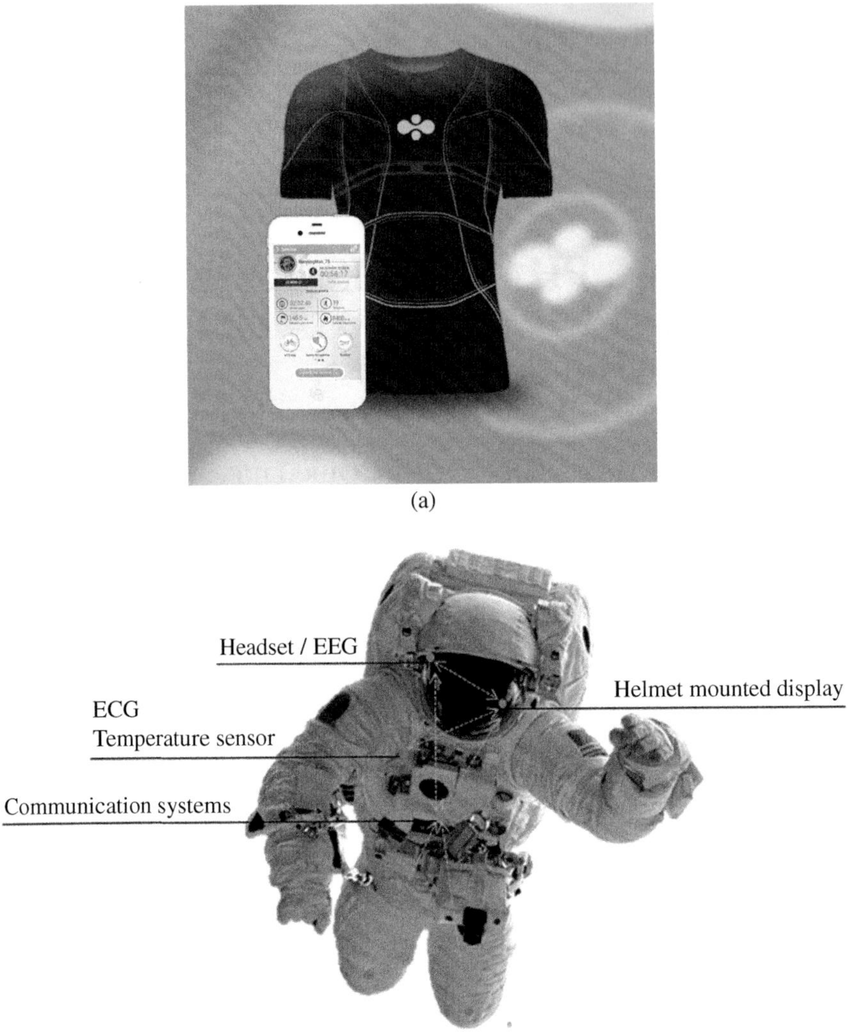

(a)

(b)

FIGURE 7.1 (a) Smart shirt with embedded microsensors monitoring temperature, heart rate, speed, location, and acceleration of the user (courtesy: Cityzen Science). (b) Sensors embedded in an astronaut's extravehicular activity suit. Reproduced with permission of Cityzen Group USA.

location tracking of firefighters during interventions. In addition, future NASA plans include launching manned interplanetary missions to Mars. These missions present new challenges that require rethinking of the extravehicular activity (EVA) suit design and associated communication systems (Figure 7.1b). Replacing the current wired biosensing within the suit by a wireless body area network (BAN) would provide benefits such as the ability to monitor multiple vital signs and easily reconfigure the sensors to suit particular mission goals.

Recently, studies carried out at millimeter-waves have grown rapidly as millimeter-waves offer many advantages compared to microwaves. First, because of the large available spectrum (7 GHz worldwide), very high data rates can be reached (beyond several Gb/s) [3]. Second, it provides a high level of security and low interference with adjacent networks [4]. Finally, compared to on-body devices operating at microwaves, the size of similar millimeter-wave systems is significantly reduced.

This chapter reports the recent progress made on antennas, phantoms, and body-centric propagation at millimeter-waves. In Section 7.2, the numerical and experimental modeling of the human body is discussed. Antennas for off-body communications and their interaction with the human body are discussed in Section 7.3. In Section 7.4, the on-body propagation is discussed, and several antenna designs specifically developed for on-body communications are described. Finally, conclusions are drawn in Section 7.5.

7.2 HUMAN BODY MODELING AND EXPOSURE GUIDELINES

7.2.1 From Microwaves to Millimeter-Waves

7.2.1.1 Tissue-Equivalent Model at Microwaves Many numerical phantoms have been reported for theoretical analysis and computations. Simple-shaped phantoms are generally used in theoretical analyses and are therefore called *theoretical phantoms*. However, more realistic human-like numerical phantoms are necessary in some particular cases. These heterogeneous phantoms are called *voxel phantoms*.

Homogeneous (Figure 7.2) and layered phantoms are the simplest ones. They can be used for dosimetric studies and antenna characterization. Whereas they usually have simplified geometry (e.g., parallelepiped, sphere, or cylinder), they can also reproduce realistic geometry of the human body [5–7].

The development of highly accurate whole-body voxel models was possible due to medical imaging technologies, such as MRI and X-ray computed tomography. A wide range of human body models is now available with various ranges of morphologies (i.e., weight, height, body mass index), age, nationality, and gender [9, 10]. However, those realistic and detailed models are not useful at millimeter-waves, where the penetration is very shallow (typically 0.5 mm at 60 GHz [8]).

7.2.1.2 Phantoms at Microwaves

7.2.1.2.1 Liquid Phantoms Liquid phantoms consist of a container filled with a liquid, having the requested average tissue-equivalent dielectric properties in the considered frequency range. These phantoms are extensively used in specific absorption rate (SAR) measurements, where direct measurement of the electric field inside the phantom is carried out using a small probe. Two examples of liquid phantoms are presented in Figure 7.3.

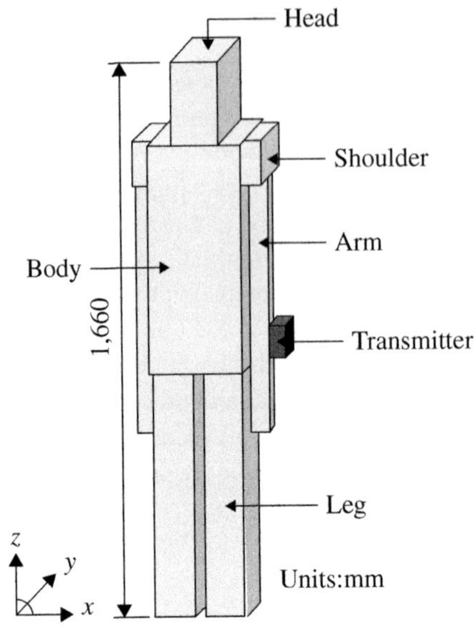

FIGURE 7.2 An example of a theoretical model with simplified shape and homogeneous dielectric properties. *Source*: Fuji et al. 2006 [6]. Reproduced with permission of IEEE.

Furthermore, the choice of tissue dielectric parameters for homogeneous tissue-equivalent liquid should be determined to avoid any over- or underestimation when compared with SAR results from heterogeneous head modeling. Therefore, when selecting the equivalent tissue parameters, detailed internal head anatomies must be considered. These parameters must be chosen carefully to ensure that the SAR measured in the liquid phantom does not underestimate what is expected in an actual human head. The dielectric parameters of most simulated tissues generally represent an average value of several different tissue layers.

To the best of our knowledge, liquid phantoms are only available up to 6 GHz [11]. In addition, due to the presence of the container shell, they do not allow measurements at the phantom surface. Therefore, liquid phantoms can hardly be used at millimeter-wave because of the very shallow penetration of millimeter-waves into the human skin.

7.2.1.2.2 Semi-Solid Phantoms It is possible to develop semi-solid phantoms capable of self-shaping using coagulant agents. Guy et al. [14] have introduced two gel phantoms simulating human fat/bone and muscle in the 0.2–1 GHz range. These authors proposed a phantom simulating high water content tissues (e.g., muscle, skin) composed of water, sodium chloride, TX-150 (a gelling agent, widely referred to as "super stuff"), and polyethylene powder. The relative permittivity can be varied over a wide range by varying the percentage of polyethylene powder, while the conductivity is controlled by the salinity of the material. For tissues with low water content such as bone and fat, a phantom composed of laminac polyester resin, catalyst,

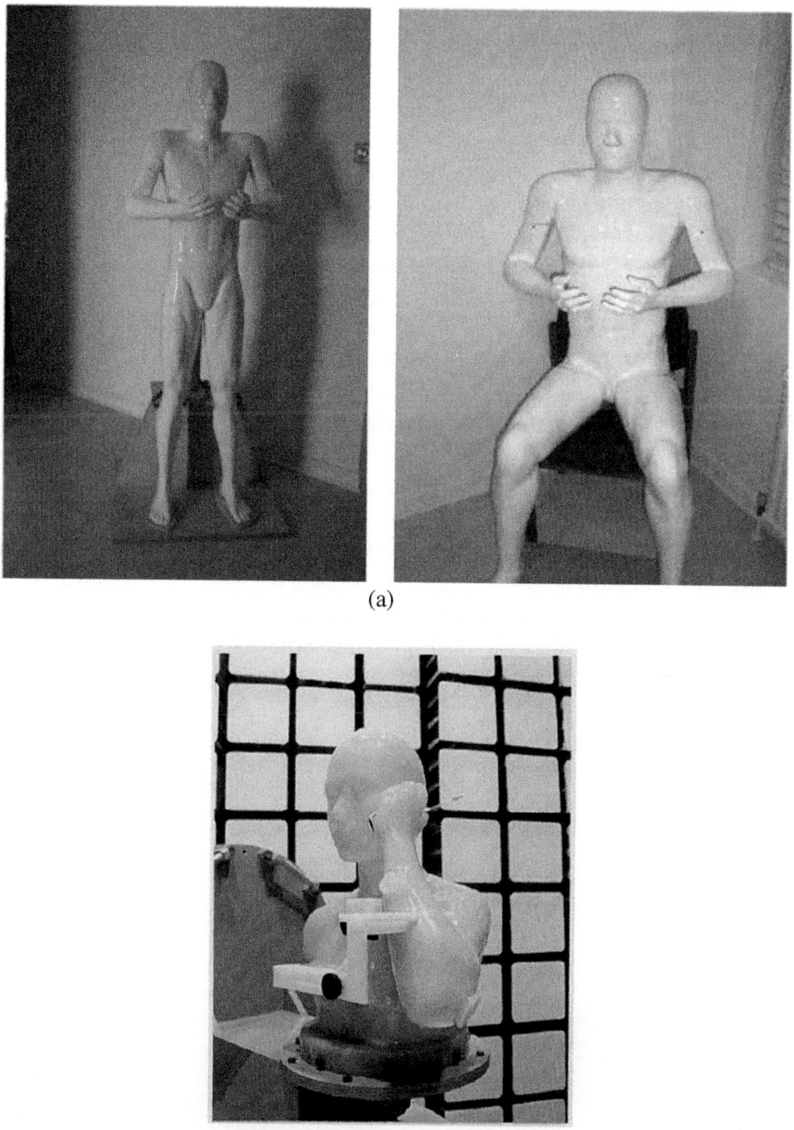

(a)

(b)

FIGURE 7.3 (a) Whole-body specific anthropomorphic mannequin (SAM) phantom in standing and seated positions. *Source*: Gabriel and Chadwick 2008 [12]. Reproduced with permission of IEEE. (b) Multipostural realistic human torso phantom. *Source*: Ogawa et al. 2001 [13]. Reproduced with permission of IEEE.

acetylene black, and aluminum powder is employed. The dielectric constant is controlled by varying the amounts of aluminum powder, and the conductivity is adjusted using acetylene black.

Ito et al. [15,16] have added sodium dehydroacetate (preservative) and agar (coagulant) to Guy's recipes (for high water content tissues). Phantom electrical properties can be adjusted over a wide frequency range. For instance, for a single composition ratio, the phantom electrical properties are almost equal to those in biological tissue over the frequency range of 0.3–2.5 GHz (i.e., muscle and brain, [15]) or 3.1–10.6 GHz (i.e., 2/3 muscle-equivalent phantom, [16]), respectively. Besides, a skin-equivalent phantom has been recently introduced at 10 GHz [17]. The dielectric properties of these phantoms degrade over time due to the loss of water and growth of fungi. Therefore the lifespan of these phantoms is limited (i.e., about 1 month, but depends on the preservative and it should be wrapped with a plastic film).

Okano et al. [18] proposed a high preservative phantom with a lifespan greater than 6 months, which is called a glycerin phantom. This phantom can simulate tissues with both high and low water content. However, the frequency range covered by this phantom is narrower than that of the Ito and Guy's phantoms.

Semi-solid phantoms are low-cost, easy to fabricate, and easy to manufacture into arbitrary shape and multiple layers. They can be used for SAR measurement on and inside the phantom. Figure 7.4a shows a 2/3 muscle-equivalent phantom representing a human arm for on-body antenna measurements in the 3.1–10.6 GHz range [19]. Figure 7.4b represents a human torso phantom of a Japanese male fabricated for SAR and antenna measurements [15]. Such phantoms are excellent candidates at millimeter-waves, since the skin consists of approximately 65% of free water [20]. The first phantom introduced at millimeter-waves is a semi-solid phantom [21], as described later in this chapter.

7.2.1.2.3 Solid Phantoms Solid phantoms (Figure 7.5) represent an interesting solution for SAR measurement at the surface of the body, antenna measurements, and

(a) (b)

FIGURE 7.4 (a) Two-third muscle-equivalent phantom (Chahat et al. 2011 [19]. Reproduced with permission of IEEE.) (b) Realistic human torso agar-based solid phantom (Courtesy Prof. Koichi Ito, Chiba University).

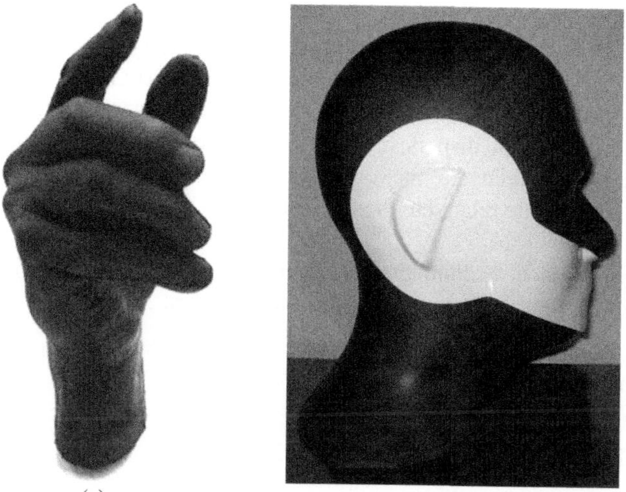

(a) (b)

FIGURE 7.5 Carbon–silicone phantoms: (a) Hand phantom, (b) Outer SAM-shaped solid head. *Source*: Gabriel and Chadwick 2008 [12]. Reproduced with permission of IEEE.

for studies of the propagation around, as well as inside, the body since they can accurately represent inhomogeneous structures of the human body. Numerous phantoms have been proposed including (1) a mixture of ceramic and graphite powder [22, 23], (2) a combination of two carbon fibers and silicone rubber [24], and (3) conductive plastic containing carbon black [25].

These phantoms can be designed to fit a wide range of complex permittivities. Besides, both tissues with low and high water content can be modeled for frequencies up to 6 GHz. Compared to liquid and semi-solid ones, solid phantoms provide longer lifespan with stable dielectric and mechanical properties over a long time. However, such phantoms are expensive and require special and expensive equipment, along with specific high-temperature (~260°C) and high-pressure manufacturing procedures. The first carbon-based solid phantom at millimeter-waves was recently presented in [26], and more details are provided in Section 7.2.3.3.

7.2.2 Dielectric Properties and Measurement Techniques for Phantoms and Human Tissues at Millimeter-Waves

7.2.2.1 Measurement Methodologies As skin covers 95% of the human body surface [8], we essentially consider the interactions with the skin. From the electromagnetic viewpoint, human skin can be considered as an anisotropic multilayer dispersive structure made of three different layers, namely, epidermis, dermis, and subcutaneous fat layer. The skin is mainly composed of 65.3% of free water, 24.6% of proteins, and 9.4% of lipids [20]. It is worthwhile to underline that the dielectric properties of human tissues can vary with age as demonstrated in [5].

To determine reflection from, transmission through, and absorption in the body, as well as for electromagnetic modeling purposes, it is necessary to assess the dielectric properties of the skin. In contrast to frequencies below 20 GHz, existing data on the permittivity of tissues in the millimeter-wave band are very limited [27–31]. In the 10–100 GHz range, the dispersive dielectric properties of the skin and biological solutions are primarily related to the rotational dispersion of free water molecules. In particular, high losses are related to the free water relaxation with the peak at 23 GHz at 33°C.

Several measurement techniques have been implemented so far to measure the skin or phantom dielectric properties at millimeter-waves:

- Open-ended coaxial probe [21],
- Heating kinetics technique [31],
- Extrapolation using a Debye model [27, 29].

7.2.2.1.1 Open-Ended Coaxial Probe The open-ended coaxial measurement setup at millimeter-waves is presented in [21]. This probe was designed by Agilent Technologies (Santa Clara, CA, USA) for permittivity measurements of lossy liquids and semi-solids up to 67 GHz.

SPEAG® company offers a dielectric probe for measurements over the 5 – 67 GHz frequency range. This open-ended coaxial probe measures the dielectric properties of liquids, solids, and semi-solids over a broad range of parameters. The measurement method is fast and non-destructive to the material under test.

7.2.2.1.2 Heating Kinetics Technique The heating kinetics technique allows performing non-destructive measurements *in situ*. The temperature kinetics resulting from millimeter-wave exposure and measured using an IR camera can be used to determine the permittivity of high water content materials including the skin and phantoms [31]. The proposed procedure consists in fitting the solution of the bioheat transfer equation (BHTE) to the measured temperature kinetics; this enables to define the incident power density (IPD) I and skin depth δ. Then, for a given kinetics, a unique combination (I,δ) is found by minimizing the standard deviation. The obtained values of δ are further used to calculate the skin or phantom permittivity spectra. It is worthwhile to note that this method can be applied only to high water content materials with the same relaxation time as water.

7.2.2.1.3 Extrapolation From a Debye Model The skin can be described by a Debye model (see Section 7.2.3.1 for more details) with a relaxation time equal to that of free water. Hence, measurement can be done at lower frequencies using well-controlled and commercially available open-ended coaxial probe and the complex permittivity can be extrapolated using a Debye model. This technique was used by Gabriel et al. [29] who reported extrapolated complex permittivity values up to 110 GHz based on human skin measurements performed below 20 GHz. Also, Gandhi

TABLE 7.1 Overview of the skin dielectric properties at 60 GHz.

Reference	Complex permittivity ε^*	T, °C	Method	Sample type
Gandhi et al. [27]	$8.89 - j13.15$	37 ± 0.5	E	*In vitro*
Alabaster et al. [28]	$9.9 - j9.0$	23	M	*In vitro*
Gabriel et al. [29] "wet skin"	$10.22 - j11.84$	37	E	*In vitro*
Gabriel et al. [29] "dry skin"	$7.98 - j10.90$	32.5 ± 0.5	E	*In vivo*
Alekseev et al. [30]	$8.12 - j11.14$	32.5 ± 0.3	M	*In vivo*
Chahat et al. [21]	$8.02 - j10.5$	32.5 ± 0.5	M	*In vivo*
Chahat et al. [31]	$8.4 - j10.96$	32.5 ± 0.5	M	*In vivo*

E, Extrapolation; M, Measurement; T, theoretical value.

et al. [27] reported skin dielectric properties obtained using a Debye model relying on measurements performed on rabbit skin at 23 GHz.

7.2.2.2 Skin Dielectric Properties Previously reported skin data at 60 GHz are summarized in Table 7.1. Clearly, the literature data vary significantly from one study to another depending on sample types and measurement techniques. Since the skin mainly consists of free water [20], its complex permittivity is strongly dispersive and temperature-dependent. Whereas the skin temperature under normal environmental conditions is around 32.5°C, the skin permittivity of *in vitro* measurements is given for different temperatures, for example [27] and [28]. Besides, water content of *in vitro* skin samples might vary depending on the measurement protocol. It is therefore preferable to perform *in vivo* measurements.

Two sets of data were recently introduced by Chahat et al. using an open-ended coaxial slim probe [21] and a new method based on heating kinetics [31]. Those results are in excellent agreement with Gabriel et al. (dry skin) [29] and Alekseev et al. [30] data as shown in Figure 7.6. Therefore, the skin permittivity data provided by Gabriel et al. [29] and Alekseev et al. [30] can be used as reference.

The attenuation of the power density in the skin at 60 GHz is shown in Figure 7.7. The very shallow penetration of millimeter-waves into the skin is clearly shown (i.e., less than typically 0.5 mm above 60 GHz) as the power decreases exponentially in the skin as a function of depth. It is worthwhile to notice that the maximum power density occurs at the surface which is not the case at microwave frequencies.

Figure 7.8 represents the power reflection and transmission coefficients at normal incidence over the 50–70 GHz band. Noticeably, the power reflection coefficient decreases with frequency, and is around 37.5% in the 57–66 GHz band dedicated to future high data rate wireless communications in V-band.

7.2.3 Numerical and Experimental Phantoms at Millimeter-Waves

7.2.3.1 Numerical Skin-Equivalent Phantom Taking into account the very shallow penetration of millimeter-waves into the skin (typically 0.5 mm at 60 GHz), using homogeneous skin-equivalent phantoms provides accurate results for the antenna/human body interaction evaluation as well as for the propagation channel

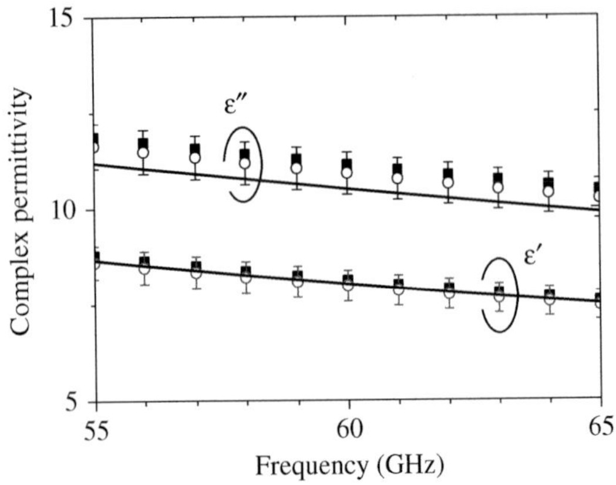

FIGURE 7.6 Comparison between experimental result for the wrist skin permittivity (—) and data provided by Gabriel et al. (dry skin) (■) and Alekseev et al. (o) models. Error bars represent ±5% deviations around Gabriel's reference values. *Source*: Chahat et al. 2012 [21]. Reproduced with permission of IEEE.

characterization [8]. For broadband analysis, dispersive models can be used. Debye model with a single relaxation time τ equal to that of free water at the same temperature was demonstrated to provide a good accuracy for modeling the experimental permittivity data in the considered frequency range [21]:

$$\varepsilon^* = \varepsilon_\infty + \frac{\Delta\varepsilon}{1 + j\omega\tau} + \frac{\sigma}{j\omega\varepsilon_0}. \tag{7.1}$$

In this equation, $\omega = 2\pi f$, f [Hz] is the angular frequency, $\Delta\varepsilon = \varepsilon_s - \varepsilon_\infty$ is the magnitude of the dispersion of the free water fraction of skin, ε_s is the permittivity

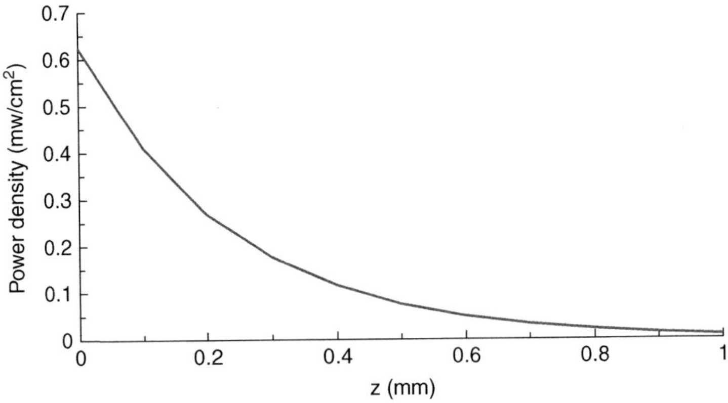

FIGURE 7.7 Attenuation of the PD in the skin at 60 GHz, normal incidence, and for an incident PD of 1 mW/cm^2.

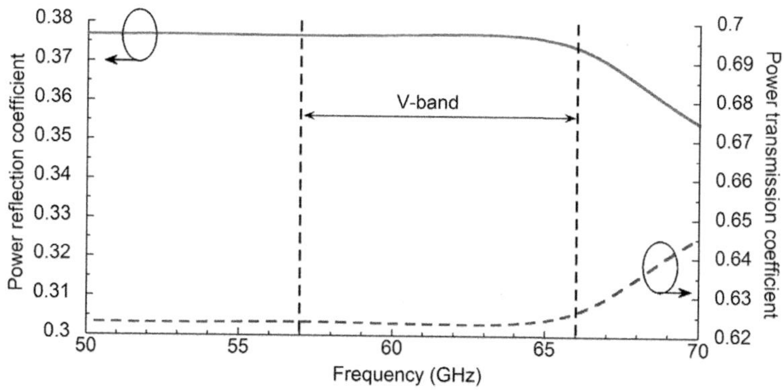

FIGURE 7.8 Power reflection and transmission coefficients in the 50–70 GHz at normal incidence.

at $\omega\tau \ll 1$, ε_∞ is the optical permittivity, $\varepsilon_0 = 8.85 \cdot 10^{-12}$, F/m is the permittivity of vacuum, and σ [S/m] is the ionic conductivity. The optimized parameters to fit the measured human skin permittivity in the 55–65 GHz range are the following: $\varepsilon_\infty = 4.1$, $\varepsilon_s = 34.8$, $\tau = 6.9 \times 10^{-12}$ s, and $\sigma = 0.7$ S/m [21]. This model allows an accurate representation of typical broadband dielectric properties of dry skin in the numerical modeling. It is important to underline that non-dispersive models can also be used to reduce computational times.

7.2.3.2 Semi-Solid Skin-Equivalent Phantom

7.2.3.2.1 Phantom Composition The first homogeneous semi-solid skin-equivalent phantom at millimeter-waves has been reported in [21]. The main components employed for its fabrication are the following:

- Deionized water. Water is the main constituent of the phantom because it is also the main skin component. It primarily determines the dispersive behavior of the phantom.
- Agar. It is employed for the retention of self-shaping, and its contribution to the phantom dielectric properties is negligible for small concentrations (typically below 4%).
- Polyethylene powder. It is used to tune the real and imaginary parts of the phantom permittivity.
- TX-151. Since the agar and polyethylene powder cannot be mixed directly, the viscosity is increased using TX-151.
- Sodium azide (NaN_3). It serves as a preservative.

The fabrication steps are the following. Deionized water, sodium azide, and agar are mixed in a kettle and heated on a stove, while the mixture is continuously stirred. When this liquid starts boiling, heating is stopped. TX-151 is sprinkled into the

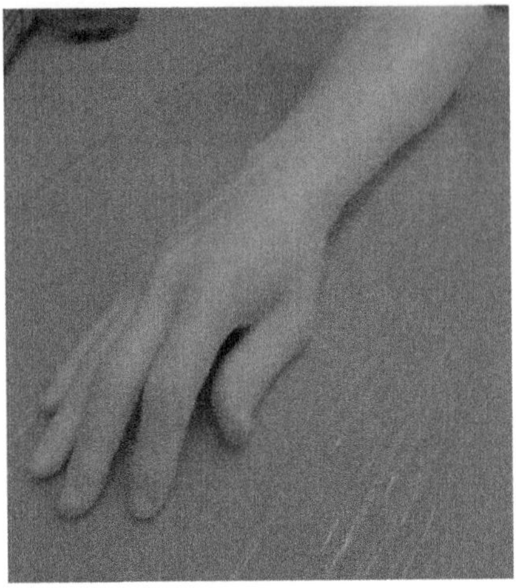

FIGURE 7.9 Skin-equivalent phantom representing a forearm and a hand.

liquid and quickly mixed. Then the polyethylene powder is added into the stirred liquid. Finally, the obtained mixture is poured into a mold and cooled in the same container for a few hours to room temperature for solidification. Using alginate gel powder, molds with realistic body-specific shapes can be manufactured for the phantom fabrication. An example of a semi-solid phantom representing a human form arm is shown in Figure 7.9.

7.2.3.2.2 Measured Dielectric Properties The measured dielectric properties of the skin-equivalent phantom and skin are compared in Figure 7.10 using an

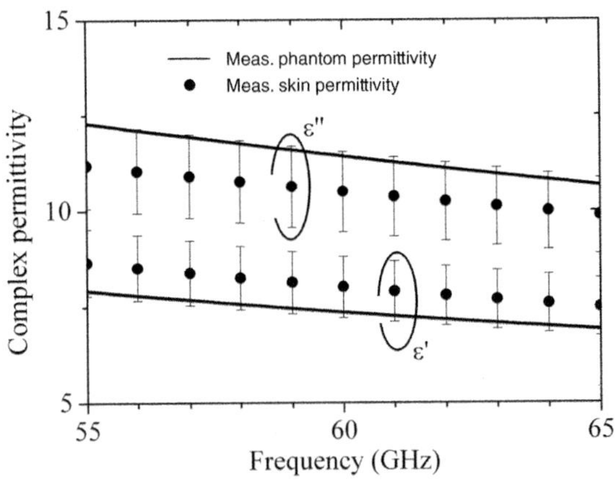

FIGURE 7.10 Dielectric properties of the skin-equivalent phantom compared to those of skin. Error bars represent ±10% of the measured skin permittivity. *Source*: Chahat et al. 2012 [21]. Reproduced with permission of IEEE.

TABLE 7.2 The dielectric properties of the proposed phantom (using two different techniques) compared to those of the reference data provided by Gabriel et al. [29]. $\Delta\varepsilon^*$ is the error relative to Gabriel et al. data.

	ε^*	$\Delta\varepsilon^*$	δ (mm)
Reference value (Gabriel et al.) [29]	$7.98 - j10.9$	–	0.48
Phantom (coaxial probe)	$7.4 - j11.4$	$7.3\% - j\,4.6\%$	0.45
Phantom (heating kinetics)	8.3 – j10.8	$4\% - j\,0.9\%$	0.49

open-ended coaxial probe. The dielectric properties of the proposed phantom are within ±10% of the measured skin permittivity. Table 7.2 compares the dielectric properties of the phantom measured using the coaxial probe and the heating kinetics technique [31] to those of the reference values provided by Gabriel et al. [29]. The measured data are in excellent agreement with the reference data and demonstrate that this phantom can be used for antenna measurement, on-body propagation, and dosimetric studies.

7.2.3.3 ***Solid Skin-Equivalent Phantom*** Currently, the existing solid phantoms are not covering the millimeter-wave band. Recently, a solid phantom has been proposed for body-centric propagation measurements in the 58–63 GHz range [26]. The phantom is designed to reproduce the electromagnetic reflection properties at the air/phantom interface similar to the ones at the air/dry skin interface. In order to satisfy this requirement, a thin lossy dielectric sheet is fabricated and metalized on one side. In the following, the composition materials, fabrication procedures and electromagnetic properties are detailed.

7.2.3.3.1 Composition and Fabrication Method The main components employed for the fabrication of a solid skin-equivalent phantom are the following:

- Polydimethylsiloxane (PDMS), a widely used silicon-based organic polymer [61].
- Carbon black glassy spherical powder used to increase losses.
- Aluminum foil used for metallization.

The first step is the preparation of PDMS where one part of curing agent is added to ten parts of silicone gel. This composite is mixed and then degassed in a vacuum chamber. Second, carbon black powder is added to the PDMS composite. The resulting composite is mixed and fully degassed. Then, the final carbon–PDMS composite is poured into a plastic container and placed into an oven for 2 hours to dry at 110°C. The final step consists of covering the lossy composite with an aluminum foil tape. Figure 7.11 illustrates a prototype of the dielectric composite and the final phantom prototype (named dielectric metallic backed phantom, i.e., DMBP).

7.2.3.3.2 Measured Dielectric Properties Phantom thickness h is optimized in order to get the same power reflection coefficient as the one at the air/skin interface.

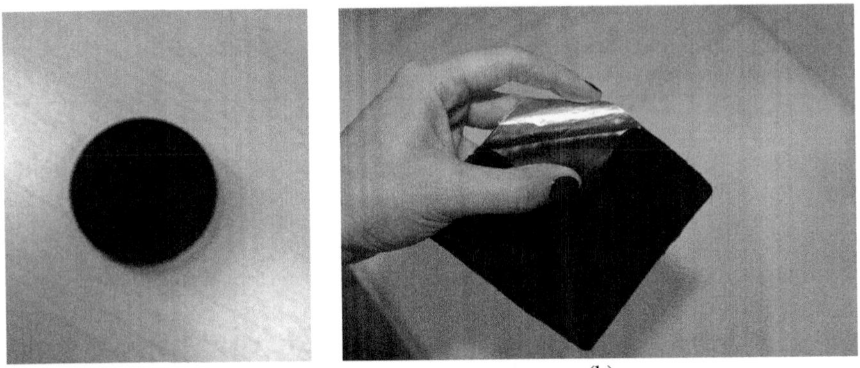

(a) (b)

FIGURE 7.11 Prototypes: (a) carbon-PDMS composite, (b) dielectric metallic backed phantom. *Source*: Guraliuc et al. 2014 [26]. Reproduced with permission of IEEE.

In Figure 7.12, the power reflection coefficient at 60 GHz is shown for a plane wave incident on a semi-infinite dry skin-equivalent phantom and on a DMBP with variable thickness. Parallel and perpendicular polarizations are considered. It is shown that an optimal solution is represented by the following combination: $\varepsilon_r = 11.6$, $\tan\delta = 0.25$, $h = 1.3$ mm. The proposed phantom can be successfully used for body-centric propagation measurements in the 58–63 GHz band.

7.2.4 Exposure Limits and Guidelines

Future millimeter-wave integrated systems must be compliant with standards (e.g., IEEE 802.11, IEEE 802.15, IEC 62209-1) based on the International Commission on Non-Ionizing Radiation Protection (ICNIRP), IEEE and European Committee for Electrotechnical Standardization (CENELEC) exposure limits and recommendations. Since at millimeter-waves the electromagnetic field exposure is superficially absorbed, the relevant dosimetric quantity is the IPD for far-field exposures. Compared to microwaves, at millimeter-waves, SAR is not used as a dosimetric quantity, but it is a useful intermediate parameter to retrieve the incident and in-tissue power density (PD). In order to protect humans exposure against possible adverse health effects, exposure limits are provided by guidelines [32, 33] and standards [34–37]. They are applied for:

- Occupational public: individuals in the course of performing their work. These individuals are fully aware of exposure, must know how to control and limit the exposure and must have exposure training knowledge. Higher exposure limits apply.
- General public: all consumer devices. There is no knowledge of exposure required. More restrictive exposure limits apply compared to occupational public restrictions.

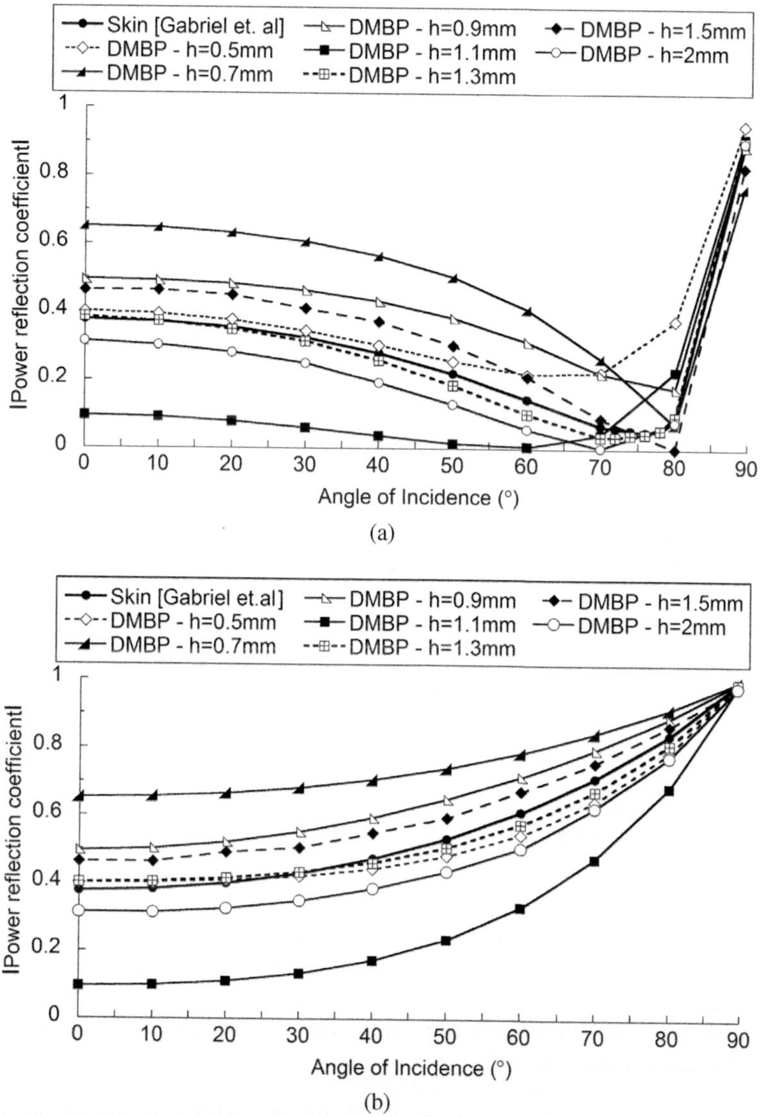

FIGURE 7.12 Power reflection coefficient at 60 GHz for parallel and perpendicular polarization. *Source*: Guraliuc et al. 2014 [26]. Reproduced with permission of IEEE.

The exposure limits are divided in:

- Basic restrictions (BRs): applied to exposure to time-varying electric, magnetic, and electromagnetic fields based directly on established health effects. The physical quantities are current density J [A/m^2], specific absorption rate SAR [W/kg], power density PD [W/m^2].

TABLE 7.3 BRs for occupational and general public [32–37].

				Averaging	
Organization	Frequency (GHz)	Public	Incident power density (mW/cm^2)	Surface (cm^2)	Time (min)
ICNIRP and CENELEC	10–300	Occupational	5	20	$68/f^{1.05}$
			100	1	
		General	1	20	
			20	1	
IEEE	30–300	Occupational	10	100	$2.524/f^{0.476}$
			100	1	
	30–100	General	1	100	$25.24/f^{0.47}$
			100	1	

f, frequency

- Reference levels (RLs) [32,33] or maximum permissible exposure (MPEs) [37] are provided for practical exposure assessment to determine whether the basic restrictions are likely to be exceeded. Are determined from the BRs using measurements or computations. The physical quantities are electric field strength E [V/m], magnetic field strength H [A/m], magnetic flux density B [Wb/m^2], power density S [W/m^2], current flowing through the limbs I_L [A], contact current I_C [A], and specific energy absorption SA [J/kg] (i.e., the time integral of SAR considered for exposures to an EM pulse).

Table 7.3 summarizes the exposure limits in terms of power density for the three main organizations and for both occupational and general public.

It should be underlined that ICNIRP and CENELEC (which adopted the ICNIRP recommendations) provide exposure limits also in terms of *E*-, *H*- and *B*-field for reference levels (RLs). These are summarized in Table 7.4.

The three organizations ICNIRP, IEEE, and CENELEC considered scientific literature data to establish the exposure limits. Some important considerations are here presented, and the particularities with respect to millimeter-waves are underlined in detail in this Review document [38].

Whole-body resonance The exposure limits for RLs/MPEs are based on well-established literature findings, which showed that the whole human body absorbs

TABLE 7.4 RLs for occupational and general public [32–36].

Organization	Frequency (GHz)	Public	*E*-field (V/m)	*H*-field (A/m)	*B*-field (T)
ICNIRP and CENELEC	2–300	Occupational	137	0.36	0.45
		General	61	0.16	0.20

different rates of electromagnetic field depending on the frequency [39–41]. It was shown that whole-body-averaged SAR values approach maximum values when the E-field vector is parallel with the longest axis of the body, and is four-tenths of a wavelength of the incident field. For example, at 2.45 GHz, a standard man (height = 1.75 m, weight = 70 kg) absorbs about half of the incident energy. For the standard man the maximum absorption is at about 70 MHz, which represents the whole-body resonance [42]. Whole-body resonance and SAR frequency dependence have been taken into account to establish the recommended RLs/MPEs. Above 10 GHz, the absorption is quasi-optical and body resonance considerations do not apply. At millimeter-waves, the penetration depth is much less than 1 mm.

Temperature considerations The BRs limits at millimeter-waves for local exposure rely on scientific published results at microwaves on temperature increase in the eye and the potential adverse health effects produced by it. A temperature threshold in the eye was established at 41°C (i.e., a temperature increase of about 3–4°C) over which the cataract may appear. Most of the studies showed that the maximum eye temperature increase (~4°C for a SAR of 10 W/kg at 2.45 GHz [43]) both in near-field and far-field exposures is below the threshold and does not induce any adverse effects.

Current studies performed at millimeter-waves on the skin showed that exposures at the recommended limits will not increase its temperature by as much as 1°C [44–49].

Time considerations Below 10 GHz, the standards used 6 min averaging time, which corresponds to a time constant for partial body heating. Above 10 GHz, the averaging time recommended is:

- $68/f^{1.05}$ (CENELEC, ICNIRP): about 0.92 min at 60 GHz, 0.74 min at 74 GHz, and 0.65 min at 84 GHz.
- $25.24/f^{0.476}$ a less restrictive value recommended by IEEE: about 3.6 min at 60 GHz, 3.25 min at 74 GHz, and 3.06 min at 84 GHz.

Spatial considerations The rationals for averaging surface area used in the exposure levels are the following:

- 20 cm^2 correspond to the surface area of a cube with edge dimension of 1.8 cm, surface correlated with the human eye, where some biological effects have been noticed (i.e., cataract). The eye has the following dimensions: at birth is about 1.6–1.7 cm in diameter; at 13 years old the eye attains the full size, from anterior to posterior, about 2.4 cm diameter, a volume of about 6 cm^3 and a mass of 7.5 g [50].
- 1 cm^2 corresponds to the surface area of the cornea, where some effects like cataract may appear. The exposure limits averaged over 1 cm^2 should not exceed 20 times the values averaged over 20 cm^2.

- As previously notices, IEEE recommendations are less restrictive. The averaging surface area is 100 cm^2, which corresponds to the surface of human face/hand area.

7.3 ANTENNAS FOR OFF-BODY COMMUNICATIONS AT MILLIMETER-WAVES

7.3.1 From Microwaves to Millimeter-Waves

At microwaves, antennas placed in close proximity to a lossy medium experience strong power absorption, radiation pattern distortion, shift in resonance frequency, and changes in input impedance, for example [1] and [51–54]. Therefore, when located close to the human body, wearable antennas need to be designed to operate in a robust way so that the influence of the body on the antenna performance is minimized.

Microstrip patch antennas have minimal sensitivity to the body proximity. They are low-profile, low-cost, easy to manufacture, and potentially conformal to the human body. Many wearable antennas have been proposed in the literature at microwaves under diverse forms: textile antennas [54–56], flexible antennas [57, 58], or stretchable antennas [59, 62]. The patch antenna shown in Figure 7.13a is made of Felt and electro textile. The patch antenna represented in Figure 7.13b is based on embroidered conductive metal-polymer fibers (*E*-fiber) on polymer–ceramic composite.

Although patch antennas are very robust on the body, they are limited in bandwidth. An alternate solution for multiple frequency operation with wide bandwidth and reduction of the human body effect is to use electromagnetic band gap (EBG) structure [63, 64]. The antenna represented in Figure 7.13c is an example of prototype developed to operate at two frequency bands [63]: from 2.4 GHz to 2.69 GHz (ISM, Bluetooth, Zigbee, WiFi, and LTE) and from 5.15 GHz to 5.875 GHz (WiFi).

At microwaves, a high level of fabrication accuracy can be achieved for textile antennas (see Figure 7.14). Two textile antennas developed at IETR, University of Rennes 1, are shown in Figure 7.14. The measured and calculated reflection coefficients of the two textile patch antennas are shown in Figure 7.14c. They are in excellent agreement which demonstrate the excellent fabrication accuracy achieved at microwave. Conductive elements of the antennas are cut from *ShieldIt*^{TM} *SUPER* using a laser cutter (ProtoLaser S, LPKF, OR, USA). This electro textile is a rugged rip-stop polyester substrate, plated with Nickel and Copper, and coated on one side with a non-conductive hot melt adhesive activated at 130°C. Hence, the metallized fabric pattern can be easily adhered to cotton or other fabrics by applying heat with a thermal press or even the laser itself. Fabrication tests were performed using a laser cutter on *ShieldIt*^{TM} *SUPER* at IETR. The laser provides fabrication conditions leading to a cutting accuracy of about ±30 μm and minimum line width of 0.5 mm using this electro textile (Figure 7.15). Obviously, this level of accuracy is not enough for millimeter-wave antennas.

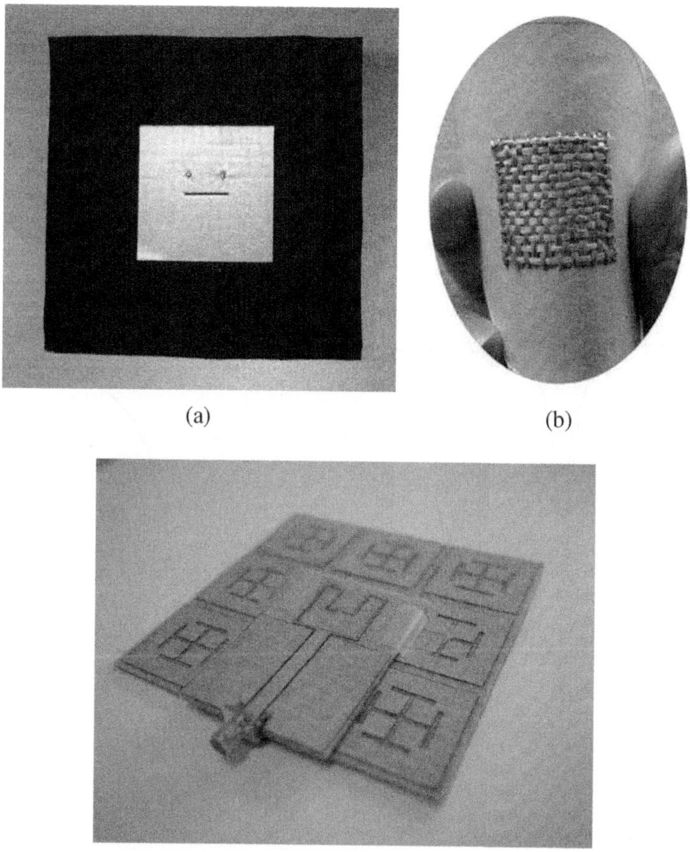

(a) (b) (c)

FIGURE 7.13 (a) Patch antenna with dual polarization operating in the 2.45 ISM band. *Source*: Vallozzi et al. 2008 [55]. Reproduced with permission of IEEE. (b) Patch antenna built on embroidered *E*-fibers and polymer composites at 2.2 GHz. *Source*: Wang et al. 2012 [57]. Reproduced with permission of IEEE. (c) Dual-band wearable textile antenna on an EBG substrate.

Improvements were recently reported for the embroidery process allowing to fabricate highly flexible antennas on PDMS as shown in Figure 7.13b. The achieved geometrical accuracy is 0.3 mm [58] which is still far from what is obtained using a laser machine on *ShieldIt* (i.e., ± 30 μm).

At millimeter-waves, the electromagnetic coupling between antennas and the human body is not expected to be the same as at microwaves. It was shown that about 40% of power is reflected at the air/skin interface for the normal incidence at 60 GHz [8]. The effect of the body on the antenna performance is studied below by considering two patch antenna arrays.

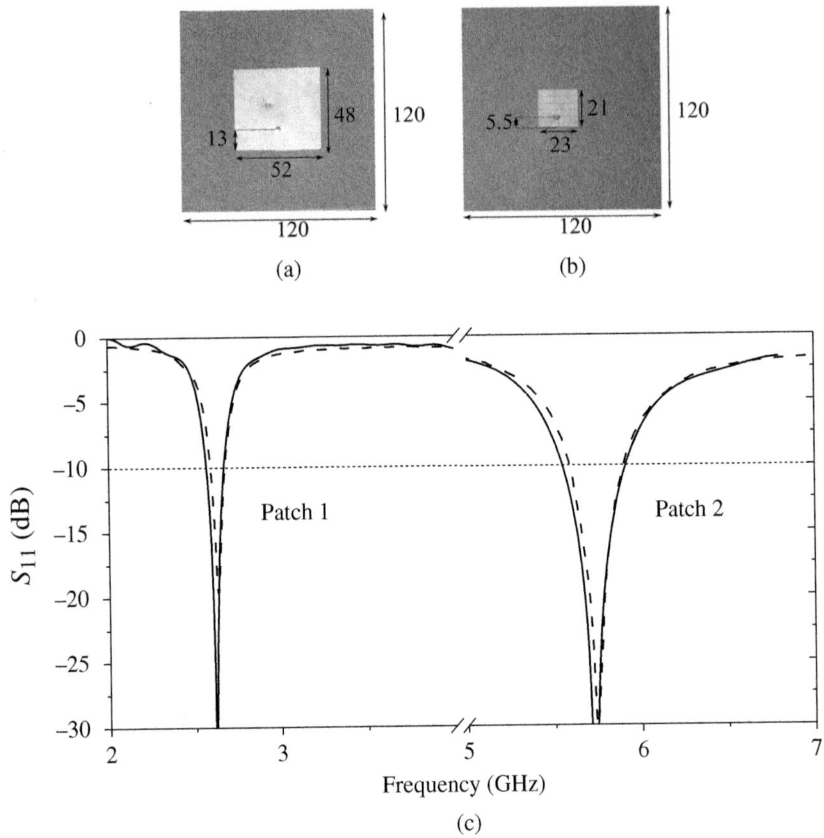

FIGURE 7.14 Textile patch antennas operating at (a) 2.6 GHz (patch 1) and (b) 5.7 GHz (patch 2). (c) Reflection coefficient demonstrating an excellent agreement between (——) simulated and (- - -) measured results.

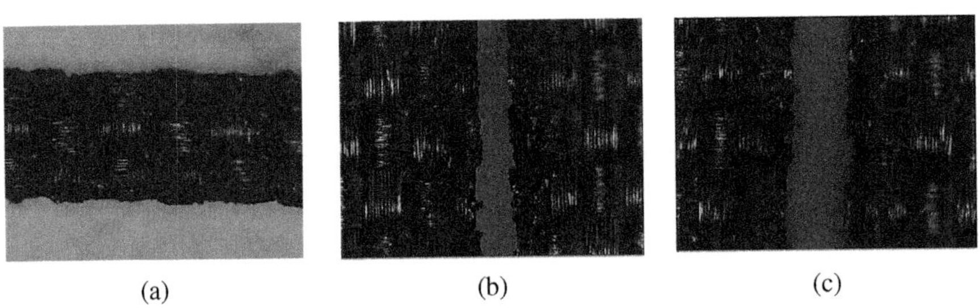

FIGURE 7.15 Test of accuracy using the laser system on *ShieldIt*^{TM} Super. (a) Minimum width line achievable (*width* = 0.52 ± 0.03 mm). (b) Minimum gap with demonstration of discrepancies due to the electro textile (*gap* = 0.15 ± 0.03 mm). (c) Larger gap with satisfactory accuracy (*gap* = 0.25 ± 0.03 mm).

7.3.2 Antenna Requirements for Off-Body Communications

Wearable antennas have to be as compact as possible and should be easily integrated with the transceiver. They need to be efficient with minimal power absorption inside the human body that behaves as a highly lossy dispersive dielectric material at millimeter-waves. They also have to be light-weight, low-cost, and preferably fully integrated in garments. Because of the high atmospheric attenuation at 60 GHz and limitations on the radiated power, medium-gain antennas (~12 dBi) are often required [65]. Whereas in controlled environments, line-of-sight (LOS) channels can be efficiently exploited using medium-gain passive antennas; directive beam steering antennas are desirable for non-line-of-sight (NLOS) channels [65–69].

7.3.3 Study of the Interaction with the Human Body of a Patch Antenna Array

7.3.3.1 Antenna Model To study the effect of the human body on the antenna performance, simple scenarios have been considered in the literature. As an example, a microstrip-fed 2 × 2 patch single-layer antenna array was developed [70]. The latter is printed on a thin RT Duroid 5880 substrate ($h = 127$ μm, $\varepsilon_r = 2.2$, $tan\delta = 0.003$). The layout is represented in Figure 7.16a. The antenna is linearly polarized along y-direction, and, for measurement purposes, it is mounted on a 3 mm-thick ground plane (Figure 7.16b) to avoid significant substrate bending and to achieve an accurate and stable placement of a V-connector. In future BAN applications, such types of antennas are expected to be directly integrated into the clothing or wearable devices.

7.3.3.2 Antenna Performance The reflection coefficients measured in free space and on a skin-equivalent phantom demonstrate insignificant variations, Figure 7.17. It is worthwhile to emphasize that at microwaves, a resonance shift is usually observed.

In addition, the radiation patterns in E- and H-planes were also measured in free space and on a skin-equivalent phantom (Figure 7.18). It can be seen in Figure 7.18

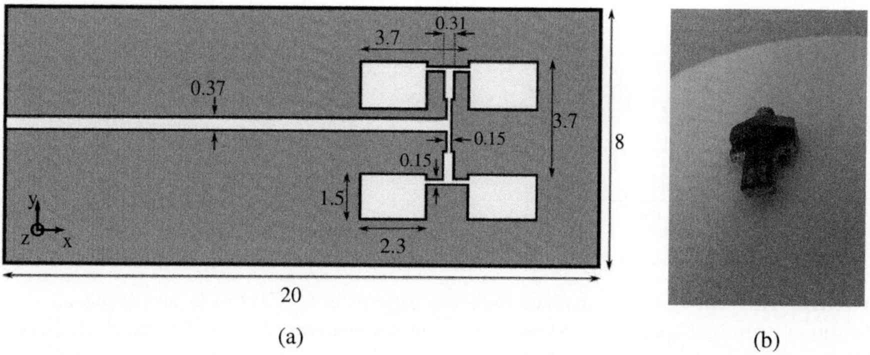

FIGURE 7.16 2 × 2 patch single-layer antenna array at 60 GHz. (a) Schematic representation of the antenna model and dimensions in millimeter. (b) Manufactured patch array antenna with a V-connector on a skin-equivalent phantom.

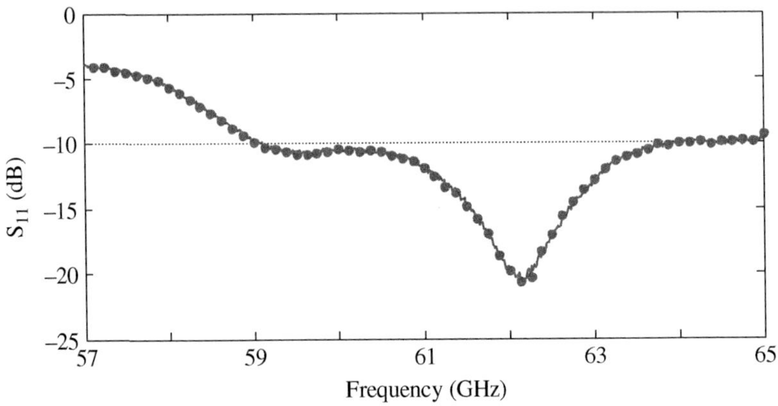

FIGURE 7.17 Measured reflection coefficients of the antenna. —— In free space. ••• On the skin-equivalent phantom.

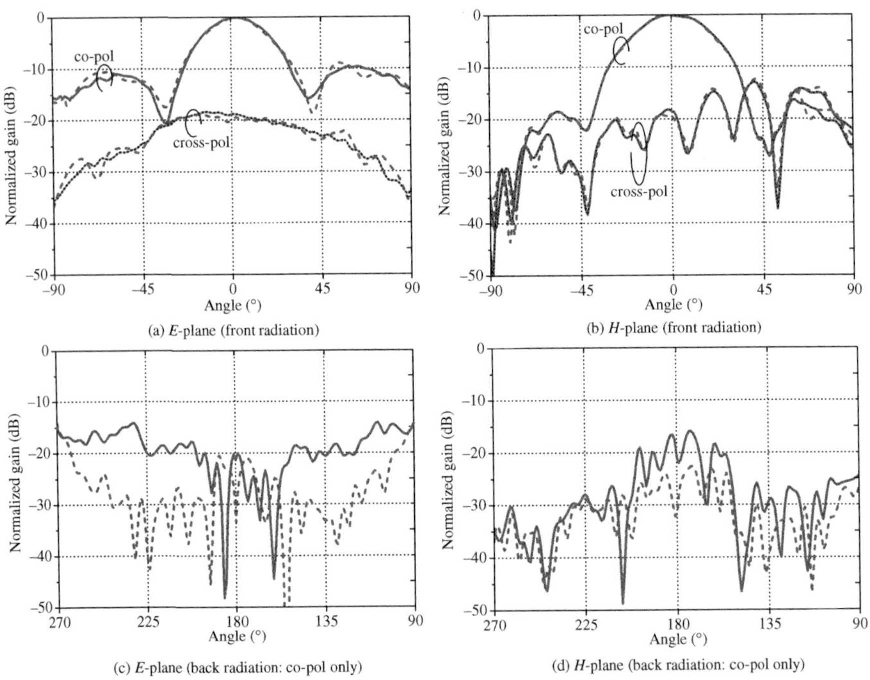

FIGURE 7.18 Measured normalized radiation patterns at 60 GHz in *E*- and *H*-planes. — Measurement in free space. - - - Measurement on the phantom. *Source*: Chahat et al. 2012 [70]. Reproduced with permission of IEEE.

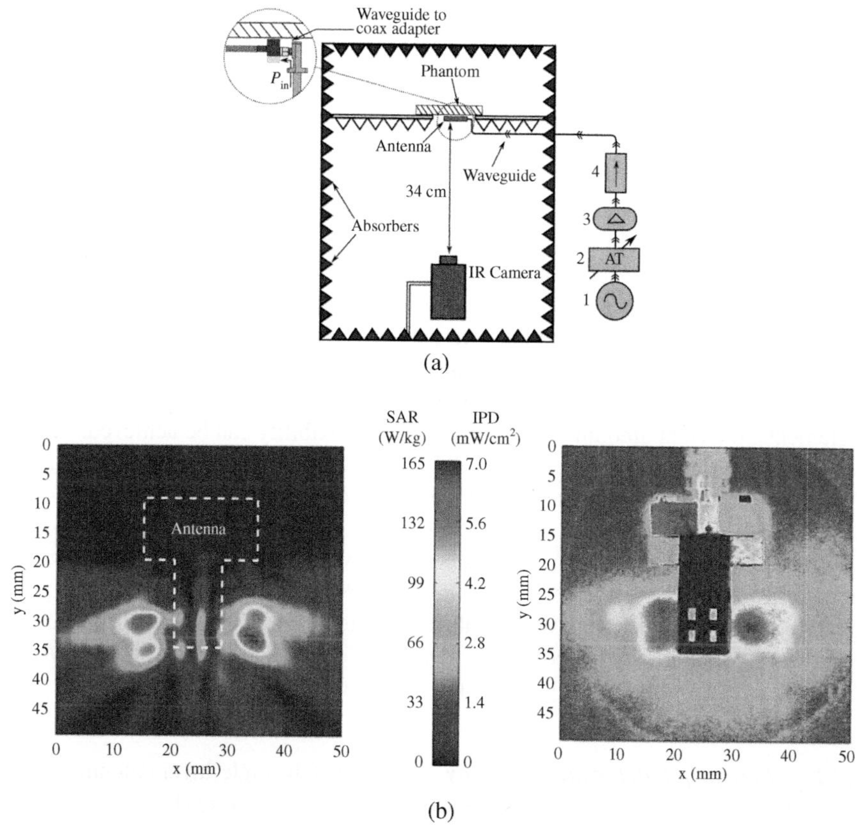

FIGURE 7.19 (a) SAR measurement setup. (b) SAR and IPD distributions at 60 GHz. (Left) Numerical results for the antenna on the skin. (Right) Measurements on the skin-equivalent phantom. P_{in} = 322 mW. *Source*: Chahat et al. 2012 [70]. Reproduced with permission of IEEE. For a color version of this figure, see the color plate section.

that front radiations remain very slightly affected. The backward radiation was measured separately in both configurations, whereas the measured level on the phantom is mainly reduced in *E*-plane (Figure 7.18c) and due to absorption and reflection, it remains very slightly affected in *H*-plane (Figure 7.18d). This is an expected result since absorption is higher when the *E*-field is parallel to the phantom surface [8]. These observations are in agreement with the calculated and measured SAR and IPD, as shown in Figure 7.19. More details regarding the measurement methodology can be found in [70].

7.3.4 Textile Antennas

Textile wearable antennas for body-centric communications have been attracting increased interest, owing to their huge potential benefits for a large set of daily life applications. However, on-textile fabrication process is very challenging at millimeter-waves, especially due to the roughness of the textile surface and the size of textile fibers and electro textiles with respect to the geometrical dimensions of the metallic patterns. Different techniques have been explored to solve the challenges of fabricating millimeter-wave antenna arrays on commercially available textiles [71,72].

The first challenge in developing a reliable fabrication process for textile antennas is finding fabric capable of supporting the tight electrical and mechanical requirements of millimeter-wave technologies. Indeed, the choice of a substrate thickness, dielectric constant ε_r, and loss tangent $tan\delta$ is essential when it comes to millimeter-waves in order to reduce losses and also to enhance the efficiency.

With the combined use of *ShieldIt* Super (a nickel- and copper-plated rip-stop nylon) as a ground plane and laser-machined copper foils for feed lines and radiating elements, the right amount of performance and flexibility can be achieved. More details about the fabrication process of millimeter-wave textile antennas are available in [71] and [72]. Examples of fabricated antennas and microstrip lines are provided in Figure 7.20.

Textile dielectric properties are key data for any antenna design. Measurement results retrieved using the open-stub technique are given in [71]. This methodology is applied to a 0.2 mm thick cotton woven fabric extracted from a shirt. The performance properties of this fabric extracted from a shirt exhibit an insertion loss of 1.6 dB/cm at 60 GHz, a relative permittivity of 2.0 and a loss tangent of 0.02.

7.3.4.1 *Microstrip Patch Antenna Array*

A four-patch single-layer antenna array printed on the 0.2 mm thick textile was introduced in the literature (Figure 7.21a) in [71]. A 15.2 mm-long microstrip line feeds the array to avoid too strong reflections from the feeding connector (Figure 7.21). Of course, as textiles are more lossy than classical substrates, it is recommended to reduce the access line length as much as possible. Whereas the antenna could be fed using a central probe, the microstrip feed line is the easiest solution to perform measurements on textile. The fabricated antenna integrated with a V-connector is shown in Figure 7.21b.

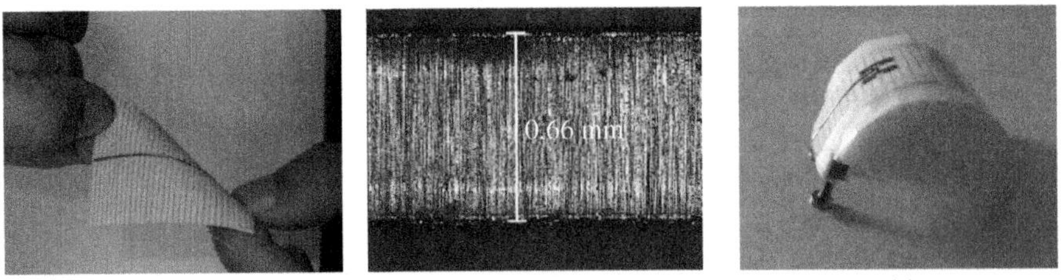

(a) (b) (c)

FIGURE 7.20 Examples of fabricated textile antennas and microstrip lines. (a) 50 Ω microstrip line. (b) Zoom on the microstrip line. (c) Textile patch array antenna. *Source*: Chahat et al. 2013 [71]. Reproduced with permission of IEEE.

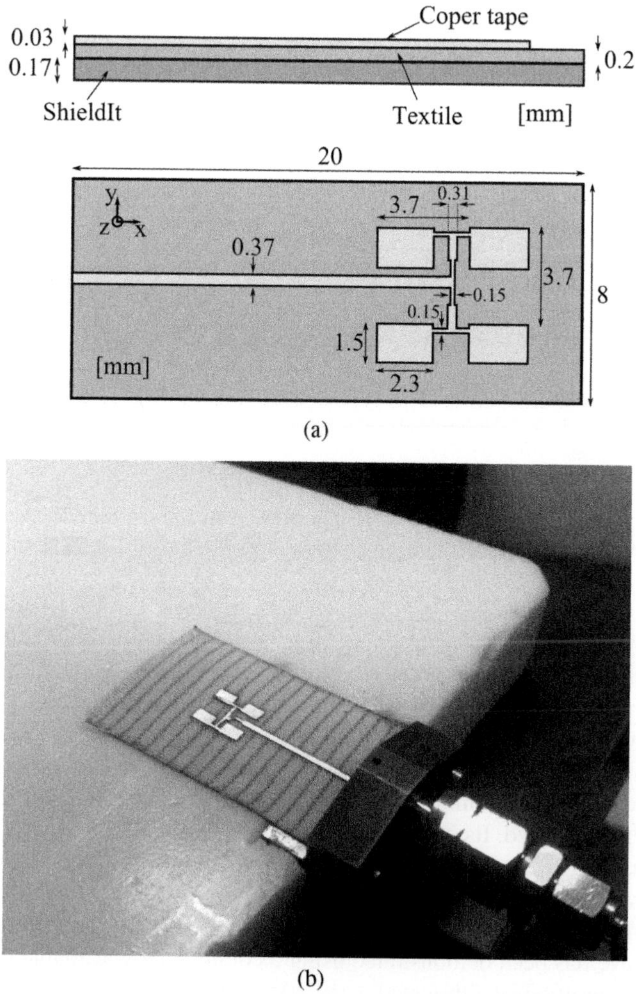

FIGURE 7.21 (a) Layout of the microstrip antenna array printed on textile. (b) Measurement set-up on the skin-equivalent phantom for a distance between the ground plane and the phantom equals to $d = 0$ mm. *Source*: Chahat et al. 2013 [71]. Reproduced with permission of IEEE.

The reflection coefficient S_{11}, measured using a 110-GHz Agilent 8510XF VNA, is plotted in Figure 7.22. Excellent agreement is obtained between simulated and measured results. As can be seen, excellent reproducibility can be achieved. In addition, the reflection coefficient is very slightly affected over the skin-equivalent phantom (Figure 7.22).

In addition, the gain, directivity, and efficiency of this textile antenna have been characterized in free space and on the skin-equivalent phantom for different antenna/body separation d (Table 7.5). High losses are experienced in the feed line

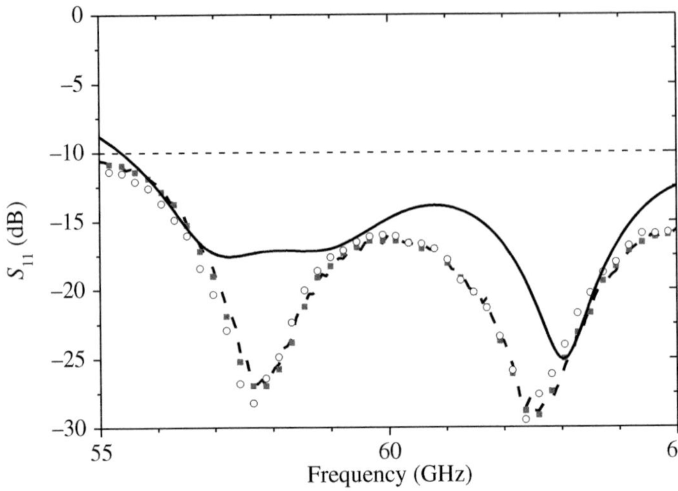

FIGURE 7.22 Effect of the phantom on the reflection coefficient. — Simulated in free space. Measured in free space - - - . Measured on phantom with $d = 5.6$ mm ■ ■ ■ and $d = 0$ mm ○○○.

(about 3.3 dB) whose length was selected to facilitate the measurement. In practice, this length should be shortened or even suppressed (i.e., using central coaxial feed probe) to increase the antenna gain and efficiency.

The antenna performances (i.e., reflection coefficient and radiation) were tested after a number of hand washing cycles. The antenna was measured before and after washing when fully dried. Its performance remained stable.

Based on computed and measured results, off-body antennas operating at millimeter-waves are very slightly sensitive to the human body. It was shown that textile antennas are feasible at millimeter-wave. A patch antenna array operating at millimeter-wave has been demonstrated using a commercially available textile. The textile antenna prototypes, fabricated using a simple and industry-compatible manufacturing process, demonstrate excellent flexibility capabilities simplifying integration into clothes.

TABLE 7.5 Comparison of antenna performances in terms of gain, directivity, and efficiency.

	Gain (dBi)		Directivity (dBi)		Efficiency (%)	
	Simulation	Measurements	Simulation	Measurements	Simulation	Measurements
Free space	8.6	8.0	12.1	11.9	45	41
On-body ($d = 0$ mm)	8.1	7.7	12.1	12.1	39.8	36.3
On-body ($d = 5$ mm)	8.3	7.8	11.9	11.7	43.6	43.7

7.4 ANTENNA AND PROPAGATION FOR ON-BODY PROPAGATION

7.4.1 State-of-the-Art

A new wireless sensors generation is under development in the area of wireless body-centric communications. One of the target applications is in the medical field, where their usefulness has been proven in replacing the existing wired telemetry systems, for example [73]. Extensive research has been performed at microwaves, for example [73–85]. In particular, in assisted living applications where it is important to monitor and recognize people's activities [74, 75]; in kinesiotherapy where physical patients therapy needs to be monitored after a stroke, joint replacements or reconstructions, spinal cord injury, motor function disability due to Parkinson's disease, and dementia [76–81]. Using wireless wearable sensors can significantly improve patient's health-care quality and efficiency. Continuous and remote patients monitoring would allow them to have a normal life, reducing regular visits to the hospital. Other applications of wearable sensors are in the military field (monitor physiological status and position of soldiers on the field), entertainment, sports, etc.

In a typical scenario, the network topology consists of wearable wireless transceivers strategically placed on the human body (on-body network) communicating with any other node located in the surrounding environment (off-body) without any wired connections [73–82]. The on-body sensors can be accelerometers, gyroscopes, vital sign sensors—-blood pressure, temperature sensors, ECG sensors, EMG (electromyogram), glucose sensor (transmitter, TX), and insulin pump sensor (receiver, RX). Some examples are shown in Figure 7.23' in particular, a "Garment integrated wireless sensor network" used for rehabilitation after a hip replacement [77]. Accelerometers and magnetic sensors are integrated on a textile pair of pants, able to measure the posture and the load placed on the operated load (necessary to gain muscle strength). Data are collected and processed by a central unit and send to a PC via Bluetooth connection. In [83], an upper limb kinesthetic garment (ULKG) able to detect the posture of the wrist, elbow, and shoulder was developed to allow patients to continue the rehabilitation training at home. A sensing glove can be used to detect posture and gesture of the hand and fingers [84]. A glucose sensor and an insulin pump sensor used by diabetic patients [85] are sketched in the same figure.

In this context, a proper propagation channel characterization is required to establish the proper placement and position of the sensors on the body. Analytical, numerical, and experimental studies have been performed at microwaves [19, 53, 86–93], and more recently at millimeter-waves [4, 8, 73, 94–106]. These studies showed that:

- electric field normal to the body surface improves the propagation;
- propagation depends on the antenna/body separation, antenna polarization, angle of arrival of the received signal;
- a proper alignment between TX and RX nodes is required;
- an LOS channel model is more appropriate, especially at millimeter-waves due to the strong channel attenuation in the presence of the human body and due to the limited distance between the communication nodes;

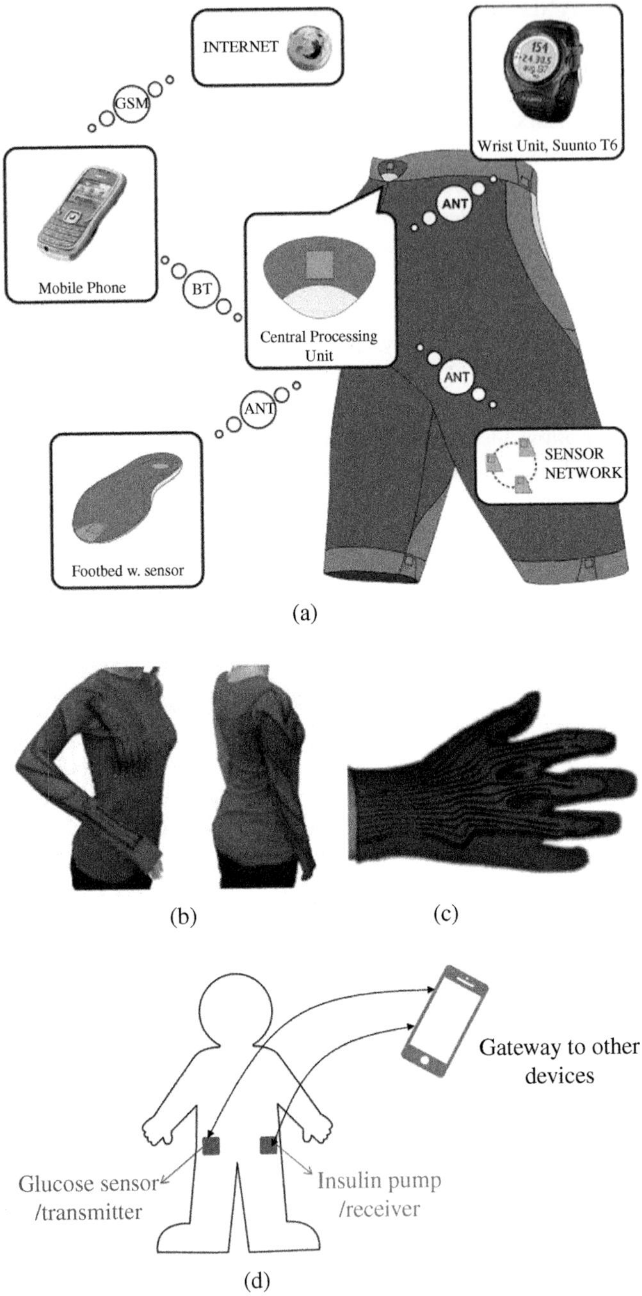

FIGURE 7.23 (a) HipGuard pants for hip patient rehabilitation (from: [77]: © *2008 IEEE*). (b) Upper limb kinesthetic garment prototype (from: [84]: © *2005 IEEE*). (c) Sensing glove prototype (from: [84]: © *2005 IEEE*). (d) Closed-loop insulin delivery.

- human body daily activities influence the propagation properties;
- attenuation at millimeter-waves is much stronger than at microwaves;
- TX-RX separation distance at millimeter-waves should be small;
- signal fading is caused by shadowing and polarization mismatch.

For upcoming body-centric systems at millimeter-waves, it is important to accurately quantify the effect of clothing on wave propagation. So far, only few studies are available in the literature. In particular, [8] provides an analytical solution for normal incidence, while in [107] a numerical investigation restricted to the transmission coefficient for a normally incident plane wave in the presence of a textile is presented. Recently, in [97] an analytical, numerical, and experimental study on the textile effect on the millimeter-wave propagation has been investigated, while in [98] it has been shown how an electro textile can improve the propagation, especially in the case of an NLOS scenario.

7.4.2 On-Body Antennas

7.4.2.1 Requirements Common requirements for any wearable antennas are the following: light weight, low profile, low cost, and ideally conformal to the body's surface. They should also be as insensitive as possible to the human body and minimize the exposure of users to radiation. On-body wearable antennas should exhibit a radiation pattern and polarization minimizing the path-link loss. At 60 GHz, the propagation loss in free space is very large and, according to on-body channel measurements performed using monopole antennas, typical path loss on the human body varies between 57 dB and 88 dB, depending on the link [99].

For on-body communications, maximum radiation should be tangential to the body surface to maximize coupling between body-worn devices. Previous studies showed that a monopole antenna gives very good results for on-body applications at lower frequencies [86], but it does not provide high enough gain, as required in the 60 GHz band. Several antenna concepts have been proposed at millimeter-wave to overcome this issue: Yagi-Uda antennas [73, 100, 101] and woodpile EBG-based antennas [102, 103]. While the latter is complex to fabricate and therefore expensive, the Yagi-Uda antenna appears to be a very attractive solution.

7.4.2.2 Yagi-Uda Antennas for On-Body Communications Yagi-Uda antennas have been proposed as an appropriate solution for on-body communications. They offer a good compromise in terms of size and gain performance compared to other end-fire antennas such as tapered-slot antennas. Figure 7.24 shows a printed Yagi-Uda array that consists of a 1×4 Yagi-Uda array printed on a 0.127 mm thick RT/Duroid 5880 substrate ($\varepsilon_r = 2.2$) [100]. It demonstrates a gain of 15 dBi at 60 GHz.

Another type of Yagi-Uda antenna was proposed for on-body communications at 60 GHz [96]. It consists of a row of vias, acting as directors, placed in front of the output of a substrate integrated waveguide (SIW) (see Figure 7.24b). These vias are loaded at the top and bottom with printed strips to guarantee that the equivalent length

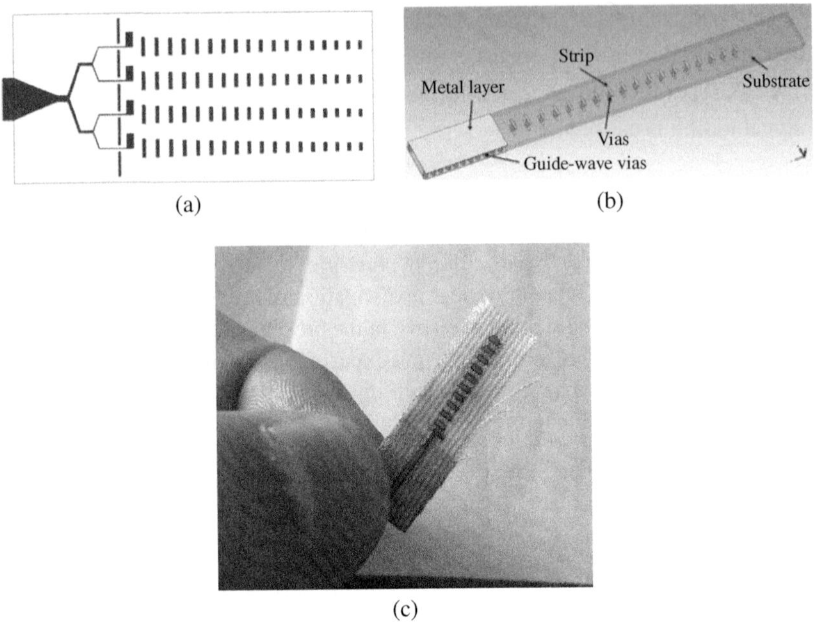

FIGURE 7.24 (a) A 1 × 4 printed Yagi-Uda array. (b) SIW Yagi-Uda antenna. *Source*: Pellegrini et al. 2013 [96]. Reproduced with permission of IEEE. (c) Textile Yagi-Uda antenna. *Source*: Chahat et al. 2012 [72]. Reproduced with permission of IEEE.

of the directors is close to a half-wavelength. The measured gain equals to 12.5 dBi at 60 GHz. The substrate used here is RT/Duroid 5880 ($\varepsilon_r = 2.2$) with a thickness of 0.787 mm.

Moreover, a textile Yagi-Uda antenna has been designed and characterized [72]. The permittivity of the textile substrate has been characterized and the Yagi-Uda antenna has been fabricated using an *ad hoc* manufacturing process. Its performance in terms of reflection coefficient, radiation pattern, gain, and efficiency has been studied in free space and on a tissue-equivalent phantom. It is shown that the antenna is matched in the 57–64 GHz band. Its measured on-body efficiency and maximum gain equal 48.0% and 11.9 dBi, respectively.

The performances of such antennas have been studied on the human body [96]. An example of on-body to off-body communications using a simple Yagi-Uda antenna is shown in Section 7.0 along with the human body effect on the antenna performance.

7.4.2.3 Woodpile EBG-Based Antennas The woodpile antenna consists of resonant cavity fed by monopole-like feed source. The woodpile structure improves the gain by more than 5 dB compared to a conventional monopole antenna. The antenna is shown in Figure 7.25 along with its radiation pattern. Such antennas

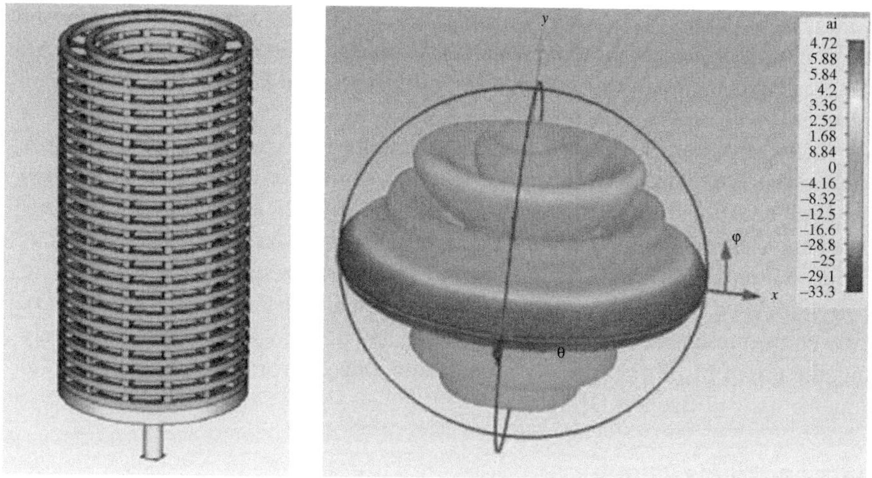

(a) (b)

FIGURE 7.25 (a) Woodpile textile antenna. (b) Radiation pattern at 60 GHz. *Source*: Brizzi et al. 2012 [103]. Reproduced with permission of IEEE.

are bulky (and expensive which does not make them ideal candidates for on-body communications.

7.4.3 Theoretical Analysis of On-Body Propagation

As already shown in the previous paragraphs, antennas for on-body applications are placed in an extremely complex scenario, where several important parameters have a strong influence on the features of the overall system. While the use of equivalent phantoms allows for a realistic testing of prototype behavior, an earlier stage analysis is always required before the prototyping phase. Effective and reliable numerical modeling tools are then of paramount importance to identify and optimize the relevant design parameters of the system.

7.4.3.1 Full-Wave Models The most accurate strategies to model on-body propagation are full-wave numerical schemes, rigorously solving Maxwell's equations for a radiating device in the presence of a human body. To this aim, an electromagnetic model of the body (describing its geometrical and physical parameters) must be available (see Section 7.0), and discretized through an appropriate mesh. These analyses, already performed at low frequencies, have recently been applied to millimeter-wave scenarios. The most common full-wave approach is probably the finite-difference-time-domain (FDTD), consisting in an iterative solution of time-domain Maxwell's equations, where partial derivatives are replaced by finite difference expressions. FDTD is particularly well-suited to study electromagnetic body interactions, characterized by complex shapes and inhomogeneous materials, thanks to its flexibility and versatility. On the other hand, due to the large electric dimensions of human bodies, the number of unknowns becomes unbearable at millimeter-waves even for modern parallel computation, thus limiting the analysis to only a part of the human body. An example of this methodology is found in [110], where the head-to-shoulder link is investigated with an FDTD approach.

An alternative numerical method, which could lead to a reduction of the number of unknowns, is the Method of Moments (MoM). MoM is based on the discretization of integral equations derived by Maxwell's partial differential equations by means of the equivalence theorem. Unknown currents are placed on suitable surfaces, thus leading to a reduction in the number of unknowns with respect to methods where the discretization is performed over the whole domain under study (as in FDTD). However, the presence of non-canonical shapes and inhomogeneous bodies severely limits the use of MoM for on-body propagation, unless hybrid methods are used to overcome these difficulties [108].

7.4.3.2 ***Simplified Analyses*** In order to simplify the complexity of these problems, semi-analytic schemes have been proposed, for example, by Dubrovka et al. [109], through a numerical synthesis of the Green's function of a dipole radiating in the presence of a human body. The Green's functions can be precalculated through full-wave schemes and re-used to compute radiation patterns for different antennas, provided that the human body considered is kept fixed. While the cited work is applied at microwave frequencies, the same approach could be in principle used at millimeter-waves.

Alternative approaches go toward further simplifications of the problem. Still relying on the full-wave solution of an electromagnetic problem, canonical models can be used for human body (e.g., a lossy circular cylinder in [110]). Other approaches are based on specific hypotheses holding at millimeter-waves, where electrical dimensions of human bodies are large, and ray tracing can be an effective approach [110].

These simplified methods are certainly attractive for their reduced complexity, but unfortunately do not grant physical insights in wave propagation mechanisms, since they still rely on the full-wave numerical solution of electromagnetic problems, however simplified they may be.

7.4.3.3 ***Analytical Methods*** For this reason, analytical models have been proposed, originally developed with reference to different applications (e.g., radio-frequency propagation along the earth). The attractiveness of these approaches is due to the availability of closed-form results, not requiring further numerical simulations and leading to an analytic estimation of important parameters (path loss, depolarization effects, etc.). Furthermore, these results can also be interpreted as a clear physical description of the wave species propagating along the body, each one being characterized by its amplitude, propagation direction, attenuation, and polarization. For these reasons, these methods had already been proposed in the framework of on-body propagation at lower frequencies, and have recently been adapted to millimeter-wave propagation.

Unfortunately, simple analytical results can be obtained only in the presence of elementary sources radiating in the presence of canonical-shape obstacles. Two main simplified body models have then been proposed: an infinite flat body or a circular-section cylinder. The first geometry can model propagation along the trunk, while the second should describe propagation around the trunk, the arms, and the legs. In the following these two cases are briefly described. Other more complex geometrical configurations have also been proposed, such as a combination of circular and elliptical cylinders modeling the arms and the trunk, respectively, yielding integral expressions for the attenuation of the relevant waves [111].

7.4.3.3.1 Skin Modeled as a Flat Shape Let us consider an arbitrarily oriented elementary dipole radiating over the human body, modeled as a flat dielectric, laterally unbounded (see Figure 7.26). The skin loss is large enough to apply the approximations proposed by Norton in his seminal papers [112–114]. Through a suitable generalization to *layered* lossy media [117], this approach has firstly been applied in the range 2–10 GHz [115, 116], since the penetration-depth values require the modeling of several skin layers at low frequencies. On the other hand, at millimeter-waves only the first skin layer is relevant, and the original results by Norton can straightforwardly be applied.

According to Norton results, the excited field, measured at more than a wavelength from the dipole, can be expressed as [118]:

$$E(r) = E^{GO}(r) + E^{N}(r), \tag{7.2}$$

where E^{GO} is the geometrical-optics field (GOF), due to the direct radiation of the source and the first reflection from the flat surface; E^{N} is the so-called Norton wave, whose contribution can be dominant over the GOF near the surface. Hence, E^{N} can be considered as the correction term to the GOF needed for the total field to respect the

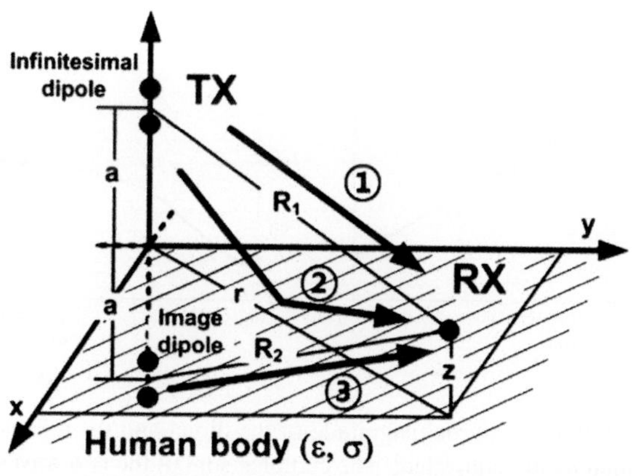

FIGURE 7.26 Geometric configuration of a dipole radiating on a flat model of skin. *Source*: Brizzi et al. 2012 [103]. Reproduced with permission of IEEE.

boundary conditions at the skin interface. Its behavior is proportional to the following expression

$$F(w)\,\frac{e^{-jk_0 R}}{4\pi R} \tag{7.3}$$

where

$$F(w) = 1 - j\sqrt{\pi w}e^{-w}\mathrm{erfc}\left(j\sqrt{w}\right) \tag{7.4}$$

erfc being the complementary error function, and

$$w \cong -\frac{jk_0 R(\varepsilon_c - 1)}{2\varepsilon_c^2} \tag{7.5}$$

In the low-frequency range, the behavior of the function F cannot be easily described, and the field values must be numerically computed. On the other hand, at millimeter-waves, electrical distances are larger, either in the horizontal or in the vertical directions, and far ranges are often of interest. According to the distance between the dipole and the skin, the GOF contribution or the Norton wave contribution is dominant in the sum (7.2).

If the dipole height is large enough (a threshold value of approximately 3 mm is found at 60 GHz), at the air/skin interface the GOF is dominant. It should be remarked that the GOF is the sum of a direct and a reflected field, both decaying as 1/R free-space terms: At large distances along the horizontal direction, the two terms tend to be of the same magnitude and of opposite phase. This is shown in Figures 7.27a and 7.27b on the human skin at 60 GHz, and it can be proven as a general property of fields along stratified media (we cannot have a nonzero far-field pattern in the horizontal

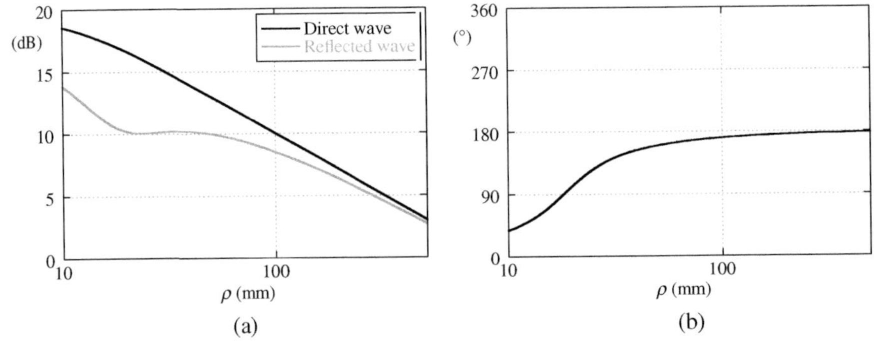

FIGURE 7.27 (a) $1/R$ decay of the magnitude of the direct and the reflected waves: Their magnitudes are asymptotically equivalent. The complex sum of the two waves gives the GOF. (b) Difference between the phases of the direct and reflected waves. The phase difference tends to 180°, meaning a cancellation between the two waves.

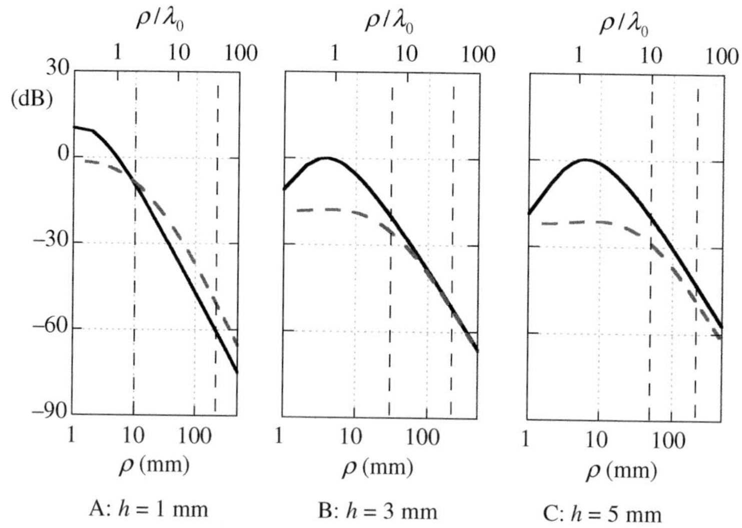

FIGURE 7.28 Geometrical-optic field (*solid black lines*) and Norton wave (*dashed red lines*) versus radial distance ρ from the vertical dipole. Different dipole heights h are considered. Both the GOF and the Norton wave decay as $1/R^2$ at a large horizontal distance from the source.

direction). For this reason, the total GOF decays faster than each free-space term and shows an asymptotic $1/R^2$ decay.

On the contrary, if the dipole is close enough to the skin (less than 3 mm in the vertical direction at 60 GHz), Norton wave (7.3) is dominant. At sufficiently large distances along the horizontal direction, also the expression (7.3) is proportional to $1/R^2$, as can be seen through simple asymptotic expressions for the *erfc* function (see Figure 7.28 for a summary of the possible wave contributions and decays).

As an important result, according to the range of distances considered from the source (with respect to the wavelength) and to the distance source-skin, different dominant waves are identified, and their attenuations can be quantified, leading to closed-form expressions for the path loss exponents [118]. Interestingly, new ranges of distances appear at millimeter-waves, at several wavelengths from the source, corresponding to the path loss exponents approaching to an asymptotic value of four, not observable at lower frequencies.

7.4.3.3.2 Skin Modeled as a Cylindrical Shape If an elementary dipole radiates in proximity of a cylinder (Figure 7.29a) whose dielectric constant is larger than unity and whose radius is greater than a wavelength, the field close to the air–skin surface can be expressed with a creeping-wave expression [111] (previously proposed at lower frequencies in [119] and [120])

$$E = C\frac{e^{-jk_0\rho_s}}{\rho_s}\sum_{i=1}^{+\infty}\frac{F(h_s)F(h_r)}{\tau_i - q^2}e^{-j\xi\tau_i} \tag{7.6}$$

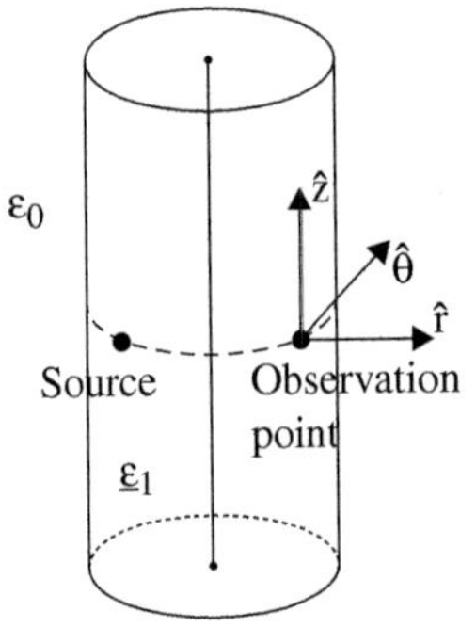

(a)

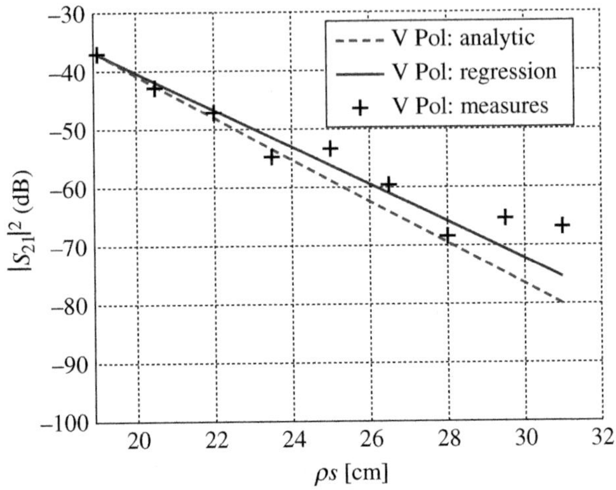

(b)

FIGURE 7.29 (a) Geometric configuration of a dipole radiating on a cylindrical model of skin. (b) Magnitude of the transmission coefficient between the radiating and the receiving dipole when the observation point is moved along the dashed line in (a). *Source*: Petrillo et al. 2014 [105]. Reproduced with permission of IEEE.

where ρ_s is the linear distance between the dipole and the observation point, measured along the circular contour of the cylinder. In (7.6) C, τ_i, ξ_i, and q are suitable coefficients depending on the dipole polarization, h_s and h_r are the heights of the dipole and of the observation point with respect to the skin surface, and the F functions are "height-gains" expressed through Fock-type Airy functions.

Also in this case, different range of distances between the dipole and the observation point can be defined. In proximity to the source, several terms in the series (7.6) must be retained. Further from the source, the $i = 1$ term is dominant and a suitable approximation of (7.6) leads to the derivation of an analytic expression for the path loss (showing an exponential field decay, see Figure 7.29b).

7.4.3.3.3 Presence of Textile Finally, the presence of textile materials deserves a specific treatment: the dielectric contrast at the interface between the air and the textile does not verify the fundamental hypotheses of large losses at the basis of the derivation of Norton wave, and Norton formalism cannot be applied. Apart from FDTD approaches [113], capable to describe generally shaped bodies but computationally expensive, a simple spectral-domain formulation has successfully been implemented [121, 122].

Under a perfectly flat body assumption, the spectral-domain electric field excited by a dipole on a general stratified medium (here made by the textile, possibly an air gap, and the skin) is known in closed form. The spatial-domain field can then be computed through a Fourier integral: for an effective integral evaluation, an asymptotic extraction has also been proposed, avoiding convergence problems if the observation and the source points lie on the same plane.

This simple approach allows a straightforward assessment of the influence of textile on the on-body propagation, and can be included in a full-wave computation tool if the resulting excited field is integrated over an equivalent current describing the presence of a realistic antenna.

7.4.4 On-Body Channel Measurements

Propagation along and around skin-equivalent model is investigated here at 60 GHz. Considering that at this frequency, the human body is electrically very large (typically 1150 λ_g), some body parts can be approximated with simple geometrical shapes, for instance torso can be approximated by a rectangle. The path gain is measured between two standard open-ended waveguides placed above a flat semi-solid and solid skin-equivalent phantom and around an elliptical cylinder phantom (Figure 7.30). The phantoms were prepared as previously introduced in Section 7.2. The two waveguides are connected through coaxial-cables and waveguide-to-coaxial-cable adaptors. A full-port calibration was performed prior to the measurements to account for the insertion loss in the cables and adaptors. The path gain is also measured when textiles like cotton, felt, and electro textile are placed above the phantom. Cotton and felt dielectric properties have been measured using the free-space technique.

7.4.4.1 Propagation Along the Phantom With and Without Textiles Figure 7.31 shows the path gain versus the separation distance between the two waveguides along the two phantoms: semi-solid and solid (DMBP) skin-equivalent phantom. A very

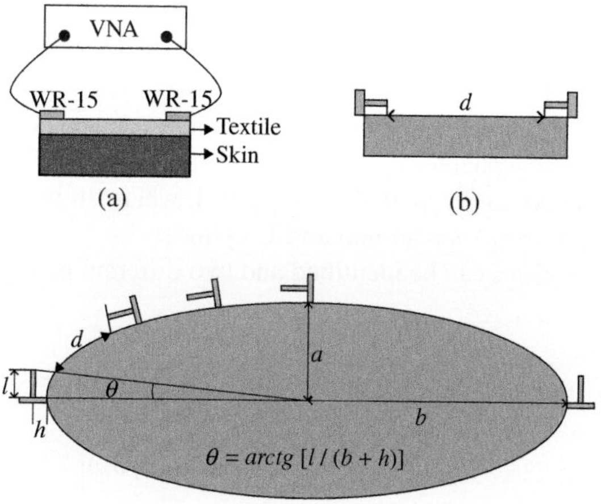

FIGURE 7.30 Measurement setup: (a) two V-polarized open-ended WR-15 waveguides placed on textile/skin-equivalent phantom. Cross-section view: (b) flat model, (c) elliptical cylinder model ($l = 21$ mm and $h = 4$ mm).

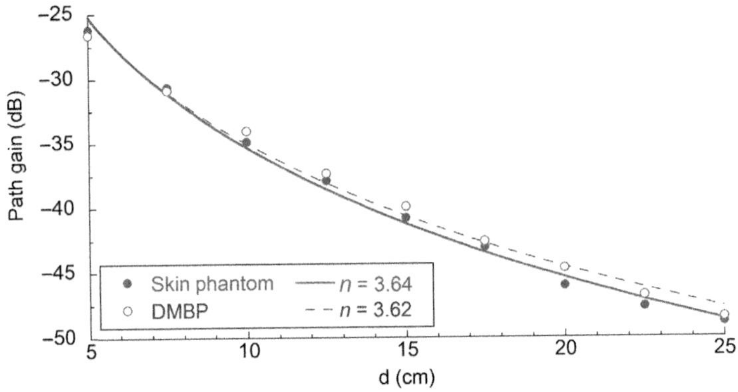

FIGURE 7.31 Measured path gain at 60 GHz versus separation distance d between two open-ended V-polarized waveguides placed above a semi-solid and solid skin-equivalent phantom.

good agreement in terms of path gain is obtained. By applying a logarithmic fit on the curves, the power decay exponent n is found to be around 3.6.

The effect of cotton (0.2 mm thick, $\hat{\varepsilon} = 2 - j0.04$) and felt (0.2 mm thick, $\widehat{\varepsilon} = 1.22 - j \cdot 0.036$) textiles placed over the semi-solid phantoms is shown in Figure 7.32a. It is shown that the presence of cotton textile decreases the transmission by 2.5–5 dB depending on the distance d. As expected, a less pronounced decrease of 1.5–3.5 dB is observed for the felt textile. It is also noticed that the presence of textiles does not affect the power decay exponent, but it significantly affects the path gain absolute values. It is also shown that an electro textile layer (0.17 mm thick ShieldIt Super® with $\sigma = 1.96 \times 10^5$ S/m) significantly increases the path gain for distances higher than 8 cm compared to the skin-equivalent phantom only (Figure 7.32b). In particular, the path gain increases more than 15 dB for distances exceeding 20 cm.

*7.4.4.2 **Propagation Around the Phantom*** In Figure 7.33, it is shown that placing an electro textile above the phantom, the path gain increases by 5–15 dB. These results are consistent with previously reported results [105], where an increase of 5–30 dB was obtained for the propagation around a PEC cylinder.

In this case, two regions can be identified and two different path gain models can be applied.

- LOS region (from 5 to 12 cm):

$$PG_{\mathrm{dB}}(d) = P_{0,\mathrm{dB}} - \left|10n\log(d/d_0)\right| \tag{7.7}$$

- NLOS region (from 12 to 30 cm):

$$PG_{\mathrm{dB}}(d) = P_{0,\mathrm{dB}} - |\alpha(d - d_0)| \tag{7.8}$$

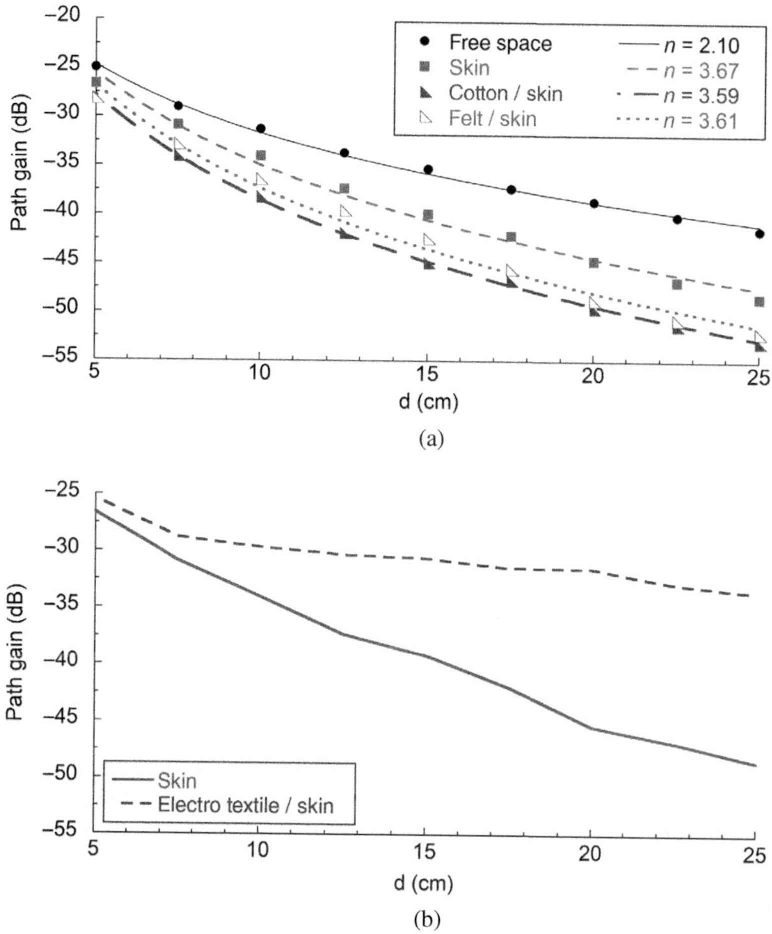

FIGURE 7.32 Measured path gain at 60 GHz versus separation distance d: (a) in free space and along the semi-solid skin-equivalent phantom with/without textiles. (b) along the skin and electro textile/skin model.

where $P_{0,\mathrm{dB}}$ is the path gain evaluated at the reference distance d_0 (5 cm in our study), n and α are the decay coefficients.

The residual sum of squares (RSS) is calculated to estimate which model better fits the path gain. It is the sum of the squared differences between the collected data y_i, and estimated attenuation law $f(x_i)$. Smaller RSS indicates a better fitting of the model to the data.

$$\mathrm{RSS} = \sum_{i=1}^{n} [y_i - f(x_i)]^2 \tag{7.9}$$

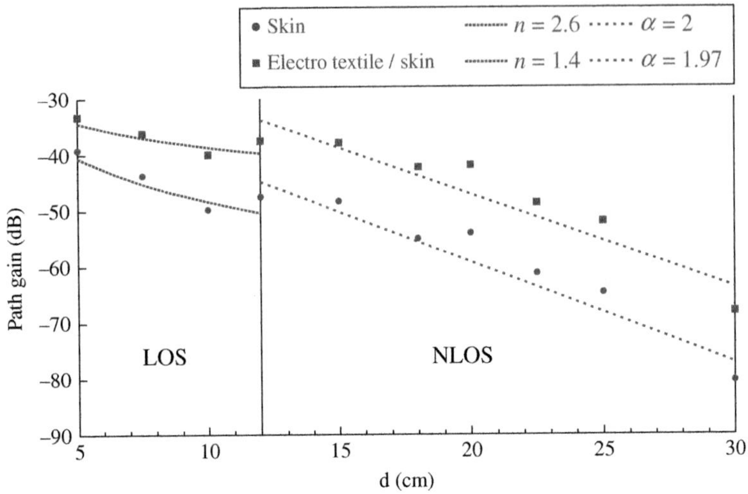

FIGURE 7.33 Measured path gain at 60 GHz versus separation distance d around the elliptical cylinder phantom with/without electro textile.

The results show a smaller attenuation coefficient compared to the flat model, due to the different distance range where the n parameter is defined (i.e., the LOS region for the flat and elliptical model). Indeed, the shape curvature affects the path gain exponent values.

Besides increasing the path gain, the electro textile behaves like a shield layer for the skin, reducing the exposure of the body. The absorbed power and the peak SAR are reduced to 0.068 mW and 0.003 W/kg (electro textile/skin model) from 4.14 mW and 1.23 W/kg (skin model), considering the averaging mass of 1 g and input power of 10 mW.

7.4.5 On-Body to Off-Body Propagation Scenario

7.4.5.1 Investigated Scenario In this section, a realistic scenario is envisioned for off-body propagation at 60 GHz. A wearable end-fire antenna [123, 124] is placed on a wrist and communicates with a transceiver placed at a certain distance in front of the hand (Figure 7.34). The antenna placed on the wrist can be integrated in a wristwatch, bracelet, or glove.

A short-range communication between a smartwatch and a base station (e.g., laptop, kiosk) is considered here, as described in Figure 7.34. Propagation is evaluated in terms of path gain, measured between the two antennas connected to the two ports of a Rohde & Schwarz® ZVA67 VNA. The waveguide placed at 5 cm in front of the hand was moved in the yz plane (along y-axis from -10 to 10 cm, and along z-axis from 0 to 4 cm). The wearable antenna is a low-profile high-gain (~12 dBi) antenna with an end-fire radiation, printed on a 0.254 mm thick RT Duroid 5880 substrate ($\varepsilon_r = 2.2$, $\tan\delta = 0.003$) [123]. The phantom was prepared as previously mentioned

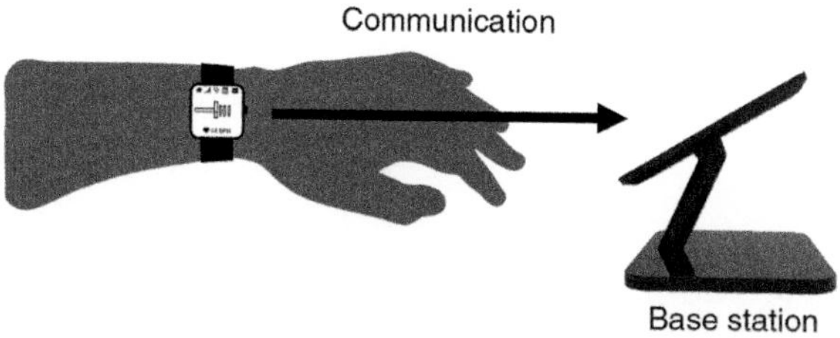

FIGURE 7.34 Off-body transceiver embedded into a smartwatch and communicating with a base station (e.g., laptop, kiosk, tablet.).

in Section 7.2, and modeled as a realistic hand shape, 23 cm length, 2.5 cm height, and 10 cm width.

7.4.5.2 Wearable End-Fire Antenna Characteristics Figure 7.35 shows the measured reflection coefficient of the end-fire antenna in free space and on a skin-equivalent phantom at a height of 5.6 mm (a height due to the presence of a V-connector used for antenna feeding). It is shown that, on the phantom the reflection coefficient is similar to the free space.

Figure 7.36 represents the measured antenna radiation patterns in *E*- and *H*-plane at 60 GHz, when the antenna/body separation distance is 5.6 mm. As previously noticed at microwaves [19, 54, 87, 125, 126], the radiation pattern is strongly affected by the presence of human body, due to the reflection from and absorption in the body. In the *H*-plane, the radiation pattern is tilted by 10° due to the reflections at the air/phantom interface.

7.4.5.3 Propagation Performance Figure 7.37 shows the path gain, with its maximum for the waveguide placed at a height of about 3–4 cm with respect to the antenna on the wrist. This particular height corresponds to the noticed 10° tilt in the *H*-plane radiation pattern (Figure 7.36b). For the waveguide moved along the *y*-axis from −5

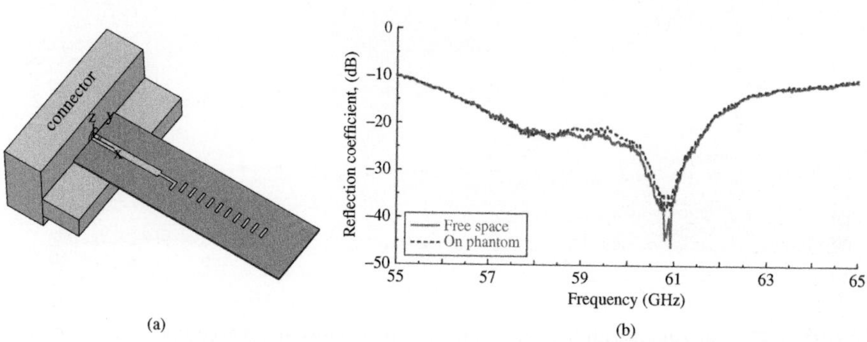

FIGURE 7.35 (a) Antenna prototype. (b) Measured reflection coefficient of the end-fire antenna in free space and on the phantom for the antenna/body spacing $h = 5.6$ mm.

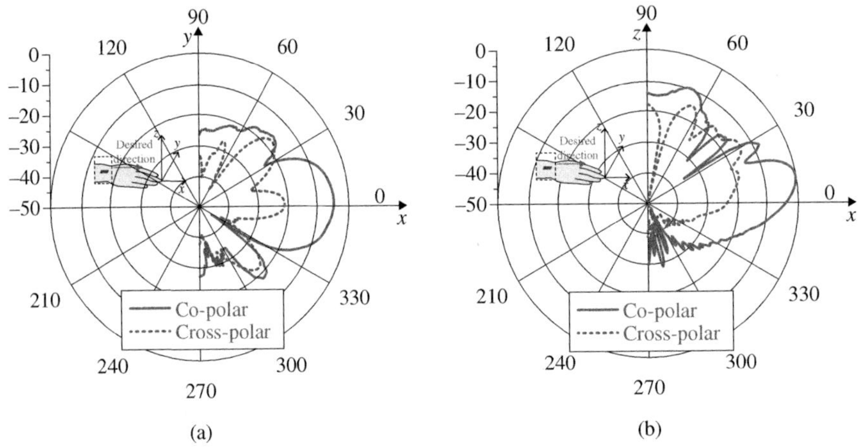

FIGURE 7.36 Measured radiation patterns at 60 GHz of the end-fire antenna (a) E- and (b) H-planes. Antenna/phantom distance is $h = 5.6$ mm.

to 5 cm, higher path gain values have been observed. This interval corresponds to an angular width of about 30°, that is, the half-power beam width of the end-fire antenna in the E-plane.

It has been experimentally shown that a wearable end-fire antenna can be used in an on-body to off-body propagation scenario. In the presence of the skin phantom the antenna radiation changes, and in particular in the H-plane it is tilted by 10°. It has been demonstrated that a better propagation occurs for the transmitter placed in front of the wearable antenna at a height corresponding to the tilt angle of the H-plane radiation pattern.

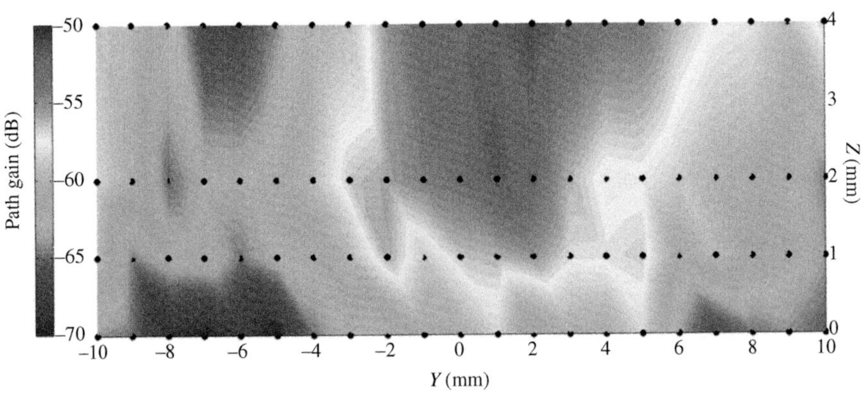

FIGURE 7.37 Measured path gain between end-fire antenna and waveguide in an off-body scenario. Distance between the hand and waveguide is $d = 5$ cm. For a color version of this figure, see the color plate section.

TABLE 7.6 Battery capacity of recent tablet, smartphone, and wearable products [127,128].

Capacity (mAh)	9000	8877	2815	2800	315	570
Product	Google Nexus	iPad Air	iPhone 6	Galaxy S5	Galaxy Gear	Google Glass

7.4.6 A Challenge to Overcome: Transceivers for Wearable Devices at Millimeter-Waves

While antenna, packaging and interconnect technology is available which enables the use of millimeter-wave communications for wearable devices and BANs, one major remaining challenge is the transceiver hardware itself. Tremendous efforts to develop low-cost, commercially compatible millimeter-wave link solutions in silicon technology are underway, and remain an ongoing research area. Progress, however, has been limited for wearable compatible solutions. The primary obstacle is that unlike conventional mobile devices (smartphones and tablets), smartwatches, smart glasses, and other wearables offer very limited battery capacities, typically a factor of 5 to 10 below their mobile counterparts. Table 7.6 compares the battery capacity of several recent products.

The low battery capacity of wearables is further exacerbated as millimeter-wave transceivers that, unlike their microwave and WiFi counterparts, consume tremendous power due to both the wider bandwidth data converters/digital signal processor (DSP), and most notably the millimeter-wave power amplifier (PA) required in the transmitter front-end. While non-commercial millimeter-wave PA technologies like InP and GaN can offer peak power-added efficiencies (PAE) well over the 50% range, silicon technology has been somewhat limited to below the 20% range. These peak PAEs are already extremely optimistic as most modern millimeter-wave is based on quadrature amplitude modulation (QAM) or other forms of non-constant envelope signaling, requiring significant back-off (3–6 dB) to preserve link error vector magnitude (EVM) and overall symbol linearity from the effects of gain compression. Table 7.7 reports several examples of recent millimeter-wave PA developments in commercial silicon and their effective power-added efficiency (PAE) when an already generous 6 dB back-off for 16 QAM operation is applied.

To balance a 60 GHz link budget considering both the multipath and fast/slow symbol fade effects for a typical indoor 1–10m scenario requires a transmitter chain on the order of 10–15 dBm of conducted power (on the order of 10 mW). Applying the

TABLE 7.7 Power-added efficiency of recently reported power amplifiers.

Peak PAE efficiency (%)	7	11	6.9	5.8	3.2	4.9	18
Operating PAE efficiency 6 dB back for 16 QAM (%)	3.3	1.7	1.9	2.3	2.5	1.4	4.3
Reference	[129]	[130]	[131]	[132]	[133]	[134]	[135]

average of the above PAE numbers (~2.5%), the device immediately consumes on the order of 400 mW DC power, just to support the PA. If we consider the overhead of all the required circuit blocks (the phase locked loop (PLL)/synthesizer, receiver chain, data converters, CDR, modulator, pulse shaping, and pre-emphasis), the overall power could easily be more than triple based on the latest reported 60 GHz transceivers [136–138]. With a power consumption at this level (3 × 400 mW = 1200 mW) and considering the 1V supply voltage typical of deep scaled CMOS, the battery in a smart watch like the galaxy gear (315 mAh) will effectively last 315 mAh/1200 mW = 15.8 minutes.

This first-order calculation is still extremely optimistic, as the battery capacity also has to support the application processor (chip that runs the user apps or "gadgets"), the display of the device, and any additional devices (sensors, USB port, microphone, camera, etc.). The latest generation application processors like the A8 processor from Apple and FD40 from Toshiba [139] can easily consume 200 mW, while portable displays can easily draw more than 300 mW due to their backlight systems. All considered, it seems that commercial millimeter-wave transceivers will need further development that offers more than an order of magnitude reduction in power consumption, before successful integration into wearable products with meaningful battery life is possible.

7.5 CONCLUSION

Tremendous progress has been done in BANs arena over the last decade, with recent focus toward BAN at millimeter-waves mainly in V-band. Indeed, the 60 GHz band offers significant advantages such as (1) high data rate capabilities (i.e., beyond several Gbps), (2) high level of security and low interference with adjacent BANs, and (3) significant reduction of on-body devices. This chapter provides a review of recent achievements on experimental and numerical body modelling, wearable antennas for off-body and on-body communications, and body-centric propagation at millimeter-waves.

As skin covers 95% of the human body surface, it constitutes the main target for the interactions of the human body with millimeter-waves due to the very shallow penetration of these waves into the skin (typically 0.5 mm at 60 GHz). From the electromagnetic viewpoint, human skin can be considered as an anisotropic multilayer dispersive structure made of three different layers, namely, epidermis, dermis, and subcutaneous fat layer. Measurement techniques have been briefly revised, and skin dielectric properties (extracted from *in vitro* or *in vivo* samples) have been provided. Experimental and numerical phantoms have been described. These phantoms play a key role in the understanding of wearable antenna performance, propagation mechanisms on the human body, and absorbed power in the human body. If many phantom types are available at microwaves, only a few numbers of phantoms have been proposed so far at millimeter-waves. In particular two phantoms recently proposed by IETR have been discussed in detail. The first one is a homogeneous semi-solid skin-equivalent phantom with realistic shapes; its composition (deionized water, agar, polyethylene powder, TX-151, sodium azide) and measured properties have been described in the 55–65 GHz band. The second one is solid phantom introduced for body-centric propagation measurements in the 58–63 GHz range.

Exposure limits and recommendations provided by the ICNIRP, IEEE, and CENELEC have been summarized. Since the electromagnetic field is absorbed superficially at millimeter-waves, the relevant dosimetric quantity for far-field exposures is not the SAR as in microwave regime, but the IPD. The basic restrictions and reference levels have been provided for general and occupational public.

An overview of available wearable antenna technologies has been reported for body-centric systems (both off- and on-body communications). Various antenna designs with broadside or end-fire radiation have been described in details; their interaction with the human body has been also discussed. As expected, it has been shown that microstrip antennas are excellent candidates as their performances remain almost insensitive to the body proximity, and power dissipated in the human body is minimized. In addition, textile antennas can properly be manufactured at millimeter-wave using a simple fabrication process, which is believed to pave the way for the future development of smart clothes at millimeter-waves.

With the increasing need of high data-rate communication in daily life, a very fast development of smart wearable devices is expected in many applications such as healthcare, entertainment, sport, smart home, and so on. Obviously, implementing on-body wireless links at millimeter-waves appears to be very challenging due to on-body wave attenuation. Nonetheless, as an example, a specific short-range communication scenario has been studied here between a Yagi-Uda antenna embedded in a smart watch and a base station located few centimeters away from the user hand. As the path gain remains in the order of 50–70dB, such a scenario seems feasible.

Although antenna, packaging and interconnect technology is already available and enables the use of millimeter-wave communications for wearable devices and BANs, one major remaining challenge is the transceiver hardware itself. Commercial millimeter-wave transceivers will need further development before successful integration into wearable products with meaningful battery life is possible.

REFERENCES

[1] P. S. Hall and Y. Hao, *Antennas and Propagation for Body Centric Communications Systems*. Artech House, Norwood, MA, 2006.

[2] D. Guha and Y. M. M. Antar, *Microstrip and Printed Antennas: New Trends, Techniques and Applications*. Wiley-Blackwell, 2011, ISBN-10: 0470681926.

[3] T. Baykas, C. S. Sum, Z. Lan, J. Wang, M. A. Rahman, and H. Harada, "IEEE 802.15.3c: the first IEEE wireless standard for data rates over 1 Gb/s," *IEEE Commun. Mag.*, vol. 49, no. 7, pp. 114–121, July 2011.

[4] S. L. Cotton, W. G. Scanlon, and P. S. Hall, "A Simulated Study of Co-Channel Inter-BAN Interference at 2.45 GHz and 60 GHz," in European Wireless Technology Conference, Paris, September 27–28, 2010, pp. 61–64.

[5] J. Wang, O. Fujiwara, and S. Watanabe, "Approximation of aging effect on dielectric tissue properties for SAR assessment of mobile telephones," *IEEE Trans. Electromagn. Compat.*, vol. 48, no. 2, pp. 408–413, May 2006.

[6] K. Fuji, M. Takahashi, K. Ito, and N. Inagaki, "Study On the Electric Field Distributions Around Whole Body Model with a Wearable Device Using the Human Body as a Transmission Channel," in First European Conference on Antennas and Propagation, Nice, November 6–10, 2006, pp. 1–4.

[7] Y. Uno, K. Saito, M. Takahashi, and K. Ito, "Characteristics of loop antennas closely-placed to different arm models," *IEICE Tech. Rep.*, vol. 110, no. 23, AP2010-13, pp. 7–10, May 2010.

[8] M. Zhadobov, N. Chahat, R. Sauleau, C. Le Quément, and Y. Le Dréan, "Millimeter-wave interactions with the human body: state of knowledge and recent advances," *Int. J. Microw. Wireless Technol.*, vol. 3, no. 2, pp. 237–247, April 2011.

[9] A. Christ, G. Schmid, R. Djafarzadeh, R. Überbacher, S. Cecil, M. Zefferer, E. Neufeld, M. Lichtsteiner, M. Bouterfas, and N. Kuster, "Numerische bestimmung der spezifischen absorptionsrate bei ganzkörperexposition von kindern: abschlußbericht", Technical Report, IT'IS Foundation, Zürich, Switzerland, July 2009.

[10] A. Christ, W. Kainz, E. G. Hahn, K. Honegger, M. Zefferer, E. Neufeld, W. Rascher, R. Janka, W. Bautz, J. Chen, B. Kiefer, P. Schmitt, H. P. Hollenbach, J. X. Shen, M. Oberle, D. Szczerba, A. Kam, J. W. Guag, and N. Kuster, "The Virtual family – development of surface based anatomical models of two adults and two children for dosimetric simulations", *Phys. Med. Biol.*, vol. 55, no. 23, January 2010.

[11] MCL-T broadband tissue equivalent liquid: 30 MHz to 6 GHz. Available at http://www.mcluk.org/pdfs/bbl.pdf

[12] C. Gabriel and P. Chadwick, "Development of Whole-Body Shell and Solid Phantoms and for SAR and Radio Propagation Measurement", in International Conference on Bioinformatics and Biomedical Engineering, Shanghai, China, May 16–18, 2008, pp. 805–807.

[13] K. Ogawa, T. Matsuyoshi, H. Iwai, and N. Hatakenaka, "A High-Precision Real Human Phantom for EM Evaluation of Handheld Terminals in a Talk Situation," in IEEE Antennas and Propagation Society International Symposium, Boston, July 8–13, 2001, pp. 68–71.

[14] A. W. Guy, "Analyses of electromagnetic fields induced in biological tissues by thermographic studies on equivalent phantom models," *IEEE Trans. Microw. Theory Techn.*, vol. 19, no. 2, pp. 205–214, February 1968.

[15] K. Ito, K. Furuya, Y. Okano, and L. Hamada, "Development and characteristics of a biological tissue-equivalent phantom for microwaves," *Electron. Commun. Japan. Pt. I-Commun.*, vol. 84, no. 4, pp. 67–77, April 2001.

[16] T. Takimoto, T. Onishi, K. Saito, M. Takahashi, S. Uebayashi, and K. Ito, "Characteristics of biological tissue equivalent phantoms applied to UWB communications," *Electron. Commun. Japan*, vol. 90, no. 5, pp. 48–55, May 2007.

[17] J. Yonebayashi, S. Takamatsu, K. Saito, M. Takahashi, and K. Ito, "Development of Dynamic Phantom for Evaluation of Breath Detection Doppler Radar," in The Bioelectromagnetics Society 32nd Annual Meeting, Seoul, Korea, June 13–18, 2010.

[18] Y. Okano, K. Ito, and H. Kawai, "Solid phantom composed of glycerin and its application to SAR estimation," *Trans. IEICE*, vol. J83-B, no. 4, pp. 534–543 (in Japanese), 2000.

[19] N. Chahat, M. Zhadobov, R. Sauleau, and K. Ito, "A compact UWB antenna for on-body applications," *IEEE Trans. Antennas Propag.*, vol. 59, no. 4, pp. 1123–1131, April 2011.

[20] F. A. Duck, "Physical properties of tissue: a comprehensive reference network," U.K.: Academic, ISBN 0122228006, 1990.

[21] N. Chahat, M. Zhadobov, and R. Sauleau, "Broadband tissue-equivalent phantom for BAN applications at millimeter waves," *IEEE Trans. Microw. Theory Techn.*, vol. 60, no. 7, pp. 2259–2266, July 2012.

[22] T. Kobayashi, T. Nojima, K. Yamada, and S. Uebayashi, "Dry phantom composed of ceramics and its application to SAR estimation," *IEEE Trans. Microw. Theory Techn.*, vol. 41, no. 1, pp. 136–140, January 1993.

[23] H. Tamura, Y. Ishikawa, T. Kobayashi, and T. Nojima, "A dry phantom material composed of ceramic and graphite powder," *IEEE Trans. Electromagn. Compat.*, vol. 39, no. 2, pp. 132–137, May 1997.

[24] Y. Nikawa, M. Chino, and K. Kikuchi, "Soft and dry phantom modeling material using silicone rubber with carbon fiber," *IEEE Trans. Microw. Theory Techn.*, vol. 44, no. 10, Part 2, pp. 1949–1952, October 1996.

[25] J. T. Chang, W. Fanning, P. M. Meaney, and K. D. Paulsen, "A conductive plastic for simulating biological tissue at microwave frequencies," *IEEE Trans. Electromagn. Compat.*, vol. 42, no. 1, pp. 76–81, February 2000.

[26] A. R. Guraliuc, M. Zhadobov, O. De Sagazan, and R. Sauleau, "Solid phantom for body-centric propagation measurements at 60 GHz," *IEEE Trans. Microw. Theory. Techn.*, vol. 62, no. 6, pp. 1376–1380, June 2014.

[27] O. P. Gandhi and A. Riazi, "Absorption of millimeter waves by human beings and its biological implications," *IEEE Trans. Microw. Theory Techn.*, vol. 34, no. 2, pp. 228–235, February 1986.

[28] C. M. Alabaster, "Permittivity of human skin in millimetre wave band," *Elec. Lett.*, vol. 39, no. 21, pp. 1521–1522, October 2003.

[29] S. Gabriel, R. W. Lau, and C. Gabriel, "The dielectric properties of biological tissues: III. Parametric models for the dielectric spectrum of tissues," *Phys. Med. Biol.*, vol. 41, no. 11, pp. 2271–2293, November 1996.

[30] S. I. Alekseev and M. C. Ziskin, "Human skin permittivity determined by millimeter wave reflection measurements," *Bioelectromagnetics*, vol. 28, no. 5, pp. 331–339, July 2007.

[31] N. Chahat, M. Zhadobov, R. Sauleau, and S. Alekseev, "New method for determining dielectric properties of skin and phantoms at millimeter waves based on heating kinetics," *IEEE Trans. Microw. Theory Techn.*, vol. 60, no. 3, pp. 827–832, March 2011.

[32] ICNIRP, "Guidelines for limiting exposure to time-varying electric, magnetic, and electromagnetic fields (up to 300 GHz)", *Health Phys.*, vol. 74, no. 4, pp. 494–522, 1998.

[33] ICNIRP, "Exposure to high frequency electromagnetic fields, biological effects and health consequences (100 kHz – 300 GHz)", ISBN 978-3-934994-10-2, 2009.

[34] 1999/519/EC, "Council recommendation of 12 July 1999 on the limitation of exposure of the general public to electromagnetic fields (0 Hz to 300 GHz)".

[35] 2004/40/EC, "Directive of the European parliament and of the Council of 29 April 2004 on the minimum health and safety requirements regarding the exposure of workers to the risk arising from physical agents (electromagnetic fields)".

[36] EN 50413 – 2008, "Basic standard on measurement and calculation procedures for human exposure to electric, magnetic and electromagnetic fields (0 Hz – 300 GHz)".

[37] IEEE Standard for safety levels with respect to human exposure to radio frequency electromagnetic fields, 3 kHz to 300 GHz, ISBN 0-7381-4835-0 SS95389, April 2006.

[38] "Beyond 2020 heterogeneous wireless networks with millimeter-wave small cell access and backhauling"- MiWaveS Project - Deliverable D1.3 "Dosimetric aspects related to the human body exposure to mm Waves," December 2014, Available at http://www.miwaves.eu/MiWaveS_D1.3_v1.0.pdf

[39] C. H. Durney, H. Massoudi, and M. F. Iskander, "Radiofrequency radiation dosimetry handbook," USAF School of Aerospace Medicine, Brooks AFB, TX, Report USAFSAM-TR-85-73, 1986.

[40] O. P. Gandhi, "Advances in Dosimetry of Radio-frequency Radiation and Their Past and Projected Impact on the Safety Standards," in Proceedings of IMTC Instrumentation and Measurement Technology Conference, San Diego, CA, April 20–22, 1988, pp. 109–113.

[41] P. W. Barber, "Electromagnetic power deposition in prolate spheroid models of man and animals at resonance," *IEEE Trans. Biomed. Eng.*, vol. 24, no. 6, pp. 513–521, November 1977.

[42] C. H. Durney, C. C. Johnson, P. W. Barber, H. Massoudi, M. F. Iskander, J. L. Lords, D. K. Ryser, S. J. Allen, and J. C. Mitchell, *Radiofrequency Radiation Dosimetry Handbook*, 2nd ed. (Report SAM-TR-78-22), Brooks Air Force Base, USAF School on Aerospace Medicine, TX 78235, 1978.

[43] NRPB, "Consultation Document — Proposals for Limiting Exposure to Electromagnetic Fields (0–300GHz)," National Radiological Protection Board, Chilton, Didcot, OX11 0RQ, 185, U.K., May 2003.

[44] A. Kanezaki, A. Hirata, S. Watanabe, and H. Shirai, "Parameter variation effects on temperature elevation in a steady-state, one-dimensional thermal model for millimeter wave exposure of one- and three-layer human tissue," *Phys. Med. Biol.*, vol. 55, pp. 4647–4659, July 2010.

[45] F. Gustrau and A. Bahr, "W-band investigation of material parameters, SAR distribution, and thermal response in human tissue," *IEEE Trans. Microw. Theory Techn.*, vol. 50, no. 10, pp. 2393–2400, October 2002.

[46] S. I. Alekseev and M. C. Ziskin, "Millimeter-wave absorption by cutaneous blood vessels: a computational study," *IEEE Trans. Biomed. Eng.*, vol. 56, no. 10, pp. 2380–2388, October 2009.

[47] S. I. Alekseev and M. C. Ziskin, "Influence of blood flow and millimeter wave exposure on skin temperature in different thermal model," *Biolectromagnetics*, vol. 30, pp. 52–58, January 2009.

[48] P. J. Riu, K. R. Foster, D. W. Blick, and E. R. Adair, "A thermal model for human thresholds of microwave-evoked warmth sensations," *Bioelectromagnetics*, vol. 18, no. 8, pp. 578–583, December 1997.

[49] G. Shafirstein and E. G. Moros, "Modelling millimeter wave propagation and absorption in a high resolution skin model: the effect of sweat glands," *Phys. Med. Biol.*, vol. 56, no. 5, pp. 1329–1339, March 2011.

[50] P. Riordan-Eva and E. T. Cunningham, *Vaughan & Asbury's General Ophthalmology*, 18th ed., eds., P. Riordan-Eva, and E.T. Cunningham, McGraw-Hill Medical, New York, ISBN 978-0071634205, Jun. 2011.

[51] K. L. Wong and C. I. Lin, "Characteristics of a 2.4-GHz compact shorted patch antenna in close proximity to a lossy medium," *Microw. Opt. Technol. Lett.*, vol. 45, no. 6, pp. 480–483, June 2005.

[52] L. Vallozzi, W. Vandendriessche, W. Vandendriessche, H. Rogier, C. Hertleer, and M. L. Scarpello, "Wearable Textile GPS Antenna for Integration in Protective Garments," in European Conference on Antennas and Propagation, EuCAP'2010, Barcelona, Spain, April 12–16, 2010, pp. 1–4.

[53] M. M. Khan, A. Alomainy, and Y. Hao, "Dual Band and Diverse Radiation Pattern Antenna for Power Efficient and Reliable On-Body and Off-Body Communications for Healthcare Applications," in IEEE International Symposium on Antennas and Propagation, Spokane, WA, July 3–8, 2011, pp. 396–399.

[54] S. J. Boyes, P. J. Soh, Yi Huang, G. A. E. Vandenbosch, and N. Khiabani, "Measurement and performance of textile antenna efficiency on a human body in a Reverberation chamber," *IEEE Trans. Antennas Propag.*, vol. 61, no. 2, pp. 871–881, February 2013.

[55] L. Vallozzi, H. Rogier, and C. Hertleer, "Dual polarized textile patch antenna for integration into protective garments," *IEEE Antennas Wireless Propag. Lett.*, vol. 7, pp. 440–443, May 2008.

[56] S. Yan, P. Soh, and G. Vandenbosch, "Compact all-textile dual-band antenna loaded with metamaterial inspired structure," *IEEE Antennas Wireless Propag. Lett.*, vol. 14, pp. 1486–1489, July 2015.

[57] Z. Wang, L. Zhang, Y. Bayram, and J. Volakis, "Embroidered conductive fibers on polymer composite for conformal antennas," *IEEE Trans. Antennas Propag.*, vol. 60, no. 9, pp. 4141–4147, September 2012.

[58] A. Kiourti and J. L. Volakis, "High-geometrical-accuracy embroidery process for textile antennas with fine details," *IEEE Antennas Wireless Propag. Lett.*, vol. 14, pp. 1474–1477, July 2015.

[59] S. Shao, A. Kiourti, R. Burkholder, and J. Volakis, "Broadband textile-based passive UHF RFID tag antenna for elastic material," *IEEE Antennas Wireless Propag. Lett.*, vol. 14, pp. 1385–1388, June 2015.

[60] A. Kiourti and J. L. Volakis, "Stretchable and flexible e-fiber wire antennas embedded in polymer," *IEEE Antennas Wireless Propag. Lett.*, vol. 13, pp. 1381–1384, July 2014.

[61] S. Hage-Ali, N. Tiercelin, P. Coquet, R. Sauleau, H. Fujita, V. Preobrazhensky, and P. Pernod, "A millimeter-wave microstrip antenna array on ultra-flexible micromachined polydimethylsiloxane (PDMS) polymer," *IEEE Antennas Wireless Propag. Lett.*, vol. 8, pp. 1306–1309, December 2009.

[62] D.-H. Kim, N. Lu, R. Ma, Y.-S. Kim, R.-H. Kim, S. Wang, J. Wu, S. M. Won, H. Tao, A. Islam, K. J. Yu, T. Kim, R. Chowdhury, M. Ying, L. Xu, M. Li, H.-J. Chung, H. Keum, M. McCormick, P. Liu, Y.-W. Zhang, F. G. Omenetto, Y. Huang, T. Coleman, and J. A. Rogers, "Epidermal electronics," *Science*, vol. 333, no. 6044, pp. 838–843, August 2011.

[63] N. Chahat, M. Zhadobov, R. Sauleau, and K. Mahdjoubi, "Improvement of the On-Body Performance of a Dual-Band Textile Antenna using an EBG Structure," in Loughborough Antennas and Propagation Conference (LAPC), Loughborough, November 8–9, 2010, pp. 465–468.

[64] L. Liu, S. Zhu, and R. Langley, "Dual-band triangular antenna with modified ground plane," *Electron. Lett.*, vol. 43, no. 3, pp. 142–143, February 2007.

[65] P. F. M. Smulders, "Impact of Regulations on Feasible Distance Between 60 GHz Devices," in Proceedings of the Fourth European Conference on Antennas and Propagation (EuCAP), Barcelona, Spain, April 12–16, 2010, pp. 1–4.

[66] S. L. Cotton, W. G. Scanlon, and B. K. Madahar, "Millimeter-wave soldier-to-soldier communications for covert battlefield operations," *IEEE Commun. Mag.*, vol. 47, no. 10, pp. 72–81, October 2009.

[67] P. F. M. Smulders, "Statistical characterization of 60-GHz indoor radio channels," *IEEE Trans. Antennas Propag.*, vol. 57, no. 10, pp. 2820–2829, October 2009.

[68] R. C. Daniels, J. N. Murdock, T. S. Rappaport, and R. W. Heath, "60 GHz wireless: up close and personal," *IEEE Microw. Mag.*, vol. 11, no. 7, pp. 44–50, December 2010.

[69] S. Collonge, G. Zaharia, and G. El Zein, "Wideband and dynamic characterization of the 60 GHz indoor radio propagation—future home WLAN architectures," *Ann. Telecommun.*, vol. 58, no. 3–4, March–April 2003, pp. 417–447.

[70] N. Chahat, M. Zhadobov, L. Le Coq, S. Alekseev, and R. Sauleau, "Characterization of the interactions between a 60-GHz antenna and the human body in an off-body scenario," *IEEE Trans. Antennas and Propag.*, vol. 60, no. 12, pp. 5958–5965, December 2012.

[71] N. Chahat, M. Zhadobov, and R. Sauleau, "60-GHz textile antenna array for body-centric communications," *IEEE Trans. Antennas and Propag.*, vol. 61, no. 4, pp. 1816–1824, April 2013.

[72] N. Chahat, M. Zhadobov, L. Le Coq, and R. Sauleau, "Wearable endfire textile antenna for on-body communications at 60 GHz," *IEEE Antennas Wireless Propag. Lett.*, vol. 11, pp. 799–802, July 2012.

[73] K. Lorincz, B. Chen, G. W. Challen, A. R. Chowdhury, S. Patel, P. Bonato, and M. Welsh, "Mercury: A Wearable Sensor Network Platform for High Fidelity Motion Analysis," in Conference on Embedded Network Sensor System, New York, November 2009, pp. 183–196.

[74] Q. Wang, W. Shin, X. Liu, Z. Zeng, C. Oh, B. K. AlShebli, M. Caccamo, C. A. Gunter, E. Gunter, J. Hou, K. Karahalios, and L. Sha, "I-Living: An Open System Architecture for Assisted Living," in Proceedings of the IEEE International Conference on Systems, Man and Cybernetics, Taipei, October 8–11, 2006, pp. 4268–4275.

[75] M. Annavaram, N. Medvidovic, U. Mitra, S. Narayanan, G. Sukhatme, Z. Meng, S. Qiu, R. Kumar, G. Thatte, and D. Spruijt-Metz, "Multimodal Sensing for Pediatric Obesity Applications," in Proceedings of the International Workshop on Urban, Community, and Social Applications of Networked Sensing Systems, Raleigh, NC, November 2008, pp. 21–25.

[76] T. Pawar, S. Chaudhuri, and S. P. Duttagupta, "Body movement activity recognition for ambulatory cardiac monitoring," *IEEE Trans. Biomed. Eng.*, vol. 54, no. 5, pp. 874–882, May 2007.

[77] P. Iso-Ketola, T. Karinsalo, and L. Vanhala, "HipGuard: A Wearable Measurement System for Patients Recovering from a Hip Operation," in Proceedings on Pervasive Computing Technologies for Healthcare Conference, Tampere, February 2008, pp. 196–199.

[78] K. D. Nguyen, I. M. Chen, Z. Luo, S. Yeo, and H. L. Duh, "A wearable sensing system for tracking and monitoring of functional arm movement," *IEEE/ASME Trans. Mechatron.*, vol. 16, no. 2, pp. 213–220, February 2010.

[79] J. Baek, G. Lee, W. Park, and B. J. Yun, "Accelerometer signal processing for user activity detection," Book Chapter, Knowledge-Based Intelligent Information and Engineering Systems, vol. 3215, pp. 610–617, 2004.

[80] M. Galli, C. Rigoldi, L. Mainardi, N. Tenore, P. Onorati, and G. Albertini, "Postural control in patients with down syndrome," *Disability Rehabil.*, vol. 30, no. 17, pp. 1274–1278, 2008.

[81] A. M. Adami, M. Pavel, T. L. Hayers, and C. M. Singer, "Detection of movement in bed using unobtrusive load cell sensors," *IEEE Trans. Inf. Technol. Biomed.*, vol. 14, no. 2, pp. 481–490, March 2010.

[82] A.R Guraliuc, P. Barsocchi, F. Potortì, and P. Nepa, "Limb movements classification using wearable wireless transceivers," *IEEE Trans. Inf. Tech. Biomed.*, vol. 15, no. 3, pp. 474–480, May 2011.

[83] F. Lorussi, E. P. Scilingo, M. Tesconi, A. Tognetti, and D. De Rossi, "Strain sensing fabric for hand posture and gesture monitoring", *IEEE Trans. Inf. Tech. Biomed.*, vol. 9, no. 3, pp. 372–381, September 2005.

[84] A. Tognetti, F. Lorussi, R. Bartalesi, M. Tesconi, G. Zupone, and D. De Rossi, "Analysis and Synthesis of Human Movements: Wearable Kinesthetic Interfaces", in Proceedings of the IEEE International Conference on Rehabilitation Robotics, June 28–July 1, 2005, pp. 488–491.

[85] K. Kumareswaran, M. L. Evans, and R. Hovorka, "Closed-loop insulin delivery: towards improved diabetes care," *Discov. Med.*, vol. 13, no. 69, pp. 159–170, February 2012.

[86] P. S. Hall, Y. Hao, Y. I. Nechayev, A. Alomainy, C. C. Constantinou, C. Parini, M. R. Kamarudin, T. Z. Salim, D. T. M. Hee, R. Dubrovka, A. S. Owadall, W. Song, A. A. Serra, P. Nepa, M. Gallo, and M. Bozzetti, "Antennas and propagation for on-body communication systems," *IEEE Antennas Propag. Mag.*, vol. 49, no. 3, pp. 41–58, August 2007.

[87] E. Reusens, W. Joseph, B. Latre, B. Braem, G. Vermeeren, E. Tanghe, L. Martens, I. Moerman, and C. Blondia, "Characterization of on-body communication channel and energy efficient topology design for wireless body area networks," *IEEE Trans. Inform. Technol. Biomed.*, vol. 13, no. 6, pp. 933–945, November 2009.

[88] A. R. Guraliuc, A. A. Serra, P. Nepa, and G. Manara, "Path gain models for on-body communication systems at 2.4 and 5.8GHz," *Ann. Telecommun.*, vol. 66, no. 3–4, pp. 205–212, April 2011.

[89] Q. H. Abbasi, A. Sani, A. Alomainy, and Y. Hao, "Numerical characterization and modeling of subject-specific ultrawideband body-centric radio channels and systems for healthcare applications," *IEEE Trans. Inform. Technol. Biomed.*, vol. 16, no. 2, pp. 221–227, March 2012.

[90] P. S. Hall and Y. Hao, *Body-Centric Wireless Communications*. Artech House, Norwood, MA, 2012.

[91] G. A. Conway and W. G. Scanlon, "Antennas for over-body-surface communication at 2.45 GHz," *IEEE Trans. Antennas Propag.*, vol. 57, no. 4, pp. 844–855, April 2009.

[92] A. A. Serra, A. R. Guraliuc, P. Nepa, G. Manara, I. Khan, and P. S. Hall, "Dual-polarization and dual-pattern planar antenna for diversity in body-centric communications," *IET Microw. Antennas Propag.*, vol. 4, no. 1, pp. 106–112, January 2010.

[93] T. S. P. See, A. Alomainy, Y. Hao, and Z. N. Chen, "On-Body Characterization of a Compact Planar UWB Antenna," in First European Conference on Antennas and Propagation (EuCAP), Nice, November 6–10, 2006, pp. 1–5.

[94] N. Chahat, G. Valerio, M. Zhadobov, and R. Sauleau, "On-body propagation at 60 GHz," *IEEE Trans. Antennas Propag.*, vol. 61, no. 4, pp. 1876–1888, April 2013.

[95] Y. I. Nechayev, X. Wu, C. C. Constantinou, and P. S. Hall, "Millimetre-wave path loss variability between two body-mounted monopole antennas," *IET Microw. Antennas Propag.*, vol. 7, no. 1, pp. 1–7, January 2013.

[96] A. Pellegrini, A. Brizzi, L. Zhang, K. Ali, Y. Hao, X. Wu, C. C. Constantinou, Y. Nechayev, P. S. Hall, N. Chahat, M. Zhadobov, and R. Sauleau, "Antennas and propagation for body centric wireless communications at millimeter wave frequencies: a review," *IEEE Antennas Propag. Mag.*, vol. 55, no. 4, pp. 262–287, August 2013.

[97] A. R. Guraliuc, M. Zhadobov, G. Valerio, N. Chahat, and R. Sauleau, "Effect of textile on the propagation along the body at 60 GHz," *IEEE Trans. Antennas Propag.*, vol. 62, no. 3, pp. 1489–1494, March 2014.

[98] A. R. Guraliuc, M. Zhadobov, G. Valerio, and R. Sauleau, "Enhancement of On-Body Propagation at 60 GHz Using Electro textiles," in IEEE Antennas and Wireless Propagation Letters, March 20, 2014, pp. 603–606.

[99] Y. I. Nechayev, X. Wu, C. C. Constantinou, and P. S. Hall, "Millimetre-wave path loss variability between two body-mounted monopole antennas," *IET Microw. Antennas Propag.*, vol. 7, no. 1, pp. 1–7, January 2013.

[100] X. Y. Wu, L. Akhoondzadeh-Asl, and P. S. Hall, "Printed Yagi-Uda array for on-body communication channels at60 GHz," *Microw. Opt. Technol. Lett.*, vol. 53, no. 12, pp. 2728–2730, December 2011.

[101] X. Y. Wu and P. S. Hall, "Substrate integrated waveguide Yagi-Uda antenna," *Electron. Lett.*, vol. 46, no. 23, pp. 1541–1542, November 2010.

[102] A. Brizzi, A. Pellegrini, L. Zhang, and Y. Hao, "Woodpile EBG-Based Antennas for Body Area Networks at 60 GHz," in 4th High Speed Intelligent Communication Forum, Nanjing, May 10–11, 2012, pp. 1–4.

[103] A. Brizzi, A. Pellegrini, and Y. Hao, "Design of a Cylindrical Resonant Cavity Antenna for BAN Applications at V Band," in IEEE International Workshop on Antenna Technology, Tucson, March 5–7, 2012, pp. 152–155.

[104] K. Ali, A. Pellegrini, A. Brizzi, and Y. Hao, "Full Wave and Ray-Based Analysis of a Body-Centric Scenario at V Band," in 7th European Antennas Propagation Conference (EuCAP), Gothenburg, April 8–12, 2013, pp. 780–783.

[105] L. Petrillo, T. Mavridis, J. Sarrazin, D. Lautru, A. Benlarbi-Delai, and P. De Doncker, "Analytical creeping waves model and measurements for 60 GHz body area networks," *IEEE Trans. Antennas Propag.*, vol. 62, no. 8, pp. 4352–4356, August 2014.

[106] T. Mavridis, L. Petrillo, J. Sarrazin, D. Lautru, A. Benlarbi-Delai, and P. De Doncker, "Theoretical and experimental investigation of a 60 GHz off-body propagation model," *IEEE Trans. Antennas Propag.*, vol. 62, no. 1, pp. 393–402, January 2014.

[107] K. Ali, A. Brizzi, A. Pellegrini, Y. Hao, and A. Alomainy, "Investigation of the Effect of Fabric in On-Body Communication Using Finite Difference Time Domain Technique at 60GHz," in Antennas and Propagation Conference (LAPC), Loughborough, November 12–13, 2012, pp. 1–3.

[108] M. De Gregorio, P. Usai, G. Tiberi, A. Monorchio, A. Brizzi, A. Pellegrini, and Y. Hao, "Hybridizing CBFM and Ray Tracing for On-Body Propagation," in Antennas and Propagation Society International Symposium (APSURSI), Orlando, July 7–13, 2013, pp. 1660–1661.

[109] R. Dubrovka, C. Parini, Y. Hao, A. Alomainy, P. Hall, C. Constantinou, and Y. Nechayev, "On-body propagation loss estimation using method of equivalent sources," *Electron. Lett.*, vol. 42, no. 9, pp. 506–508, April 2006.

[110] Fort, F. Keshmiri, G. Roqueta Crusats, C. Craeye, and C. Oestges "A body area propagation model derived from fundamental principles: Analytical analysis and comparison with measurements," *IEEE Trans. Antennas Propag.*, vol. 58, no. 2, pp. 503–504, February 2010.

[111] R. Chandra and A. J. Johansson, "An analytical link-loss model for on-body propagation around the body based on elliptical approximation of the torso with arms' influence included," *IEEE Antennas Wireless Propag. Lett.*, vol. 12, pp. 528–531, April 2013.

[112] K. A. Norton, "The propagation of radio waves over the surface of the earth and in the upper atmosphere—Part I ground-wave propagation from short antennas, "*Proc. IRE*, vol. 24, no. 10, pp. 1367–1387, October 1936.

[113] A. Norton, "The physical reality of space and surface waves in the radiation field of radio antennas," *Proc. IRE*, vol. 25, no. 9, pp. 1192–1202, September 1937.

[114] A. Norton, "The propagation of radio waves over the surface of the earth and in the upper atmosphere—Part II The propagation from vertical, horizontal, and loop antennas over a plane earth of finite conductivity," *Proc. IRE*, vol. 25, no. 9, pp. 1203–1236, September 1937.

[115] Lea, P. Hui, J. Ollikainen, and R. G. Vaughan, "Propagation between on-body antennas," *IEEE Trans. Antennas Propag.*, vol. 57, no. 11, pp. 3619–3627, November 2009.

[116] J. Bae, H. Cho, K. Song, H. Lee, and H.-J. Yoo, "The signal transmission mechanism on the surface of human body for body channel communication," *IEEE Trans. Microw. Theory Techn.*, vol. 60, no. 3, pp. 582–593, March 2012.

[117] J. R. Wait, "The ancient and modern history of ground wave propagation," *IEEE Antennas Propag. Mag.*, vol. 40, no. 5, pp. 7–24, October 1998.

[118] N. Chahat, G. Valerio, M. Zhadobov, and R. Sauleau, "On-body propagation at 60 GHz," *IEEE Trans. Antennas Propag.*, vol. 61, no. 4, pp. 1876–1888, April 2013.

[119] G. A. Conway *et al.*, "An Analytical Path-Loss Model for On-Body Radio Propagation," in Proceedings of URSI International Symposium on Electromagnetic Theory (EMTS), Berlin, August 16–19, 2010, pp. 332–335.

[120] T. Alves, B. Poussot, and J. M. Laheurte, "Analytical propagation modeling of BAN channels based on the creeping-wave theory," *IEEE Trans. Antennas Propag.*, vol. 59, no. 4, pp. 1269–1274, April 2011.

[121] A. Guraliuc, M. Zhadobov, G. Valerio, N. Chahat, and R. Sauleau, "Effect of textile on the propagation along the body at 60 GHz," *IEEE Trans. Antennas Propag.*, vol. 62, no. 3, pp. 1489–1494, March 2014.

[122] A. Guraliuc, M. Zhadobov, G. Valerio, and R. Sauleau, "Enhancement of on-body propagation at 60 GHz using electro textiles," *IEEE Antennas Wireless Propag. Lett.*, vol. 13, pp. 603–606, March 2014.

[123] N. Chahat, C. Leduc, M. Zhadobov, and R. Sauleau, "Antennas and Interaction with the Body for Body-Centric Wireless Communications at Millimeter-Waves", in European Conference on Antennas and Propagation(EuCAP), Gothenburg, April 8–12, 2013, pp. 757–760.

[124] A. R. Guraliuc, N. Chahat, C. Leduc, M. Zhadobov, and R. Sauleau, "End-fire antenna for BAN at 60 GHz: impact of bending, on-body performances, and study of an on to off-body scenario", *Electronics*, vol. 3, no. 2, pp. 221–233, April 2014.

[125] K. L. Wong, and C. I. Lin, "Characteristics of a 2.4-GHz compact shorted patch antenna in close proximity to a lossy medium", *Microw. Opt. Technol. Lett.*, vol. 45, no. 6, pp. 480–483, June 2005.

[126] M. R. Islam, N. H. Chamok, and M. Ali, "Switched parasitic dipole antenna array for high-data-rate body-worn wireless applications", *IEEE Antennas Wireless Propag. Lett.*, vol. 6, pp. 693–696, June 2012.

[127] Apple Product Brief. Available at www.apple.com

[128] Samsung Product Brief. Available at www.samsung.com

[129] Jenny Y. Liu, Adrian Tang, Ning-Yi Wang, Qun Jane Gu, R. Berenguer, H. Hsieh, P. Wu, C. Jou, and M. F. Chang, "A V-band Self-Healing Power Amplifier with Adaptive Feedback Bias Control in 65 nm CMOS", in IEEE Radio Frequency Integrated Circuits Symposium (RFIC 2011), Baltimore, June 5–7, 2011, pp. 1–4.

[130] K. Raczkowski, S. Thijs, W. D. Raedt, B. Nauwelaers, and P. Wambacq, "50-to-67 GHz ESD-protected power amplifiers in digital 45 nm LP CMOS", in IEEE International Solid-State Circuits Conference–Digest of Technical Papers, San Francisco, February 8–12, 2009, pp. 382–383.

[131] W. L. Chan, J. R. Long, M. Spirito, and J. J. Pekarik, "A 60 GHz-band 1 V 11.5 dBm Power Amplifier with 11% PAE in 65 nm CMOS", in IEEE International Solid-State Circuits Conference–Digest of Technical Papers, San Francisco, February 8–12, 2009, pp. 380–381.

[132] N. Kurita and H. Kondoh, "60 GHz and 80 GHz Wide Band Power Amplifier MMICs in 90 nm CMOS Technology", in IEEE Radio Frequency Integrated Circuits Symposium, Boston, June 7–9, 2009, pp. 39–42.

[133] S. Aloui, E. Kerherve, R. Plana, and D. Belot, "RF-Pad, Transmission Lines and Balun Optimization for 60 GHz 65 nm CMOS Power Amplifier", in IEEE Radio Frequency Integrated Circuits Symposium, Anaheim, May 23–25, 2010, pp. 211–214.

[134] B. Martineau, V. Lnopik, A. Siligaris, F. Gianesello, and D. Belot, "A53-to-68 GHz 18 dBm Power Amplifier with an 8-Way Combiner in Standard 65 nm CMOS", in IEEE International Solid-State Circuits Conference–Digest of Technical Papers, San Francisco, February 7–11, 2010, pp. 428–429.

[135] M. Boers, "A 60 GHz Transformer Coupled Amplifier in 65 nm Digital CMOS", in IEEE Radio Frequency Integrated Circuits Symposium, Anaheim, May 23–25, 2010, pp. 343–346.

[136] Arun Natarajan, Scott K. Reynolds, Ming-Da Tsai, SeanT. Nicolson, Jing-Hong Conan Zhan, Dong Gun Kam, Duixian Liu, Yen-Lin Oscar Huang, Alberto Valdes-Garcia, and Brian A. Floyd, "A fully-integrated 16-element phased-array receiver in SiGe BiCMOS for 60-GHz communications", *IEEE J. Solid-State Circuits*, vol. 46, no. 5, pp. 1059–1075, May 2010.

[137] A. Tang, F. Hsiao, D. Murphy, I-N. Ku, J. Liu, S. D'Souza, N.-Y. Wang, H. Wu, Y.-H. Wang, M. Tang, G. Virbila, M. Pham, D. Yang, Q. J. Gu, Y.-C. Wu, Y.-C. Kuan, C. Chien, and M.-C. F. Chang, "A Low-Overhead Self-healing Embedded System for Ensuring High Yield and Long-Term Sustainability of 60GHz 4Gb/s Radio-on-a-Chip",

in IEEE International Solid-State Circuits Conference (ISSCC 2012), February 19–23, 2012, pp. 316–318.

[138] T. LaRocca, Y.-C. Wu, R. Snyder, J. Patel, K. Thai, C. Wong, Y. Yang, L. Gilreath, M. Watanabe, H. Wu, and M.-C. Frank Chang, "A 45GHz CMOS Transmitter SoC with Digitally-Assisted Power Amplifiers for 64 QAM Efficiency Improvement", in IEEE Radio Frequency Integrated Circuits Symposium (RFIC 2013), Seattle, June 2–4, 2013, pp. 359–362.

[139] Yu Kikuchi, Makoto Takahashi, Tomohisa Maeda, Hiroyuki Hara, Hideho Arakida, Hideaki Yamamoto, Yousuke Hagiwara, Tetsuya Fujita, Manabu Watanabe, Takayoshi Shimazawa, Yasuo Ohara, Takashi Miyamori, Mototsugu Hamada, Masafumi Takahashi, Yukihito Oowaki, " A 222mW H.264 Full-HD Application Processor with x512b Stacked DRAM in 40nm", in IEEE International Solid-State Circuits Conference, San Francisco, February 7–11, 2010, pp. 326–327 .

8

WEARABLE ACTIVE ANTENNA MODULES FOR ENERGY-EFFICIENT RELIABLE OFF-BODY COMMUNICATION SYSTEMS

PATRICK VAN TORRE, LUIGI VALLOZZI, AND HENDRIK ROGIER
Department of Information Technology, Ghent University, Ghent, Belgium
iMinds, Ghent, Belgium

8.1 INTRODUCTION

The term *off-body wireless communication* identifies a particular type of wireless communication that is performed between the human body and any device in the external world. Such a class of communication is in contrast to the so-called *on-body wireless communication*, that takes place between several devices worn on a human body, communicating with each other. Together, these two classes of communication are commonly indicated as *body-centric communications* [1].

A *wearable textile system* can be defined as a garment, realized by means of *smart textiles*, which extends the classic concept of clothing, providing a number of new functionalities, in addition to the classic protective function of clothing. *Smart textiles* can be viewed as special textiles that, in addition to the classical function of protection of the human body from the adversities of the environment, possess additional functionalities [2]. Briefly, smart textiles provide additional *sensing* or/and *actuating functions*. A wearable textile system is thus composed of several parts, each one implementing a certain functionality, which can be listed as *sensors*, *actuators*, *data processing unit*, *power supply*, *interconnections*, and *hardware for the communication system*.

Electromagnetics of Body-Area Networks: Antennas, Propagation, and RF Systems, First Edition.
Edited by Douglas H. Werner and Zhi Hao Jiang.

In the context of *wearable textile systems* and *off-body wireless communication*, *energy efficiency* and *reliability* of the communication link, which are two strongly related problems, represent a crucial matter, especially in human-life related applications such as rescue workers interventions or hospital patients monitoring. In particular, an *energy-efficient* wearable antenna system must be designed in such a way in order to require the lowest possible power consumption to achieve a given level of reliability (quantified by a given receive *Bit Error Rate—BER*). Equivalently, a *reliable* wearable antenna system is designed with the aim of ensuring a BER lower than a given maximum limit for Signal-to-Noise Ratio (SNR) larger than a specified value (e.g., BER $< 10^{-5}$ for SNR > 20 dB). In multipath off-body wireless communication, this represents a challenging task, due to a number of detrimental effects introduced by the channel. More specifically, in such a scenario, transmitter and receiver are typically in relative motion which, combined with the multipath nature of the propagation, results in well-known signal fluctuations phenomena at the receive side, such as *fading* and *shadowing*, *path loss* and *Doppler spread*, as explained in more detail in Section 8.2.1. As known from wireless communication theory [3], when additive noise is present on the receive signal, the detector is capable to correctly detect the received symbols, if the signal level is sufficiently larger than the additive noise. On the contrary, when signal fluctuations phenomena produce a particularly low signal level, this becomes indistinguishable from additive noise, resulting in errors in symbol detection, that is, an increase of BER, that means a decrease of reliability.

The use of multiple wearable textile antennas was proven to be an effective technique to counter the negative effects of channel fluctuation, that is, to increase reliability or, equivalently, energy efficiency. This can be done by implementing *receive/transmit diversity*, or more general Multiple-Input Multiple-Output (*MIMO*) communication [4–6], described in Section 8.2.2, and by means of appropriate signal-processing techniques on the transmitted and received signals, presented in Section 8.2.3.

Another recently developed technique that can be particularly effective to improve reliability, described in Section 8.4.3, is *spatial modulation (SM)*. SM allows to enhance spectral efficiency of the communication link by means of a low-complexity algorithm based on switching of symbols' transmission among multiple antennas.

Concerning novel antenna types, *active wearable antennas* are proposed as valid candidates to enhance the reliability of a communication link, by simply providing a remarkably higher gain, when compared to classical passive wearable antennas. A novel active antenna, developed by our research group, is presented in Section 8.3, which represents a very interesting solution to achieve remarkable higher gains in wearable antennas, allowing to overcome the absorption by the wearer's human body and guarantee higher reliability in off-body communication links.

Next to the implementation solutions summarized above, during the design phase of an off-body communication system, it is also very important, for the design engineer, to be able to model and simulate the communication before its real implementation, and predict thus its performance in terms of reliability and power efficiency. In the present chapter we present two modeling frameworks recently developed by the authors. The first one, described in Section 8.4.1, models the complex gain of a MIMO link affected by shadowing and fading. The model is based on the assumption that shadowing and fading are two mutually independent stochastic processes, each one characterized by its own correlation properties, expressed by a correlation matrix. Such a model has been proven to be capable to reproduce experimentally obtained BERs and capacity. Moreover, it can be used as a useful tool to generate channel gain realizations with given correlations for shadowing and fading. The second one, detailed in Section 8.4.2, is a modeling framework, based exclusively on deterministic methods, capable to reproduce and simulate an entire narrowband off-body MIMO communication link. In particular, the framework models the multipath propagation channel by means of *ray tracing*, and the on-body or stand-alone transmit and receive antennas, including the human body, by means of three-dimensional (3D) full-wave EM solvers. Moreover, the framework simulates the link level operation of the link, by means of a Monte Carlo simulation of a digital data transmission between transmit and receive sides, giving as results BER curves, SNRs, channel correlations and statistical distributions. The framework has been validated by modeling an off-body communication employing fourth-order diversity inside an office environment. The results in terms of BER curves obtained by the framework showed an excellent agreement with the experimental ones, demonstrating its validity.

Finally, two examples of applications are presented in Section 8.5, being an *on-body sensor network* and a *distributed exposimeter*, recently realized by our research group. The on-body sensor network is composed of several autonomous, energy-efficient 2.45 GHz wearable wireless sensor nodes, represented by a transceiver implemented on a very thin and flexible substrate, fully integrated onto a wearable patch antenna, which can be conveniently integrated in a garment worn on a human body. The use of multiple cooperating nodes on a human subject was successfully employed to realize a power-efficient and reliable on-body sensor network composed of four transmitting body-worn nodes, sending data detected by an accelerometer to a remote base station. The main purpose of such a sensor network is to increase the communication reliability of the link with the base station. This is realized by means of mutual cooperation between the nodes, realized by enabling *data forwarding*. In this technique, the data transmitted by one node to the receiver are forwarded and stored by the other TX nodes as well. Such a redundancy allows to reduce the rate of lost packets at RX, that means an increased reliability. The achievable performance of the 4-sensors network was experimentally verified and proven to be comparable with the one of a fourth-order diversity system. The sensor network has also been experimentally tested with three single nodes worn by three different human subjects performing independent movements, detected and transmitted by the nodes, demonstrating a very high link reliability in this case as well.

The distributed exposimeter consists of four wearable sensor nodes, similar to the ones used for the sensor network, employing a linearly polarized patch antenna operating in the GSM-900 frequency band. Such a system represents an improvement with respect to existing, commercially available exposimeters, which are composed of a single on-body sensor antenna, therefore suffering of inaccuracies due to body shadowing. In the proposed exposimeter, the use of several on-body sensor antennas, distributed on the human body (front, back, sides) results in a nearly omnidirectional total radiation pattern, making the reception independent on the direction of arrival of the incoming EM waves. The performance of such a system was tested during a measurement with a human subject walking in an urban environment, with the system detecting the received EM-field strength from GSM base stations located in the surroundings, demonstrating superior performance with respect to existing commercial products.

8.2 DIVERSITY AND MIMO TECHNIQUES FOR OFF-BODY WIRELESS CHANNELS

8.2.1 The Indoor Off-Body Radio Propagation Channel

In an indoor radio propagation scenario, the received signal can vary according to several phenomena, being *path loss*, *fading*, and *shadowing* [7]. For a moving transmitter and/or receiver, the signals also experience a variable *Doppler shift*, according to the mobile speed. In conventional radiowave propagation theory, the mechanisms behind path loss, fading, and shadowing are described separately. However, in practice, the three effects appear superimposed and are difficult to distinguish.

Fading causes the fastest signal variations and also produces the largest amplitude fluctuations over small displacements. The environment has a large influence on the amplitude of these fluctuations. If the displacement is more than a few wavelengths, a significant influence of shadowing is often present in the resulting signal, as different obstacles in the environment contribute differently to the radiowave propagation as the mobile user moves along.

8.2.1.1 Signal Variation Phenomena The different signal variation phenomena can be described as follows:

8.2.1.1.1 Path Loss Path loss results from the attenuation of the power density transported by an electromagnetic wave due to the geometrical increase of its wave front surface. The path loss experienced by a radio signal increases with the distance between the receiver and transmitter and can be expressed in dB by

$$PL(d) = PL(d_0) + 10 \cdot n \cdot log_{10}\left(\frac{d}{d_0}\right) \tag{8.1}$$

with PL the path loss in dB, d the distance in meter, $PL(d_0)$ the path loss in dB at a distance $d_0 = 1$ m, and n the path loss exponent.

8.2.1.1.2 Small-Scale Fading Small-scale fading refers to the rapid signal fluctuation that is caused by the alternating points of constructive and destructive interference experienced in the environment. The interfering signals consist of reflected and/or scattered components and potentially also a direct signal.

Figure 8.1 illustrates the behaviour of the received SNR at $f = 2.45$ GHz due to small-scale fading for a mobile receiver moving along a non line-of-sight path in an

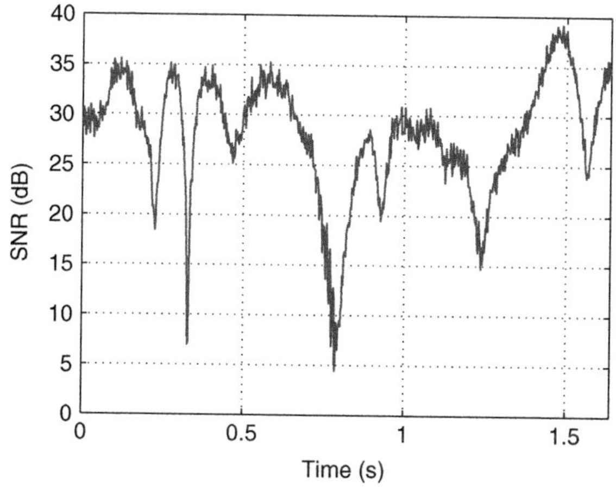

FIGURE 8.1 Time-varying received signal-to-noise ratio for a receiving antenna worn on a firefighter, walking in an indoor environment.

indoor environment, with fluctuation over a 34 dB range and several signal minima per second.

8.2.1.1.3 Shadowing Shadowing is a phenomenon where radio signals are attenuated by obstacles on their propagation path, causing an envelope fluctuation that varies more slowly than for small-scale fading, typically over distances larger than 10 λ. In the indoor environment, shadowing will always be accompanied by fading and, hence, the distribution of the signal envelope results from both effects.

8.2.1.1.4 Doppler Shift Radio signals experience a Doppler shift when the receiver and/or transmitter are moving, due to a superposition of reflected and scattered signals, all having different angles of arrival (AOA). In free space, the Doppler shift, expressed in Hz, is proportional to the relative speed between the transmitter and receiver and is given by $f_d = v \cdot f_c \cdot cos\phi/c$, with f_c the carrier frequency in Hz, v (in m/s) the relative speed between the transmit and receive antennas (positive when they are approaching each other), and ϕ the angle of arrival of the electromagnetic wave, relative to the direction of movement. The maximum Doppler shift expected at a given speed is $f_m = v \cdot f_c/c$. Hence, a superposition of signal components with various Doppler shifts is received, resulting in a Doppler spread in the range $f_c - f_m$ to $f_c + f_m$. In the demodulation of the received signals, the Doppler shift should be frequently estimated to compensate for the movements of the rescue worker.

8.2.2 Diversity and MIMO Systems

The described signal fluctuation phenomena, due to the multipath nature of a wireless off-body communication channel, result in errors in the data stream detected at the receive side of the link. Such errors occur when the received signal experiences a deep fade in amplitude, causing the additive noise to be larger than the signal itself, resulting in erroneous symbols detection at the receiver. The first well-known technique used to mitigate this negative effect is called *diversity*. The main concept is that by relying on multiple replicas of the same signal (differing from each other for a characteristic such as space, polarization, frequency, etc.), the probability that all of them experience a deep fade at the same time is much lower than having a deep fade on a single signal, that is, in absence of diversity. Diversity is practically implemented by deploying multiple antennas at the transmit or receive side, resulting in *transmit* or *receive* diversity. The simultaneous use of diversity at both sides of the link, that is, transmit and receive diversity, represents the generalization of diversity known as MIMO communication [8]. We briefly describe these techniques as follows.

8.2.3 Signal Processing Techniques in Diversity and MIMO Systems

In diversity and, more generally, MIMO communication systems, the different available replicas of the signals are processed using several techniques, aiming at the mitigation of errors on the received data stream. They can be classified according to the type of diversity (*transmit* or *receive*).

8.2.3.1 Receive Diversity Signal Processing Techniques

8.2.3.1.1 Selection Combining Selection Combining (SC) is the technically simplest form of receive diversity. The received signals are compared and the strongest signal is switched to the receiver chain. The comparison of signal strengths is possible using low-cost logarithmic detectors, producing an output voltage proportional to the signal strength in dBm. As only one full receiving chain is needed, selection combining provides a low-cost diversity solution.

8.2.3.1.2 Maximum Ratio Combining Maximum ratio combining (MRC) [9] is a receiver diversity technique where the different signals are combined constructively to achieve a resulting SNR that is higher than any of the separate channels. Theoretically, the resulting SNR is the sum of the SNRs for the individual receiving channels. Although the performance of MRC is better than for SC, the technique requires full receiving chains for each channel to be combined. An accurate channel estimation is necessary to optimally combine the signals with a correct phase relationship. MRC results in a higher SNR at the detector, with a gain depending on the distribution of the received signals and the gain equality for these channels. MRC is realized by constructive addition of the received signals, each scaled according to their power.

Using complex channel estimates, performing MRC for the received signals

$$\begin{aligned} r_0 &= h_0 \cdot s_0 + n_0 \\ r_1 &= h_1 \cdot s_0 + n_1 \end{aligned} \tag{8.2}$$

means that the received symbol is estimated as

$$\tilde{s}_0 = h_0^* \cdot r_0 + h_1^* \cdot r_1 \tag{8.3}$$

8.2.3.2 Transmit Diversity Signal Processing Techniques

8.2.3.2.1 Antenna Selection Diversity can be easily applied at the transmitting side by switching the output signals of the transmitter alternatingly to a number of transmit antennas. The difference with selection combining is that channel feedback from the receiving side is necessary if the transmitter has to be able to select the antenna that produces the strongest signal at the receiver.

Often channel feedback is not available and, in that case, the transmit power can be switched to all transmit antennas, sequentially. In this case the data rate is reduced by the number of transmit antennas as the same set of data symbols has to be repetitively transmitted on each antenna. If the reduction of data rate is acceptable, antenna selection provides a low-cost solution for transmit diversity.

8.2.3.2.2 Space-Time Coding Space-time block coding was first introduced by Siavash Alamouti [9] in 1998 for a synchronous transmission of data on two antennas operating at the same frequency. Without the need for channel feedback, the signals can be constructively combined at the receiver, with a performance similar to receiver MRC and without a reduction of the data rate.

To perform Alamouti space-time coding (STC) using two transmit antennas and one receiving antenna, the following symbols are transmitted in two subsequent time slots:

	Antenna 0	Antenna 1
time t	s_0	s_1
time t+T	$-s_1^*$	s_0^*

The transmitted symbols are then estimated as

$$\begin{aligned}\tilde{s}_0 &= h_0^* \cdot r_0 + h_1 \cdot r_1^* \\ \tilde{s}_1 &= h_1^* \cdot r_0 - h_0 \cdot r_1^*\end{aligned} \tag{8.4}$$

Note that, although the signal combination is performed differently compared to MRC, the result for the signals is equivalent. A scalar channel is created, multiplying the symbol amplitude by a factor $(\alpha_0^2 + \alpha_1^2)$ and preserving the phase.

8.2.3.3 MIMO Systems A MIMO system combines transmit and receive diversity to achieve the corresponding gains at both sides of the wireless communication link. A MIMO system is often labeled as $N_T \times N_R$ MIMO, with N_T the number of transmit antennas and N_R the number of receive antennas.

With orthogonal STC at the transmitter, an $N_T \times N_R$ MIMO system is equivalent to MRC receiver diversity of $(N_T \cdot N_R)$-th order, except for the potential absence of transmit array gain if the total transmitted power for all antennas is the same as for a similar Multiple-Input Multiple-Output (SISO) system. Figure 8.2 illustrates the cumulative distribution function (CDF) for measurements with 1×1 SISO, 2×2 MIMO, and 4×4 MIMO configurations [10], expressed as a function of the E_b/N_0

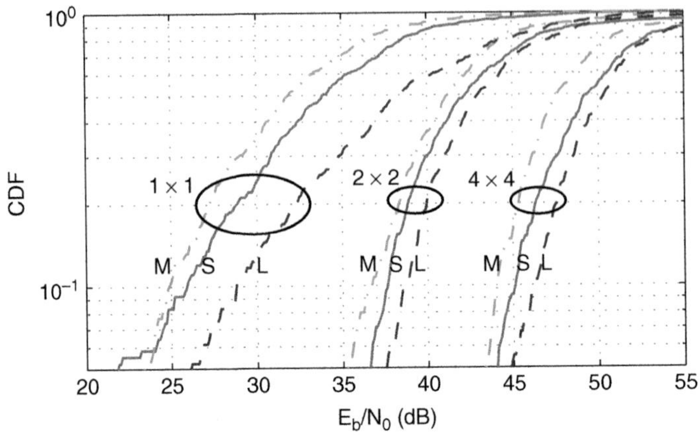

FIGURE 8.2 Cumulative distribution function of signal envelopes for persons equipped with front and back dual-polarized antennas for small, medium, and large body sizes measured for various MIMO orders.

in dB. E_b/N_0 refers to the ratio of the received energy per bit to the one-sided spectral noise density. For the Quadrature Phase-Shift Keying (QPSK) modulation used, SNR $= 2E_b/N_0$. The measurements are performed in non line-of-sight conditions with persons of substantially different body sizes. From the CDF, the gain achieved by employing MIMO techniques is clear.

- The average SNR is higher for a higher-order MIMO system.
- The slope of the CDF is steeper for a higher-order MIMO system, illustrating a reduced SNR variance.

Figure 8.3 illustrates the time-domain behavior of the SNR and BER for a walking firefighter [10]. Considering the SNR, both the higher average and reduced variance are visible. The bar graphs at the bottom show the bit error rates (BERs) after demodulation and detection of the QPSK symbols, with a significantly reduced BER for the higher-order MIMO systems. The large reduction in BER results from the combination of a higher average SNR and also a reduced variance of this SNR.

Defining $E_b/N_0 = SNR/2$ as the ratio of the energy transmitted per data bit to the one-sided spectral noise density, the bit error probability $P_b(i)$ for frame i with QPSK modulation is given by

$$P_b(i) = \frac{1}{2}\text{erfc}\left(\sqrt{\frac{E_b(i)}{N_0(i)}}\right). \tag{8.5}$$

Owing to this relationship, the BER is not only reduced by a larger average E_b/N_0, but also by a smaller variance. Clearly, the minima have the largest impact on the average BER, which is why the reduction of the signal variance obtained by employing MIMO techniques is so important.

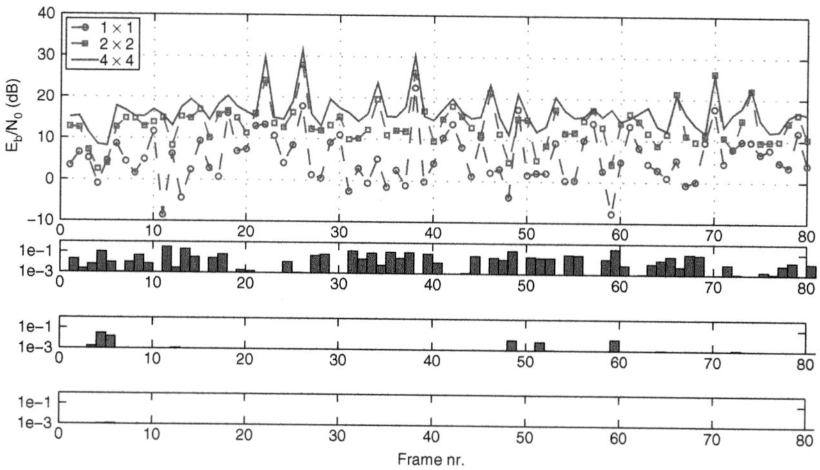

FIGURE 8.3 SNR and BER evolution for subsequent frames transmitted in 1 × 1, 2 × 2, and 4 × 4 configurations for the medium-sized person.

8.3 ACTIVE WEARABLE ANTENNAS: EFFICIENT DESIGN AND IMPLEMENTATION

In the recent past, the development and integration of unobtrusive wearable antennas and RF devices into clothing for body-centric communications gained increasing importance in the scientific community. This integration of such devices into garments known as *wearable textile systems* will offer additional functionalities to the wearers, improving their quality of life without impeding movements. This novel technology is mainly oriented to applications such as medical patients' health monitoring and protective clothing. For this reason, numerous studies were reported in literature, focusing on the design of wearable antennas, for both *off-body* [11] and *on-body* communications [12]. Despite the limitations imposed on transmit power levels in wireless personal area networks due to presence of the human body, communication links need to be highly reliable to ensure safe operation in emergency scenarios and health-care applications. Moreover, the proximity of the human body significantly affects the wearable antenna performance. Among observed effects there are, for example, higher losses due to power absorption by the human body, distortion of radiation patterns, resonance frequency variations, and input impedance mismatch due to capacitive coupling with the human body [13]. In order to avoid power absorption by the human body and to maximize the antenna gain, wearable antennas on Electromagnetic Band Gap (EBG) substrates were proposed in [14].

Active wearable antennas are very appealing solutions to overcome performance degradation in body-centric communication, due to the proximity of the human body. In such a technology, remarkably high gains are achieved by integrating active circuitry onto a conventional wearable antenna, allowing to overcome absorption by the human body of the wearer.

In this section, we present a comprehensive design paradigm for the design of energy-efficient active wearable antennas, based on the combination of circuit and full-wave optimization [15].

8.3.1 Full-Wave/Circuit Co-design Paradigm for Energy-Efficient Active Antennas in Professional Garments

8.3.1.1 Introduction In recent years, the increase of operational safety and efficiency of police, army, and rescue services has been object of extensive scientific research. More specifically, wearable electronic systems largely extend the potential of rescue worker protective garments by adding new functionalities such as sensing, localization, and wireless communication. Smart fabrics and interactive textiles (SFIT) [16] integrated into professional clothing do not impede movements during interventions and, at the same time, offer constant monitoring of activity, life parameters, and environmental conditions, relaying this information in a wireless way to a remote supervision post for operations' coordination. Moreover, alarms and specific instructions can be fed back to each subject in action, in case of hazardous situations. *Reliability* and *autonomy* of SFIT are two key design requirements, since such systems are deployed for harsh and critical conditions of operation. The electronic components must be highly energy-efficient, in order to ensure sufficient autonomy and avoiding, at the same time, the use of heavy battery units. Since most of the required power is employed to establish the wireless link, it is very preferable that textile antennas exhibit large gain and high radiation efficiency. A large amount of space is available in garments, enabling to implement antenna systems with these characteristics. Moreover, by backing the antenna with a large ground plane, it is possible to drastically reduce the absorption of radiated power by the human body. However, during the design phase of wearable flexible antennas, particular care has to be taken to counter antenna performance degradation caused by bending, wrinkling, and crumpling [17, 18], resulting by wearer's movements. Moreover, an appropriate material selection is required, favoring the choice of substrate materials that exhibit low losses, due to absorption of humidity by the substrate fabric. Also to protect the antenna from washing, a thermoplastic polyurethane (TPU) coating is often advisable. Next to the increase in power efficiency in SFIT systems, energy scavenging can be performed by adding harvesters that are able to pick energy from the different available sources that can be located in proximity of the body, resulting in increased operational autonomy. In this section, a dedicated comprehensive design paradigm is described, to implement energy-efficient active wearable antennas, exhibiting stable performance when integrated in rescue workers' professional garments. To increase autonomy and reliability of the active antenna modules, electronics were integrated on the feed plane of the textile antenna underneath its ground plane and energy harvesters positioned directly on the antenna plane. In order to preserve the antenna performance after the integration of the additional components, specific measures were taken, which are outlined as follows. In particular, in the following subsection we describe a novel comprehensive design paradigm for the design of autonomous active wearable antennas, relying on full-wave/circuit co-design and co-optimization. A compact communication module, as described in detail in [15], can be obtained by integrating active electronic circuitry underneath the wearable antenna, avoiding fragile connections that can be easily damaged when under stress during activity. In addition, the placement of the active electronic circuits directly on the feed plane allows the minimization of the length of the radio-frequency (RF) connections, resulting in smaller signal attenuation, reduced electromagnetic compatibility issues, and therefore a higher signal integrity. In a subsequent phase, the active antennas can be equipped with energy harvesters, which increase their autonomy. This can be done, for example, by integrating a set of solar cells directly above the radiating patch of a wearable module, which do not influence its radiation characteristics. The described co-design strategy was successfully used for the design of a wearable low-cost *through-wall Doppler radar* for rescue workers and a *wearable active Galileo E1-band antenna*, which are described in detail in [15].

8.3.1.2 Circuit/Full-Wave Co-design Paradigm In the proposed technique, the active antenna design is performed by means of a joint circuit/full-wave optimization for simultaneous antenna and active electronics design to achieve a multitude of design goals [19, 20]. Typically, impedance matching and circular polarization represent the desired characteristics for the antenna, as well as high gain, input and output matching, together with a low-noise figure (NF) for a low-noise amplifier (LNA) connected to the antenna output. We have developed two ad-hoc strategies that keep the design process manageable for a large number of available optimization variables [19].

The first type of design flow, displayed in Figure 8.4, aims at the design of a module with optimized noise characteristics. In the first stage of the full-wave/circuit co-optimization procedure, an optimal complex radiation impedance for the passive textile antenna is determined, which produces the minimum NF at the output of the LNA. Such an impedance is computed for the active electronics only, resulting in a preliminary optimal LNA design. A full-wave simulator is employed in the process, generating an N-port S-matrix which expresses the interconnections of the active electronics circuit. In a subsequent circuit simulation, the previously obtained S-matrix is combined with the N-port descriptions of all active and passive lumped

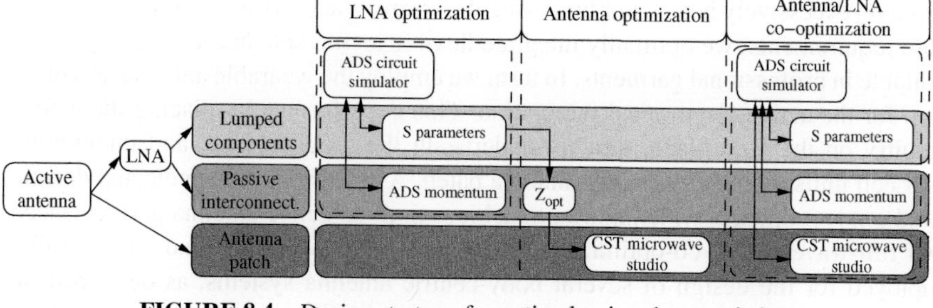

FIGURE 8.4 Design strategy for optimal noise characteristics.

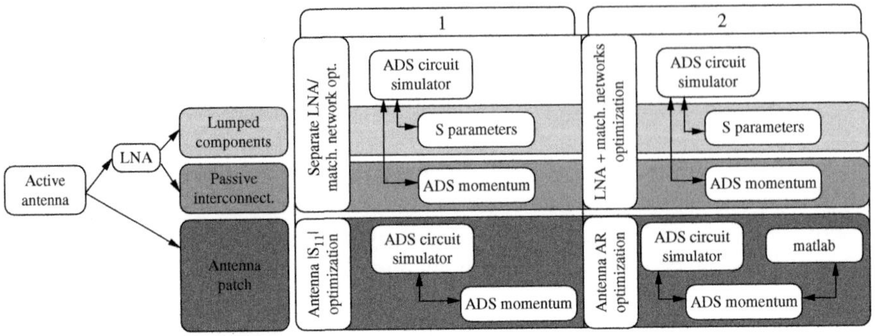

FIGURE 8.5 Design strategy for optimal impedance matching.

components that are present in the circuit. In the second stage, the dimensions of the passive antenna are optimized by means of a full-wave EM solver, to determine the values that provide the optimal radiation impedance. In the third and final step, the co-optimization is performed to jointly maximize the performance of the entire system composed of antenna and active electronics. In Figure 8.4, the design flow is displayed for optimal noise characteristics without the need of a matching network. It guarantees short RF connections as well as no excess losses due to additional matching-network components. The strategy employs the most appropriate simulators for each part of the active textile antenna. More specifically, the 3D full-wave frequency-domain simulator of CST Microwave Studio was used to model the passive antenna, which allows to take into account the finite conductivity of the antenna plane and ground plane made out of the electrotextile Flectron. Since the LNA circuit is realized on a thin polyimide flex, the planar 3D full-wave simulator ADS Momentum is more appropriate to model the circuit's interconnections. We refer to [19] for further details.

A second type of design flow, aiming at the joint optimal impedance matching of passive antenna and the LNA is displayed in Figure 8.5. Only the planar full-wave simulator ADS Momentum is used in such a design process, since all the conductive layers are made out of copper patterns on polyimide flex. More details can be found in [19, 21].

Smart electronic systems for localization, sensing, and wireless communications are required to be robust, reliable, and to possess a high degree of autonomy while not being excessively heavy, nor hindering the wearer's activity and comfort. To meet these requirements, we optimally integrate flexible textile antennas on the large area available in professional garments. In turn, we employ the wearable antenna as a platform for the active electronics' integration. This is performed by placing the active circuitry on the feed plane, directly underneath the ground plane, and integrating solar cell units directly above the antenna patch. A comprehensive dedicated design paradigm is proposed, which allows to achieve optimal active antenna performance. Two full-wave/circuit co-optimization strategies were developed and successfully employed for the design of several body-centric antenna systems, as described in [15]. We exploit the large area available in garments in order to implement a multiple antenna system, which improves the signal quality by means of diversity and MIMO communication schemes.

8.4 BODY-CENTRIC MIMO CHANNELS

Wearable textile antenna systems conveniently allow rescue workers to communicate vital information captured by sensors, employing a fully garment-integrated system. The body provides a large area to deploy multi-antenna systems for spatial diversity. Additionally, dual-polarized antennas allow to double the number of signals for an equal number of antenna patches. In an indoor propagation environment, the signals are influenced by fading and shadowing. Shadowing occurs not only due to objects in the environment, but also by the human body, carrying the antennas.

For the development of off-body communication systems, it is important to have models of the propagation to allow the prediction of the system's behaviour and performance. Two models are described here, a *stochastical model*, relying on the correlation properties of fading and shadowing, and a *deterministic model*, relying on ray tracing based on a detailed description of the propagation environment.

The specific properties of body-centric communication allow the use of a new modulation technique, known as *spatial modulation*, specifically exploiting those properties to provide more data throughput.

8.4.1 Off-Body Stochastic Modeling

8.4.1.1 The Wireless Channel Indoor body-centric communication is influenced by fading and shadowing, of which the statistical distributions are Rayleigh and log-normal, respectively. In the correlated shadowing model, fading and shadowing are considered to be independent and can hence be modeled by means of separate correlation matrices.

It is shown that by separating shadowing from small-scale fading, the main mechanisms of the off-body communication channel are accurately captured by the model:

- Correlated fading and correlated shadowing are considered mutually independent, therefore the channel is accurately characterized by means of the shadowing and fading correlation matrices.
- Experimentally determined BER and channel capacity characteristics are accurately reproduced by the model.
- Random channel realizations with correlation properties corresponding to the modeled environment can be generated for simulating new modulations and codings.

The model's correlation parameters are determined based on channel measurements in non line-of-sight conditions with constant path loss [22]. The rescue worker with the antennas deployed into his garment walks at approximately 17 m from the transmitter, with the direct signal path blocked by brick walls. Channel measurements are performed for a 4×4 MIMO configuration with the rescue worker equipped with front and back dual-polarized textile antennas [23]. The base station uses similar antennas, separated by a center-to-center distance of either 10λ (122 cm) or 1.5λ (18 cm), to allow the comparison of the behaviour depending on the spatial antenna separation. All antenna patches are mounted for $+45°/-45°$ slanted cross-polarization.

A 3/4-rate 4×4 orthogonal space-time code [24, pp. 194 (5.143)] is employed for the MIMO transmission. In a system with N_R receive and N_T transmit antenna ports, the received signal corresponding to a code word $\mathbf{C}$ is represented by

$$\mathbf{R} = \mathbf{H}\mathbf{C} + \mathbf{W}. \tag{8.6}$$

In our 4×4 MIMO transmission, $\mathbf{R}$, $\mathbf{H}$, $\mathbf{C}$, and $\mathbf{W}$ are all matrices of dimensions 4×4.

The value $h_{m,n} = (\mathbf{H})_{m,n}$ represents the complex channel gain corresponding to the mth receive and nth transmit antenna port. $\mathbf{C}$ is a space-time matrix with orthogonal rows, whose elements are linear functions of QPSK modulated information symbols, with variance σ_s^2, and their complex conjugates. The noise matrix $\mathbf{W}$ is composed of i.i.d. complex-valued Gaussian random variables, with independent real and imaginary parts having equal variances $N_0/2$. The variable $r_{m,k} = (\mathbf{R})_{m,k}$ represents the signal received by the mth receive antenna during the kth time slot of the space-time code.

The channel matrix $\mathbf{H} = \mathbf{A}_{sh} \circ \mathbf{H}_{ss}$ is the Hadamard product ($\circ$) of a shadowing matrix $\mathbf{A}_{sh}$ and a fading matrix $\mathbf{H}_{ss}$, both of dimensions 4×4. We define $\mathbf{h}$, α_{sh} and $\mathbf{h}_{ss}$ as vectors of dimension 16×1, obtained by stacking the columns of the matrices $\mathbf{H}$, $\mathbf{A}_{sh}$ and $\mathbf{H}_{ss}$, respectively. Hence, $\mathbf{h}$ can be expressed as $\mathbf{h} = \mathbf{D}_{sh}\mathbf{h}_{ss}$, where $\mathbf{D}_{sh}$ is a 16×16 diagonal matrix with $(\mathbf{D}_{sh})_{i,i} = \alpha_{sh,i}$. Assuming small-scale Rayleigh fading, $\mathbf{h}_{ss}$ is complex Gaussian distributed with zero average and correlation matrix $\mathbf{R}_{ss} = E[\mathbf{h}_{ss}\mathbf{h}_{ss}^H]$.

8.4.1.2 Separation of Shadowing and Fading

In case of lognormal shadowing distribution, the real quantities $\alpha_{\mathrm{dB},i} = 20log(\alpha_{sh,i})$ are Gaussian distributed with zero mean and correlation matrix $\mathbf{R}_{\mathrm{dB}} = E[\alpha_{\mathrm{dB}}\alpha_{\mathrm{dB}}^T]$. If fading and shadowing are statistically independent, the correlation matrices $\mathbf{R}_{ss}$ and $\mathbf{R}_{\mathrm{dB}}$ fully characterize the statistics of the MIMO channel. Accepting that fading and shadowing are statistically independent, the resulting channel correlation matrix $\mathbf{R}_h = E[\mathbf{h}\mathbf{h}^H]$ is determined by

$$(\mathbf{R}_h)_{i,j} = (\mathbf{R}_{sh})_{i,j}(\mathbf{R}_{ss})_{i,j}, \tag{8.7}$$

where the correlation matrix $\mathbf{R}_{sh} = E[\alpha_{sh}\alpha_{sh}^T]$ can be computed from $\mathbf{R}_{\mathrm{dB}}$ as follows:

$$(\mathbf{R}_{sh})_{i,j} = E[\alpha_{sh,i}\alpha_{sh,j}] = \exp\left(\frac{1}{2} \cdot \frac{\ln(10)^2}{10} \cdot \frac{(\mathbf{R}_{\mathrm{dB}})_{i,i} + (\mathbf{R}_{\mathrm{dB}})_{j,j} + 2(\mathbf{R}_{\mathrm{dB}})_{i,j}}{4}\right). \tag{8.8}$$

8.4.1.3 Bit Error Rate Calculation The BER for a space-time code over the MIMO channel, fully characterized by $\mathbf{R}_{\mathrm{dB}}$ and $\mathbf{R}_{ss}$, is now calculated as follows. The conditional BER for QPSK modulation with a given $\mathbf{h}$ is known to be [25]

$$BER(\mathbf{h}) = Q\left(\sqrt{\frac{\sigma_s^2}{N_0}|\mathbf{h}|^2}\right). \tag{8.9}$$

Averaging over the small-scale fading $\mathbf{h_{ss}}$ and shadowing α_{sh} results in the average BER. First the averaging is performed over $\mathbf{h_{ss}}$ for a given $\mathbf{D}_{sh}$. For a given $\mathbf{D}_{sh}$, $\mathbf{h}$ represents Rayleigh fading with correlation matrix $\mathbf{D}_{sh}\mathbf{R}_{ss}\mathbf{D}_{sh}$.

The result of averaging $BER(\mathbf{h})$ over the fading for a given shadowing realization is a function of σ_s^2/N_0 and $\mathbf{D}_{sh}\mathbf{R}_{ss}\mathbf{D}_{sh}$:

$$E_{\mathbf{h}_{ss}}[BER(\mathbf{D}_{sh}\mathbf{h}_{ss})] = g(\sigma_s^2/N_0; \mathbf{D}_{sh}\mathbf{R}_{ss}\mathbf{D}_{sh}). \tag{8.10}$$

An analytical expression for $g(.;.)$ can be found in [26–28]. Subsequently, averaging is performed over the shadowing:

$$BER_{avg,th} = E_{\mathbf{D}_{sh}}[g(\sigma_s^2/N_0; \mathbf{D}_{sh}\mathbf{R}_{ss}\mathbf{D}_{sh})]. \tag{8.11}$$

$BER_{avg,th}$ is the average BER corresponding to the theoretical channel model, characterized by $\mathbf{R}_{\mathrm{dB}}$ and $\mathbf{R}_{ss}$. The averaging operator in equation (8.11) is calculated by means of Monte Carlo integration:

$$BER_{avg,th} = \frac{1}{N}\sum_{n=1}^{N} g(\sigma_s^2/N_0; \mathbf{D}_{sh}(n)\mathbf{R}_{ss}\mathbf{D}_{sh}(n)). \tag{8.12}$$

In this equation, $\mathbf{D}_{sh}(n)$ is a diagonal matrix defined as $(\mathbf{D}_{sh}(n))_{i,i} = 10^{\alpha_{\mathrm{dB},i}(n)/20}$, with $\alpha_{\mathrm{dB}}(n) = \mathbf{A}\mathbf{u}(n)$. The lower triangular matrix $\mathbf{A}$ results from the Cholesky decomposition of $\mathbf{R}_{\mathrm{dB}} = \mathbf{A}\mathbf{A}^T$. The set $\{\mathbf{u}(n), n = 1, \ldots, N\}$ consists of N statistically independent realizations of a real Gaussian vector, with zero average and uncorrelated components with unit variance.

The performance indicator $BER_{avg,th}$ will be calculated as a function of E_b/N_0, with E_b the average energy per bit per receive antenna. For QPSK modulation and the considered space-time codes,

$$E_b = \frac{\sigma_s^2 \cdot E[|h|^2]}{2N_R} = \frac{\sigma_s^2 \cdot Tr(\mathbf{R}_h)}{2N_R}, \tag{8.13}$$

with $\mathbf{R}_h$ calculated based on $\mathbf{R}_{\mathrm{dB}}$ and $\mathbf{R}_{ss}$.

The theoretical error performance $BER_{avg,th}$ from equation (8.12) will now be compared to the $BER_{avg,meas.}$ obtained from the channel measurements.

$$BER_{avg,meas.} = \frac{1}{M}\sum_{m=1}^{M} BER(\mathbf{h}(m)), \tag{8.14}$$

with $BER(.)$ the conditional BER (8.9) and $\{\mathbf{h}(m), m = 1, \ldots, M\}$ a set of M channel measurements. The average received bit energy per receive antenna that corresponds to the measurements is given by

$$E_b = \frac{\sigma_s^2}{2N_R} \frac{1}{M} \sum_{m=1}^{M} |\mathbf{h}(m)|^2. \tag{8.15}$$

8.4.1.4 Shadowing Parameter Extraction: Standard Deviation The 4×4 MIMO channels consists of 16 SISO channels. The shadowing standard deviation $\sigma_{\mathrm{dB},i}$ for the ith SISO channel is determined by considering a trial value $\sigma_{\mathrm{dB}}^{(i)}$. The corresponding small-scale fading variance $(\mathbf{R}_{ss})_{i,i} = (\sigma_{ss}^{(i)})^2$ that is consistent with the measurements $\{h_i(m), m = 1, \ldots, M\}$ is determined from

$$(\mathbf{R}_{sh})_{i,i}(\mathbf{R}_{ss})_{i,i} = \frac{1}{M} \sum_{m=1}^{M} |h_i(m)|^2, \tag{8.16}$$

where $(\mathbf{R}_{sh})_{i,i} = E[\alpha_{sh,i}^2]$ can be computed from $\sigma_{\mathrm{dB}}^{(i)}$ (8.8).

The shadowing standard deviation estimates, found for the SISO channels composing the measured 4×4 MIMO links with 10λ and 1.5λ TX antenna separation [22], are shown in Table 8.1.

TABLE 8.1 Shadowing factors $\sigma_{\mathrm{dB}}^{(i)}$ for the 16 SISO channels composing the 4×4 MIMO links with 10λ and 1.5λ TX antenna separation. Unequal average channel gain illustrated by $10log(E[|h_i|^2])$ values, normalized to 0 dB for the strongest channel. *Source*: Van Torre et al. 2012 [22]. Reproduced with permission of IEEE.

		10λ		1.5λ	
TX ant.	RX ant.	$10log(E[\|h_i\|^2])$	$\sigma_{\mathrm{dB}}^{(i)}$	$10log(E[\|h_i\|^2])$	$\sigma_{\mathrm{dB}}^{(i)}$
1	1	−1.6059	2.1584	−2.4898	1.4385
1	2	−1.9405	1.7713	−1.4675	3.3456
1	3	−4.6638	2.0962	−7.1399	3.8046
1	4	−2.8610	1.5815	−2.9732	1.2774
2	1	−0.4957	2.3589	0	3.6634
2	2	−0.8068	0.5304	−0.9383	2.2557
2	3	−4.5796	1.5501	−7.2706	2.5974
2	4	−0.5549	1.7198	−0.9591	3.9298
3	1	−0.1340	2.0013	−0.3374	1.2791
3	2	0	1.155	−1.4767	1.0524
3	3	−3.4440	1.7413	−7.8415	1.1525
3	4	−0.2739	1.6356	−2.5750	0.7179
4	1	−1.4143	0.2036	−3.4653	0.7280
4	2	−1.6455	0.6695	−1.5893	1.5102
4	3	−5.0189	1.3132	−8.4081	0.9396
4	4	−2.4016	0.0998	−4.1365	1.1217

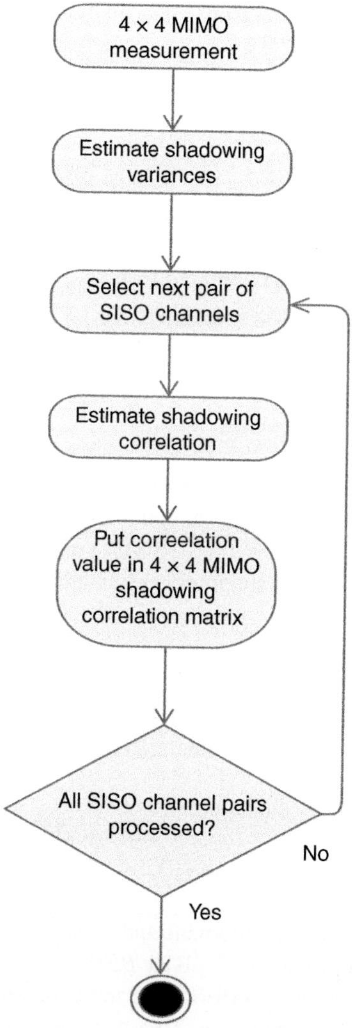

FIGURE 8.6 Flowchart of the shadowing estimation procedure determining the correlation coefficients for all 120 possible combinations of 2 out of 16 measured SISO channels.

8.4.1.5 Shadowing Parameter Extraction: Cross Correlation Values After estimating the shadowing variances of the SISO channels, the shadowing correlation coefficients between different SISO channels are determined. Figure 8.6 displays a flowchart of the estimation procedure.

Consider a pair (h_i, h_j) of SISO channels $(i \neq j)$. For i and j ranging from 1 to 16, the number of pairs to be considered equals $(16 \times 15)/2 = 120$. The shadowing correlation $\rho_{\mathrm{dB}}^{(i,j)}$ and the fading correlation $\rho_{ss}^{(i,j)}$ between the considered channels are defined through the relations $(\mathbf{R}_{\mathrm{dB}})_{i,j} = \sigma_{\mathrm{dB}}^{(i)}\sigma_{\mathrm{dB}}^{(j)}\rho_{\mathrm{dB}}^{(i,j)}$ and $(\mathbf{R}_{ss})_{i,j} = \sigma_{ss}^{(i)}\sigma_{ss}^{(j)}\rho_{ss}^{(i,j)}$.

Let us denote by $\rho_{\text{dB}}^{(i,j)} \in (-1, 1)$ a trial value of the shadowing correlation. The corresponding fading correlation $\rho_{ss}^{(i,j)}$ that is consistent with the measurements $\{h_i(m), h_j(m), m = 1, \ldots, M\}$ is determined from

$$(\mathbf{R}_{sh})_{i,j}(\mathbf{R}_{ss})_{i,j} = \frac{1}{M}\sum_{m=1}^{M} h_i(m)h_j^*(m), \tag{8.17}$$

where $(\mathbf{R}_{sh})_{i,j} = E[\alpha_{sh,i}\alpha_{sh,j}]$ can be computed from $\sigma_{\text{dB}}^{(i)}$, $\sigma_{\text{dB}}^{(j)}$ and $\rho_{\text{dB}}^{(i,j)}$ [25].

After determining the values of $\sigma_{\text{dB}}^{(i)}$ and $\sigma_{ss}^{(i)}$ for all 16 SISO channels and $\rho_{\text{dB}}^{(i,j)}$ and $\rho_{ss}^{(i,j)}$ for all 120 combinations of two SISO channels, we construct the shadowing correlation matrix (of dimension 16×16) that corresponds to the MIMO setup with four transmit and four receive antennas.

8.4.1.6 *Measured Versus Calculated MIMO BER Characteristics* The bit error calculations are now applied to the experimental datasets, each containing 2500 measured frames of the 4×4 MIMO link (i.e., 16 SISO channels have been simultaneously measured). The number N of shadowing realizations used in the BER calculation (8.12) is:

- $N = 10^4$ for the estimation of $\sigma_{\text{dB}}^{(i)}$ values per SISO channel.
- $N = 10^4$ for the estimation of $\rho^{(i,j)}$ per pair of SISO channels.
- $N = 10^5$ for the generation of the calculated MIMO BER characteristics (8.12).

Figure 8.7 displays the BER characteristics $BER_{avg,th}$ from equation (8.12) and $BER_{avg,meas}$ from equation (8.14) for the 4×4 MIMO link, with two dual-polarized antenna patches on both sides of the link (TX1-4, RX1-4). TX antenna spacings of 1.5λ and 10λ are considered [22].

The BER curves related to measurement and model agree very well. The BER performance of the link improves by 1 dB at $BER = 10^{-3}$ when going from 1.5λ to 10λ transmit antenna spacing, a performance increase which is mainly attributed to the increased shadowing decorrelation. This performance increase is reproduced accurately by the model.

8.4.1.7 *Measured Versus Calculated MIMO Capacity* Channel capacity characteristics are also accurately reproduced by the model. Ergodic capacity (in bit/s/Hz) is defined as [3]:

$$\overline{C} = E\left[log_2\left(det\left[\mathbf{I}_{N_R} + \frac{\rho}{N_T}\mathbf{H}\mathbf{H}^H\right]\right)\right], \tag{8.18}$$

where ρ is the ratio of total transmit power to noise power, $det(.)$ the determinant, and $E[.]$ the expectation over the channel statistics. Figure 8.8 displays a good match of the capacity characteristics for measurement and model [22].

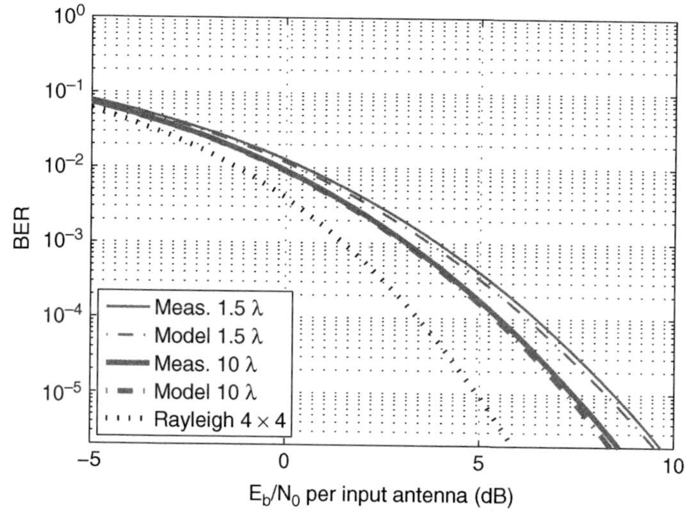

FIGURE 8.7 BER characteristics for the two 4×4 measurements, with the TX antennas spaced 1.5λ and 10λ apart, versus theoretical BER curves based on the fading and shadowing correlation matrices. The dotted curves represent the theoretical BER for i.i.d. Rayleigh fading channels.

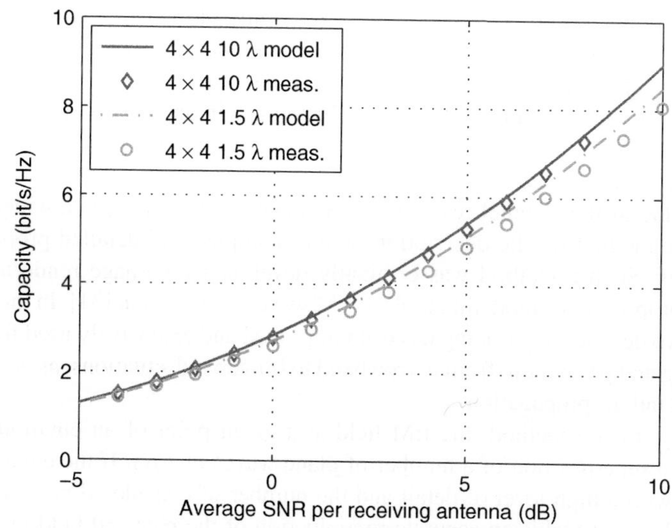

FIGURE 8.8 Ergodic capacity characteristics for model and measurement for different MIMO orders, transmit antenna spacings, and diversity types. Upper set of curves for 4×4 MIMO at 10λ or 1.5λ. Lower set for 2×2 MIMO at 10λ with spatial (sp.) diversity or polarization (pol.) diversity.

8.4.2 Off-Body Deterministic Modeling

8.4.2.1 ***Introduction*** In order to characterize an off-body wireless communication link, models and simulation tools are needed. Even though real measurements are the most accurate way to analyze such communication links, simulation models may represent a valuable alternative, allowing to predict the link performance with an acceptable accuracy, as well as in a relatively fast and inexpensive way. In recent times, a large number of tools have been developed for modeling of antennas, human body, and communication channel, but few work has been done to develop a unique simulation tool that accurately models the complete off-body communication link. Therefore, we developed a novel fully-deterministic simulation framework for this purpose [29, 30]. In particular, by combining a 3D full-wave EM solver for body-worn antenna modeling, a ray tracer for channel modeling and Matlab for digital data transmission simulations, the complete off-body communication link can be modeled and its performance in terms of SNR, BER, and channel correlations can be predicted. The validation of the proposed approach was performed by modeling the measurement campaign scenario described in [5] and comparing the simulated and measured performance in terms of BER, SNR, and correlations. A good agreement was obtained, demonstrating the effectiveness of the proposed approach.

During the last decades, many methods and tools were developed for modeling each element of a communication system separately. As far as antennas are concerned, full-wave 3D EM solvers are largely available nowadays, enabling to simulate every kind of radiator. Moreover, 3D electromagnetic human body models were developed, and they can be included in close proximity of the antennas. Concerning the wireless propagation channel, a high number of models are available, that can be classified as *analytical*, *empirical*, *stochastic* and *deterministic*.

Deterministic methods model the channel response in an *exact* way. This means that, given the exact geometrical and dielectric description of the propagation environment, together with the characteristics and positions of transmitter and receiver, the exact value of the channel response is calculated. *Ray tracing* is a widely used deterministic method for the description of very complex and detailed propagation environments. Such a method was originally developed for image rendering, and afterwards proposed as a modeling tool for EM waves propagation [31]. In the course of the last two decades, ray tracing has been improved and extensively used for channel modeling and prediction, from narrowband to broadband situations, as well as for outdoor and indoor propagation.

In the ray tracing method, the EM field at a given point of an environment is obtained as a superposition of a number of plane waves or *rays*. If the environment is modeled with a high level of detail and the number of considered rays is chosen sufficiently large, it yields an accurate reproduction of the received field in the real situation. The drawback of such a method lies in the trade-off between accuracy and computational cost, which are inversely proportional to each other.

Ray tracing was chosen as a channel modeling technique for the present work, because indoor environments may be characterized by highly variable characteristics and propagation mechanisms, making it very difficult to model them by stochastic or any other above-mentioned methods. In this section, we build a framework for off-body communication systems modeling, which is capable of describing the complete communication link, including wearable antennas, wearer's human body, and wireless channel. Therefore, we make use of deterministic methods, rather than analytical, stochastic, or other non-deterministic methods. Moreover, the proposed framework gives the possibility to simulate a digital base-band data transmission for determination of performance parameters such as *received* SNRs, BER, *channel statistics*, and *channel correlations*.

8.4.2.2 Description of the Modeling Framework In order to build a model responding to the above mentioned characteristics, we applied several software tools, for each of the main blocks composing the complete system, and combined their results. A conceptual block scheme of a most general MIMO off-body communication system, between M fixed antennas at the base station and N wearable antennas on a human body, is depicted in Figure 8.9. The corresponding tools, used to model the main elements of such an off-body communication system, together with the corresponding outputs, used by the model to construct the channel response, can be shortly described as follows:

- *Transmitting and receiving antennas* are completely characterized by their radiation vectors $\bar{F}_{tx}(\theta, \phi)$, $\bar{F}_{rx}(\theta, \phi)$. In the proposed framework they are obtained by 3D EM solver simulations or by analytical expressions for basic types of antenna such as dipoles. The use of the radiation vector was chosen because it contains information about the *polarization* and *phase* of the antenna, in contrast to the gain. Knowledge of the polarization of transmitting and receiving antennas is very important since it will have an effect on the system performance, especially when exploiting polarization diversity.

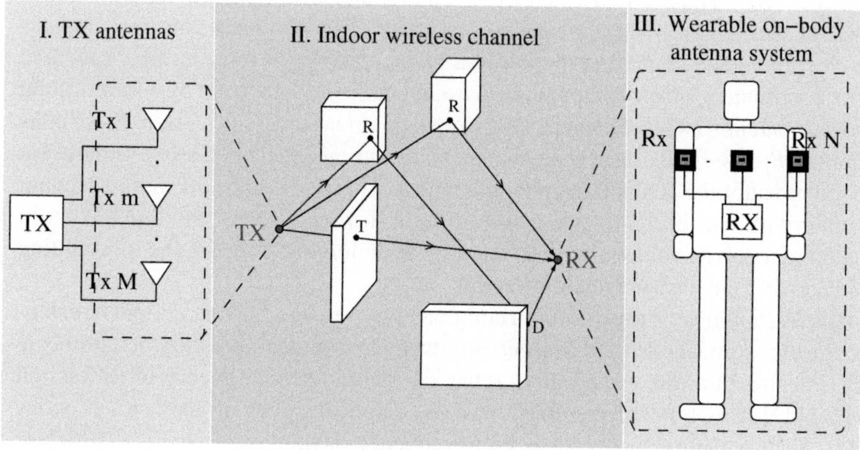

FIGURE 8.9 Schematic overview of an off-body indoor wireless communication system.

- *The propagation channel response* is determined by ray tracing, which calculates the *vectorial complex channel responses*, indicated as $\bar{g}^{(V)}, \bar{g}^{(H)}$. These complex vectors represent the fields radiated by transmitting vertically and horizontally polarized isotropic antennas, respectively. They are computed for a given input power at the isotropic antenna terminals, and for given transmitter and receiver positions in the considered propagation environment. The vectorial channel responses are computed as a superposition of L components, where L is the number of propagating rays considered in the ray tracing.

  Such responses do not take into account the radiation patterns of the actual transmitting and receiving antennas yet, that will be included in a following stage in order to obtain the *scalar channel response h*, defined as the narrowband scalar response between the input terminal of the transmitting antenna and the output terminal of the receiving antenna.

8.4.2.3 Modeling the Indoor Environment A commercially available ray tracing software called *WinProp*™ (by "AWE communication") was employed to model the indoor environment and perform the ray tracing computation of the channel response.

8.4.2.3.1 CAD Tool for Environment Drawing and Database Calculation The first block of the software, called WallMan, is a CAD-drawing software tool that is used to make a technical drawing of the indoor environment in which the simulated communication system operates. The tool possesses a user interface in which all the constitutive elements of the considered environment can be drawn such as walls, floor, ceiling, door and windows, desks, cupboards, and other furniture elements. Each object is defined by entering the coordinates of its vertexes together with its dielectric parameters, that is, ε_r and σ. The program puts the constructed model in an appropriate format to be given as input to the following block that represents the actual ray tracing program.

8.4.2.3.2 Ray Tracing and Plane-Wave Approximation The second constitutive block of the WinProp software is the ray tracer. This computer routine computes the electromagnetic field distribution in a given rectangular receiver area, specified by its coordinates, given as input the environment model, produced by a transmitting *isotropic antenna* (with vertical and horizontal polarization), with a given power at its terminals P_t. We clarify once again that the transmitting antenna we are considering at this point is not the actual transmitting antenna. This one will be taken into account together with the receiving antenna(s) in the following stage of the framework, in which the received field will be corrected by including the effect of the transmitting and receiving antenna(s) radiation vectors.

An *electromagnetic* ray tracer is a computer program used to *render*, that is, calculate the electromagnetic field in a certain given environment in a fully deterministic way. For this purpose, the program relies on the geometrical theory of diffraction (GTD) [32], which is a generalization of the classical theory of geometrical optics (GO). Such a theory represents a high-frequency asymptotic technique, which is valid when the dimensions of the scattering objects are many wavelengths. In this circumstance, the rigorous solution of Maxwell's equations would be an intractable mathematical problem, which justifies the use of such an approximate technique. GO is a high-frequency approximated method for determining wave propagation for incident, reflected, and refracted rays and GTD extends this theory by introducing another mechanism called *diffraction*, resulting in diffracted rays. The high-frequency approximation considers the EM field at a given point as the superposition of several plane-wave components, which propagate along rectilinear paths, normally indicated as *rays* through the propagation environment. Under such circumstances, each plane-wave component radiated from a transmitting antenna can experience several propagation mechanisms when propagating through the environment, being *reflection, transmission* and *diffraction*.

8.4.2.4 Matlab Simulation Model

8.4.2.4.1 Channel Response Let us consider the communication link between one particular couple of TX-RX antennas placed at fixed positions in the propagation environment space between which a scalar propagation channel response h can be defined, as shown in Figure 8.10. The scalar channel response indicated as $h = |h|e^{j\varphi(h)}$ represents the narrowband, static channel response between the input terminals of the actual transmitting and receiving antennas. This is the quantity which we are interested in, for use in simulations of base band transmission of data. We now explain how the scalar channel response can be calculated starting from the output of the ray tracer.

As previously introduced, the ray tracer program provides, as an output, the two so-called *vectorial channel responses* $\bar{g}^{(V)}$, $\bar{g}^{(H)}$, representing the components of the electric field at the given receiver antenna position for vertically and horizontally oriented transmitting isotropic antennas, respectively. Such a situation is schematized in Figure 8.11. These two responses are calculated by the ray tracer as a superposition of L rays. In particular, the lth-ray component of the vectorial channel response can

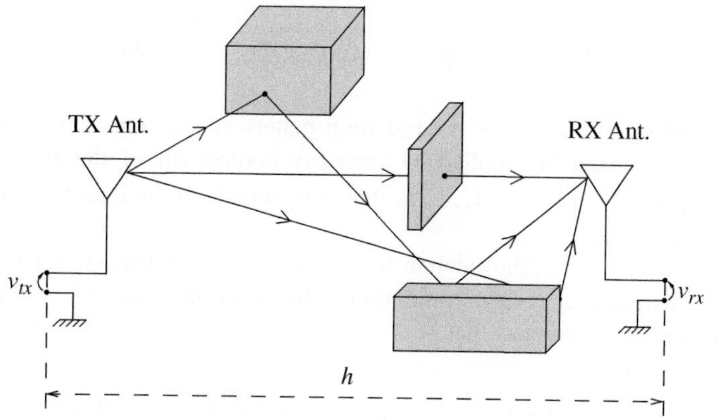

FIGURE 8.10 Scalar channel response between the terminals of TX and RX antennas.

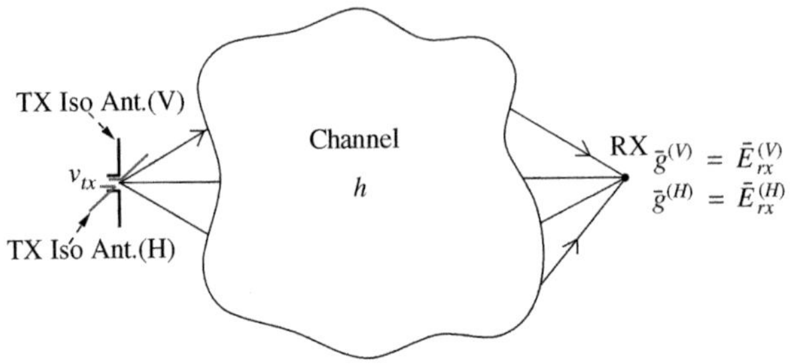

FIGURE 8.11 Vectorial channel responses between the terminals of isotropic horizontal and vertical TX antennas, and field components at a given RX antenna position.

be written in matrix notation, as the product of a tensor, representing the channel response for the lth ray and the radiation vector of the isotropic antenna:

$$\bar{g}_l^{(pol)} = \bar{\bar{G}}_l \cdot \bar{F}_{iso}^{(pol)}, \qquad pol = \{H, V\} \tag{8.19}$$

where the channel response tensor, independent from the orientation of the considered transmitting isotropic antenna (pol), is the product of reflection, transmission, diffraction tensors, and the propagation factor experienced by the lth ray.

The scalar channel response h can be obtained from the vectorial channel responses $\bar{g}_l^{(H)}$ and $\bar{g}_l^{(V)}$ ($l = 1, \ldots, L$), by means of an appropriate weighted sum of the L components of the two responses with proper coefficients calculated on the basis of the radiation-vector pattern of the actual transmitting antenna, and after inclusion of the receiving antenna radiation-vector pattern.

8.4.2.4.2 Inclusion of the TX Antenna Radiation Pattern The TX antenna is completely described by its 3D radiation-vector pattern, obtained from analytical expressions or from 3D full-wave simulation and can be written as

$$\bar{F}_{tx}(\theta, \phi) = F_{x,tx}(\theta, \phi)\,\hat{x} + F_{y,tx}(\theta, \phi)\,\hat{y} + F_{z,tx}(\theta, \phi)\,\hat{z}\,.$$

In order to include the effect of the radiation pattern of the transmitting antenna, weighting factors must be applied to each ray component of the two channel responses, being $\bar{g}_l^{(H)}$, $\bar{g}_l^{(V)}$, $l = 1, \ldots, L$, that corresponds to the direction of departure of the lth ray, $(\theta_{D,l}, \phi_{D,l})$.

For a given orientation of the antenna, the radiation vector of the actual TX antenna can always be written as a linear combination of the radiation vectors of the horizontal and vertical isotropic antennas, that is,

$$\bar{F}_{tx}(\theta_{D,l}, \phi_{D,l}) = C_l^{(H)}\,\bar{F}_{iso}^{(H)}(\theta_{D,l}, \phi_{D,l}) + C_l^{(V)}\,\bar{F}_{iso}^{(V)}(\theta_{D,l}, \phi_{D,l}) \tag{8.20}$$

where $C_l^{(H)}, C_l^{(V)} \in \mathbb{C}$ are the weighting factors for the lth ray, for the horizontal and vertical vectorial channel responses, respectively. Such weighting factors are easily calculated from $\bar{F}_{tx}(\theta_{D,l}, \phi_{D,l})$, $\bar{F}_{iso}^{(H)}(\theta_{D,l}, \phi_{D,l})$ and $\bar{F}_{iso}^{(V)}(\theta_{D,l}, \phi_{D,l})$, which are known quantities.

At this point, the weighting factors are multiplied with the corresponding vectorial channel responses, obtaining the field components of the actual received field for the two cases $pol = \{H, V\}$, that is,

$$\begin{cases} \bar{E}_{rx,l}^{(H)} = C_l^{(H)} \, \bar{g}_l^{(H)} = C_l^{(H)} \, \bar{\bar{G}}_l \cdot \bar{F}_{iso}^{(H)} \\ \bar{E}_{rx,l}^{(V)} = C_l^{(V)} \, \bar{g}_l^{(V)} = C_l^{(V)} \, \bar{\bar{G}}_l \cdot \bar{F}_{iso}^{(V)} \end{cases} \tag{8.21}$$

Finally, the total electric field for the lth ray, accounting the actual TX antenna radiation pattern, is obtained by superposition of the two components for the H and V cases.

$$E_{rx,l} = \bar{E}_{rx,l}^{(H)} + \bar{E}_{rx,l}^{(V)} = \bar{\bar{G}}_l \cdot \bar{F}_{tx}(\theta_{D,l}, \phi_{D,l}) \tag{8.22}$$

8.4.2.4.3 Inclusion of the RX Antenna Radiation Pattern Once the field at the receiver position is known, accounting for the actual TX antenna, it is straightforward to include the receiving antenna radiation pattern and finally obtain the *scalar channel* $h = |h| e^{j\varphi(h)}$ between the TX and RX antenna terminals.

It is known that the voltage at the receiving antenna terminals, the received field, and the RX antenna radiation vector are related by means of

$$v_{rx} = \alpha \, \bar{E}_{rx} \cdot \bar{F}_{rx} \, , \tag{8.23}$$

where $\bar{F}_{rx}$ is the receiving antenna's radiation vector, assuming a unit current $I_t = 1$ A at its terminals, and

$$\alpha = -\frac{2j\lambda}{Z_c}$$

The inclusion of the RX antenna pattern and the determination of the scalar channel response h proceeds as follows, for each ray $l = 1, \ldots, L$.

From the knowledge of the the 3D radiation pattern of the RX antenna (obtained by analytical expressions or full-wave simulations) and the AOA of the lth ray, $(\theta_{A,l}, \phi_{A,l})$, $l = 1, \ldots, L$ (obtained as an output from the ray tracer), the value of the actual RX antenna radiation vector is calculated for each of its three components, that is,

$$\bar{F}_{rx,l} = F_{x,rx}(\theta_{A,l}, \phi_{A,l}) \, \hat{x} + F_{y,rx}(\theta_{A,l}, \phi_{A,l}) \, \hat{y} + F_{z,rx}(\theta_{A,l}, \phi_{A,l}) \, \hat{z} \tag{8.24}$$

Next, the voltage at the RX antenna terminals v_{rx} is obtained as the superposition of L rays' contributions as

$$v_{rx} = \alpha \sum_{l=1}^{L} \bar{E}_{rx}(\theta_{A,l}, \phi_{A,l}) \cdot \bar{F}_{rx}(\theta_{A,l}, \phi_{A,l}). \tag{8.25}$$

Finally, the scalar channel response h can be easily obtained as

$$h = \frac{v_{rx}}{v_{tx}}. \tag{8.26}$$

It is important to remark that the scalar channel response h expresses the narrowband, static channel response, referred to given positions of the transmitter and receiver. Actually, for the performed simulations, the scalar channel calculation was repeated for several positions of the mobile receiver, organized on a square grid or a rectilinear path of receiver points. Thus, if the total number of receiver points is N, the scalar channel response, over all points, can be expressed as $h(k),\ k = 1, \ldots, N$. Moreover, the scalar channel calculation was repeated, in general, for a given number of antennas or antenna ports at the receiving side, denoted by n_r. Hence, in general we obtain n_r scalar channel responses indicated as $h_j(k),\ j = 1, \ldots, n_r,\ k = 1, \ldots, N$.

8.4.2.5 Baseband Simulations of Data Transmission The scalar channel responses provide the necessary information to simulate the complete communication link. Such simulations were performed by means of Matlab, and aimed at determining the performance of a certain communication scheme by the computation of performance parameters such as *received* SNRs, BERs (E_b/N_0), *channels correlation*, and *channels statistical distribution*. Our work focused on simulation of communication schemes using *receive diversity* only, also known as single-input multiple-output (SIMO) schemes, with the single-input single-output scheme being a particular case, when the number of receiving antennas is $n_r = 1$. A scheme of such a base-band data transmission simulation, with one transmitting antenna and n_r receiving antennas, using MRC at the receiver, is depicted in Figure 8.12. The scheme of the base-band communication link simulation, for each symbol, can be explained as follows: at the TX side, first log_2M bits, with M being the order of the chosen symbol modulation, are generated by a random source. These bits are then mapped into one symbol $c(k)$. The symbol is transmitted across the communication channels, whose channel gains $h_1(k), \ldots, h_{n_r}(k)$ $(k = 1, \ldots, N)$ are obtained by ray tracing as previously explained. At the RX side, n_r received signals are present to which Additive White Gaussian Noise (AWGN) is added each with given power spectral density $N_{0,i}$. In particular, the n_r received signals, plus AWGN, can be written as $r_i(k) = c(k)\, h_i(k) + n_i(k)$, with $i = 1, \ldots, n_r$, where each noise sample $n_i(k)$ is characterized by a given power spectral density $N_{0,i}$.

Then, the received signals undergo combining that, in our particular case, is Maximal Ratio Combining (MRC). For this purpose, we assume perfect channel

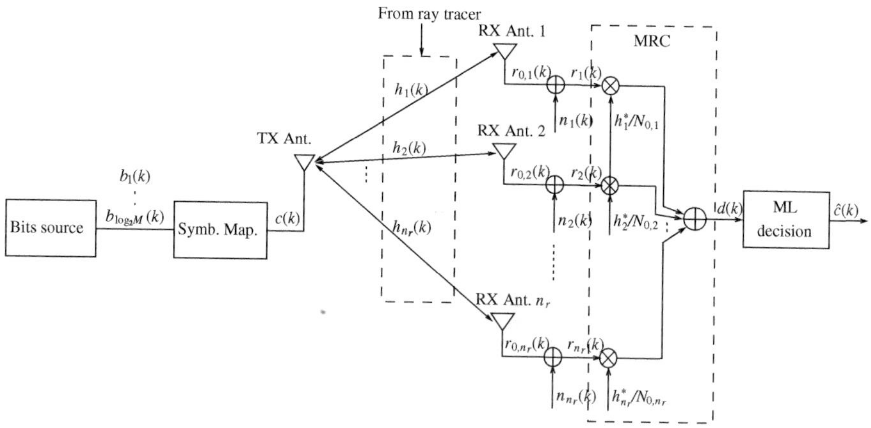

FIGURE 8.12 Scheme of the realized SIMO base-band data transmission simulation, with MRC at receiver.

gain knowledge at the receiver. In particular, the signal obtained after MRC can be written as $d(k) = \sum_{i=1}^{n_r} r_i(k) \frac{h_i^*(k)}{N_{0,i}}$.

The estimate of the transmitted symbol $\hat{c}_k$ is then decided from $d(k)$ according to the maximum likelihood criterium. Finally, $\hat{c}(k)$ is compared with the transmitted symbol $c(k)$: in case $\hat{c}(k) \neq c(k)$, an error counter is incremented. The simulation is repeated for $k = 1, \ldots, L_D$. In our case binary phase-shift keying (BPSK) was used, so that $M = 1$, that is, one symbol corresponds to one bit. In this case, if N_{err} is the total number of bit (or equivalently symbol) errors, the final BER value is calculated as $BER = N_{err}/L_D$. Then, the estimation of the average BER, on a certain receiver space composed of N points, consists of simulating a transmission of L_D bits (typically $L_D = 10^5$ or 10^6), for each of the N scalar channel response points $h_i(k)$, $k = 1, \ldots, N$, $i = 1, \ldots, n_r$, and for a given number of E_b/N_0 points (with N_0 being the average of the N_i, $i = 1, \ldots, n_r$). N AWGN BER$_i(E_b/N_0)$ curves are thus obtained, one for each of the N points composing the receiver space. After that, the average BER(E_b/N_0) curve on the considered receiver space is calculated as the average of the above mentioned N AWGN BER(E_b/N_0) curves.

8.4.2.6 *Case Study and Simulation Results*

8.4.2.6.1 Simulations of a **1 × 4** *Link with on-Body Dual Polarized Antennas* The simulations results presented in this section aimed to reproduce the results of the measurement campaign shown in [5]. Therefore, a rectilinear path is chosen as receiver space, having the same length and location of the path followed by the test person in the above mentioned measurement campaign. A layout showing the position of the receiver path space in the considered simulated environment is displayed in Figure 8.13.

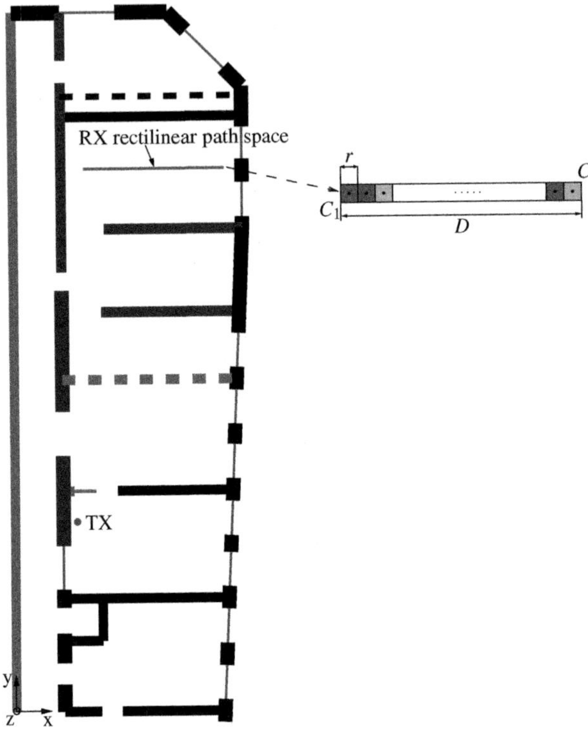

FIGURE 8.13 Positions of TX and RX rectilinear path space in the simulated indoor environment.

The used diversity scheme is a combination of *pattern diversity*, given by the opposite orientation of the radiation pattern of front and back antenna and *polarization diversity* between the two port signals of each antenna.

A baseband simulation of digital data transmission was performed for this case. In Figure 8.14 the simulated BER(E_b/N_0) resulting from the data transmission simulation is shown using BPSK symbols transmitted through the four considered channels and employing maximal ratio combining at the receiving side. One notices that the performance of the fourth-order diversity system, using both front and back Dual-Polarized Patch Antennas (DPPAs), shows a considerable improvement with respect to the case of using only one DPPA worn at the front or at the back side of the human body, in good agreement with the measurement campaign results described in [5].

We verified the effectiveness of the proposed model in reproducing the real measurement scenario from a BER point of view. In order to make a comparison, we need to point out some differences between the real measurement scenario and its simulated model.

As shown in Figure 8.15, one can see that there is a good agreement between the simulated and measured BER curves, both for the 1×4 and the *average* 1×2 *polarization diversity* case, with the performance being slightly better in the simulation

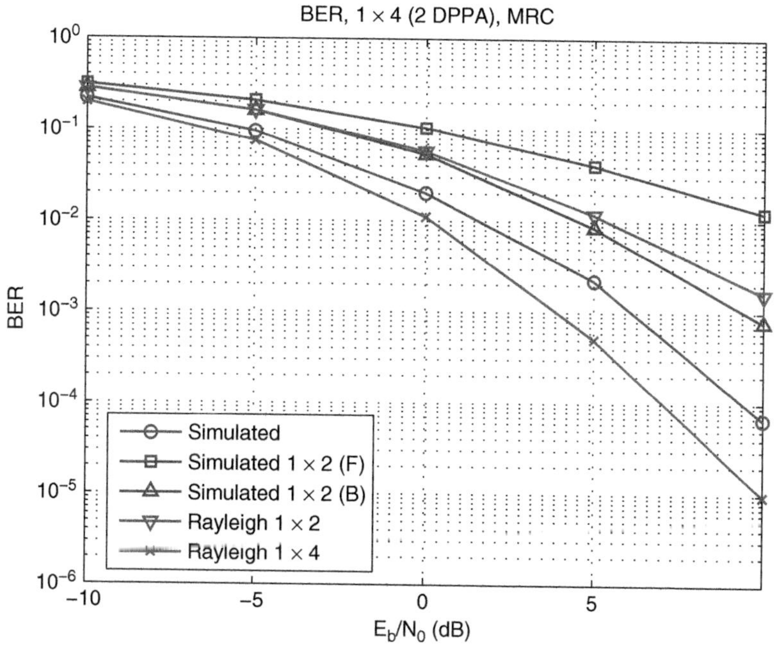

FIGURE 8.14 Simulated bit error rate curve, for the 1 × 4 link, with two RX dual polarized textile patch antennas, worn at front and back of rescue worker's body, compared to the BER curves of the 1 × 2 cases for each antenna.

than in the measurement. This is probably due to the limited number of propagation rays considered in the simulation, resulting in a more favorable fading distribution in the simulation than in the reality.

Given the agreement between both simulation and measurement, it can be stated that an effective deterministic model for an entire off-body communication link was developed.

In particular, modeling the fourth-order diversity off-body communication link between a fixed base station and a human-body equipped with two DPPAs, and simulating its performance in terms of BER yields results which are in sufficiently good agreement with the one obtained in the measurement campaign described in [5], proving the effectiveness of the proposed modeling method.

8.4.3 Spatial Modulation

8.4.3.1 Introduction Spatial modulation (SM) is a novel transmission mode, allowing to increase the data throughput of a wireless link by means of the simple addition of an RF switch. To perform SM, multiple transmit antennas are switched onto a single hardware transmit chain. The antenna switch is controlled symbol by symbol, based on a series of additional information bits to be transmitted.

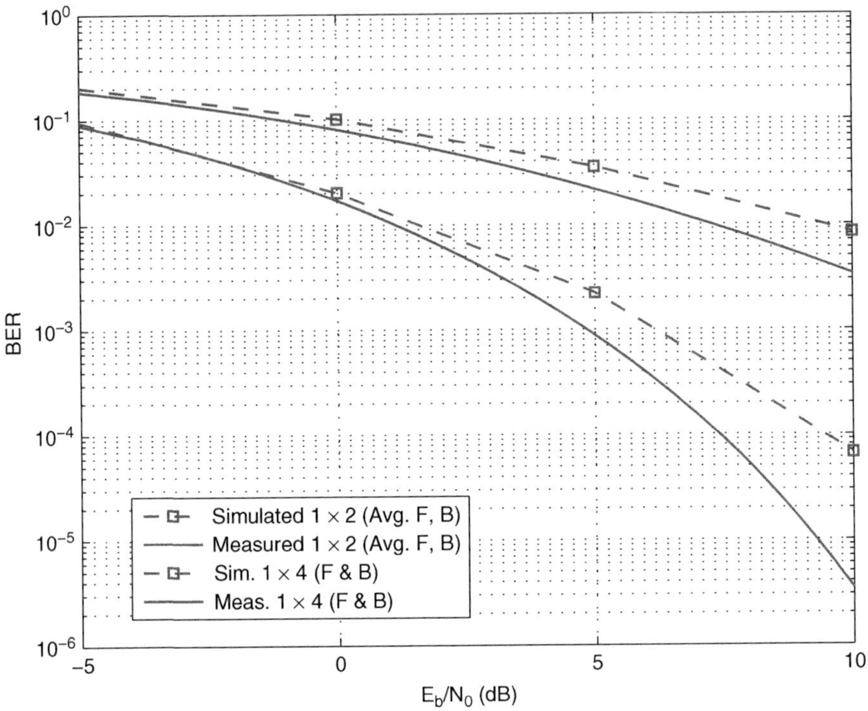

FIGURE 8.15 Simulated and measured bit error rate curves for the *average* 1 × 2 link with polarization diversity and the 1 × 4 link with combined pattern-polarization diversity.

Therefore, the selected antenna number for each transmitted symbol provides N extra information bits for a set of 2^N transmit antennas. The absence of separate transmit chains per antenna significantly reduces the complexity, size, and cost of a multi-antenna transmitter.

In literature, only a few papers treat measurements for this modulation form [33–36]. Further experiments are necessary to study the practical performance of SM [36]. For off-body transmissions from mobile users, shadowing by the body substantially reduces the antenna correlation for both LoS and NLoS scenarios. Therefore, SM seems a promising modulation technique for off-body communication.

In the following experiment, SM is employed for a two-antenna system with BPSK modulation, effectively doubling the spectral information density by switching between two transmit antennas. Measurements are performed to assess the practical performance of SM for off-body wireless communication from a walking person toward a base station in LoS as well as NLoS conditions. The person is equipped with textile antennas integrated into the front and back sections of his garment. For comparison to well-known modulation methods, spatial multiplexing (SMX) and space-time code (STC) transmissions are performed within the same transmission frame, evaluating the performance of the three modulation forms included under

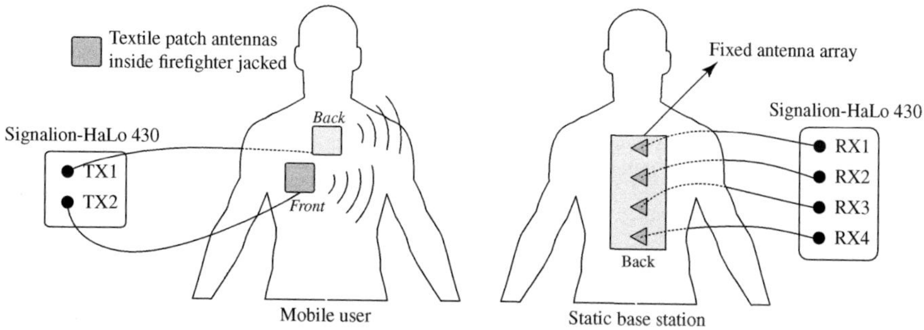

FIGURE 8.16 Setup of the measurement system.

equal channel conditions. Here, we discuss only the results for Non Line-of-Sight (NLoS), as this represents the most hostile propagation environment. The full results of the measurement campaign are documented in [36].

8.4.3.2 Measurement Setup For the measurement in the 2.45 GHz band, a person is equipped with dual-polarized textile antennas [37] integrated into the front and back sections of his jacket. The static base station receiver employs a wearable vertical uniform linear array (ULA) composed of four antennas described in detail in [38]. The measurement setup is illustrated in Figure 8.16. The transmit antennas are directly connected to the Signalion HaLo430 wireless testbed, producing all considered modulations. SM is achieved by transmitting symbols only from one TX port at any given symbol time, generating the same SM signal as when employing a TX switch. The mobile user is performing a random walk including reorientation. The measurements are performed at a distance of approximately 10 m between transmitter and receiver in NLoS conditions. The sum of the transmitted powers on front and back antennas is 10 mW.

The transmissions are performed on the front (TX1) and back (TX2) antennas, at 2 bits/s/Hz for each of the following transmission modes:

- SM: Spatial Modulation in BPSK
- SMX: Spatial Multiplexing in BPSK
- STC: Alamouti space-time block coding in QPSK

When employing SM, two bits are transmitted per symbol. One information bit is conveyed by the phase modulation of the carrier, and another bit by means of the selected antenna number, which can differ for each symbol. In SMX, two independent BPSK data streams are transmitted on both antennas. For STC, the data is Alamouti encoded and QPSK modulation is chosen in order to obtain the same spectral efficiency of 2 bits/s/Hz for all modulations tested within this transmission frame. At the receiving side, a Signalion HaLo430 wireless testbed is connected to the four ports of the textile antenna array. Signals are sampled on all ports and transferred to Matlab for processing.

STC is detected by the "Two-Branch Transmit Diversity with M Receivers" method proposed by Alamouti [39]. SM and SMX symbols are recovered by means of the optimal maximum-likelihood (ML) receiver,

$$[\hat{l}_t, \hat{s}_t] = \underset{\substack{l\in\{1,2,\ldots N_t\} \\ s\in\{s_1,s_2,\ldots s_M\}}}{\operatorname{argmin}} \left\{ \|\mathbf{y} - \mathbf{h}_l s\|_F^2 \right\}, \tag{8.27}$$

where $\mathbf{y}$ is the $N_r \times 1$ received signal vector, $\mathbf{h}_l$ the channel vector corresponding to the lth transmit antenna and s is the transmitted symbol. The estimated values $[\hat{l}_t, \hat{s}_t]$ provide the detected transmit antenna number and symbol, respectively.

8.4.3.3 Statistics of Measured NLoS Signals Based on the detected pilot symbols, a statistical analysis is performed documenting the channel behavior. The CDF and signal envelope correlation properties are displayed in Figure 8.17. The signal powers are normalized, in order to make the average received power, averaged over all channels, equal to 0 dBm. Note that the same scaling factor is used for all channels in this normalization process. The CDF characteristics largely overlap, illustrating limited channel imbalance. This is the case for the different receive channels as well as for the front (TX1) and back (TX2) transmit antennas. Fading occurs over a 35 dB range, a well-known phenomenon in NLoS communications under Rayleigh fading conditions.

The envelope correlation for the received signals is reported in [36, Table I], from which one can see that the absolute value of the envelope correlation coefficient

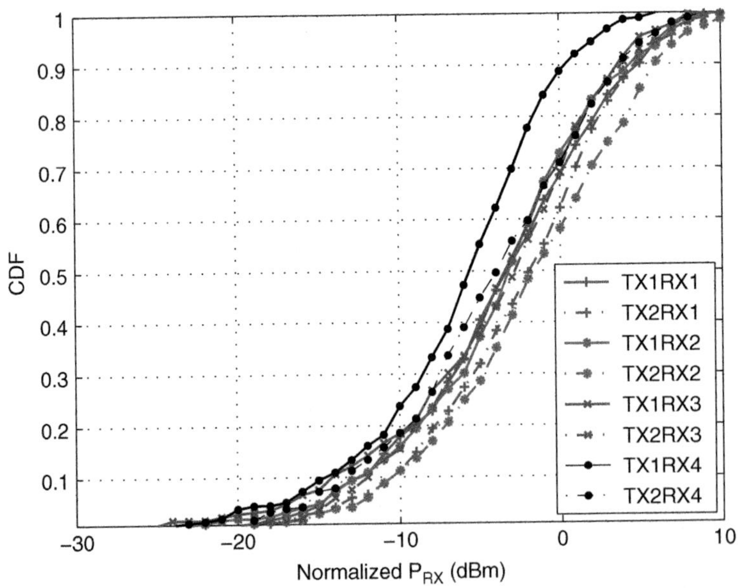

FIGURE 8.17 CDFs for the normalized received power per NLoS channel.

between two different channels remains less or equal to 0.45. According to Alamouti [39], envelope correlation coefficients below 0.7 promise a significant diversity gain. All correlation coefficients are well below that level, with the highest values corresponding to signals originating from the same transmit antenna. All correlation values between signals originating from front and back transmit antennas are negative, illustrating the complementarity of those antennas and their ability to counter shadowing by the body when their signals are combined.

8.4.3.4 Bit Error Rate for SM, SMX, and STC In contrast to network-analyzer based measurements described in [34] and [35], the experiment discussed here realizes a true wireless data link requiring amplitude, phase and frequency-offset estimation. The off-body measurement with a walking person results in serious channel variation, requiring a re-estimation of the channel parameters for each received frame. The accuracy of the channel estimates is important for a reliable demodulation.

8.4.3.5 2×1 ***MISO Link*** Figure 8.18 displays BER characteristics as a function of the received SNR on receive channel RX1, connected to the top antenna element of the array. The spectral efficiencies are equal to 2 bits/s/Hz for all transmissions, because QPSK modulation is employed for STC, whereas BPSK is used for SMX and SM. In indoor fading propagation conditions, without employing any diversity, a very high SNR is necessary for SM and SMX links. In contrast, STC provides second-order transmit diversity, resulting in a much better BER performance. To obtain an

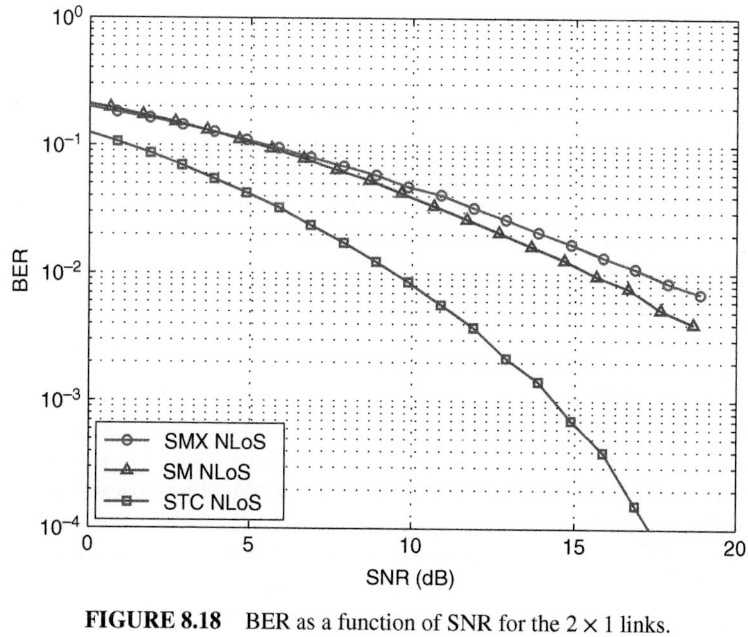

FIGURE 8.18 BER as a function of SNR for the 2×1 links.

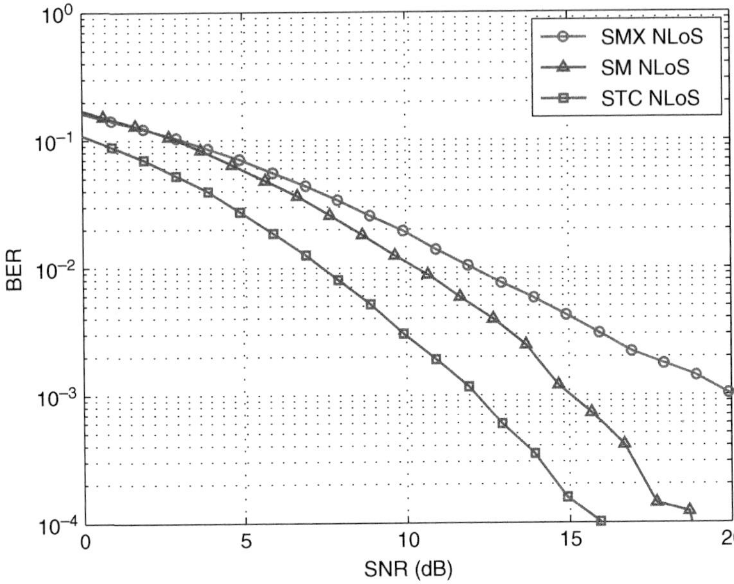

FIGURE 8.19 BER as a function of SNR for the 2 × 2 links.

acceptable BER at a reasonable SNR for SM or SMX, receiver diversity must be employed.

8.4.3.6 2 × 2 ***MIMO Link*** Figure 8.19 displays BER characteristics for identical transmissions, but now with second- order receive diversity, employing the two outer antenna elements of the array, corresponding to receive channels RX1 and RX4. The characteristics are shown as a function of the average received SNR per input antenna, visualizing diversity gain as well as array gain.

For SM, a BER $\approx 10^{-3}$ is obtained at an SNR = 15 dB. The slope of the characteristic corresponds to second-order diversity for uncorrelated Rayleigh fading channels. SMX needs an SNR $\approx$ 20 dB, whereas STC obtains the same BER already at an SNR $\approx$ 12.2 dB.

8.4.3.7 2 × 4 ***MIMO Link*** An extra increase in performance results for the 2 × 4 link, as illustrated in the BER characteristics in Figure 8.20. Employing all four receiving elements provides more array and diversity gain, as well as richer channel information to correctly detect the transmit antenna in SM, or to separate the parallel streams in SMX.

A BER $\approx 10^{-3}$ is now achieved by SM at an SNR = 8.9 dB, corresponding to a gain of 6.1 dB compared to the 2 × 2 configuration. SMX achieves the same BER at SNR $\approx$ 11.5 dB, whereas STC requires only an SNR = 7.1 dB.

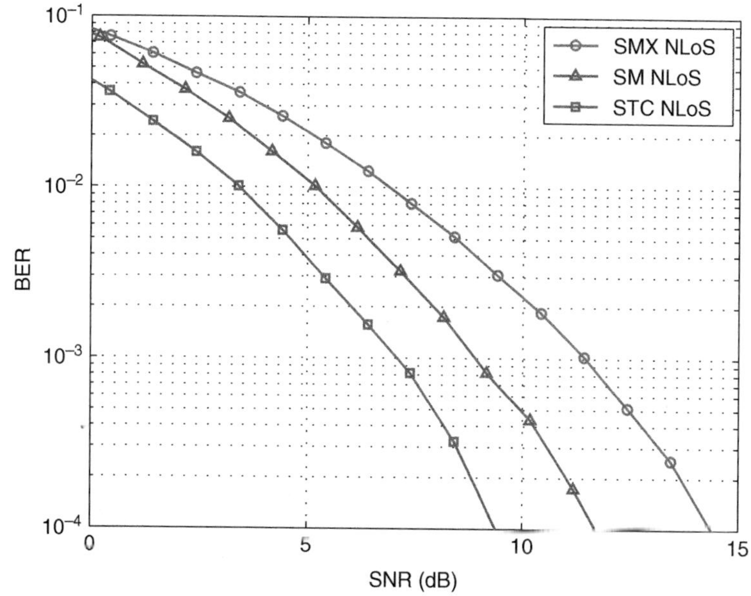

FIGURE 8.20 BER as a function of SNR for the 2×4 links.

8.4.3.8 Discussion A comparison of the performance for the different 2 bit/s/Hz transmission schemes for the MIMO links in an NLoS propagation environment at BER = 10^{-3}, indicates that SM performs 5 dB better than SMX and 2.8 dB worse than STC for the 2×2 link. For the 2×4 link, SM performs 2.6 dB better than SMX and only 1.8 dB worse than STC.

SM is clearly a useful modulation scheme for off-body communication, especially in a 2×4 MIMO configuration. Although STC already provides an equal BER at 1.8 dB lower SNR, we note that an STC transmitter requires two full QAM (Quadrature Amplitude Modulating) transmit chains, whereas the SM transmitter only needs one transmit chain. The reduced complexity of the SM hardware motivates the choice of spatial modulation for off-body communications.

SMX requires an SNR that is 2.6 to 5 dB higher than for SM. The performance of SMX is degraded due to the negative correlation between front and back transmit antennas, each transmitting independent data streams in a scenario where one signal is often much stronger than the other.

8.5 APPLICATIONS

8.5.1 Wearable Distributed Sensor Network

8.5.1.1 Introduction A fully flexible autonomous wearable 2.45 GHz wireless sensor node [40] with all its functionality fully integrated onto a dual-polarized textile patch antenna [37, 41] was recently developed. The electronic circuits on the polyimide substrate include a microcontroller, an ADF7242 diversity transceiver, a flash memory, and sensors. The textile ground plane of the wearable antenna limits the interaction with the human body, stabilizing the radiation characteristics and limiting the user's exposure. The embedded software on the microcontroller provides wireless *ad hoc* network functionality. The autonomous wearable wireless nodes cooperate, forming a modular, low-power, body-worn system with receive diversity. Besides sensor network functionality, the autonomous units can also be deployed as a channel measurement device, synchronously recording signal levels received on different autonomous nodes into their local flash memories.

8.5.1.2 Wearable Node Design The heart of the transceiver system is formed by the ADF7242 integrated circuit from analog devices, supporting IEEE 802.15.4-2006 2.4 GHz Physical layer (PHY) requirements, as well as Gaussian Frequency-Shift Keying (GFSK), Frequency-Shift Keying (FSK), Gaussian Minimum Shift Keying (GMSK) and Minimum Shift Keying (MSK) modulation. Depending on the range and data rate required, one of these modulations is chosen. The IEEE 802.15.4-2006 standard is excellent for reliable data packet communication in low data-rate, wireless sensor networks. The ADF7242 device features a dual-port radio-frequency (RF) interface, directly connected to a dual-polarized textile antenna's ports for $\pm 45°$ slanted quasi-linear polarizations [37]. This architecture enables second-order receive polarization diversity by means of selection combining, improving the reliability of the wireless communication.

A low-power Silicon Laboratories C8051F921 microcontroller runs embedded software, which is developed in the C programming language. The embedded software controls the transceiver chip, reads the sensor data, and organizes data transfer with the flash memory. The actions of the firefighter wearing the network are monitored by on-board 3-axis accelerometers, which are present on each node. The energy-efficient, wearable node is powered by a small battery, ensuring the autonomy of the system during the course of a full rescue operation.

The antenna with the polyimide substrate circuit is shown in Figure 8.21. This cross section of the electronic system clearly displays the system integration onto the feed plane of the textile antenna.

The maximum output power of the transceiver is +4.8 dBm. Taking into account the antenna gain and circuit losses, the maximum output power of the system is 8.9 dBm effective isotropic radiated power (EIRP), which complies with the ETSI EN 300 328 standard for wide band transmissions.

To guarantee the flexibility of the unit, all integrated-circuit components are chosen in small packages and are placed at sufficiently large center-to-center distances in the printed circuit board (PCB) layout, allowing a small bending radius for the full antenna-circuit assembly. To confirm the flexibility, the node is tested in bent conditions, attached to a plastic tube with a curvature radius of 7.5 cm, as shown on Figure 8.23. The performance measurements in bent conditions confirm a correct operation with only a minor degradation of the antenna's radiation pattern.

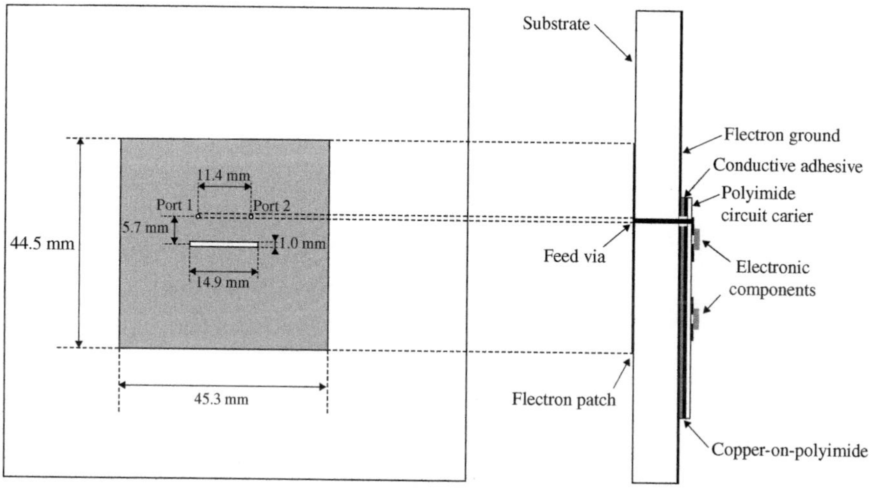

FIGURE 8.21 Cross-section of the dual-polarized textile antenna with electronic circuitry mounted onto the feed plane. *Source*: Vanveerdeghem et al. 2014 [42]. CC BY 4.

8.5.1.3 On-Body Sensor Network Patients, health-care professionals, and rescue workers can be monitored by a distributed sensor network. In such a network, sensor data are transmitted from each sensor node to a central access point. Regardless of whether the nodes are all worn by the same person or spread over different users, other nodes in the network cooperate to forward the packets.

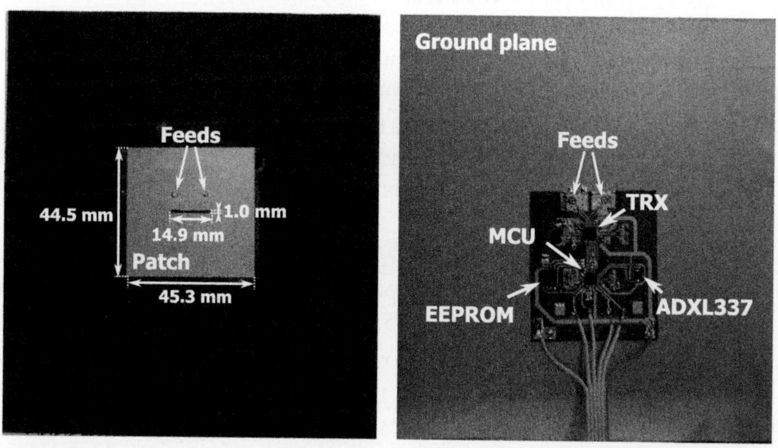

FIGURE 8.22 (a) Front side of the wireless sensor node. (b) Back side of the wireless sensor node with the electronic components mounted on the flexible substrate. *Source:* Vanveerdeghem et al. 2014 [42]. CC BY 4.0. For a color version of this figure, see the color plate section.

FIGURE 8.23 The integrated system bent around a plastic tube with a radius of 7.5 cm. For a color version of this figure, see the color plate section.

Synchronous sensor data acquisition and packet transmission is implemented for an N-node body-worn network by means of the following protocol:

- Power-on cycle and node enumeration: during subsequent power-on of the battery-operated nodes, the nodes detect each other's presence in the network and are automatically enumerated, acquiring their unique soft-ID's, which determine their time slots in the transmission cycle.
  - When powered on, each node listens for packets from other nodes. If no packets are received within a one-second time-out period, this node is assigned soft-ID = 1 and autonomously starts transmitting sensor data, including this soft-ID number.
  - Following the final packet of the transmission cycle, an extra receive time slot is preserved to listen for other packets from new nodes that can join the network, as displayed in Figure 8.25.
  - At the moment an extra node is powered-on, it first listens, receiving data packets from all active nodes. The new node responds with its own sensor data and dynamically determined soft-ID number, during the receive time slot. All nodes in the network adapt to this new configuration, always leaving an open receive time slot after the last transmission in the cycle.
  - Each node also transmits a fixed hard-coded ID, allowing unambiguous identification of a node's serial number by the base station. The hard-coded ID is

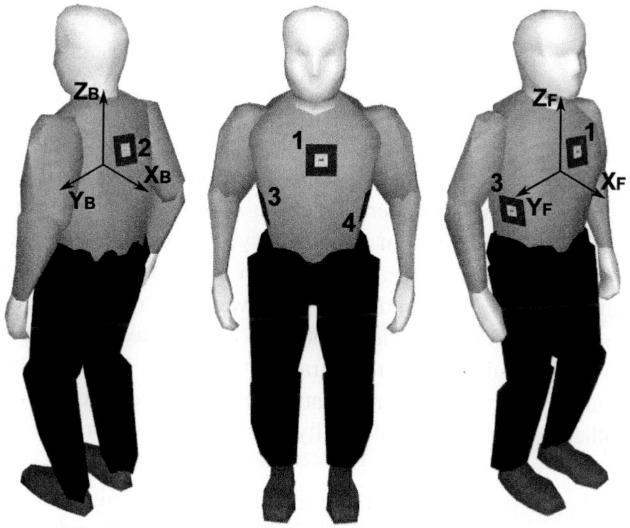

FIGURE 8.24 Wearable wireless body sensor network composed of four fully autonomous and wireless textile nodes, synchronously capturing, preprocessing, and relaying sensor data to a central access point. *Source:* Vanveerdeghem et al. 2014 [42]. CC BY 4.0.

also used to avoid re-enumeration of already active nodes in case of a temporary communication breakdown.

- Base station transmissions that should occur during the receive time slot, are recognized by a hard-coded ID > 127, and are not enumerated in the network. The base station transmits data packets containing alert levels, setting different alarm conditions at the rescue worker's side. The alert level value is forwarded by the wireless nodes themselves in their own transmit time slot together with the sensor data to ensure all rescue workers receive the alert status. Alerts produce audible alarm signals on the wearable nodes via a beeper.

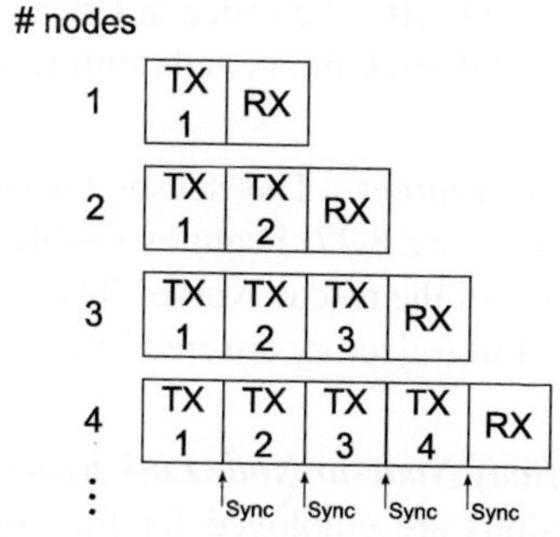

FIGURE 8.25 Power-on cycle and time slot structure of the autonomous body-centric network. *Source:* Vanveerdeghem et al. 2014 [42]. CC BY 4.0.

Hard node ID	Soft node ID	Node ID_{max}	Base station alert level	Time stamp + Packet #	Sensor data node 1 + class.	...	Sensor data node N + class.

FIGURE 8.26 Data packet structure. *Source:* Vanveerdeghem et al. 2014 [42]. CC BY 4.0.

- The network nodes continuously monitor each others' transmissions to allow data forwarding and measurement synchronization. The sensor data from any transmitting node are stored in the memory of each receiving node, which will forward the collected data in its own transmit time slot. Thanks to the redundancy created by this mode of data forwarding, a significantly increased performance results, comparable to the gain achievable by means of transmit diversity. All nodes determine their transmit time slot intelligently, based on the time stamps and network enumeration values of the received data packets from the other nodes. This enables fully synchronous sensor data capturing on all nodes.
- The transmission packets are organized as shown in Figure 8.26. In IEEE 802.15.4 mode, an error-detecting code is automatically appended, ensuring the correctness of the received data.

8.5.1.4 Measurements for Four On-Body Nodes The performance of the wearable sensor network is assessed in realistic conditions, deployed in a protective garment worn by a rescue worker performing a number of movements and postures. Real-world situations are included for an indoor office environment.

Four wireless nodes are integrated into a firefighter garment as displayed in Figure 8.24. Only the front (1) and rear (2) nodes are initially taken into consideration, avoiding the graphs to be too cluttered with data. Given the relative orientation of the two nodes, the vectors of the x- and y-axes of both nodes are opposite. The measured gravity vector provides absolute information about the rescue worker's orientation. The central access point consists of similar hardware as the mobile nodes, but now with two omnidirectional monopole antennas connected.

The base station node is located at 2 m height above the office floor (RX1) and connected to a PC running MATLAB. The office at Ghent University is located on the first floor and consists of solid brick floors and reinforced concrete walls.

8.5.1.5 Accelerometer Measurement The sensor measurements in the indoor environment are displayed in Figure 8.27. Features visible in the graph are caused by the movements and postures of the rescue worker. The on-body network provides reliable connectivity over the full trajectory covered by the test person.

8.5.1.6 Evaluation of On-Body Node-to-Node Link Reliability In the body-worn network, off-body transmissions are employed for the node-to-node communication on the same body, relying on reflections in the environment. An experiment is performed illustrating the received signal levels as well as the packet loss for the node-to-node links, demonstrating the link reliability. Even between front and

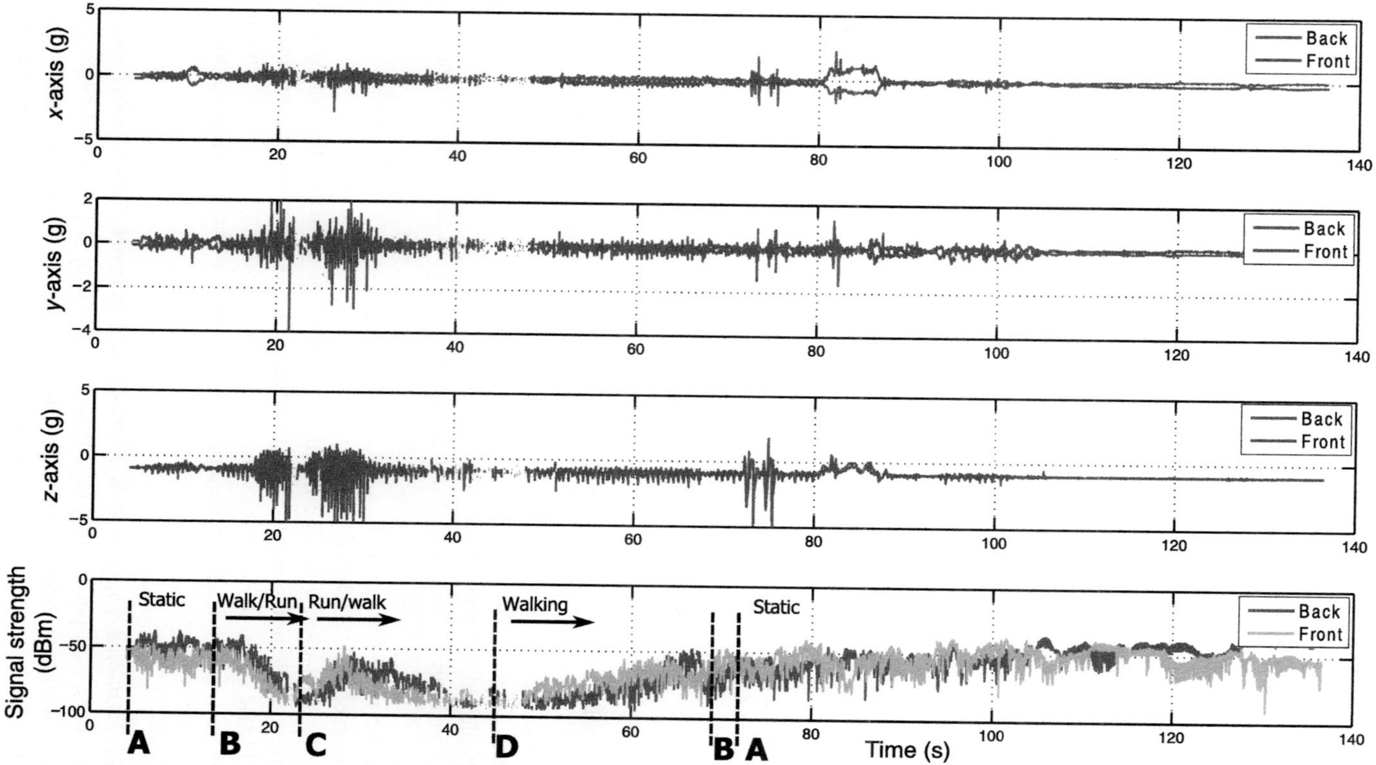

FIGURE 8.27 Measurement results along the indoor path followed by some activities performed in the neighborhood of A (*x*,*y*,*z*-axes + signal strength). *Source:* Vanveerdeghem et al. 2014 [42]. CC BY 4.0.

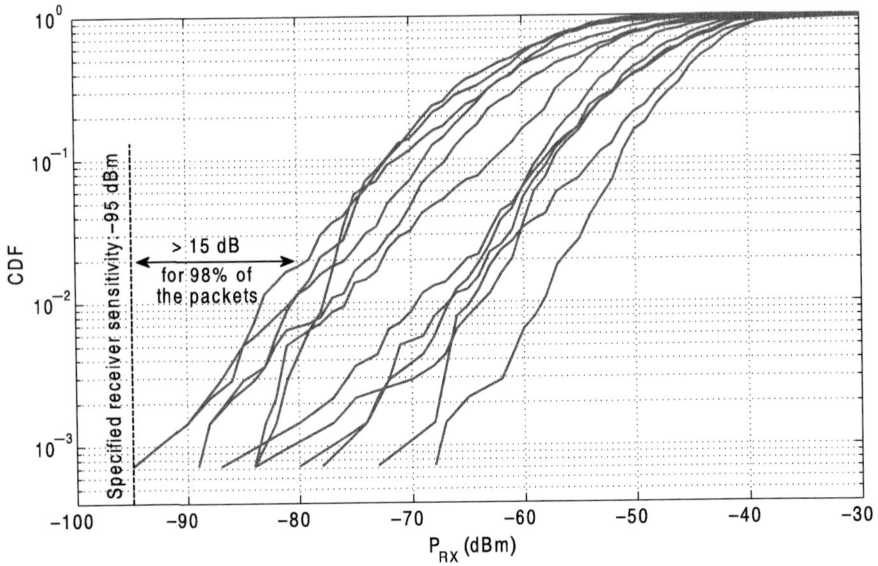

FIGURE 8.28 Cumulative distribution function of the received signal levels for the 12 on-body node-to-node links in a front, back, left, and right configuration. *Source:* Vanveerdeghem et al. 2014 [42]. CC BY 4.0.

back nodes, which are dramatically shadowed by the human body, a reliable connection exists.

A person equipped with four nodes is walking randomly, including maneuvers such as jumping and lying down. The nodes labeled 1 to 4 are located in the front, back, left, and right of the body in that order. The total number of 5940 received packets leads to the CDF of the received signal powers for the 12 node-to-node links, as displayed in Figure 8.28. For 98% of the received packets the signal level is at least 15 dB above the ADF7242 transceiver's specified receiving threshold of −95 dBm.

In total, 12 node-to-node links exist in a four-node network. The nodes corresponding to each link are listed in Table 8.2, including the median received power and packet loss for each link. The maximum number of subsequently lost packets on any node-to-node link is three, corresponding to a maximum link interruption of 120 ms. Please note that, owing to the packet forwarding, the link reliability for the full network is significantly better than for the individual node-to-node links.

*8.5.1.7 **Performance Analysis for a Three-Person Network*** As a further assessment of the person-to-person link quality, a measurement is performed employing three persons walking randomly within the same area, separated by varying distances of up to 20 m. Each person is wearing one network node on the chest. The experiment illustrates that

TABLE 8.2 Median received power and packet loss for the on-body node-to-node links. Node numbers correspond to the following locations: 1, front; 2, back; 3, left; 4, right.

TX node	RX node	Median P_{RX}	Packet loss (%)	Max # subsequently lost packets
1	2	−59	0.43541	1
1	3	−55	3.4833	1
1	4	−48	4.4993	3
2	1	−60	3.7736	1
2	3	−45	3.701	2
2	4	−54	3.7736	1
3	1	−50	3.3382	1
3	2	−47	0.21771	1
3	4	−61	3.701	2
4	1	−47.5	3.5559	1
4	2	−44	0.072569	1
4	3	−58	3.4833	1

- Node-to-node communication is very reliable between nodes mounted on different persons operating in a team.
- Owing to the packet forwarding, not a single measurement is lost even when only one unit is installed per person.

The CDFs of the received signal powers for the six node-to-node links are displayed in Figure 8.29. The signal level is at least 20 dB above the ADF7242 transceiver's specified receiving threshold of −95 dBm for 98% of the received packets. Although the persons are constantly reorienting and changing distance during the random walk, the spread between different CDF curves is much more limited for the person-to-person links, in comparison to CDFs for the on-body links. Statistically, the propagation conditions are similar between each pair of nodes, with each node worn on the front of a person performing a random walk over a 4-minute time span. For the on-body node-to-node communication, the nodes are in fixed locations on the same body, causing systematic differences in the CDF curves for the different on-body links. The packet-loss values, as shown in Table 8.3, are significantly lower for the person-to-person links, confirming the results observed in the CDF characteristics.

Distributing the nodes over several persons does not compromise the wireless communication between nodes. The wearable wireless sensor network successfully measures three-axis accelerations and transmits these measurements to the base station without a single missing measurement in all experiments performed. On-body links and person-to-person links are always very reliable, thanks to the low packet loss for each link, with more than 98% of the transmitted packets received at a power level that was at least 15 dB higher than the specified receiving threshold for the

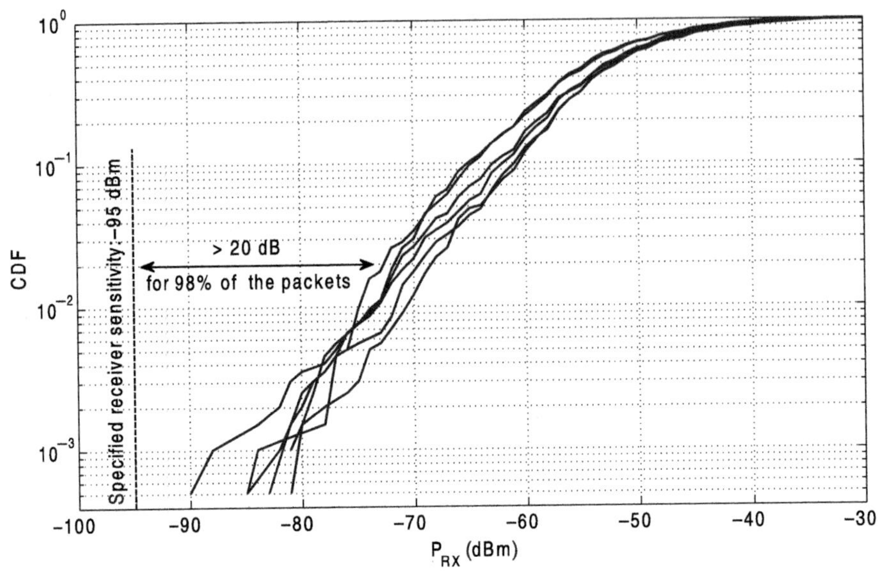

FIGURE 8.29 Cumulative distribution function of the received signal levels for the six off-body person-to-person links. *Source:* Vanveerdeghem et al. 2014 [42]. CC BY 4.0.

transceiver chip. Combined with the redundancy introduced by packet forwarding, a very high link reliability is realized by the full network.

8.5.2 Distributed Exposimeter

The specific absorption rate (SAR), in Watts per kilogram body mass, is a generally accepted unit to express the absorption of radio frequency (RF) radiation by the human body. The most commonly studied SAR value is the whole-body averaged SAR (SARwb) [43], which is generally evaluated using numerical simulations. To allow practical measurements, reference levels have been defined on the incident

TABLE 8.3 Median received power and packet loss for the person-to-person links.

TX person	RX person	Median P_{RX}	Packet loss (%)	Max # subsequently lost packets
1	2	−51	0.55866	1
1	3	−54	0.8126	1
2	1	−52	0.76181	1
2	3	−54	0.8126	1
3	1	−51	0.71102	1
3	2	−51	0.25394	1

FIGURE 8.30 Top- and bottom-view of the patch antenna. On the bottom view, the exposimeter electronics and antenna are shown separately.

power density, as this is a measurable quantity which can be compared to the international guidelines issued by organizations such as INCIRP [44].

Exposure measurements are often performed employing personal exposimeters (PEMs) [43, 45], which are commercially available. Such measurements are relatively inaccurate, due to the influence of the body on the electromagnetic fields impinging on the PEM's antenna. Body shadowing in particular, contributes to a considerable underestimation of the RF radiation to which a person is exposed. A fully integrated wearable on-body PEM enables continuous monitoring of the long-term RF exposure, without hindering the test person's daily activities. An on-body personal distributed exposimeter (PDE), composed of several PEMs, increases the measurement accuracy, as demonstrated in [46]. A fully autonomous on-body PDE is proposed, composed of multiple independent RF-exposure modules integrated on the feed planes of the textile antennas [47]. The proposed exposimeter is configured for the Global System for Mobile Communications (GSM) downlink band, ranging from 925 MHz to 960 MHz.

8.5.2.1 Antennas Due to the large wavelength of the signals in the GSM downlink band, an aperture-coupled shorted patch antenna, as displayed in Figure 8.30, is chosen. This antenna topology allows a more compact implementation, while excellent antenna performance in proximity of the human body is preserved. Fragile probe feed connections [48, 49] are also avoided, making this antenna highly suitable for garment integration. The top view of the antenna and its dimensions are shown in Figure 8.30, with the electronic circuitry integrated onto the feed plane of the antenna.

8.5.2.2 System Design The block diagram of the proposed exposimeter module is presented in Figure 8.31. The top view of one of the prototypes is displayed in Figure 8.32. Several nodes form a PDE. The specifications of a single PEM node are listed in Table 8.4.

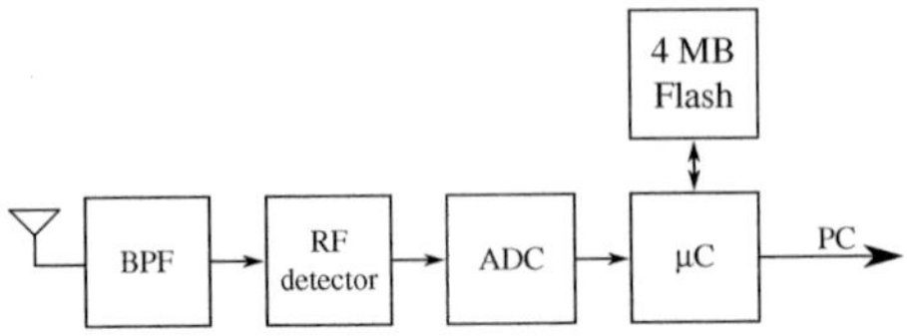

FIGURE 8.31 Block diagram of single autonomous exposimeter module with integrated band-pass filter (BPF), logarithmic RF detector, and temperature compensation (Temp Comp). The PC connection is only used to configure the microcontroller (uC) and to download logged data from the 4 MB on-board flash memory.

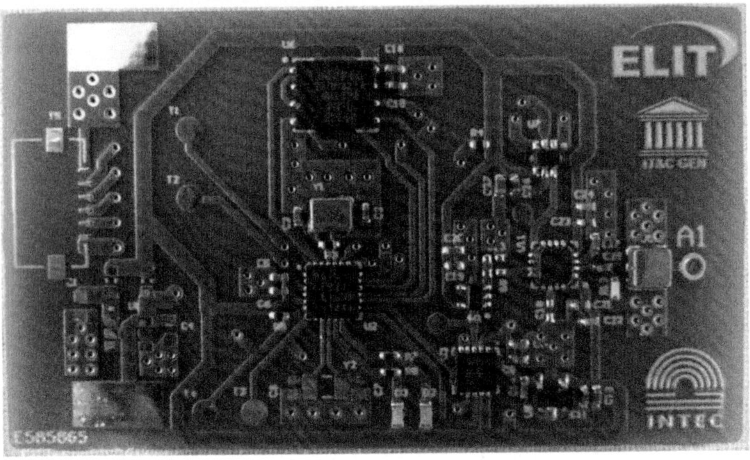

FIGURE 8.32 Top view of the system (57 mm × 33 mm).

8.5.2.3 On-Body Performance The PDE is configured for the GSM-900 down link frequency band, which is exploited worldwide. Four separate PDE nodes are distributed at optimal positions over the front and rear sections of the torso of a test person, as shown in Figure 8.33. The polarizations of the individual nodes are chosen for optimal complementarity.

TABLE 8.4 Specifications of the single PEM node. *Source*: Vanveerdeghem et al. 2015 [47]. Reproduced with permission of IEEE.

Specification	Value
Dynamic range	80 dB
Minimum input power	−70 dBm
Maximum input power	10 dBm
Operating supply voltage	3.5 –12 V
Size antenna	100 mm × 125 mm
Size PCB	35 mm × 55 mm
Frequency range	1 MHz –4 GHz

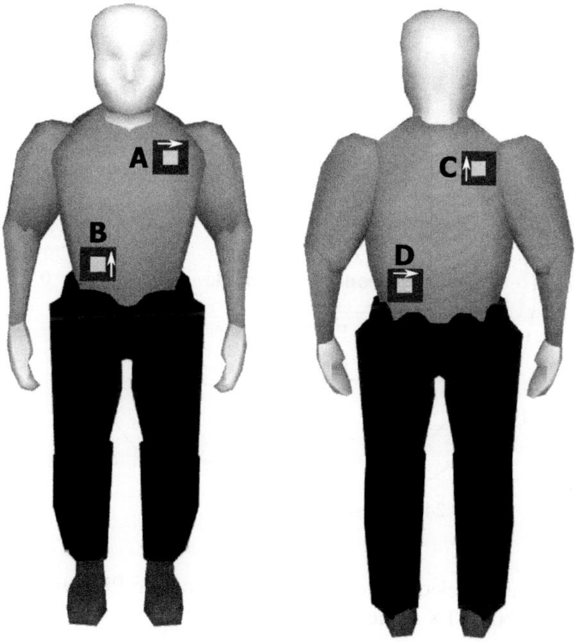

FIGURE 8.33 Positions of the 4 personal exposimeter (PEM) modules composing the personal distributed exposimeter (PDE).

The person equipped with the PDE system is mechanically rotated in the anechoic room, in the far-field (at 4.34 m distance) of a standard gain horn, radiating at 942.5 MHz, being the center frequency of the GSM-900 down-link band. The transmit power of the signal generator is 10 dBm (cable losses = 3.75 dB). The person wearing the PDE is rotated over 360° in the azimuth plane. The measurements are performed for horizontal (H) as well as vertical (V) TX polarizations. During these measurements, for comparison, a commercially available secondary exposimeter (EME Spy 140) is worn using its belt clip. In order to determine the polarization dependence of the body-worn system, the standard deviation σ_{pol} is calculated based on the difference in the received field strength for both polarizations. The polarization dependence is determined for both the "EME Spy 140" and the prototyped PDE. The following results are averaged for the full 360° rotation in the azimuth plane. Clearly, the PDE is significantly less polarization dependent than the "EME Spy 140".

$$\sigma_{\text{PDE diff H/V}} = 2.32\text{ dB}$$

$$\sigma_{\text{EME Spy 140 diff H/V}} = 3.73\text{ dB}$$

To compare the PDE and the "EME Spy 140" in an on-body scenario more in detail, the standard deviation σ of the field strength for varying azimuth angles is determined for both horizontal as well as vertical polarizations. σ indicates how circular the pattern is in the azimuth plane, with a pattern that is perfectly

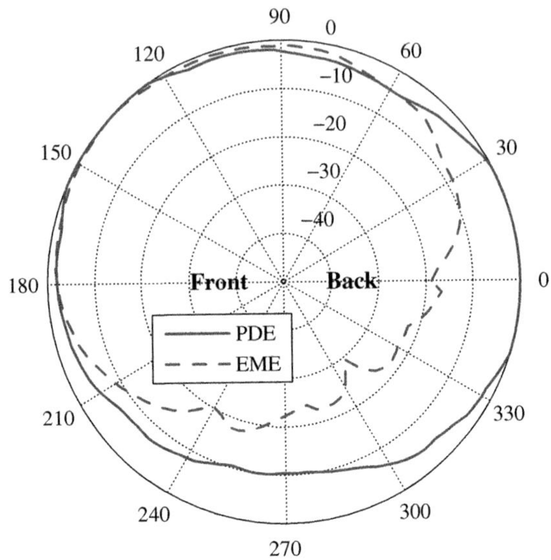

FIGURE 8.34 Normalized electric field strength (dB) on the PDE and “EME spy 140”. 180° = front side of the body, TX vertical polarization.

omnidirectional resulting in a standard deviation of 0 dB. σ_ϕ is derived from the (logarithmic) field strength on the four nodes, resulting in

$$\sigma_{\text{PDE Hor}} = 2.70 \text{ dB}$$
$$\sigma_{\text{PDE Vert}} = 3.35 \text{ dB}$$
$$\sigma_{\text{EME Spy 140 Hor}} = 6.94 \text{ dB}$$
$$\sigma_{\text{EME Spy 140 Vert}} = 9.13 \text{ dB}$$

We clearly obtain a more omnidirectional response for the PDE in comparison to the commercial exposimeter.

The normalized average field strength over the four nodes is now plotted on a logarithmic scale in Figures 8.34 and 8.35, for each azimuth angle φ, for both horizontal and vertical polarizations. This visually illustrates the above results, with the PDE clearly achieving a more uniform distribution of the field strength in the azimuth plane.

8.5.2.4 Calibration The power received by the PDE is almost independent from the polarization or the azimuth angle. However, in order to acquire accurate measurements of the actual RF field strengths at the human body, the PDE is calibrated in the anechoic room. The calibration eliminates the influence of the body on the measurement. $P^H_{geom}(\varphi)$ and $P^V_{geom}(\varphi)$ are the geometric average received powers for the horizontal and vertical transmit polarizations as a function of the azimuth angle,

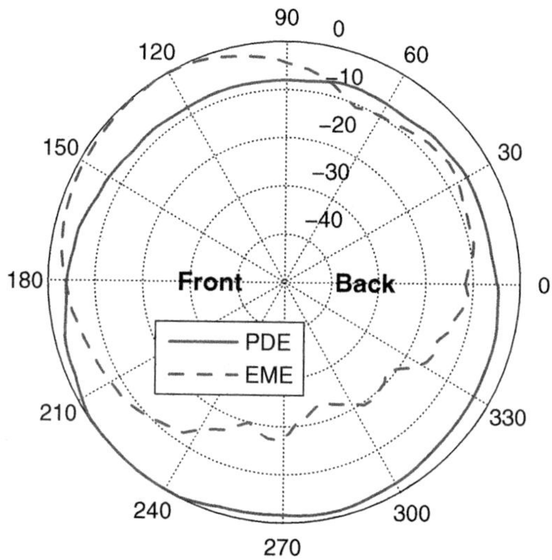

FIGURE 8.35 Normalized electric field strength (dB) on the PDE and "EME spy 140". 180° = front side of the body, TX horizontal polarization.

respectively. The incident powers S_{inc}^{H} and S_{inc}^{V} are measured employing the NBM-550 broadband probe for both polarizations.

After calibration, the geometric average antenna aperture (AA_{geom}) of the full body-worn PDE system is determined by

$$AA_{geom}(\varphi,\psi) = \frac{P_{geom}^{H}(\varphi)}{S_{inc}^{H}} cos^2(\psi) + \frac{P_{geom}^{V}(\varphi)}{S_{inc}^{V}} sin^2(\psi)$$

where $S_{inc}^{V} = 0.541\ \mathrm{mW/m^2}$ and $S_{inc}^{H} = 0.394\ \mathrm{mW/m^2}$, and with ψ the polarization of an incident electric field. $AA_{geom}(\varphi,\psi)$ is calculated for 10^3 ψ-samples, located in the interval $[0\text{–}2\pi]$ radians. The ψ-samples are drawn from a Gaussian distribution in an "Urban Macro cell" scenario [50], in order to take into account a realistic polarization of the incident electric field. The calibration is performed 100 times, averaging the results, to improve accuracy. The full calibration process and its post processing provides the distribution of AA_{geom} for realistic AOA. The median of this distribution is chosen as the value of the Antenna Aperture (AA_{geom}) for the full PDE system, which is found to be 6.58 cm^2.

This PDE is now ready for field measurements and the incident power received on the body of the test person can be determined by

$$S_{inc} = \frac{P_{geom}}{AA_{geom}}, \tag{8.28}$$

with P_{geom} equals to the geometric average of the received power on the four nodes of the PDE.

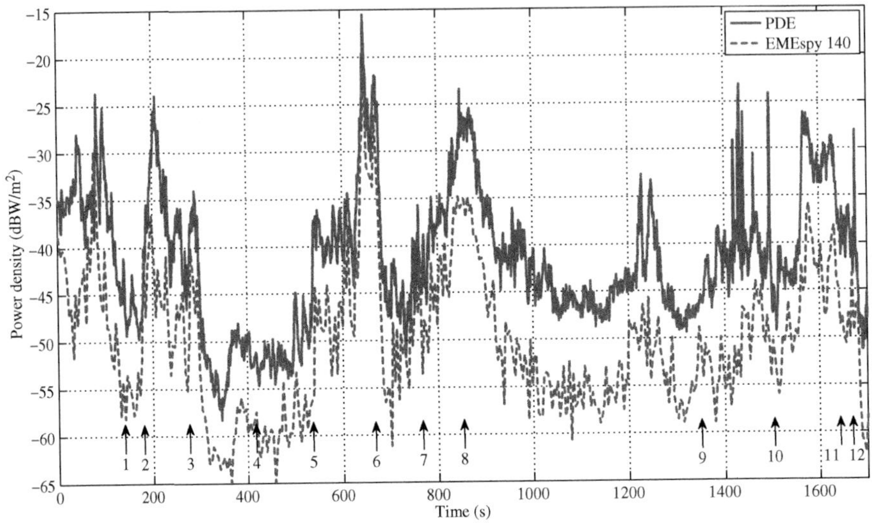

FIGURE 8.36 Received power density (dBW/m^2) on the body of the test person measured by the active nodes of the PDE versus the commercially available exposimeter. The position markers correspond to the numbers in Figure 8.37.

8.5.2.5 Real-World Measurement For the outdoor measurement campaign, a male person of 1.85 m length and a weight of 80 kg is equipped with the PDE. For comparison to current measurement techniques available on the market, the "EME spy 140" is worn using its belt clip. Walking along a path in the city-center of Ghent (Belgium), the received power is recorded each second on all the exposimeter nodes as well as each 4 s on the "EME spy 140". Figure 8.36 shows the power density received with the PDE during the complete walk as well as the power density on the "EME spy 140". In Figure 8.37, the outdoor trajectory is shown with the position markers corresponding to the numbers in Figure 8.36. The positions of the nearby GSM-900 base stations shown in the map are found in an official database of the Belgian frequency regulation organization [51].

The varying power density impinging on the human body during the walk is determined based on the measurements logged in the local memories of the exposimeter nodes of the PDE, combined with the calibration values. The power density for the "EME spy 140" is extracted directly from the measurement logging file.

The measured power density is highly correlated for both measurement devices, but with short-term differences in power density levels. Shadowing by the human body has a significant influence on the measurement results of the "EME spy 140", as the device is worn on a belt clip. Therefore, the signals received on the "EME spy 140" are more dependent on the angle of arrival of the signals in the azimuth plane. In comparison, the omnidirectional receive pattern which is obtained with the PDE ensures a more accurate assessment of the power density levels. The maximum power density measured by the PDE during the experiment is 28 mW/m^2. The average

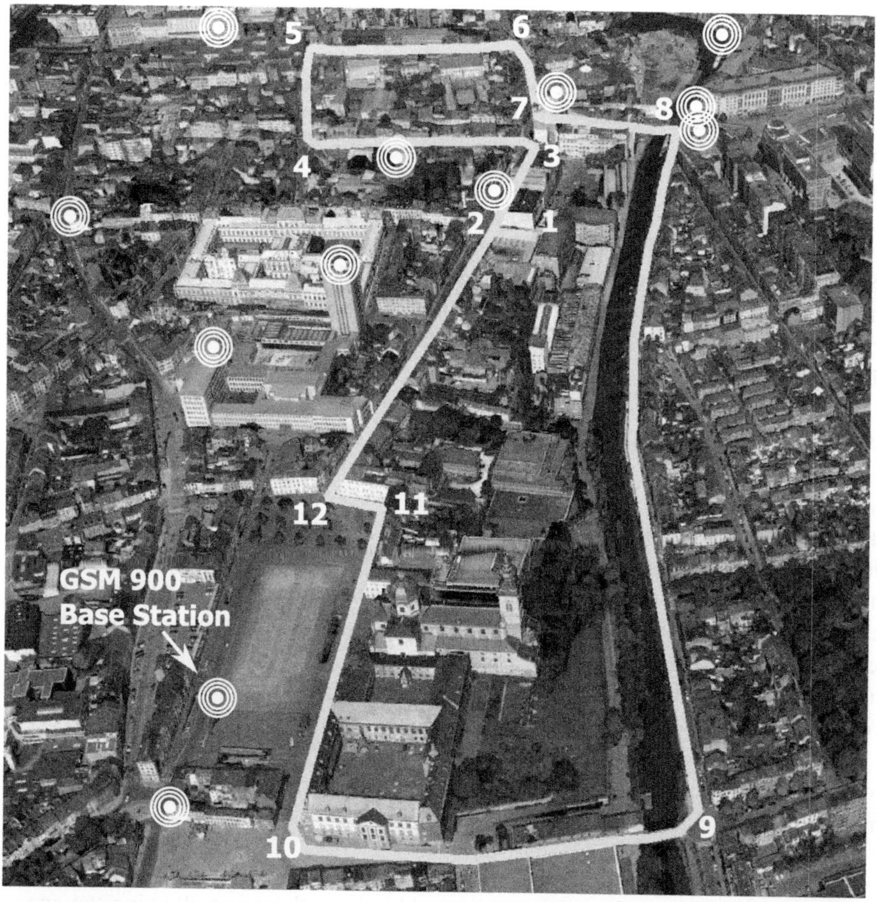

FIGURE 8.37 Map of Ghent with the path walked by the test person with numbered position markers and GSM-900 base station symbols (source: Google Earth).

power density levels over a 6-minute time frame are well below the ICNIRP reference level of 4.8 W/m^2 [44]. From these measurements the SARwb can be determined, as described in [52].

The wearable PDE increases the measurement accuracy compared to conventional single-unit PEMs. In particular, the dependency on the polarization of the signals and their angles in the azimuth plane is reduced. The proposed compact exposimeter unit is a step forward toward garment-integrated highly accurate PDEs.

8.6 CONCLUSIONS

In this chapter, an overview has been given about different state-of-the-art techniques and devices, capable of improving reliability – or equivalently power efficiency – of body-centric wireless communication systems. We can summarize the main conclusions as follows.

It was shown how the enhancement of energy efficiency or, equivalently, reliability - in terms of BER - of wireless off-body communication systems, is particularly challenging in multipath channels due to phenomena like fading, shadowing, and Doppler spread.

We have proven that a possible way to improve reliability consists in implementing diversity and MIMO schemes and by employing appropriate processing techniques. Experiments performed on 1×1, 2×2, and 4×4 links between a fixed base station and a subject with wearable antennas walking indoor indicated a significant reduction of signal variance and BER for the higher-order MIMO systems, with respect to the SISO (1×1) case.

Reliability of wireless off-body communications can as well be improved by using active wearable antennas, providing remarkably higher gains when compared to classical passive wearable antennas, obtained by combining a passive antenna plus an LNA in a single compact wearable module. A novel co-design strategy, relying on full-wave/circuit co-design and co-optimization, specifically developed for such a class of antennas was presented for this purpose. The proposed method was successfully applied to two different situations, the first one leading to a module with optimized noise characteristics and the second one to an optimal impedance matching between a passive antenna and LNA. The described co-design strategy was successfully used for the design of a wearable low-cost through-wall Doppler radar for rescue workers and a wearable active Galileo E1-band antenna.

For the efficient design and prediction of reliable and power-efficient communication systems, two modeling frameworks were successfully developed and presented in this chapter, in particular:

- A correlated channel model constructed by the authors and allowing the analytical description of the relationship between shadowing and fading, treated as two mutually independent stochastic processes with their correlation proprieties and the complex gain of a MIMO communication link. The model allows the calculation of BER of the MIMO system, as well as the extraction of shadowing parameters such as standard deviation and cross correlation values. Validation was performed by comparison between measured and calculated BER that were in very good agreement.
- The problem of characterization by means of prediction models, a crucial task for design engineers to asses "a priori" the reliability—or power efficiency—of a wireless off-body communication link was also tackled. To this aim, a novel deterministic modeling and simulation framework were developed, allowing the prediction of the reliability of the system mainly in terms of BER curves. This was constructed by combining different tools: a 3D full-wave EM solver for the (body-worn) antennas, ray-tracing for multipath indoor channel modeling, and Monte Carlo simulations for digital data transmission. The framework was successfully validated by modeling and simulating a 1×4

off-body indoor communication link and comparing the BER curves obtained by the tool with the ones obtained experimentally by a measurement campaign on the real system, obtaining a very good agreement between the two.

As a way to improve reliability of off-body communications, an innovative modulation technique was introduced known as spatial modulation (SM). This consists in a low-complexity algorithm based on switching the transmission of symbols among different antennas, resulting in an enhancement of spectral efficiency. Results of experimental tests, which compared performance of SM to those of space-time coding (STC) and SMX, showed that SM is a useful modulation scheme for off-body communications, especially in a 2×4 MIMO link. SM requires a lower SNR with respect to SMX to achieve the same BER and provides almost the same performance of STC with a much simpler transmit configuration.

Finally, two real applications employing the presented devices and techniques were presented.

- A distributed on-body sensor network composed of four autonomous wireless textile nodes capable of capturing, preprocessing, and relaying on-body sensor data (such as the ones obtained by a body-worn accelerometer) to a central remote access point. This is particularly useful for rescue worker and health-care applications. The network nodes continuously monitor transmissions between the nodes and allow data forwarding and measurement synchronization, which result in a significant increase of performance similar to the one obtainable using diversity. In particular, by means of a performance analysis carried out on a three-person network, it was shown that node-to-node communication is very reliable and packet forwarding makes the loss of sensor data extremely unlikely.
- A fully integrated on-body PDE for the GSM downlink band was developed, representing a more accurate alternative to commercially available PEMs. Our PDE consists of several wearable PDEs with GSM textile antennas deployed on both front and back side on the human body, and counters inaccuracies due to body shadowing occurring in classical PEMs. The superior accuracy of the proposed PDE compared to commercial systems was demonstrated by an experimental outdoor campaign using a PDE prototype on a moving test subject.

REFERENCES

[1] P. S. Hall and Y. Hao, *Antennas and Propagation for Body-Centric Wireless Communications*. Artech House, Boston/London, 2006.

[2] R. Chapman, ed., *Smart Textiles for Protection*. Woodhead Publishing, 2012.

[3] A. Goldsmith, *Wireless Communications*. Cambridge University Press, 2005.

[4] Y. Ouyang, D. J. Love, and W. J. Chappell, "Body-worn distributed MIMO system," *IEEE Trans. Vehicular Technol.*, vol. 58, no. 4, pp. 1752–1765, May 2009.

[5] L. Vallozzi, P. Van Torre, C. Hertleer, H. Rogier, M. Moeneclaey, and J. Verhaevert, "Wireless communication for firefighters using dual-polarized textile antennas integrated in their garment," *IEEE Trans. Antennas Propag.*, vol. 58, no. 4, pp. 1357–1368, April 2010.

[6] P. Van Torre, L. Vallozzi, C. Hertleer, H. Rogier, M. Moeneclaey, and J. Verhaevert, "Dynamic link performance analysis of a rescue worker's off-body communication system using integrated textile antennas", *IET Sci. Meas. Technol.*, vol. 4, no. 2, pp. 41–52, March 2010.

[7] R. Janaswamy, *Radiowave Propagation and Smart Antenna for Wireless Systems.* Kluwer Academic Publishers, Dordrecht, 2001.

[8] C. Oestges and B. Clerckx, *MIMO Wireless Communications: from Real-World Propagation to Space-Time Code Design.* Academic Press, 2007.

[9] S. M. Alamouti, "A simple transmit diversity technique for wireless communications," *IEEE J. Sel. Areas Commun.*, vol. 16, no. 8, pp. 1451–1458, October 1998.

[10] P. Van Torre, L. Vallozzi, H. Rogier, M. Moeneclaey, and J. Verhaevert, "Reliable MIMO Communication Between Firefighters Equipped with Wearable Antennas and A Base Station Using Space-Time Codes," in Proceedings of the 5th European Conference on Antennas and Propagation (EUCAP), April 2011, pp. 2690–2694.

[11] C. Hertleer, H. Rogier, L. Vallozzi, and L. Van Langenhove, "A textile antenna for off-body communication integrated into protective clothing for firefighters," *Spec. issue Antennas Propag. Body-Centric Wireless Comm.*, vol. 57, no. 4, pp. 919–925, April 2009.

[12] P. S. Hall, Y. Hao, Y. I. Nechayev, A. Alomainy, C. C. Constantinou, C. Parini, T. Z. Kamarudin, M. R. abd Salim, D. T. M. Hee, R. Dubrovka, A. S. Owadally, W. Song, A. Serra, P. Nepa, M. Gallo, and M. Bozzetti, "Antennas and propagation for on-body communication systems," *IEEE Antennas Propag. Mag.*, vol. 49, no. 3, pp. 41–58, June 2007.

[13] P. Salonen, Y. Rahmat-Samii, and M. Kivikoski, "Wearable Antennas in the Vicinity of Human Body," in Proceedings of Antennas and Propagation Society International Symposium, June 20–25, 2004, pp. 467–470.

[14] S. Zhu and R. Langley, "Dual-band wearable textile antenna on an EBG substrate," *IEEE Trans. Antennas Propag.*, vol. 57, no. 4, pp. 926–935, April 2009.

[15] A. Dierck, S. Agneessens, F. Declercq, B. Spinnewyn, G.-J. Stockman, P. Van Torre, L. Vallozzi, D. Vande Ginste, T. Vervust, J. Vanfleteren, and H. Rogier, "Active textile antennas in professional garments for sensing, localisation and communication," *Int. J. Microw. Wireless Technol.*, vol. 6, no. 3– 4, pp. 331–341, June 2014.

[16] D. Curone, E. L. Secco, L. Caldani, A. Lanatà, R. Paradiso, A. Tognetti, and G. Magenes, "Assessment of sensing fire fighters uniforms for physiological parameter measurement in harsh environment," *IEEE Trans. Inf. Technol. Biomed.*, vol. 16, no. 3, pp. 501–511, January 2012.

[17] C. Hertleer, H. Rogier, L. Vallozzi, and L. Van Langenhove, "A textile antenna for off-body communication integrated into protective clothing for firefighters," *IEEE Trans. Antennas Propag.*, vol. 57, no. 4, pp. 919–925, April 2009.

[18] J. Lilja, P. Salonen, T. Kaija, and P. de Maagt, "Design and manufacturing of robust textile antennas for harsh environments," *IEEE Trans. Antennas Propag.*, vol. 60, no. 9, pp. 4130–4140, July 2012.

[19] A. Dierck, F. Declercq, and H. Rogier, "Review of Active Textile Antenna Co-design and Optimization Strategies," in 2011 IEEE International Conference on RFID-Technologies and Applications (RFID-TA), September 2011, pp. 194–201.

[20] D. Masotti, A. Costanzo, M. Prete, and V. Rizzoli, "Genetic-based design of a tetra-band high-efficiency radio-frequency energy harvesting system," *IET Microw. Antennas Propag.*, vol. 7, no. 15, pp. 1254–1263, December 2013.

[21] A. Dierck, H. Rogier, and F. Declercq, "A wearable active antenna for global positioning system and satellite phone," *IEEE Trans. Antennas Propag.*, vol. 61, no. 2, pp. 532–538, January 2013.

[22] P. Van Torre, L. Vallozzi, L. Jacobs, H. Rogier, M. Moeneclaey, and J. Verhaevert, "Characterization of measured indoor off-body MIMO channels with correlated fading, correlated shadowing and constant path loss," *IEEE Trans. Wireless Commun.*, vol. 11, no. 2, pp. 712–721, February 2012.

[23] L. Vallozzi, H. Rogier, and C. Hertleer, "Dual polarized textile patch antenna for integration into protective garments," *IEEE Antennas Wireless Prop. Lett.*, vol. 7, pp. 440–443, May 2008.

[24] C. Oestges and B. Clerckx, *MIMO Wireless Communications: From Real-World Propagation to Space-Time Code Design.* Academic Press, 2007.

[25] J. G. Proakis, *Digital Communication.* Osborne-McGraw-Hill, 2001.

[26] I. M. Kim and D. Kim, "Closed-form exact BER and optimization of generalized orthogonal STBCs," *IEEE Trans. Wireless Commun.*, vol. 7, no. 9, pp. 3323–3328, September 2008.

[27] I.-M. Kim, "Exact BER analysis of OSTBCs in spatially correlated MIMO channels," *IEEE Trans. Commun.,* vol. 54, no. 8, pp. 1365–1373, August 2006.

[28] L. Jacobs and M. Moeneclaey, "BER analysis of square OSTBCs with LMMSE channel estimation in arbitrarily correlated rayleigh fading channels," *IEEE Commun. Lett.*, vol. 14, no. 7, pp. 626–628, July 2010.

[29] L. Vallozzi, "Wearable Textile Antennas for Wireless Off-Body Communication between Rescue-Workers: Prototyping, Modeling and Testing," Ph.D. Dissertation, Ghent University, October 2010.

[30] L. Vallozzi and H. Rogier, "Off-body Wireless Link Simulation Framework Using Deterministic Channel Modeling," in 2014 8th European Conference on Antennas and Propagation (EuCAP), The Hague, The Netherlands, 6–11 April, 2014, pp. 2289–2293.

[31] J. W. McKown and R. L. Hamilton Jr., "Ray tracing as a design tool for radio networks,", *Network, IEEE*, vol. 5, no. 6, pp. 27–30, November 1991.

[32] C. A. Balanis, *Advanced Engineering Electromagnetics*, 1st ed. John Wiley & Sons, New York, 1989.

[33] N. Serafimovski, A Younis, R. Mesleh, P. Chambers, M. Di Renzo, C.-X. Wang, P. M. Grant, M. A Beach, and H. Haas, "Practical implementation of spatial modulation," *IEEE Trans. Vehicular Technol.*, vol. 62, no. 9, pp. 4511–4523, November 2013.

[34] J. Zhang, Y. Wang, L. Ding, and N. Zhang, "Bit error probability of spatial modulation over measured indoor channels," *IEEE Trans. Wireless Commun.*, vol. 13, no. 3, pp. 1380–1387, March 2014.

[35] A. Younis, W. Thompson, M. Di Renzo, C.-X. Wang, M. A Beach, H. Haas, and P. M. Grant, "Performance of Spatial Modulation Using Measured Real-world Channels," in

2013 IEEE 78th Vehicular Technology Conference (VTC Fall), Las Vegas, NV, September 2–5, 2013, pp. 1–5.

[36] P. Van Torre, T. Castel, and H. Rogier, "Realistic Performance Measurement for Body-Centric Spatial Modulation Links," in 9th European Conference on Antennas and Propagation, accepted for presentation at EuCAP 2015, Lisbon, Portugal, April 2015, pp. 1–5.

[37] L. Vallozzi, H. Rogier, and C. Hertleer, "Dual polarized textile patch antenna for integration into protective garments," *IEEE Antennas Wireless Prop. Lett.*, vol. 7, pp. 440–443, May 2008.

[38] M. L. Scarpello, D. V. Ginste, and H. Rogier, "Design of a low-cost steerable textile antenna array operating in varying relative humidity conditions," *Microw. Opt. Technol. Lett.*, vol. 54, no. 1, pp. 40–44, January 2012.

[39] S. Alamouti, "A simple transmit diversity technique for wireless communications," *IEEE J. Sel. Areas Commun.*, vol. 16, no. 8, pp. 1451–1458, October 1998.

[40] P. Vanveerdeghem, P. Van Torre, C. Stevens, J. Knockaert, and H. Rogier, "Flexible dual-diversity wearable wireless node integrated on a dual-polarised textile patch antenna", *IET Sci., Meas. Technol.*, vol. 8, no. 6, pp. 452–458, November 2014.

[41] L. Vallozzi, P. Van Torre, C. Hertleer, H. Rogier, M. Moeneclaey, and J. Verhaevert, "Wireless communication for firefighters using dual-polarized textile antennas integrated in their garment," *IEEE Trans. Antennas and Propag.*, vol. 58, no. 4, pp. 1357–1368, April 2010.

[42] P. Vanveerdeghem, P. Van Torre, C. Stevens, J. Knockaert, and H. Rogier, "Synchronous wearable wireless body sensor network composed of autonomous textile nodes," *Sensors*, vol. 14, no. 10, pp. 18583–18610, October 2014.

[43] W. Joseph, G. Vermeeren, L. Verloock, and L. Martens, "Estimation of whole-body SAR from electromagnetic fields using personal exposure meters," *Bioelectromagnetics*, vol. 31, no. 4, pp. 286–295, May 2010.

[44] International Commission on Non-Ionizing Radiation Protection, "Guidelines for limiting exposure to time-varying electric, magnetic, and eletromagnetic fields (up to 300 GHz)," *Health Phys.*, vol. 74, pp. 494–552, October 1998.

[45] A. Thielens, S. Agneessens, L. Verloock, E. Tanghe, H. Rogier, L. Martens, and W. Joseph, "On-body calibration and processing for a combination of two radio-frequency personal exposimeters," *Radiat. Prot. Dosim.*, vol. 163, no. 1, pp. 58–69, April 2014.

[46] A. Thielens, H. De Clercq, S. Agneessens, J. Lecoutere, L. Verloock, F. Declercq, G. Vermeeren, E. Tanghe, H. Rogier, R. Puers, L. Martens, and W. Joseph, "Personal distributed exposimeter for radio frequency exposure assessment in real environments," *Bioelectromagnetics*, vol. 34, no. 7, pp. 563–567, October 2013.

[47] P. Vanveerdeghem, P. Van Torre, A. Thielens, J. Knockaert, W. Joseph, and H. Rogier, "Compact personal distributed wearable exposimeter," *IEEE Sensors Journal*, vol. 15, no. 8, pp. 4393–4401, Aug. 2015.

[48] F. Declercq, A. Georgiadis, and H. Rogier, "Wearable Aperture-coupled Shorted Solar Patch Antenna for Remote Tracking and Monitoring Applications (invited)," in Proceedings of the 5th European conference on Antennas and propagation EUCAP, IEEE, Rome, April 11–15, 2011, pp. 2992–2996.

[49] C. Hertleer, A. Tronquo, H. Rogier, L. Vallozzi, and L. Van Langenhove, "Aperture-coupled patch antenna for integration into wearable textile systems," *IEEE Antennas Wireless Propag. Lett.*, vol. 6, pp. 392–395, 2007.

[50] A. Thielens, G. Vermeeren, W. Joseph, and L. Martens, "Stochastic method for determination of the organ-specific averaged SAR in realistic environments at 950 MHz," *Bioelectromagnetics*, vol. 34, no. 7, pp. 549–562, October 2013.

[51] Belgian Institute for Postal services and Telecommunications, "Antenna site register," 2014.

[52] A. Thielens, P. Vanveerdeghem, S. Agneessens, P. Van Torre, G. Vermeeren, H. Rogier, L. Martens, and W. Joseph, "Whole-body Averaged SAR Assessment Using a Personal, Distributed Exposimeter," in Proceedings of BioEM2014, The Bioelectromagnetics Society (BEMS), 2014, p. 6.

9

MORE THAN WEARABLE: EPIDERMAL ANTENNAS FOR TRACKING AND SENSING

SARA AMENDOLA, CECILIA OCCHIUZZI, AND GAETANO MARROCCO

Department of Civil Engineering and Informatics,
University of Rome Tor Vergata, Rome, Italy

9.1 INTRODUCTION

Epidermal or *skin electronics* [1] is a currently emerging research trend combining multi-disciplinary backgrounds such as material science, mechanics, classic electronics and electromagnetics, and communications. Recently published works [2] demonstrate the possibility to fabricate skin-like electronics and sensors that are suitable to be deployed over thin and biocompatible conformable membranes for direct placement over living organisms (e.g., tissues), environmental substrates (plants or stems), and foods [3]. The direct "on-skin" sampling of surface temperature, humidity, pH, deformations, skin impedance, and electrophysiological potentials by means of plaster-like complex devices may enable a host of new applications to consumer electronics, human health and wellness as well as to nutrition.

Most of the scientific efforts presented so far have been oriented to high-tech fabrication methods for epidermal devices (spin-coating, photolithography, dry etching, and transfer printing), for example, to the design of engineered thin, lightweight, stretchable conformal silicone, and polyvinyl alcohol (PVA) substrates as well as to the deposition of conducting traces. Most of the described prototypes relied on wired interconnection for data transfer [4] or, at most, by inductive coupling among coils [5,6] that permit only near-contacting links.

Electromagnetics of Body-Area Networks: Antennas, Propagation, and RF Systems, First Edition.
Edited by Douglas H. Werner and Zhi Hao Jiang.

In this scenario, the virtuous synergy of epidermal electronics with the well-assessed passive radiofrequency identification (RFID) technology [7] could boost the usability of skin devices in the real world, providing a further brick of the quickly emerging Internet of Things [8]. The communication through electromagnetic backscattering, as involved in the RFID architectures, requires just a small battery-less IC transponder that is fully compatible with thin and flexible devices. Furthermore, thanks to the intrinsic sensing capabilities of RFID platforms [9], it could be possible to exploit the simple interaction between the tag and the skin to retrieve data and information about the state of the human body without the need of dedicated and complex electronics integrated into the epidermal device. However, the cohabitation of passive antenna elements with the human skin, or with living matters in general, represents an intrinsic challenge due to the high losses of tissues which strongly degrade the radiation efficiency of the device and accordingly the length of the communication link. This problem has been widely addressed by the research on wearable antennas wherein the goal is however to decouple at most the antenna radiation from the body loss by means of ground planes (patch-like antennas) [10, 11] or multi-layer spacers placed between the body and a dipole [12], a slot [13], or a loop antenna [14]. Conversely, an RFID tag for epidermal applications has to act both as an antenna and as a sensor; therefore, unlike wearable tags, it has to be placed in direct contact with the body. Very thin insulating materials need to be used, mostly acting as an adhesive interface between the conductor/electronics and the skin. The epidermal tag, moreover, has to be designed in order not to interfere with the local metabolism of the skin. Accordingly, the substrate is required to be not only biocompatible but also breathable (in order not to perturb the transpiration and the local temperature of the skin). Shielding planes, as in the case of patch antennas, have to be avoided, and the amount of antenna conductor has to be minimized.

Some early and pioneering examples of epidermal antennas, in the form of shaped dipoles, have been very recently proposed in [15–17], and consider temporary tattoo-like RFID tags. Conductive paints, profiled by stencils and inkjet printing, were used to fabricate the antenna element for direct application over the skin and over the tongue to detect the motion of impaired people. The reported read distances are of the order of 1 m. The performance of several types of thin single-layer antennas (e.g., slot, self-complementary and wire antenna) for plaster-like sensing applications at 2.4 GHz were extensively investigated in [18] to search for the antenna layout that was best suited to operate in very close proximity of the human body.

This chapter introduces the performance, the natural bounds, and some possible applications to body-centric systems of epidermal antennas that are suitable to be integrated with the RFID technology. Starting from basic principles, the maximum efficiency and gain are related to the geometrical and electrical parameters of the epidermal device. Then, a reference antenna layout is extensively used to discuss the issues arising from the placement of the antenna over different parts of the body and to introduce a practical on-body impedance retuning mechanism. Specialized membranes useful to host epidermal antennas and to achieve sensing capabilities are introduced as well.

The chapter ends with the description of two promising applications of epidermal tags to the wireless measurement of body temperature and to the monitoring of fluid exchanges at the skin interface.

9.2 RFID TECHNOLOGY

An RFID system [7] is composed of two main components: the remote transponder or tag, including an antenna and a microchip transmitter, located on the object to be identified, and the local querying system or reader which can collect the data transmitted from the tag and eventually perform a preliminary processing job. Various kinds of data and, first of all, a unique identification code (ID), can be wirelessly transferred to the reader by means of radiofrequency electromagnetic signals.

The tags could be *passive*, harvesting energy from the interrogating system, *semi-active* when a battery is included only to feed embedded sensors or to increase the sensitivity of the receiver inside the microchip, or *fully active* where a local source directly feeds a microcontroller as well as the transmitting radio. Battery less tags have an almost unlimited life, are very low cost, lightweight, and need no maintenance. Because of the absence of a local power source, passive transponders provide the greatest flexibility in seamlessly integrating the RF devices with the human skin as required by epidermal applications.

RFID systems can be classified in terms of operating frequencies: LF (low frequency, 120–145 KHz), HF (high frequency, 13.56 MHz), UHF (ultra-high frequency, 866–956 MHz). UHF devices, although strongly influenced by high dielectric targets, as in the case of the human body, may in principle promise larger activation ranges which enable a real ease of interrogation of epidermal tags thus making the UHF-RFID standard particularly attractive over the HF frequency.

This chapter is hence focused only on passive battery-less UHF systems.

The power budget of the UHF-RFID radio channel, that is, the power $P_{R\rightarrow T}$ transmitted by the reader and collected by the tag (direct link), and the power $P_{R\leftarrow T}$ backscattered by the tag toward the reader (backward link), are commonly characterized in far field by means of the Friis formula and the radar equation [7]

$$P_{R\rightarrow T} = \left(\frac{\lambda_0}{4\pi d}\right)^2 P_{in} G_R G_{Tag} \tau \cdot \eta_p \tag{9.1}$$

$$P_{R\leftarrow T} = \frac{1}{4\pi}\left(\frac{\lambda_0}{4\pi d^2}\right)^2 P_{in} G_r^2 rcs_T \eta_p \tag{9.2}$$

where $\hat{G}_{Tag} = G_{Tag} \cdot \tau$ is the realized gain of the tag, that is, the radiation gain G_{Tag} of the tag antenna reduced by the power transfer coefficient τ between the tag antenna and the microchip, P_{in} is the power emitted by the reader, G_R is the reader antenna gain, η_p is the polarization factor between the reader and the tag, d is the distance between reader and tag antennas, λ_0 is the free-space wavelength, and rcs_T is the radar cross section of the tag.

The power transmission coefficient, can be expressed in terms of tag antenna and tag IC impedance, $Z_A = R_A + iX_A$ and $Z_{chip} = R_{chip} + iX_{chip}$ by

$$\tau = \frac{4R_{chip}R_A}{|Z_{chip} + Z_A|^2} < 1 \tag{9.3}$$

The performance indicator of an on-body tag for reliable indoor monitoring is the maximum reading range d_{max} between the reader and the tag. The read range is dependent on the antenna radiation characteristics of the antenna $(\hat{G}_{Tag})$ and inversely dependent on the IC sensitivity (P_{chip}), for example, the power required to activate the IC. The maximum theoretical read range can be estimated from (9.1) as

$$d_{max} = \left(\frac{\lambda_0}{4\pi}\right)\sqrt{\frac{\mathrm{EIRP}\cdot\hat{G}_{Tag}\cdot\eta_p}{P_{chip}}} \tag{9.4}$$

where $\mathrm{EIRP} = G_R \cdot P_{in}$ is the equivalent isotropic radiated power emitted by the reader and is generally fixed to 3.2 W EIRP or 4 W EIRP, according to different countries' regulations. Depending on the size of the tag and on the object on which it is attached, the realized gain of a tag (see Table 9.1) may span from 3 dB (low-loss things) down to −30 dB (humans) or even less. Consequently, a UHF link may be established from less than 1 m (implantable tags) up to 20 m in the best configuration provided that ICs with very high sensitivity are considered in the design (Figure 9.1). Epidermal UHF tags are expected to be read up to a distance of a few meters.

9.3 RADIATION PERFORMANCE OF EPIDERMAL ANTENNAS

It is well known that the radiating properties of an antenna are strictly correlated with its physical size [19]. Electrically large antennas generally provide a better gain than smaller ones. For instance, the maximum gain of a straight dipole radiating in free space monotonically increases as function of its electrical length (Figure 9.2) and similar curves can be obtained for other layouts such as loops and slots.

When antennas radiate in the close proximity of a lossy medium like the human body, the radiation efficiency is instead severely degraded because of the high loss

TABLE 9.1 Typical realized gains and read ranges of three families of tags achievable by means of 2014 RFID technology (P_{chip} = −22 dBm).

Placement	Realized gain (dB)	Read range (m)
Free-space tag	0 to 3	10 to 19
Wearable tag	−10 to 0	4 to 10
Epidermal tag	−20 to −8	1 to 4
Implanted tag	<−20	<1

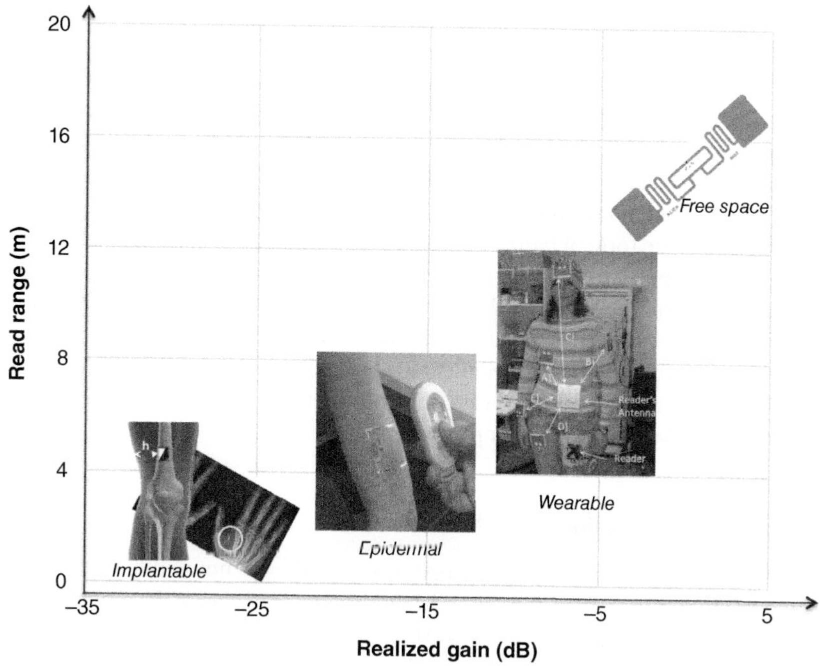

FIGURE 9.1 Typical read distances of implantable, epidermal, wearable, and free-space tag assuming a microchip sensitivity of −22 dBm.

of the underlying tissues. Accordingly, the gain-size relationship might be non-trivial and not necessarily monotonic.

In the past, investigation on antennas operating near the interface between air and lossy regions were the topic of extensive theoretical efforts, starting with the famous essay of Sommerfeld in 1909 [20]. Those studies considered simple wire antennas (usually infinitesimal electric or magnetic dipoles) above an infinite lossy half-space

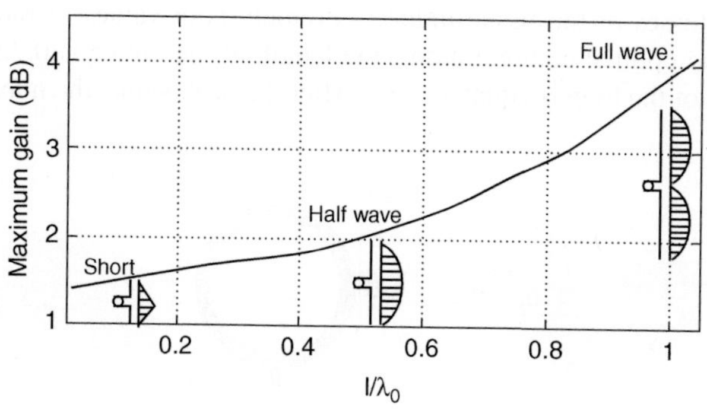

FIGURE 9.2 Strip dipole in free space: maximum gain as a function of the electrical length.

and the derived equations and knowledge background had practical concern in a broad variety of applications such as location of buried objects, geophysical prospecting, and over-horizon propagation. Soils and ground generally have a permittivity much lower than the tissues of the human body, while instead a higher conductivity, and the interesting frequencies are rather different than in the UHF-RFID communications.

Among the very few efforts of the RFID community to tackle the challenge of designing passive antennas above lossy medium, the early examples of epidermal transponders are in the form of tattoo-like nested-slot dipoles [15] or dual square loops [21]. The loop shape is traditionally expected, or believed, to outperform dipoles when placed over the body due to the dominant magnetic near field that should be less influenced by the presence of human tissues.

Before introducing possible implementations of epidermal RFID antennas and their applications, it is useful to clarify the role of the geometrical and electrical parameters of dipoles and loops for optimal placement over the skin.

9.3.1 Efficiency and Gain versus Antenna Size

Dipole and loop radiators made by a copper trace of $W = 1$ mm width are assumed to be stuck onto a muscle-like box as in Figure 9.3 (size 30 cm × 30 cm × 20 cm, electromagnetic parameters $\varepsilon = 43$, $\sigma = 0.9$ at 870 MHz) through a thin air layer which emulates a thin adhesive membrane. The profiles of maximum gain and radiation efficiency, estimated by the Method of Moments (FEKO solver [22]), are shown in Figure 9.4 with respect to the increase of the overall size of the antennas. More precisely, the length of the dipole and the outer diameter of the loop have been progressively increased up to the effective wavelength $\lambda = \lambda_0/\sqrt{\varepsilon_{\text{eff}}}$ where λ_0 is the free-space wavelength and ε_{eff} is the effective permittivity evaluated as the average of the permittivity of the body phantom and free space.

Unlike the case of free-space radiation, the efficiency of antennas over lossy media exhibits a bell-shaped asymmetrical behavior. After an initial monotonic region with a nearly linear relationship between maximum gain (efficiency) and size, there are peak values, rather similar for the two antennas ($\eta = 0.3\%$), occurring for an antenna size of 30 mm and 50 mm for the loop and the dipole, respectively. In both cases, the significant losses of the tissue induce extremely poor values of radiation gain (−20 dB). In the optimal configurations, the length of the dipole is 0.7λ, while the outer perimeter of the loop is equal to 1.3λ. After the peak value, the maximum gain

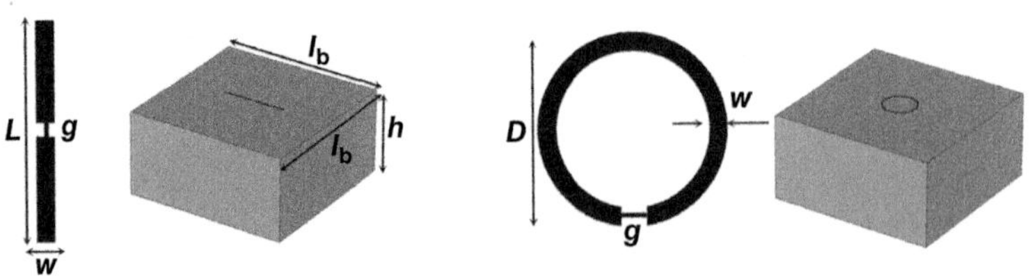

FIGURE 9.3 Dipole and loop radiators over muscle-like box ($l_b = 30$ cm, $h = 20$ cm).

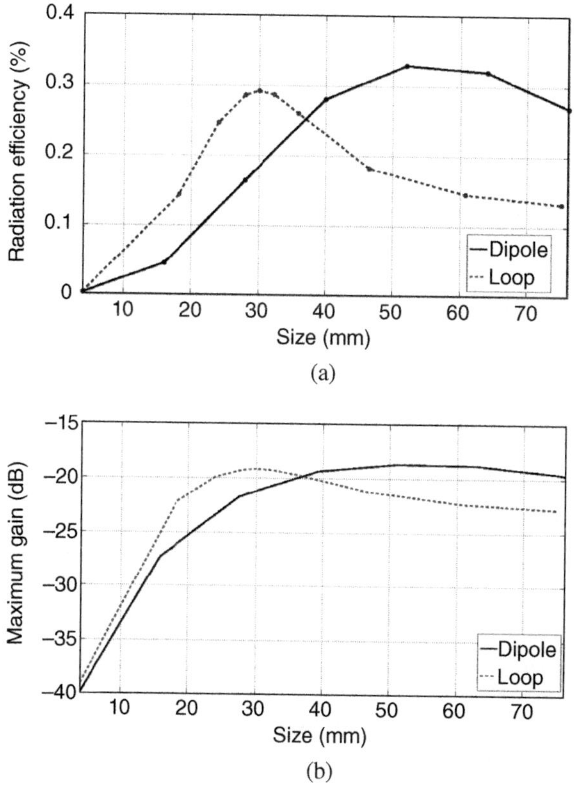

FIGURE 9.4 Dipole and loop antennas over lossy dense medium: Radiation efficiency (%) and maximum gain at 870 MHz as a function of maximum size.

remains rather stable while, instead, the radiation efficiency significantly degrades. The peak gains occur along the normal direction to the skin, which acts as both an absorber as well as a reflector.

The non-monotonic relationship between the efficiency/gain and physical size of antennas can be explained by considering the presence of two counteracting phenomena. The radiation efficiency can be expressed as the ratio

$$\eta = \frac{R_{\text{R}}}{R_{\text{R}} + R_{\text{o}} + R_{\text{B}}}$$

where R_{R} is the radiation resistance of the antenna, R_{O} is the ohmic loss of the antenna conductor, and R_{B} accounts for the power dissipated by the electromagnetic near field of the antenna into lossy body tissue. The initial rise of the efficiency in Figure 9.4 is mostly due to the increase of the radiation resistance that is strictly proportional to the overall length of the antenna. However, a further enlargement of the antenna will produce more intense power dissipation by the high conductivity of the hosting medium.

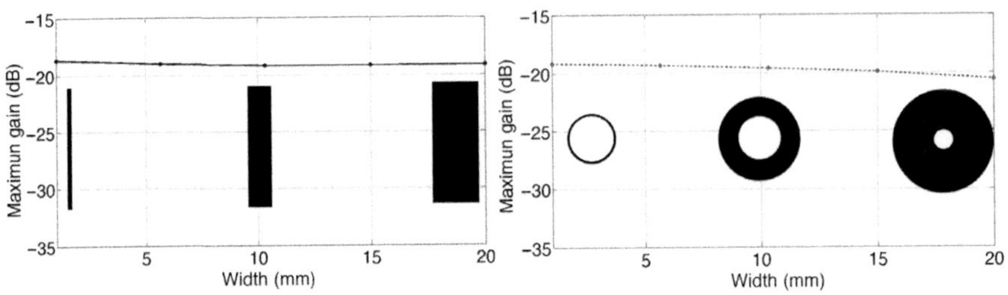

FIGURE 9.5 Simulated maximum gain at 870 MHz versus the width of the conductor for dipole ($l = 50$ mm) and loop ($r = 15$ mm) over lossy medium.

Thus, enlarging the antenna beyond its optimal size is not only undesirable for epidermal applications demanding small and unobtrusive devices; it even produces a negative impact on radiation properties.

9.3.2 Gain versus the Trace Width

Another geometrical parameter that is involved in the design and fabrication of epidermal antennas is the width of its conducting traces. Figure 9.5 shows maximum gain, obtained by simulations, when the copper trace is increased from $W = 1$ mm (as in the previous case) up to 20 mm.

The radiation parameters turned out to be roughly independent of the conductor width, apart from a negligible decrease for the loop. Therefore, wider layouts provide the same performance as the narrow ones at the expense of a major invasiveness of the epidermal antennas.

9.3.3 Radiation Performance versus the Trace Conductivity

Epidermal devices demand an easy deposition of the conductor traces on thin membranes, and several experiments considered different options than bulk copper, such as inkjet printing or stencil-based techniques which make use of conducting paints and nanoparticle inks [23]. These materials provide superior mechanical performance to the bulk copper, such as flexibility and robustness to cracks, but at the price of a reduced conductivity of one order of magnitude or even more [23].

For antennas radiating in free space, a non-negligible efficiency loss is expected when using bad conductors, while a rather different behavior occurs in the case of epidermal antennas.

Figure 9.6 shows the efficiency and gain of a dipole of fixed length ($L = 50$ mm) and trace thickness $W = 1$ mm when its conductivity is decreased from 10^8 S/m (good conductor) down to 10^2 S/m (bad conductor). There is no visible variation in the radiation performance for $\sigma > 10^4$ S/m. When further reducing the conductivity down to 1000 S/m, the gain drops by less than 2 dB. This result is not trivial and it is a consequence of the dominant power losses of the hosting human body that, for

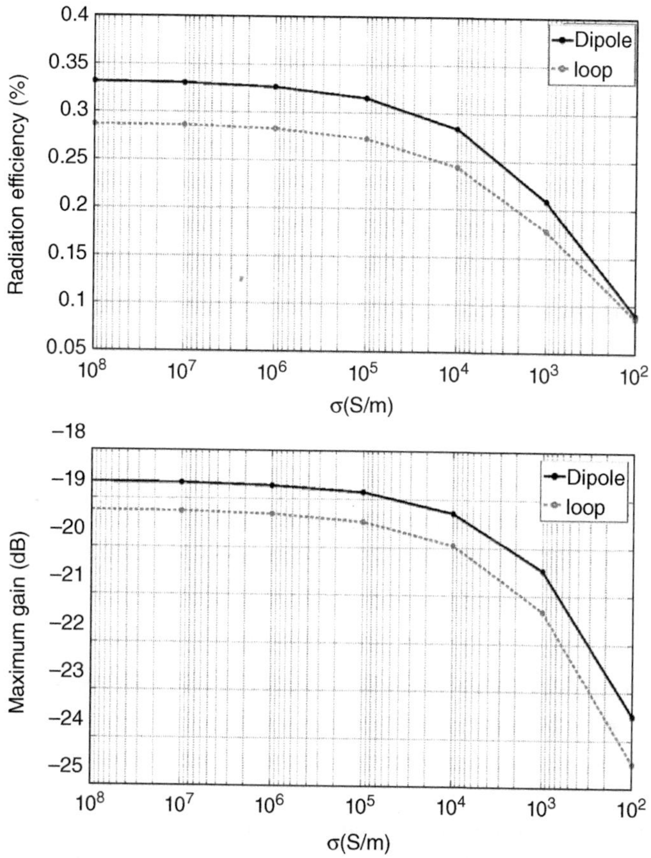

FIGURE 9.6 Efficiency and maximum gain at 870 MHz versus the finite conductivity of metal conductor.

epidermal antennas, is much more effective than in the case of wearable radiators which are usually much better decoupled from the body.

9.3.4 Radiation Performance versus the Spacing From the Skin

Epidermal antennas are stuck onto the skin by ultrathin materials (fractions of millimeters). The membrane could even be functionalized to act as a sensitive material thus enabling the epidermal antenna to act as a sensor. The thickness of the membrane has a remarkable influence on the radiation performance of the antenna since, in case of thick membranes, the electric near field diffused into the epidermis is reduced and, accordingly, the power loss is decreased.

Figure 9.7 shows the radiation efficiency of a dipole of size ($L = 50$ mm, $W = 1$ mm, $\sigma = 10^8$ S/m) placed on the same muscle box as before when the skin–dipole

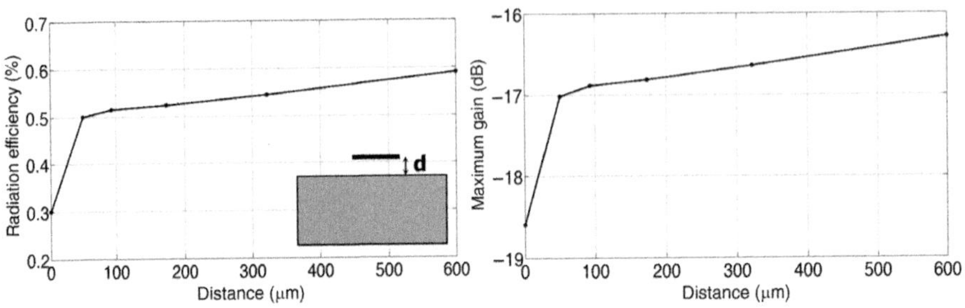

FIGURE 9.7 Efficiency and maximum gain at 870 MHz as a function of the air gap between the dipole and the muscle box.

gap is increased from $d = 0$ (direct contact) up to 600 μm. There is a sharp discontinuity when the separation distance exceeds $d = 0.1$ mm while the efficiency and gain increase is less appreciable for thicker substrates.

To summarize, the close interaction of an antenna (here loops and dipoles) with the lossy human body, and the corresponding strong power loss, induces the following typical phenomena to be accounted for, during design and manufacturing.

i) There is an upper bound in the achievable maximum gain and efficiency, roughly independent of the antenna shape while partly dependent on the thickness of the attaching membrane.
ii) As a consequence, there is an optimal size of the antenna depending on the particular layout.
iii) Dipoles and loops provide the same performance. The optimal external size of a loop is however shorter than that of the dipole.
iv) The quality of the antenna conductor plays a minor role in the radiation performance. Materials with conductivity $\sigma > 10^4$ S/m provide the same results of bulk copper.
v) The radiation performance sensibly improves when the membrane is thicker than 0.1 mm.

9.4 PERFORMANCE OF EPIDERMAL RFID DUAL-LOOP TAG

Performances of skin-mounted tags are moreover expected to be quite dependent on the specific region of application due to the high variability of the tissue composition [24] according to different body parts.

In this section, a reference UHF tag in the form of dual loop is introduced in order to clarify by means of both simulations and extensive experimentations the dependence of the electromagnetic response on the position over the body and also on the user's body mass. Thanks to the small amount of conductive material and the direct contact with human skin, this tag layout matches the requirements of biocompatibility and low invasiveness of epidermal devices, as well as it is suitable to act as a chemical–electrical transducer of skin conditions.

9.4.1 Tag Layout

The layout of the epidermal tag consists of a one-wavelength rectangular-loop connected to the RFID microchip through a loop-match impedance transformer [25].

A similar geometry has been previously proposed for on-body applications [14]. Unlike the wearable tag, the conductor of the epidermal radiator is insulated from the body by only a thin silicone membrane (thickness = 600 μm, $\varepsilon = 2.5$, $\sigma = 0.005$ S/m) so that the tag is forced to work in very close proximity with the skin. The flexible silicone substrate permits the comfortable application of the tag over any body curvatures and can be carved just around the antenna profile to ensure physiological skin transpiration (Figure 9.8). The size of the outer loop is fixed to 5 cm × 5 cm so that the physical area of the tag is compatible with that of a medical plaster. The internal loop is instead properly sized by numerical simulations (moment method throughout FEKO solver [22]) for the optimal impedance tuning within the worldwide UHF-RFID band. The terminals of the inner loop are finally connected to the RFID IC.

The initial design of the tag is performed by exploiting a simplified model of the human body, consisting of stratified planar layers of defined thickness [13], and dielectric properties [26] (see the inset in Figure 9.9). The maximum realized gain of the epidermal tag in this reference configuration (by considering NXP–G2X-TSSOP-8 IC, with $Z_{chip} = 16 - j148\ \Omega$, and $P_{chip} = -15$ dBm) occurs around 940 MHz along the broadside (frontal) direction (Figure 9.9). The corresponding theoretical read range is hence 2.5 m for the considered IC and up to 5.5 m in case the state-of-the-art COTS are used ($P_{chip} = -22.1$ dBm, [27]).

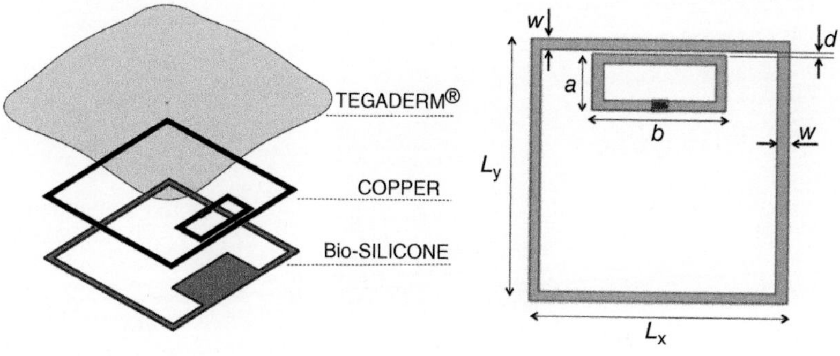

FIGURE 9.8 Layout of the epidermal tag. Size (mm): $L_1 = L_2 = 50$, $w = 2$, $d = 0.5$, $a = 26$, $b = 10$.

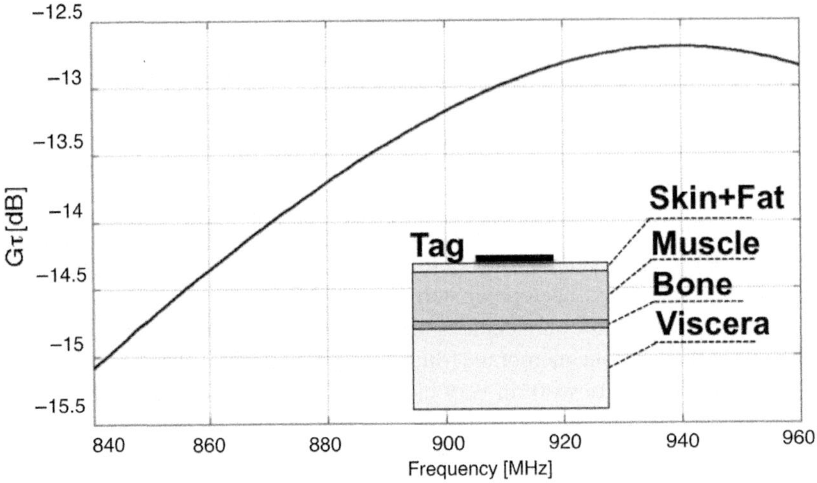

FIGURE 9.9 Simulated frequency-dependent realized gain of the epidermal tag over the reference planar model as in [13]. *Source:* Amendola et al. 2015 [26]. Reproduced with permission of IEEE.

9.4.2 Prototype and On-Skin Performance

Figure 9.10 shows a prototype of the epidermal tag. It is made of adhesive copper (thickness 35 μm) that was carved by a two-axis digital-controlled cutter. First, the loop was transferred over a thick (600 μm) biocompatible non-breathable silicone layer by means of an adhesive transfer tape with the purpose to have a robust substrate which could preserve the integrity of the antenna layout. Then, the membrane was shaped just around the conductors, and the material in excess was carved out in order to improve transpiration. The resulting gummy loop is finally stuck over a commercial

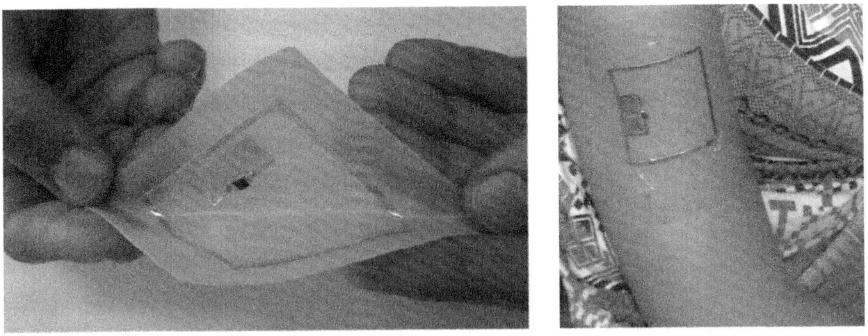

(a) (b)

FIGURE 9.10 (a) Prototype of the epidermal RFID tag sensor over biocompatible silicone membrane. (b) Epidermal RFID tag applied over an arm through Tegaderm™ fixing. *Source:* Amendola et al. 2015 [26]. Reproduced with permission of IEEE.

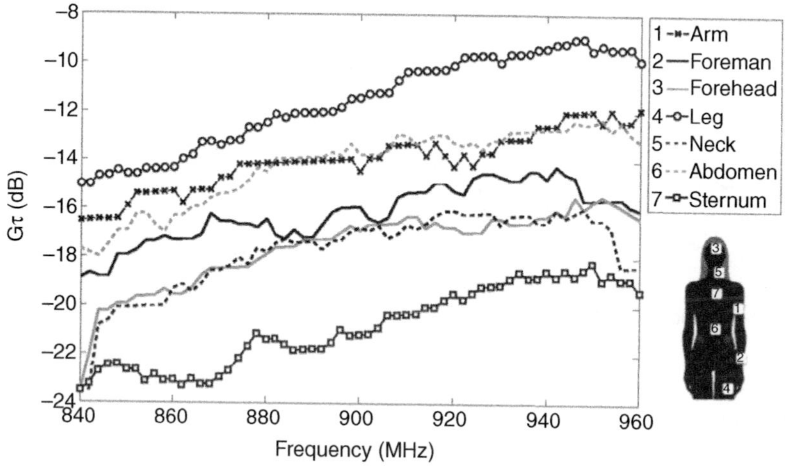

FIGURE 9.11 Variability of the measured realized gain of the epidermal tag when attached onto different body parts of a female volunteer. *Source:* Amendola et al. 2015 [26]. Reproduced with permission of IEEE.

very thin 22 μm transparent adhesive film (Tegaderm™). This second membrane is a waterproof and breathable dressing that is commonly used in clinical practice for the treatment of wounds. The plaster is still partially adhesive, with the copper placed between the two membranes, and it is finally ready for a comfortable and nearly invisible placement over the skin.

The ratio between the total area of the conductor and the size of the resulting tag is roughly 20%, with great benefit to skin transpiration. The residual surface of the skin could be, moreover, used to host other devices such as chemical sensors and even drug delivery mechanisms.

The performance of the epidermal tag is strictly related to the specific locus of placement over the human body. Figure 9.11 shows the measured realized gain of the tag placed onto different body parts of a female volunteer, such as arm, forearm, abdomen, sternum, hand, neck, and forehead.

The profiles are mutually scaled (due to change in the power absorption) and partially shifted by the effect of impedance detuning. The placement over the leg provides the highest realized gain. In particular, the realized gain is higher than −10 dBi at 950 MHz and more than −13 dBi in the whole UHF-RFID band. The worst case was instead the sternum position with a degradation of about 10 dB.

Tags placed on different subjects suffer from human body variability as visible in Figure 9.12 where the epidermal tag is attached onto arms, legs, and abdomen of three volunteers: a *"normal" male*, a *"muscular" male,* and a *female* (height and weight reported in Table 9.2).

The resulting realized gains are sensibly different in scale and peak position. Tags placed onto normal male and female volunteers exhibit, on average, better performance in comparison with the muscular man. Differences are particularly visible for

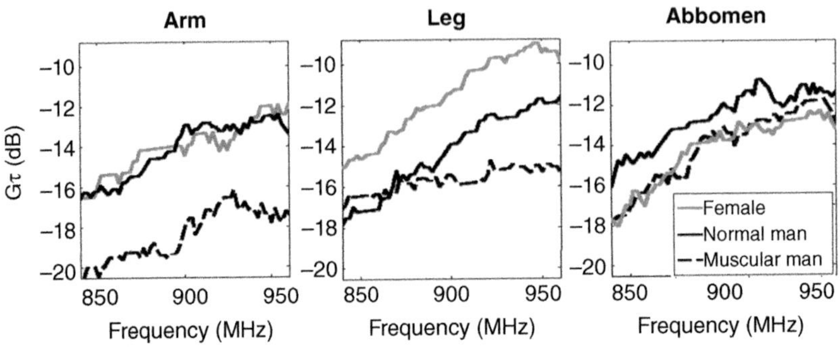

FIGURE 9.12 Variability of the measured realized gain of the epidermal tag when it is attached onto corresponding body parts of the three volunteers. *Source:* Amendola et al. 2015 [26]. Reproduced with permission of IEEE.

arm and leg placement while in the case of the abdomen there is no apparent difference between the normal and the muscular man since in both cases water-rich tissues dominate. The placement over the leg, for the same reason as above, provides instead the highest gain that, in case of the woman, may approach −9 dBi.

9.4.3 On-Skin Retuning

Previous experiments demonstrated how the anatomical characteristics of different body regions as well as human body variability (gender, physique) sensibly affect the tag radiation performances. It is hence required to provide the antenna with some post-fabrication and, possibly, post-placement retuning capabilities in order to obtain a true communication-reliable epidermal device which is independent of the frequency, of the user's mass, and of the specific position.

Since the considered tag consists of two loops, it is not possible to achieve tuning by trimming a portion of the conductor length, as typically done with dipoles, since the galvanic continuity of the loop has to be preserved. A viable retuning mechanism relies instead on the form factor of the internal coil. In fact, by analyzing the current distribution over the inner loop (Figure 9.13a), it is apparent that the surface currents are more intense in the inner perimeter, so that a perturbation of such a profile is expected to cause a change of the equivalent self-impedance of the loop and of the mutual inductance between the two coupled loops (recall the equivalent circuit model in [14]), and accordingly the input impedance of the epidermal tag will be modified.

TABLE 9.2 Data of the four volunteers for epidermal tag measurements.

Volunteer	Height (cm)	Weight (Kg)	Body mass index
Female	155	45	18.7
Normal male	174	64	21.4
Muscular male	184	85	25.4

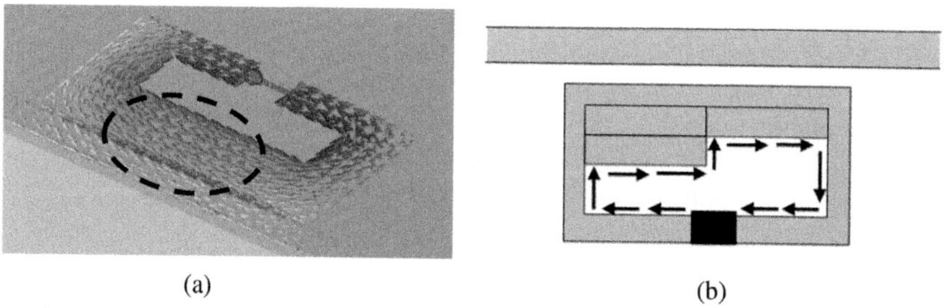

FIGURE 9.13 (a) Zoomed view of the simulated current of the feeding loop. (b) The black arrows show how the path of the excitation currents can be modified by removing/adding equal-sized strips made of conductor. *Source:* Amendola et al. 2015 [26]. Reproduced with permission of IEEE.

A practical way to implement this idea in a controllable manner is to partition the segment of the inner coil, which is closely coupled with the external loop, into some equal-size pre-cut strips (Figure 9.13b). The strips can be easily removed, even after placement over the body, to modify the loop current distribution without affecting the tag integrity.

Figure 9.14 shows the simulated changes of the power transfer coefficient for the epidermal tag when the strongly coupled subregion of the inner loop (a = 24 mm, b = 12 mm) was enlarged and partitioned into six strips of size 20 mm × 1.9 mm and one strip was removed at a time. The tag was again assumed to be placed onto the

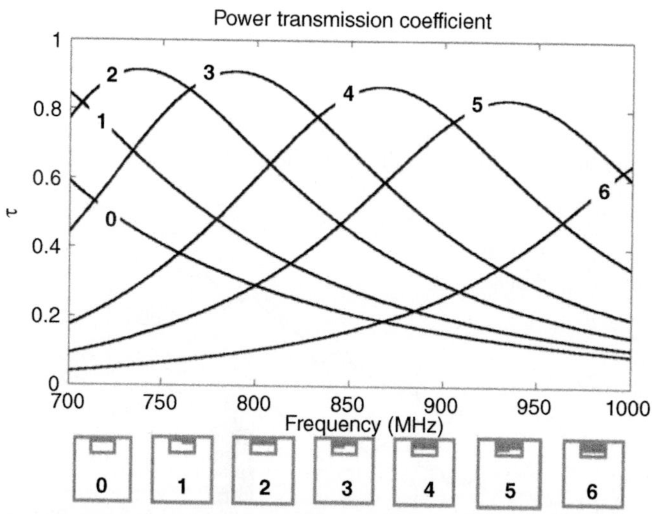

FIGURE 9.14 Parametric exploration of the simulated power transmission coefficient of the epidermal tag (placed over the reference layered medium of Figure 9.9) that is tuned by modifying the size of the feeding inner loop through removal of pre-carved strips of conductors. *Source:* Amendola et al. 2015 [26]. Reproduced with permission of IEEE.

layered geometry as in Figure 9.9. The peak of the power transfer coefficient τ shifts toward a lower frequency, with some improvement in the maximum value.

9.4.3.1 ***Examples*** Some examples of frequency retuning are given for application of the epidermal tag onto the volunteers. Experiments considered the sensible improvement of performance achievable through the skin retuning in two relevant cases:

1. the realized gain of the tag placed at a given position is initially rather poor;
2. the same tag needs to be used in different RFID bands (USA or EU).

The unperturbed tag (e.g., with all the six strips connected) was stuck onto the hip of the female volunteer. The peak of the realized gain occurred at 950 MHz, with a poor maximum value close to −20 dBi (a worse case than those experimented before). An improvement of more than 4 dB was achieved by removing just a single strip and the read distance increased about 1.5 times, as shown in Figure 9.15.

Starting from its initial "full" configuration, a tag can be properly tuned to work in any worldwide RFID band by just removing a well-defined number of strips. For example, to have a tag placed on a female arm, tuned in the EU band, three strips need to be removed (realized gain at 870 MHz equal to −13 dBi, Figure 9.16) while to shift the frequency of operation from Japan to US (realized gain close to −12 dBi) for a tag placed on the leg of a normal man, the removal of two strips is (Figure 9.17) required.

The results of previous experiments may be summarized as in Table 9.3 to better understand the effectiveness of the on-body retuning. The application of the same tag on different users and on different regions of the body (hip, arm, leg) really benefits at 950 MHz of the retuning (5, 6, 4 strips, respectively) with an overall nearly stable realized gain. The tag with 4 strips is suitable to multiple applications over the hip and arm of the female volunteer at 870 MHz and at 900 MHz as well as to the leg of the normal man but at 950 MHz. In all cases, the read ranges are reasonably similar.

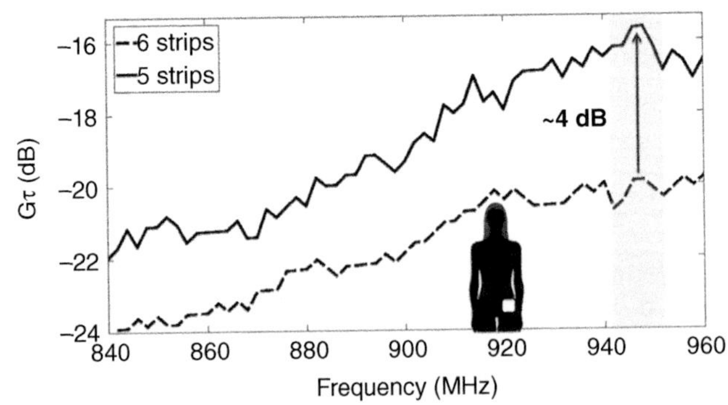

FIGURE 9.15 On-skin retuning from US to EU band of the tag placed onto a female hip. *Source:* Amendola et al. 2015 [26]. Reproduced with permission of IEEE.

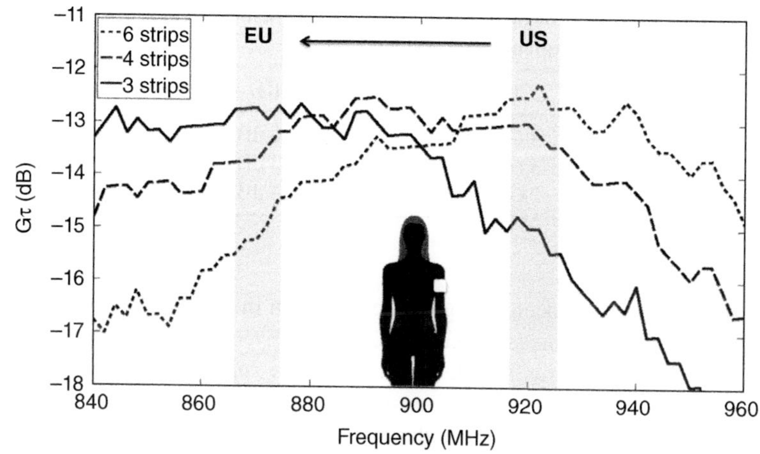

FIGURE 9.16 On-skin retuning from US to EU band of the epidermal tag placed onto the arm of the female volunteer. *Source:* Amendola et al. 2015 [26]. Reproduced with permission of IEEE.

9.5 SPECIAL (FUNCTIONALIZED) EPIDERMAL MEMBRANES

Epidermal devices must intimately integrate with human skin and enable a fully transparent, non-invasive "on-skin" sampling of bioparameters. Membranes to be used as substrates for epidermal tags play a crucial role as they represent the interface between the radiators and the human skin. Their mechanical and thermal properties must closely match to the skin itself, so as to permit an effective adhesion to the skin surface with minimum constraints on natural processes. From an electromagnetic point of view, the presence of a thin separating layer helps to mitigate the loss

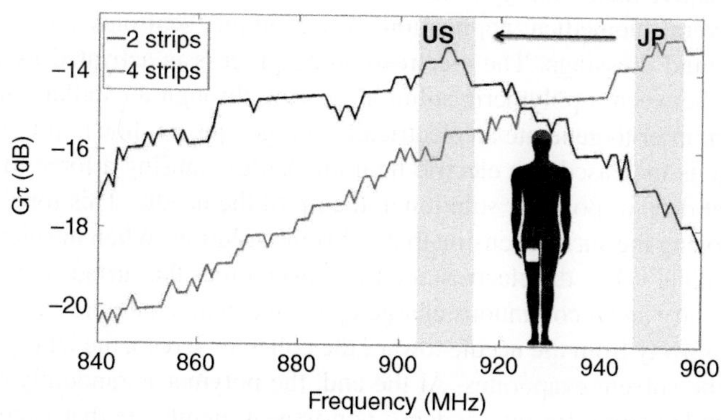

FIGURE 9.17 On-skin retuning from US to EU band of the epidermal tag placed onto the leg of the male volunteer. *Source:* Amendola et al. 2015 [26]. Reproduced with permission of IEEE.

TABLE 9.3 Summary of the strip number and (measured realized gain) for some placement and frequencies.

Placement	870 MHz	900 MHz	950 MHz
Tiny female, hip	4 (−17 dBi)	4 (−16 dBi)	5 (−16 dBi)
Tiny female, arm	3 (−13 dBi)	4 (−14 dBi)	6 (−14 dBi)
Normal man, leg	2 (−15 dBi)	2 (−14 dBi)	4 (−15 dBi)

from the body tissue by concentrating the near field in the low loss region between the antenna and the epidermis (see Section 9.3.4).

These materials can act as simple physical means supporting the antenna element or may be instead exploited for sensing purposes in case some of its dielectric and/or geometrical parameters undergo significant variations along with the physio/pathological process under monitoring.

Two classes of unconventional membranes have been proposed as substrates for RFID epidermal tags and sensors, a bioresorbable and transpiring polymeric scaffold, namely *Poli(ε-caprolactone)* (PCL) [28] and a *hydrogel* [29] particularly suited for absorbing and releasing body fluids and drugs, and hence for acting as sensitive substrate.

9.5.1 Scaffold Membranes: Poli(ε-Caprolactone)

PCL is a synthetic membrane [28], composed by semicrystalline bioresorbable *poly α-hydroxyester* with a slow degradation rate due to its hydrophobic nature and the high crystallinity degree. This membrane is produced by the electrospinning technique by using PLC granules dissolved in $CHCl_3$ and THF:DMF (1:1) solutions. Electrostatic fiber-spinning [30], also denoted as "electrospinning", is a straightforward, cost-effective methodology to fabricate non-woven micro- and/or nanofibrous fabrics for several biomedical applications, for example, scaffolds for tissue engineering or wound dressings. The electrospinning process is activated by applying high voltages between a polymeric solution flowing through a capillary and a collector plate, in order to generate an electrically charged jet, as shown in Figure 9.18. As the voltage is increased, the electric field intensifies, causing a force to build up on the pendant drop of polymer solution at the tip of the needle. This force acts in a direction opposing the surface tension that holds the solution. When the electric field reaches the critical value, the electrostatic force overcomes the surface tension at the tip of the capillary and a continuous charged jet is ejected. When the energized suspension moves away from the needle toward the collector screen, the jet rapidly thins and dries as the solvent evaporates. At the end, the polymer is randomly deposited onto the grounded target forming a dense non-woven membrane that is completely biocompatible, that is, free of toxic residues. PCL is moreover, flexible and stretchable for a comfortable application over any body curvature and, thanks to its fibrous structure, the natural transpiration of the skin is preserved (Figures 9.18 and 9.19).

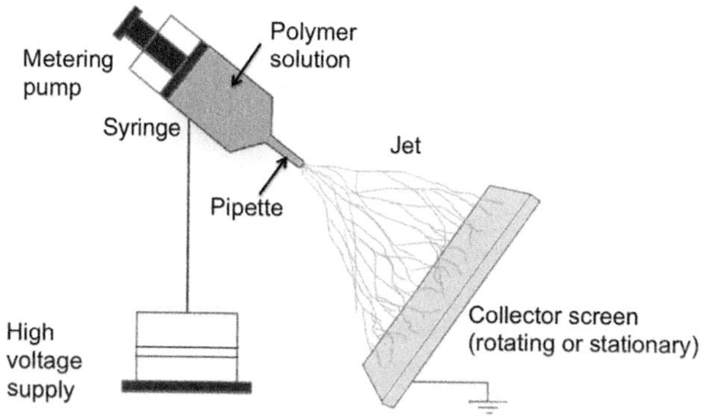

(a)

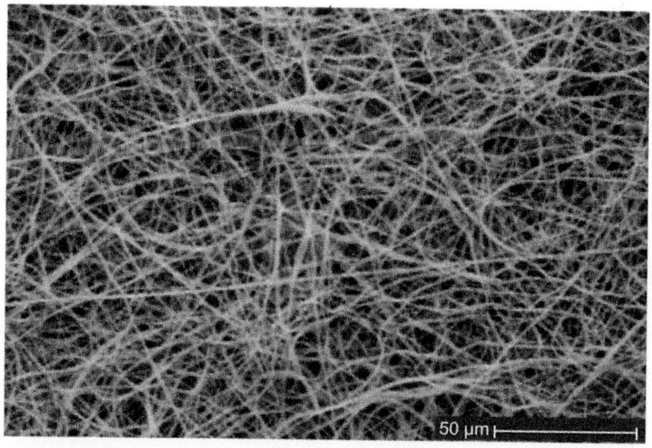

(b)

FIGURE 9.18 The PCL membrane used as substrate of the epidermal tag. (a) Fabrication scheme by electrospinning. *Source:* Del Gaudio et al. 2009 [28]. Reproduced with permission of John Wiley and Sons, Inc. (b) SEM image of the membrane. *Source:* Milici et al. 2014 [21]. Reproduced with permission of IEEE.

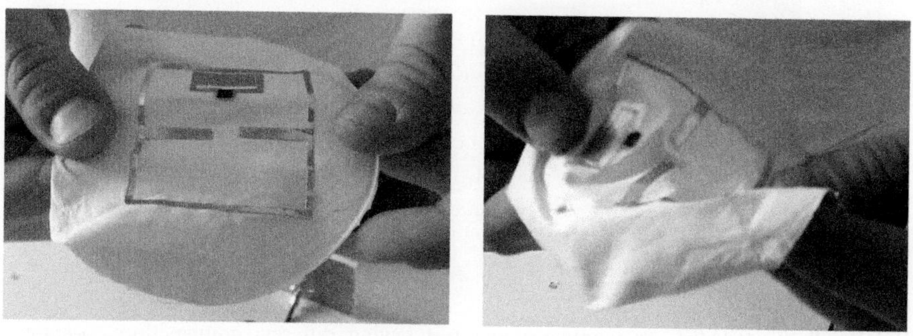

FIGURE 9.19 Prototype of the stretched epidermal RFID temperature sensor over PCL membrane. *Source:* Milici et al. 2014 [21]. Reproduced with permission of IEEE.

9.5.2 Hydrogel Membranes

Hydrogel dressings are being extensively investigated for wound treatment as they are able to absorb secretions from exuding wounds, to cool the infected areas, and recover the dehydrated tissues thus creating and maintaining the ideal environment for the wound healing process [29]. These materials can be moreover functionalized by adding doping components to selectively detect biochemical compounds or activate programmable drug release (e.g., antibacterial agents, growth factors). These features, combined with good mechanical properties (flexibility and stretchability), make hydrogels very promising substrates to fabricate RFID-based Smart Plasters.

A hydrogel membrane which has been already experimented for RFID applications [31] is a stretchable substrate made of a PVA/xyloglucan (XG). PVA is a water-soluble biocompatible synthetic polymer while xyloglucan is a hemicellulose that occurs in the primary cell wall of every vascular plant. PVA/XG hydrogel membranes is prepared starting from the aqueous solutions of the two polymers, properly mixed at different concentrations. Specific plasticizers can also be added to provide the required flexibility to the membranes, the desired mechanical strength, and the bioresorbable degree. Films of different sizes and shapes can be easily produced by natural or forced drying of the mixture (Figure 9.20).

Similar to conventional hydrogels, the PVA/XG membranes are also able to absorb body fluids (such as wound exudates) and release water and drugs. Their structure hence undergoes swelling and drying processes according to the skin conditions (Figure 9.21) being able to absorb in the specific case up to 40% of fluid.

The complex permittivity of the hydrogel membrane in the UHF band is intermediate between that of the PVA powder ($\varepsilon = 1.6$, $\sigma = 4.8 \times 10^{-3}$ S/m [33]) and

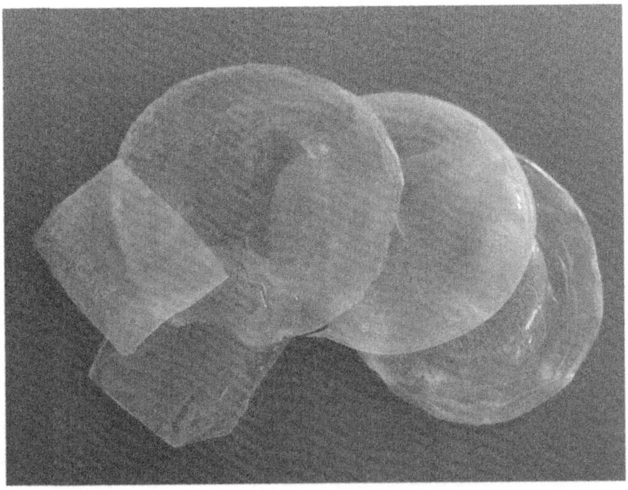

FIGURE 9.20 PVA/XG membranes under test having different compositions. *Source:* Amendola et al. 2015 [32]. Reproduced with permission of IEEE.

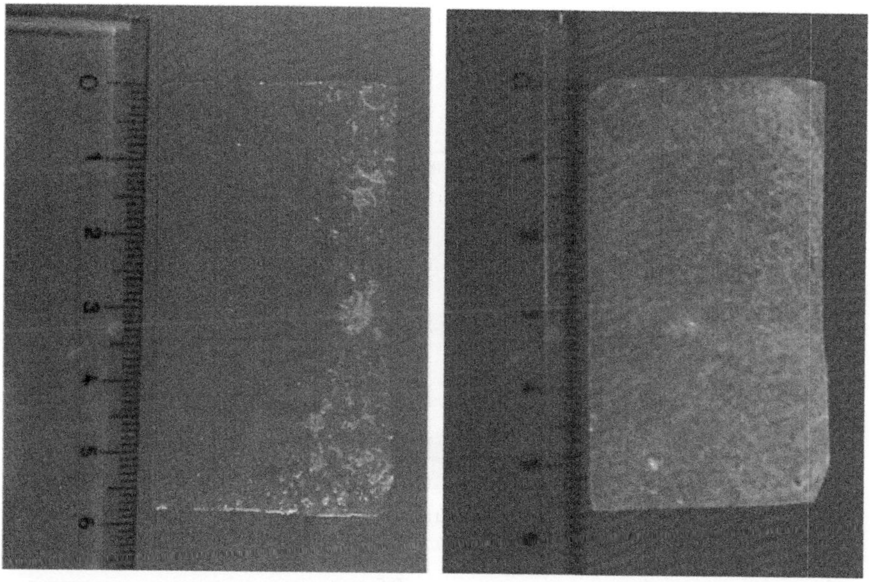

FIGURE 9.21 Samples of the fabricated hydrogel membrane. (Left) Swollen. (Right) Dried. *Source:* Occhiuzzi et al. 2015 [31]. Reproduced with permission of IEEE. For a color version of this figure, see the color plate section.

the much higher permittivity of the absorbed water. Furthermore, great variations are expected during the absorbing and releasing process. A first electromagnetic characterization of such membranes, in both static and dynamical configurations, can be found in [32]. A non-disruptive dielectric probe, composed of a multi-layer ring resonator, was properly customized with the purpose to correctly measure thin films in variable environmental conditions (Figure 9.22a). The releasing and swelling processes have been then induced by exposing the samples to an electromagnetic liquid phantom resembling the body fluid (saline solution 0.5 g/l ($\varepsilon_r = 79, \sigma = 0.26$ S/m)). To enrich the analysis, an additional commercial hydrogel-based plaster, *MySkin*® [34] was also measured.

In static conditions, all the membranes exhibit a low permittivity that is associated with a non-negligible conductivity (Table 9.4) mainly related to the amount of water/fluid absorbed from the environment or residual from the fabrication. The

TABLE 9.4 Retrieved dielectric properties.

	M1	M3	M6	M2	M4	M5	*MySkin*®
ε	3.75	4.4	3.7	2.4	2.6	2.5	5.5
σ (S/m)	0.06	0.05	0.08	0.02	0.03	0.02	0.13

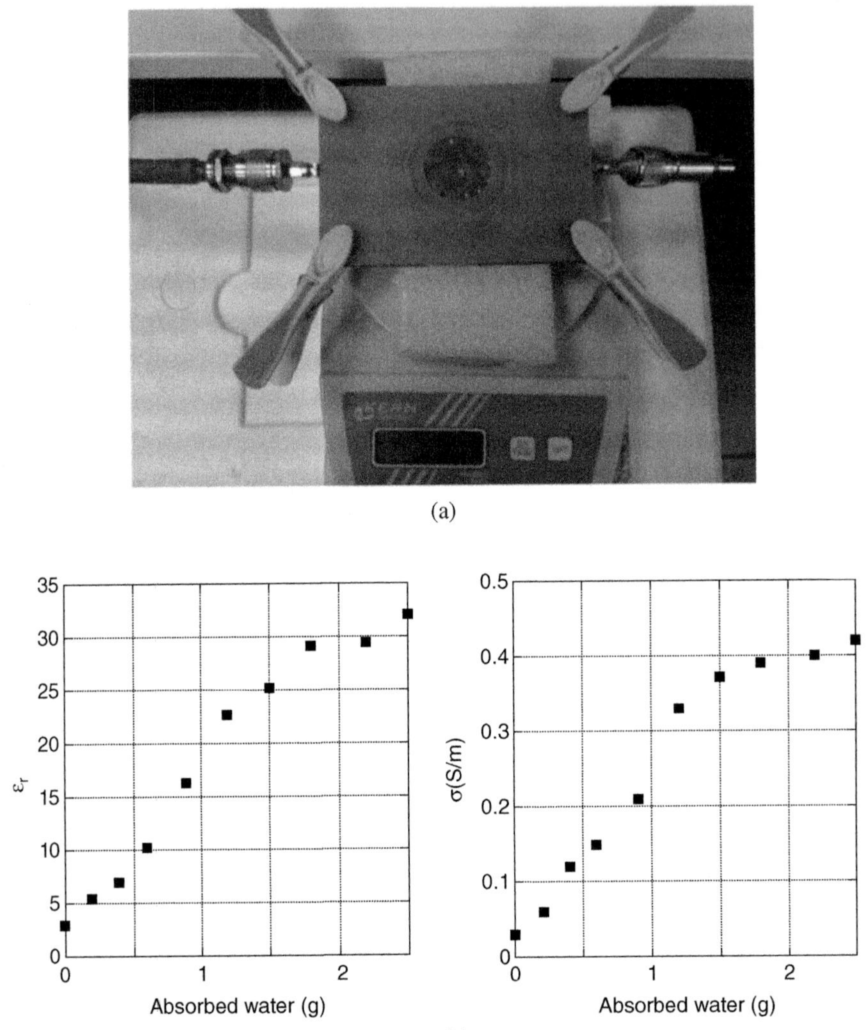

FIGURE 9.22 (a) Measurement setup with a modified version of the ring resonator in [30] with a hole for the dynamic exposition of the sample. (b) Measured dielectric properties of the hydrogel membrane related to the amount of dispersed body fluid. *Source:* Amendola et al. 2015 [32]. Reproduced with permission of IEEE.

effects are particularly evident in the case of the commercial membrane, which contains, as explicitly declared by the producer, a well-defined amount of water to create a wet environment around the wound.

During the exposition, the RF response of the membranes significantly changes along with the amount of fluid absorbed in the bulk hydrogel. A remarkable and almost linear variation of permittivity and conductivity (Figure 9.22a) is experienced that sensibly impacts the matching and radiation performance of any RFID epidermal sensors if properly designed and integrated with the membrane.

9.6 SENSING APPLICATIONS

Two examples are given of battery-less passive RFID epidermal tags based on the layout described in Section 9.4 and on the membranes introduced in Section 9.5. The first one is a thermometer suitable to be used in pathological conditions as well as in sports applications, and critical access control [21]. The second one is a sensitive plaster for wound monitoring and healing [31].

9.6.1 Epidermal RF Thermometer

The temperature sensing mechanism is based on a new family of RFID microchips (EM4325 [35]) equipped with an integrated temperature sensor able to directly provide the measured data in digital form once properly calibrated by means of standard thermocouples [36] (Figure 9.23a). Several independent laboratory tests demonstrated the good accuracy and precision of the RFID sensor, with a standard deviation lower than 0.25°C (Figure 9.23b).

The calibrated dual-loop epidermal sensor, integrated with the PCL membrane, was applied to the measurement of body temperature in both rest condition and under physical stress. The sensor was stuck onto the right side of the abdomen (Figure 9.24) of a male volunteer wearing his usual fitness clothing. A first measurement was conducted for 5 minute in rest condition, while during the second experimental session, the volunteer rode an exercise bike for about 1 hour.

As expected, at rest, the detected temperature was 35°C, which is slightly lower than the core body temperature (about 37°C). During the physical activity instead, the temperature followed a slightly cyclic profile due to thermoregulation mechanisms (e.g., sweating changes in skin blood flow) [37]. A substantial agreement with the thermocouple data can be appreciated in both the experiments.

The sensor offers the non-negligible advantage to enable a full wireless interrogation up to 30 cm when using a low power (0.5 W EIRP) interrogating device, which can be virtually integrated into mobile phones and handheld devices, or inside the exercise bike itself. But instead using fixed infrastructures, for example, long-range readers emitting up to 3.2/4 W EIRP, the read distances could increase up to 1 m and become compliant with typical access control systems. Passive temperature sensor plasters could indeed open interesting scenarios in the field of health control procedures by sensibly reducing costs and times and by improving their effectiveness and applicability. Great advantages could be offered to the automatic monitoring of people and animals in crowded and transit places (Figure 9.25), for example, in airports, stations, hospitals, and rescue camps where the possibility of epidemic events must be kept under strict control. Furthermore, since temperature is one of the most important health indicators, the passive plasters could be used to unobtrusively monitor the

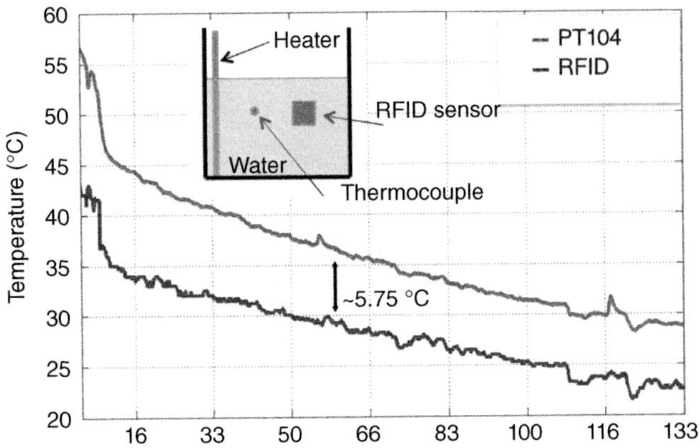

(a)

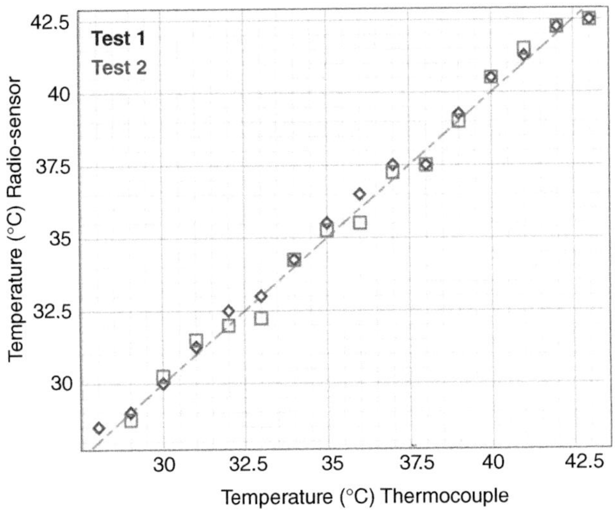

(b)

FIGURE 9.23 Sensor calibration by a reference thermocouple measurement: (a) identification of the temperature offset. (b) RFID versus true (thermocouple) measured temperature profile in further tests by using the calibrated RFID sensor. *Source:* Milici et al. 2012 [21]. Reproduced with permission of IEEE.

psychological and physical state of operators involved in high risk operations, such as militaries, firemen, and health workers.

9.6.2 Smart-Plaster for Wound Healing

A clinical application that could sensibly benefit epidermal electronics is wound care, where sensors could be used as diagnostic tools in the healing process [29].

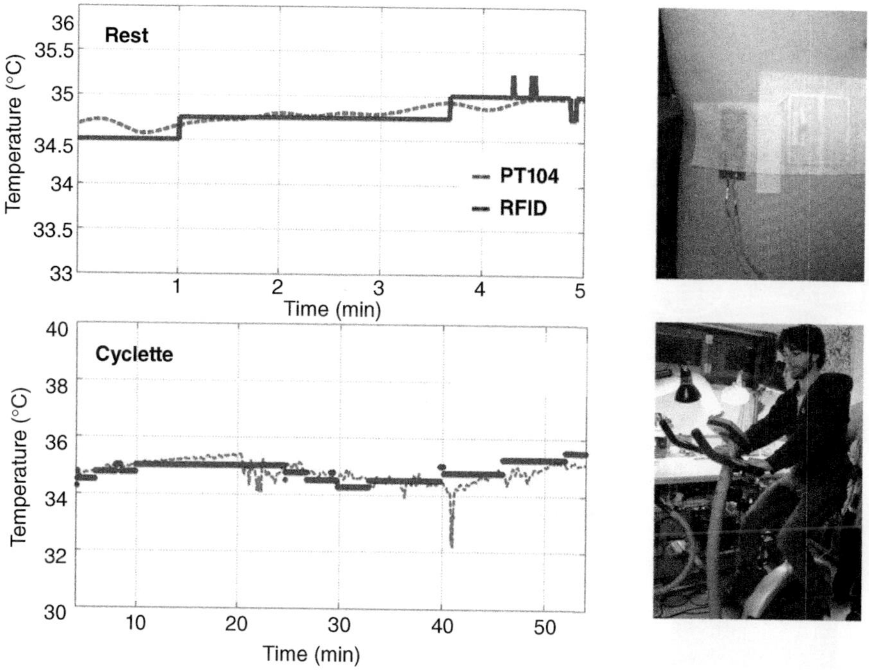

FIGURE 9.24 Temperature measurements over the body in rest condition (top) and during physical activity with exercise bike (bottom). *Source:* Milici et al. 2012 [21]. Reproduced with permission of IEEE.

Temperature, pH, and fluid loss could be locally detected by a plaster sensor and remotely transmitted by using devices either incorporated into or near wound dressings. Furthermore, the possibility to directly modify the clinical therapy, for example, by modulating the local release of drugs, according to the state of the wound, could open a promising opportunity in the management of diseases such as chronic ulcers, infected acute wounds, and large full-thickness burns.

9.6.2.1 Idea With the purpose to monitor the loss of fluids or the release of drugs through a body wound, the plaster exploits the absorbing/releasing capabilities of the hydrogel and their effects on tag radiation and matching performances. Indeed, as demonstrated in the previous section, the dielectric properties of the hydrogel strictly depend on the materials composing the membrane and on the amount of absorbed fluid.

With $\Psi(t)$ being the amount of body fluids that are embedded into the hydrogel, it is possible to consider the dependence

$$\Psi(t) => \{\varepsilon_R(\Psi), \sigma(\Psi)\}$$

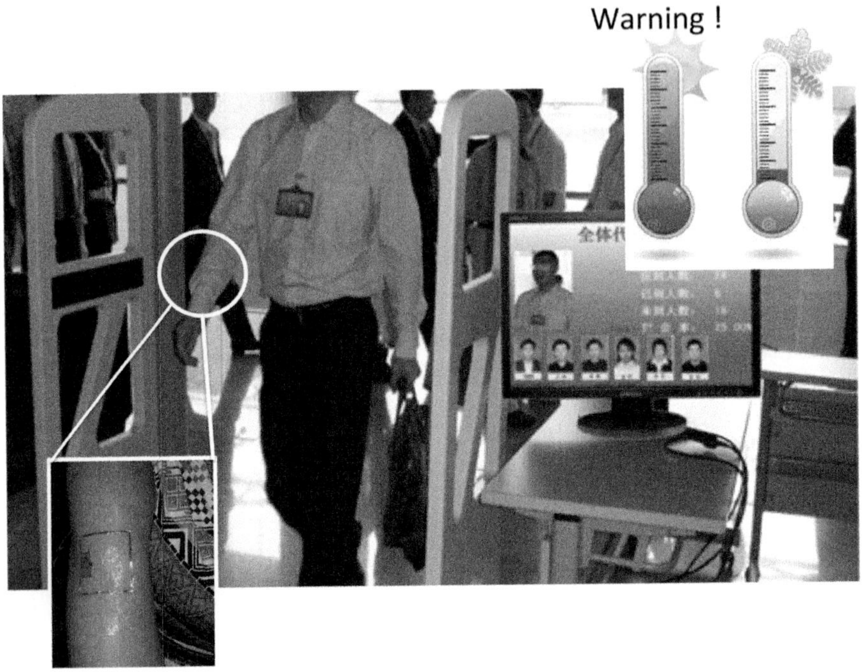

FIGURE 9.25 Possible setup for the automatic health monitoring of in-transit people.

Accordingly, a change in the fluid percentage will modify the antenna response in terms of input impedance and radar cross section and definitely the activation power of the chip as well as of the backscattered power collected by the reader. By following the formulations in Section 9.2, different sensing metrics can be easily derived from data measurable by the reader, such as the *realized gain* and the *analog identifier* [9].

The realized gain

$$G_{\tau}(\theta, \varphi)[\Psi] = \left(\frac{4\pi d}{\lambda}\right)^2 \frac{P_{\mathrm{chip}}}{G_{\mathrm{R}}(\theta, \varphi)\, P_{\mathrm{in}}^{\mathrm{to}}[\Psi]\, \eta_{\mathrm{p}}} \tag{9.5}$$

can be derived by measuring the *turn-on power* $P_{\mathrm{in}}^{\mathrm{to}}[\Psi]$, while the analog identifier AID $[\Psi]$ is a non-dimensional indicator independent of the environment and the reader–tag mutual position, and can be derived by properly combining the forward and backward powers at the turn-on condition as

$$\mathrm{AID}[\Psi] = K \frac{P_{\mathrm{chip}}}{\sqrt{P_{\mathrm{in}}^{\mathrm{to}}[\Psi] \cdot P_{\mathrm{bs}}^{\mathrm{to}}[\Psi]}} \tag{9.6}$$

The indicators in equations (9.1) and (9.2) can be hence used as *data inversion curves* $\xi[\Psi]$ between the measured data and the evolution, $\Psi(t)$, of the fluid loss or delivered to the wound.

9.6.2.2 *Prototype and Test* A prototype of the plaster (total weight 2.9 g), fabricated on adhesive copper carved by a two-axis digital-controlled cutter, is shown in Figure 9.26b. By considering a commercial RFID reader emitting 3.2 W EIRP, the tag is expected to be readable in a 0.5 – 2.5 m range according to the amount of absorbed fluid.

The sensing capabilities of the RFID plaster has been experimentally characterized by real-time measurements of the realized gain and the analog identifier during both loading (absorption) and recovery processes.

The absorbing process was induced by placing the plaster on a synthetic sponge floating on a plastic box which contains a saline water solution whose electromagnetic parameters resemble those of body fluids. The plaster was continuously interrogated by the reader for an overall measuring time of about 7 hours. As expected, the fluid absorption sensibly impacts the hydrogel features and hence on the tag radiation performances, leading to monotonic and quite linear variation of the sensing curves of G_τ [Ψ] and AID [Ψ] with 8 dB and 3 dB of sensing range, respectively (Figure 9.26 Right).

Similar to other hydrogels, the recovery, that is, the process of releasing the fluid, is in general sensibly slower than the uptake one and it is more dependent on the environmental conditions.

Figure 9.27 shows the realized gain of the fully loaded plaster after 1 and 2 days of recovery, compared with the realized gain of the unloaded membrane. The

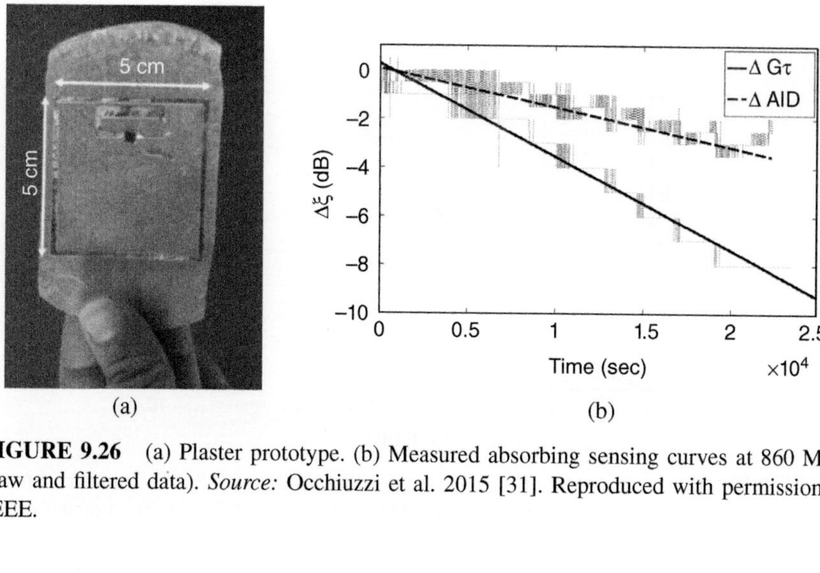

(a) (b)

FIGURE 9.26 (a) Plaster prototype. (b) Measured absorbing sensing curves at 860 MHz (raw and filtered data). *Source:* Occhiuzzi et al. 2015 [31]. Reproduced with permission of IEEE.

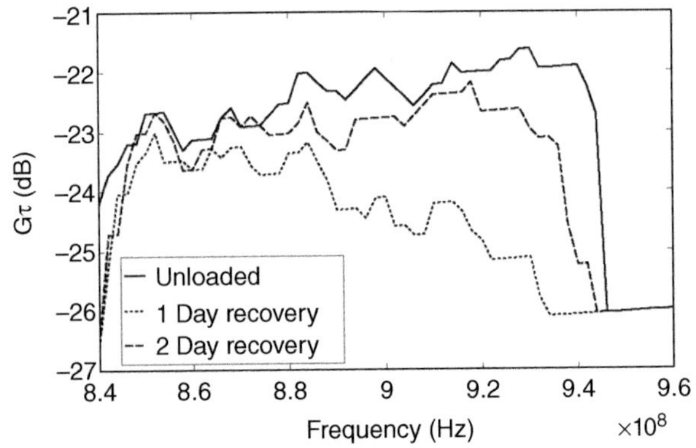

FIGURE 9.27 Realized gain of the fully loaded plaster 1 and 2 days of recovery with environmental conditions. *Source:* Occhiuzzi et al. 2015 [31]. Reproduced with permission of IEEE.

hydrogel requires more than 2 days to release all the absorbed fluid. The hysteresis looks negligible in the lower part of the UHF band while it is more evident in the upper part. An insight view on the recovery process is provided in Figure 9.28 where a partially loaded membrane (amount of absorbed fluid similar to the one produced during body perspiration) has been considered. The profiles are opposite to the ones in Figure 9.27, demonstrating again, the reversibility of the process and the effectiveness of the electromagnetic reaction of the epidermal sensor for smaller amounts of absorbed fluid thanks to the unsaturated sensing capabilities in the initial condition. Such observation could open interesting scenarios in the drug delivery area.

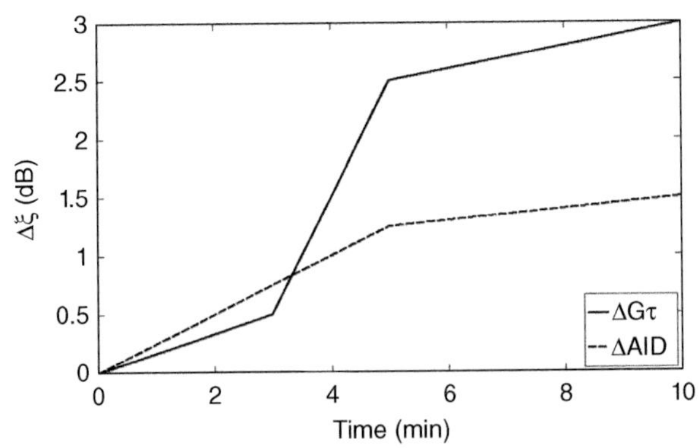

FIGURE 9.28 Recovery curves of the electromagnetic response of the RFID plaster when exposed to a small amount of body fluid. *Source:* Occhiuzzi et al. 2015 [31]. Reproduced with permission of IEEE.

9.7 CONCLUSION

The fertile research on skin electronics has paved the way to a novel approach for the continuous assessment of people's health and well-being by pushing the technologies from clothes and personal accessories directly to the human skin. Despite the early studies mostly focused on the high tech fabrication of engineered thin films with sensor capabilities, there is nowadays an increasing attention to the powering and communication issues related to the epidermal devices, which are the current bottlenecks for the real spreading of skin-worn technologies.

Numerical investigations on the electromagnetic performance of simple radiators such as dipoles and loops in direct touch with a lossy medium revealed a bell-shaped profile of the radiation efficiency with respect to the overall size of the antenna. Depending on the specific layout, there is an optimal size of the antenna (smaller in the case of loop) while the achievable efficiency and, accordingly, the maximum gain have upper bounds roughly independent of the antenna shape.

Inkjet printing or stencil-based methods which make use of conducting paints and nanoparticle inks with low conductivity may be a viable technique for the fabrication of epidermal radiators since the quality of the antenna conductor plays a minor role in the radiation performance.

Unlike shielded wearable tags, epidermal antennas are significantly sensitive to the placement conditions. Simulations and experiments demonstrated that loop-like epidermal tags, insulated from the human skin by only a thin membrane of biocompatible silicone, provide a realized gain ranging between −17 dBi and −9 dBi depending on the body part where it is placed and on the body mass of the user. Limbs and arms are the most suitable targets for what concerns communication properties and the epidermal tag may be used for general purpose applications involving just the identification of the user or even wound monitoring and controlled drug delivery. Placement over sternum, forearm, and forehead, which are relevant, for instance, for the purpose of temperature measurement, are instead more critical concerning the RFID link and hence may greatly benefit from on-body retuning.

Substrate materials of epidermal antennas have a crucial role, as they are the interface between the conductor and the epidermis whose local metabolism must be preserved. The radiation performance sensibly improves when the membrane thickness is more than 0.1 mm but a trade-off is required for extremely low profile devices.

First experimentations of epidermal tags fabricated over special membranes demonstrated the feasible application of such devices not only for tracking but also for sensing purposes. By providing an epidermal-like tag with an IC embedding digital thermometer, the temperature of the human body was continuously sampled in a wireless and passive mode from a distance up to 1 m when using conventional long-range readers, and up to 30 cm when using a low power interrogating device, (e.g., handheld reader, smartphone). The feasibility of integrating passive RFID sensor tags into medical hydrogel membranes was also experimentally demonstrated. Indeed, the hydrogel-powered epidermal-tag was revealed to be extremely sensitive to the dynamic exchanges of body fluids at the skin interface thus stimulating innovative solutions for wound care treatment. Conjugate capabilities of both temperature and fluid (e.g., sweat loss) monitoring can be realistically envisaged in the perspective of developing multifunctional RFID Smart Plasters.

REFERENCES

[1] D-H. Kim, N. Lu, R. Ma, Y.-S. Kim, R.-H. Kim, S. Wang, J. Wu, S. M. Won, H. Tao, A. Islam, K. J. Yu, T. Kim, R. Chowdhury, M. Ying, L. Xu, M. Li, H.-J. Chung, H. Keum, M. McCormick, P. Liu, Y.-W. Zhang, F. G. Omenetto, Y. Huang, T. Coleman, and J. A. Rogers, "Epidermal electronics," *Science*, vol. 333, no. 6044, pp. 838–843, August 2011.

[2] W. H. Yeo, Y. S. Kim, J. Lee, A. Ameen, L. Shi, M. Li, S. Wang, R. Ma, S. H. Jin, Z. Kang, Y. Huang, and J. A. Rogers, "Multifunctional epidermal electronics printed directly onto the skin," *Adv. Mater.*, vol. 25, no. 20, pp. 2773–2778, May 2013.

[3] Hu Tao, M. A. Brenckle, M. Yang, J. Zhang, M. Liu, S. M. Siebert, R. D. Averitt, M. S. Mannoor, M. C. McAlpine, J. A. Rogers, D. L. Kaplan, and F. G. Omenetto, "Silk-based conformal, adhesive, edible food sensors," *Adv. Mater.*, vol. 24, no. 8, pp. 1067–1072, January 2012.

[4] J.-W. Jeong, M. K. Kim, H. Cheng, W.-H. Yeo, X. Huang, Y. Liu, Y. Zhang, Y. Huang, and J. A. Rogers, "Capacitive epidermal electronics for electrically safe, long-term electrophysiological measurements," *Adv. Healthcare Mater.*, vol. 3, pp. 642–648, May 2014.

[5] D. P. Rose, M. Ratterman, D. K. Griffin, H. Linlin, N. Kelley-Loughnane, R. K. Naik, J. A. Hagen, I. Papautsky, and J. Heikenfeld "System-Level Design of an RFID Sweat Electrolyte Sensor Patch," in 2014 36th Annual International Conference of the IEEE on Engineering in Medicine and Biology Society (EMBC), Chicago, IL, August 26–30, 2014, pp. 4038–4041.

[6] J. Kim, A. Banks, H. Cheng, Z. Xie, S. Xu, K.-I. Jang, J. W. Lee, Z. Liu, P. Gutruf, X. Huang, P. Wei, F. Liu, K. Li, M. Dalal, R. Ghaffari, X. Feng, Y. Huang, S. Gupta, U. Paik, and J. A. Rogers, "Epidermal electronics with advanced capabilities in near-field communication," *Small,* vol. 11, no. 8, pp. 906–912, February 2015.

[7] D. M. Dobkin, *The RF in RFID: UHF RFID in Practice*. Newnes, 2012.

[8] S. Amendola, R. Lodato, S. Manzari, C. Occhiuzzi, and G. Marrocco, "RFID technology for IoT-Based personal health-care in smart spaces," *IEEE Internet Things J.*, vol. 1, no. 2, pp. 144–152, April 2014.

[9] C. Occhiuzzi, S. Caizzone, and G. Marrocco, "Passive UHF RFID antennas for sensing applications: principles, methods, and classifications," *IEEE Antennas Propag. Mag.*, vol. 55, no. 6, pp. 14–34, December 2013.

[10] S. Manzari, C. Occhiuzzi, and G. Marrocco, "Feasibility of body-centric systems using passive textile RFID tags," *IEEE Antennas Propag. Mag.*, vol. 54, no. 4, pp. 49–62, August 2012.

[11] C. Occhiuzzi, S. Cippitelli, and G. Marrocco, "Modeling, design and experimentation of wearable RFID sensor tag," *IEEE Trans. Antennas Propag.*, vol. 58, no. 8, pp. 2490–2498, August 2010.

[12] T. Kellomaki, "On body performance of wearable single layer RFID tag," *IEEE Antennas Wireless Propag. Lett.*, vol. 11, pp. 73–76, January 2012.

[13] G. Marrocco, "RFID antennas for the UHF remote monitoring of human subjects," *IEEE Trans. Antennas Propag.*, vol. 55, no. 6, pp. 1862–1870, June 2007.

[14] M.-C. Tsai, C.-W. Chiu, H.-C. Wang, and T.-F. Wu, "Inductively coupled loop antenna design for UHF RFID on-body applications," *Prog. Electromagnetics Res.*, vol. 143, pp. 315–330, 2013.

[15] M. A. Ziai and J. C. Batchelor, "Temporary on-skin passive UHF RFID transfer tag," *IEEE Trans. Antennas Propag.*, vol. 59, no. 10, pp. 3565–3571, October 2011.

[16] O. O. Rakibet, J. C. Batchelor, and S. W. Kelly, "RFID tags as passive enabling technology," in 2013 Loughborough Conference on Antennas and Propagation (LAPC), Loughborough (England), November 11–12, 2013, pp. 350–353.

[17] O. O. Rakibet, C. V. Rumens, J. C. Batchelor, and S. J. Holder, "Epidermal passive RFID strain sensor for assisted technologies," *IEEE Antennas Wireless Propag. Lett.*, vol. 13, no., pp. 814–817, April 2014.

[18] R. M. Makinen and T. Kellomaki, "Body effects on thin single-layer slot, self-complementary, and wire antennas," *IEEE Trans. Antennas Propag.*, vol. 62, no. 1, pp. 385–392, January 2014.

[19] C. A. Balanis, *Antenna Theory: Analysis and Design*. John Wiley & Sons, 2012.

[20] A. Sommerfeld, "The propagation of waves in wireless telegraphy," *Ann. der Phys.*, vol. 28, pp. 665–736, March 1909.

[21] S. Milici, S. Amendola, A. Bianco, and G. Marrocco, "Epidermal RFID Passive Sensor for Body Temperature Measurements", in 2014 IEEE International Conference on RFID Technologies and Applications (RFID-TA), Tampere, September 8–9, 2014, pp. 140–144.

[22] FEKO Suite 7.0, EM Software and Systems, 2014. Available at http://www.feko.info

[23] V. Sanchez-Romaguera, M. A. Ziai, D. Oyeka, S. Barbosa, J. S. R. Wheeler, J. C. Batchelor, E. A. Parker, and S. G. Yeates, "Towards inkjet-printed low cost passive UHF RFID skin mounted tattoo paper tags based on silver nanoparticle inks," *J. Mater. Chem. C*, vol. 1, pp. 6395–6402, 2013.

[24] C. Gabriel and S. Gabriel, "Compilation of the Dielectric Properties of Body Tissues at RF and Microwave Frequencies," Internet document. Available at http://niremf.ifac.cnr.it/docs/DIELECTRIC/home.html

[25] H. W. Son and C. S. Tyo, "Design of RFID tag antennas using an inductively coupled feed," *Electron. Lett.*, vol. 41, no. 18, pp. 994–996, 2005.

[26] S. Amendola, S. Milici, and G. Marrocco, "Performance of epidermal RFID dual-loop tag and on-skin retuning," *IEEE Trans. Antennas Propag.*, vol. 63, no. 8, pp. 3672–3680, August 2015.

[27] Monza R6 RFID Tag Chip. Available at http://www.impinj.com/products/monza/monza-r6-rfid-tag-chip

[28] C. Del Gaudio, A. Bianco, M. Folin, S. Baiguera, and M. Grigioni, "Structural characterization and cell response evaluation of electrospun PCL membranes: micrometric versus submicrometric fibers," *J. Biomed. Mater. Res. Part A*, vol. 89A, no. 4, pp. 1028–1039, June 2009.

[29] T. R. Dargaville, B. L. Farrugia, J. A. Broadbent, S. Pace, Z. Upton, and N. H. Voelcker, "Sensors and imaging for wound healing: a review," *Biosens. Bioelectron.*, vol. 41, no. 15, pp. 30–42, March 2013.

[30] Z.-M. Huang, Y.-Z. Zhang, M. Kotaki, and S. Ramakrishna, "A review on polymer nanofibers by electrospinning and their applications in nanocomposites," *Compos. Sci. Technol.*, vol. 63, no 15, pp. 2223–2253, November 2003.

[31] C. Occhiuzzi, A. Ajovalasitb, M. A. Sabatino, C. Dispenza, and G. Marrocco, "RFID Epidermal Sensor including Hydrogel Membranes for Wound Monitoring and Healing," in Proceedings of 9th Annual IEEE International Conference on RFID, San Diego, CA, April 15–17, 2015, pp. 182–188.

[32] S. Amendola, C. Occhiuzzi, A. Ajovalasit, M. A. Sabatino, C. Dispenza, and G. Marrocco, "Dielectric Characterization of Biocompatible Hydrogels for Application to Epidermal RFID Devices," in Proceedings of 2015 European Microwave Conference (EuMC), Paris, September 7–10, 2015, pp. 379–382.

[33] S. Ebnesajjad, *Handbook of Biopolymers and Biodegradable Plastics*, 1st ed. Elsevier, 2012.

[34] http://myskin.picsolution.com/en/index.shtml

[35] EM4325 IC. Available at http://www.emmicroelectronic.com

[36] https://www.picotech.com/data-logger/pt-104/high-accuracy-temperature-daq

[37] J. E. Hall, *Guyton and Hall Textbook of Medical Physiology*, 12th ed. Saunders, 2011.

10

INKJET-PRINTED SMART SKINS AND WIRELESSLY-POWERED SENSORS FOR WEARABLE APPLICATIONS

JOHN KIMIONIS AND MANOS (EMMANOUIL) M. TENTZERIS

School of Electrical and Computer Engineering, Georgia Institute of Technology, Atlanta, GA, USA

10.1 INTRODUCTION

A main challenge for the fabrication of wearable devices consists of their fabrication cost and prototyping time. These two affect both the time-to-market of a product and the final cost for the consumer during the full production phase. Moreover, conventional photolithography techniques used for circuit board fabrication consist *subtractive* manufacturing processes; a portion of the metallization material is removed from the parts of the board where it is not needed. These processes usually require a high cost for building masks and generate a large amount of waste.

In sharp contrast, inkjet printing is an *additive* manufacturing technique with which circuitry, high frequency structures, and antennas can be built by depositing conductive ink (e.g., silver nanoparticle solutions) on a plethora of substrates at a very low material cost, without compromising high frequency performance [1]. Since the total volume of ink required to build planar conductive structures is very small (on the order of micro- to milliliters), the fabrication time is greatly reduced, especially in scenarios where the largest area of a circuit board is non-conductive.

Currently, inkjet-printing technology has become a lot more mature, allowing the printing of more demanding structures, such as multilayer circuit elements, by overlaying different materials one-after-one in an all-printed fashion. Apart from fully

Electromagnetics of Body-Area Networks: Antennas, Propagation, and RF Systems, First Edition.
Edited by Douglas H. Werner and Zhi Hao Jiang.

conductive metal-based materials like silver nanoparticle ink, conductive polymers or semiconductive materials such as graphene oxide and carbon nanotubes (CNTs) have been recently demonstrated to be inkjet-printable, broadening the possibilities of inkjet manufacturing technologies. Exploiting the conductive or semi-conductive properties of these materials, sensors can be used for humidity or temperature level monitoring or for detecting chemicals and gases.

Inkjet printing also facilitates the fabrication of structures for demanding scenarios such as microfludic sensors for biochemical applications. State-of-the-art microfluidics typically require cleanroom fabrication and chemical treatment techniques, which are of relatively high cost. Recently, the feasibility of printing microfluidic structures has been demonstrated, even for high frequency structures such as microwave filters or antennas, significantly reducing fabrication costs. Moreover, material deposition with inkjet printing allows for the use of several substrate options, including flexible ones. Polyimide films or liquid crystal polymer (LCP) sheets can be used as substrates; the main advantage of these materials is that they are very flexible, thus they can be easily used for wearable applications. Conductive traces on these substrates can be realized either via printing silver nanoparticle lines on dielectric substrates, or by inkjet-printing photoresist inks to form masks on copper-clad flexible materials. Because inkjet printers can achieve high feature resolutions of a few tens of microns, the latter technique is a low-cost way of implementing masking for etching, without compromising accuracy.

The above are discussed in the following sections to showcase the possibilities of inkjet printing and how it can be used to realize flexible, wearable circuits, antennas, and sensors. Moreover, since another challenge is to provide power to wearable sensing systems, the realization of wireless power systems on flexible substrates using inkjet-printing techniques is discussed.

10.2 MULTILAYER INKJET PRINTING—CONDUCTORS AND DIELECTRICS

Along with its additive nature, another advantage of using inkjet printing for the fabrication of versatile, wearable electronics is the wide range of printing materials. The printing of both metallic conducting and polymer dielectric materials allows for the realization of vertically integrated multilayer structures through a purely additive process. These two general material types have the ability to create a wide variety of electronic structures, such as RF capacitors and inductors, passive filters, antennas, and sensors.

Conducting materials are typically composed of noble metal nanoparticle solutions, such as silver and gold. The suspension of these nanoparticles, often in the order of tens of nanometers in diameter, allow for the deposition and patterning of metallic materials with relatively low processing temperatures (lower than 200° C) while still providing a conductivity close to that of bulk metal (1.1×10^7 S/m) [2]. Another effective method of metallic material patterning is through the use of a printed catalyst ink. Catalyst inks, such as the palladium(II) chloride catalyst solution ($PdCl_2$),

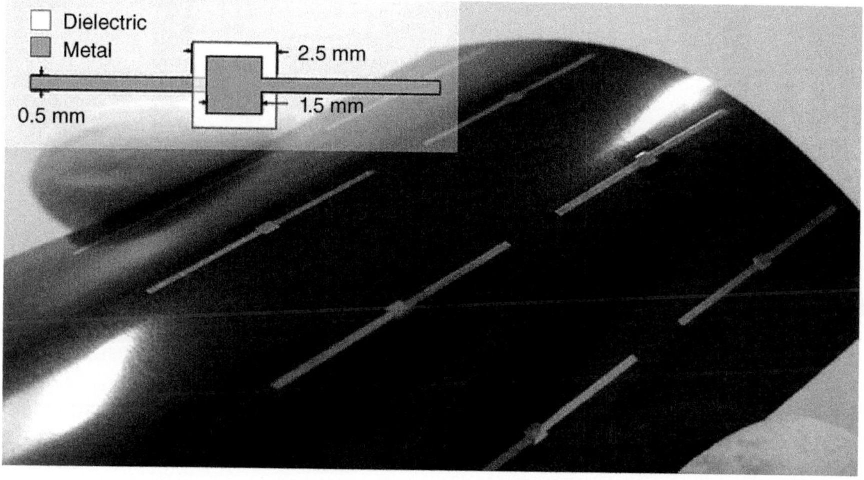

FIGURE 10.1 Fully inkjet-printed MIM capacitors using a PVP-based dielectric ink. *Source:* Cook et al. 2013 [4]. Reproduced with permission of IEEE.

allow for latent patterning of metals that do not behave efficiently in nanoparticle form, where high oxidation is a detrimental issue [3]. The printed catalyst pattern is exposed to a chemical bath where the metal pattern is electrolessly plated onto the catalyst, offering another low temperature method of inkjet-printing high conductivity metallic patterns.

Dielectric materials within inkjet printing can be separated into two categories based on their printed profile: thin film and thick film. Thin film dielectric inks exhibit printed layer thicknesses within the range of hundreds of nanometers to several micrometers. These low-profile polymeric ink materials, such as PVP and PMMA, have been demonstrated to be useful in the fabrication of purely additive passive components, such as metal-insulator-metal (MIM) capacitors and interconnects [4–6]. Figure 10.1 shows an array of fully inkjet-printed MIM capacitors using a PVP-based dielectric on a flexible polyimide Kapton substrate [4]. The printing process includes three steps: first the printing of metallic trace and bottom capacitive plate, followed by the printing of a single layer 300 nm thick polyvinylpyrrolidone (PVP) dielectric insulator, and concluded with a top capacitive plate and metallic trace.

Many electronic applications call for the utilization of thick dielectric materials, such as multilayer system interconnects and RF substrates. While many works have demonstrated inkjet-printed dielectrics, the greatest thickness achieved in literature has been limited to less than 20 μm [7]. Through the utilization of a new polymeric thick film ink, recent works have yielded printed dielectric patterns with thickness greater than 100 μm [8]. This thick film dielectric ink is formulated using an SU-8 photoresist, which is a UV-cross-linkable polymer that yields a per-layer thickness of 4 to 6 μm upon printing. The repeated deposition of these layers leads to the realization of not only thick dielectric patterns, but also dielectric patterns with an excellent control over film thickness. Figure 10.2 shows a set of inkjet-printed microstrip

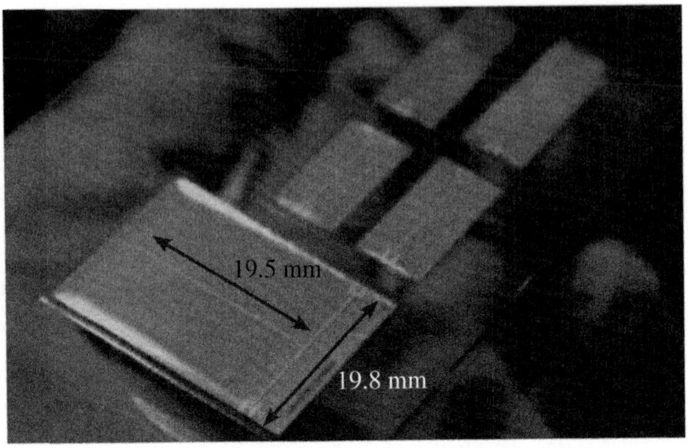

FIGURE 10.2 Inkjet-printed microstrip structures with a printed thick SU-8 dielectric substrates. *Source:* Tehrani et al. 2014 [9]. Reproduced with permission of IEEE.

structures with 80 μm thick SU-8 dielectric substrates [9]. The printing process for these devices is similar to that of the PVP capacitors presented previously: first a bottom metallic ground plane is printed, followed by a thick SU-8 polymer substrate, and concluded with a final microstrip topology on top of the printed substrate. With this thick inkjet-printed dielectric material, multi-laminate electronic designs can be transformed into vertically integrated multilayer stack ups through a purely additive fabrication process.

10.3 MULTILAYER INKJET PRINTING—ANTENNA EXAMPLES

With the current push in wireless technology to reach wider bandwidths and higher data rates, extensive attention is being drawn to the field of millimeter-wave wireless technology, referring to wireless systems with operational frequencies in the range of ten to several hundred gigahertz. The physical nature of this technology is ambitious both in its increased wireless throughput and its reduced antenna footprint size, where the wavelength of a free-space 60 GHz wave is only 5 mm, being advantageous for miniaturized wearable technologies. The small footprint of millimeter-wave antennas has led to interest in package and chip integration with wireless radios, thus making antenna size reduction and isolation from harmful active-circuitry, a popular area of research [10]. One technique typically used to reduce the footprint of an antenna is to use a multilayer design of metals and dielectrics. With standard electronic fabrication schemes, this design method requires either the use of multiple, stacked laminate substrates or the masked patterning of thin insulating layers.

Inkjet printing offers an alternative to these subtractive fabrication methods, providing a purely additive patterning process of both metallic and thick dielectric materials in order to create multilayer antenna designs. One simple application of this fabrication technology is shown in the demonstration of an inkjet-printed millimeter-wave, proximity-coupled patch antenna array [11]. This antenna utilizes

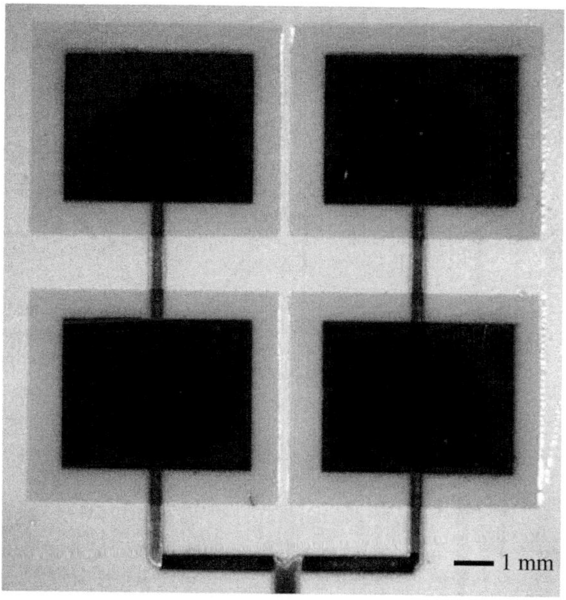

FIGURE 10.3 Inkjet-printed 24.5 GHz proximity-coupled patch antenna array utilizing dielectric SU-8 ink. *Source:* Cook et al. 2013 [11]. Reproduced with permission of IEEE.

the SU-8 polymer ink to realize 40 μm thick dielectric spacers between microstrip feeding lines and resonating patch elements on a flexible liquid crystalline polymer (LCP) substrate. An image of the fabricated antenna array is shown in Figure 10.3. S-parameter measurements of two fabricated samples, displayed in Figure 10.4, are shown to match well between each other as well as computer simulations, highlighting the reliability and extensive characterization of the inkjet-printing process.

In addition to patterning electronic structures on an existing RF substrate, this multilayer printing process can also be used to create fully printed RF structures completely isolated from base substrates. This method of fabrication is extremely advantageous for applications where an antenna might be placed close to a wirelessly destructive material, such as lossy silicon or the human body. In these circumstances, an antenna structure can be fabricated in its entirety from the ground up, such as the fully printed microstrip patch antenna shown in Figure 10.5 [8]. This 30 GHz patch antenna is fabricated directly onto a wireless integrated circuit (IC) package in a post-processed fashion, including an inkjet-printed metallic ground plane, 120 μm thick SU-8 substrate, and metallic microstrip patch antenna. Printed vias are also included with the design, bridging the metallic patterns on the top and bottom of the printed dielectric substrate.

Inkjet printing gained its popularity from its ability to rapidly prototype planar structures on bulk substrates at a very low cost, yet recent advancements in ink materials have led to the realization of multilayer structures through the incorporation of thin and thick dielectric inks. These vertically integrated structures

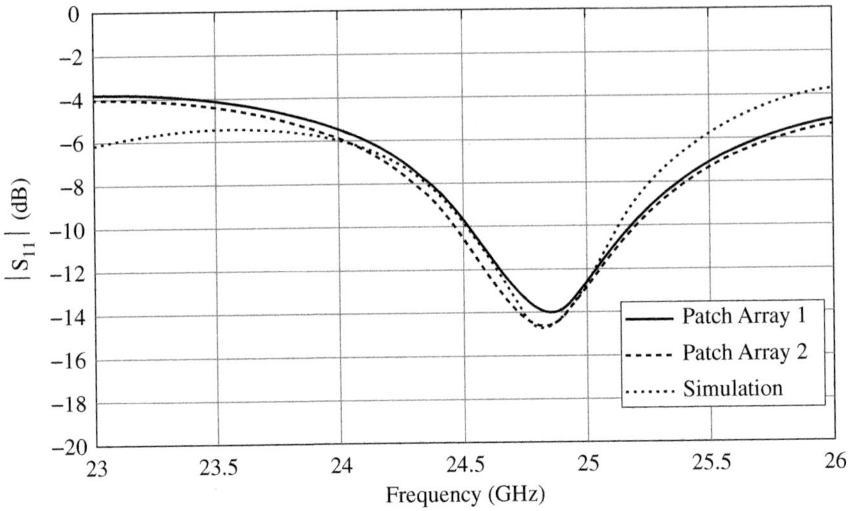

FIGURE 10.4 Measured return loss for inkjet-printed proximity-coupled patch arrays. *Source:* Cook et al. 2013 [11]. Reproduced with permission of IEEE.

reduce the material requirements of typical laminate-based designs as well as add an element of versatility with selective three-dimensional patterning. The nature of post-process antennas fabrication with inkjet printing allows for both isolated and integrated antenna designs, where antennas can be incorporated with wireless wearable electronics in any desired application.

10.4 INKJET-PRINTED SENSORS

The use of inkjet printing for the fabrication of wearable chemical sensors offers tremendous advantages with its low-cost, low-environmental footprint and compatibility with flexible material substrates. However, overly complex sensing systems

FIGURE 10.5 Fully inkjet-printed package-integrated millimeter-wave microstrip patch antenna. *Source:* Tehrani et al. 2015 [8]. Reproduced with permission of IEEE.

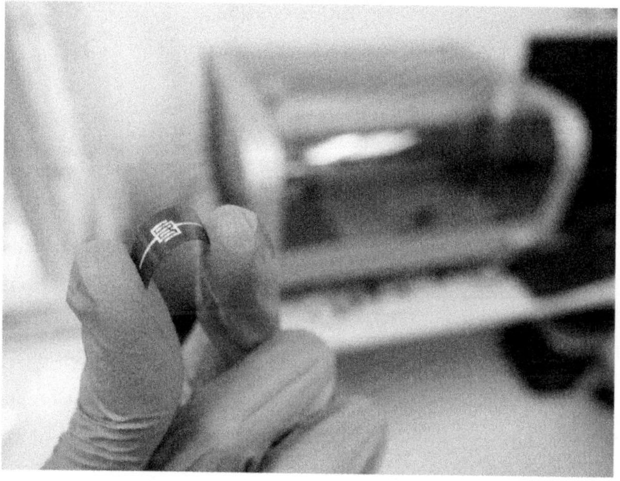

FIGURE 10.6 Fully inkjet-printed flexible chemical sensor.

and technologies are ill-suited for this type of fabrication method as their manufacturing can quickly become overly expensive and incompatible with the capabilities of inkjet printing. Fortunately, some novel and rapidly growing technologies are very well-suited for these applications. Indeed, a wide variety of conductive polymers, as well as nanomaterials, can be printed in thin films whose electrical properties change after exposure to and subsequent absorption of detected chemical species (analytes). Due to their small sizes, inkjet nozzles generally cannot be used with inks containing particles larger than 1 μm. Both conductive polymers and nanomaterials can be small enough and usually offer the possibility of being dispersed in a solvent to make printable inks. Once printed, the ink is dried and the particles are deposited on a substrate to additively manufacture a (thin) and commonly flexible sensor, as shown in the prototype in Fig.10.6.

10.5 CONDUCTIVE POLYMER-BASED SENSORS

Conductive polymers are molecular chains where electronic orbitals are conjugated along the whole length of the molecule, creating a molecular orbital that is delocalized along the entire molecule and therefore allowing the transport of electrons in the whole structure. Absorption of a chemical species by one of these polymers has a strong influence on its molecular electronic structure and can therefore be monitored and used for sensing purposes. A rather common implementation of these materials for gas sensing is the use of polyaniline (PANI) for ammonia (NH_3) sensing. In [12], a remarkably sensitive and responsive ammonia sensor has been fabricated by inkjet printing PANI-polystyrene sulfonate (PSS) nanoparticles on photopaper. This sensor can detect ammonia concentrations of as low as 10 ppb in 20 s, as shown in Figure 10.7. Ammonia sensing is a particular application

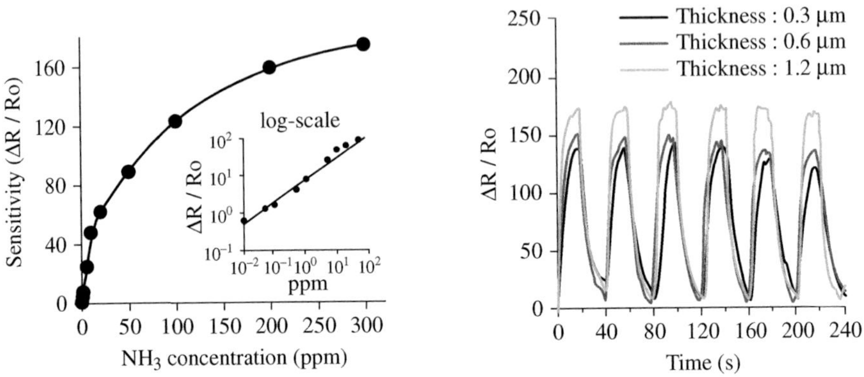

FIGURE 10.7 Plots of the maximum sensitivity versus ammonia concentration (left) and time response sensitivity to 160 ppm of ammonia (right) of the PANI-PSS based ammonia sensor. *Source:* Jang et al. 2007 [12]. Reproduced with permission of IEEE.

where conductive polymers (PANI in particular) really shine, but it becomes a lot more difficult to specifically detect most other gases. Repeatability and long-term chemical stability of these sensors are other issues that currently need to be tackled [13]. In parallel, other materials who offer better chemical stability, such as carbon nanomaterials, are also considered for these applications.

10.6 CARBON NANOMATERIAL-BASED SENSORS

Carbon nanomaterials have been in the spotlight since their recent discovery, due to the unique properties conferred to them by their 2D (or 2D-like, in the case of CNTs) structures. Two of these properties make them very attractive for sensing applications. A combination of conductivity or semi-conductivity with high surface area gives them the capacity of significantly interacting with surrounding analytes to produce a change in their observed electrical properties, such as conductivity and bandgap [14]. This potential has already been applied in a wide range of resistometric sensors, where the change of resistivity is monitored, transistor configurations, where the turn-on bias voltage can be used as sensing information [15] and, more recently, in capacitive configurations, where significant changes in capacitance are observed after interaction with the measured analyte [16].

However, most of the efforts toward building these sensors have remained very confined to stand-alone sensor design, without any consideration for the constraints of system integration, using expensive processes on non-flexible materials, which are not ideal for wearable implementations. These usual fabrication methods involve the growth of the carbon nanomaterials by using processes that require temperatures of 1000° C ([17]), which is not a temperature that flexible substrates can withstand. Inkjet printing, on the other hand, assuming that these particles are small enough and dispersible into a solvent can practically deposit them at room temperature and

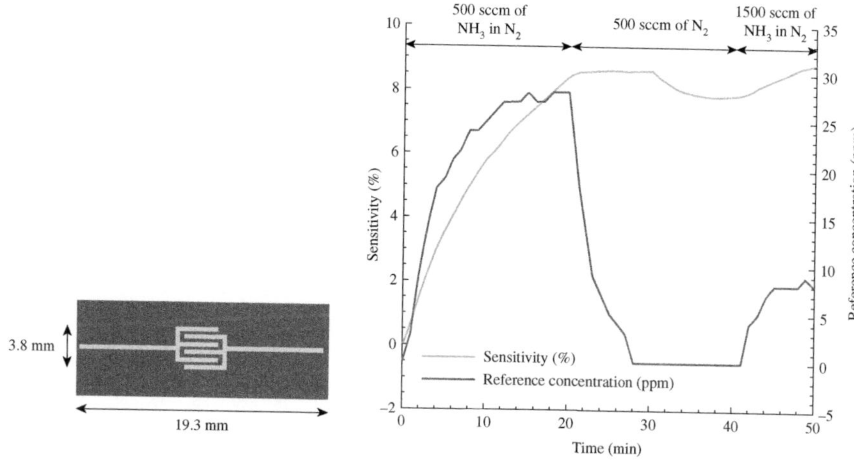

FIGURE 10.8 Inkjet-printed IDEs (left) and plot of the sensitivity over time of the rGO-based ammonia sensor, in light color, along with the response of a commercial ammonia sensor, in dark color (right). *Source:* Hester et al. 2015 [18]. Reproduced with permission of IEEE.

therefore on virtually any substrate. One of the materials commonly used for inkjet-printed sensors is graphene oxide (GO), because it is very easily dispersed in water, contrary to graphene, which is very difficult to disperse in most solvents. In [18], an inkjet-printed reduced graphene oxide (rGO) based sensor that is fully inkjet-printed on a flexible Kapton polyimide substrate is presented. The inter-digitated electrodes (IDE) were realized by inkjet printing patterns of silver nanoparticle ink (Figure 10.8-left). Then, GO nanoflakes and polystyrene nanospheres were dispersed in a water-based solution to make an ink that was subsequently printed on top of the electrodes. The printed film was then cleaned with acetone in order to remove the nanospheres, leaving pores behind them in the material, and the GO was finally reduced to obtain an rGO film. As shown in Figure 10.8-right, the sensor shows a response of about 8% to an exposition of 30 ppm of ammonia and a response time that is very comparable to that of a commercial sensor. Although not nearly as sensitive to ammonia as the PANI-PSS sensor mentioned previously, this sensor demonstrates the use of graphene as a sensing material for inkjet-printed flexible sensors.

One of the drawbacks of these materials is their propensity to interact with many chemical species and therefore their lack of selectivity, which is an essential property of a sensor in any practical implementation. However, it is possible to attach (covalently or not) carefully chosen molecules in order to coat these nanomaterials with functional groups that preferentially interact with some analytes rather than others, therefore "functionalizing" the nanomaterials and inducing selectivity [19].

This is demonstrated through the example of the first fully inkjet-printed, flexible CNT-based DMMP sensor, that was reported in [18]. The electrodes were printed

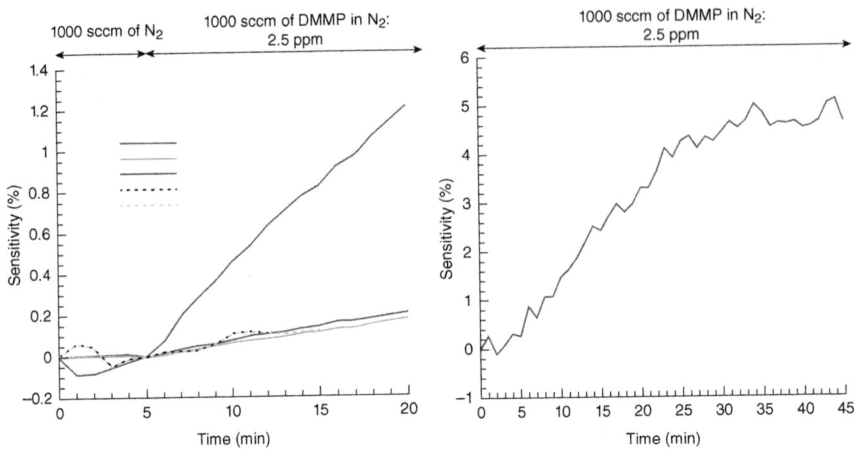

FIGURE 10.9 Compared plots of measured sensitivities of differently functionalized CNT sensors (left) and plot of the response of a HFA functionalized CNT sensor to 2.5 ppm of DMMP (right). *Source:* Hester et al. 2015 [18] Reproduced with permission of IEEE.

exactly the same way as for the rGO sensor. Some CNTs were then dispersed in Dimethylformamide (DMF) before the ink was inkjet printed on the electrodes. After being dried, the resulting CNT film was functionalized with a wide range of functional groups through a series of chemical processes.

Sensors with different functional groups, and otherwise completely identical, were compared to identify the functional group best-suited for detection of the targeted gas, Dimethyl methylphosphonate (DMMP) (Figure 10.9-left). The hexafluoroalcohol (HFA) functional group was recognized as best-suited for that purpose, by an order of magnitude in sensitivity, and was used to functionalize the final sensor (Figure 10.9-right).

10.7 INKJET-PRINTED MICROFLUIDICS

As the health-care economics increases exponentially these years and the growing global population is aging, we have reached to a point where remote monitoring is overtaking alongside the traditional hospital-focused model and has become the predominant way of real-time tracing patient progress [20, 21]. For instance, the real-time monitoring of bodily fluids such as tears, sweat, urine, and blood could be an excellent way to track patient health status and alert the patient to prevent any dangerous condition in time. On the one hand, monitoring on this level was not really realized due to difficulties associated with sample generation, collection and delivery issues in the current hospital-focused model [22]. On the other hand, the body fluids monitoring task can be easily undertaken by using wearable fluid sensors with characteristics such as miniaturization, reusability or disposability, low-cost production, and flexibility. Micro-Total-Analysis (μTAS) or Lab-on-a-Chip (LOC) is an important technical tool to realize this kind of wearable sensors [23]. In the aspect of fluid monitoring, *microfluidics* is an important concept for development of wearable fluid sensors. Due to its capability of manipulating extremely small quantities of liquid as well as its compact size and capability to be embedded into various devices, microfluidics technology has been widely used in biomedical sensing, manufacturing control, chemical assay, and other lab-on-chip applications in the past decades [24].

The first microfluidic device was fabricated in silicon and glass by photolithography and etching, a conventional planar manufacturing technique, which is in general expensive and environment unfriendly. Recently, flexible microfluidics using organic polymers have managed to be fabricated and largely reduce the manufacturing cost with soft lithography techniques [25]. However, these mold-based methods still highly rely to photolithography, which is of relatively high manufacturing cost. Recent years have witnessed an increasing number of novel and low-cost fabrication approaches that have been proposed to address these issues. Inkjet printing, a low-cost additive manufacturing technique, is a promising technology that has been recently utilized in microfluidics fabrication [26, 27]. By integrating inkjet printing and laser etching techniques, a rapid, low-cost, low-temperature, and zero-chemical treatment process has been developed. Poly (methylmethacrylate) (PMMA), a low-cost, transparent material, and regular photo paper are used in this process. An inkjet printer is used to inkjet-print conductive silver traces, as well as a polymer bonding and isolation layer on top of the metallization. A laser is then used to etch the channels and feed holes into a block of PMMA. This fabrication process successfully eliminates any photolithography need, or any other "cleanroom process." The fabricated microfluidics are lightweight, disposable, and robust, which makes them suitable for wearable applications.

In our natural environment, fluids have a wide permittivity distribution at microwave frequency due to their very different microstructure [28–30]. Also, the same fluids will show variations of permittivity at different temperatures or frequencies [31]. Moreover, with a varying mixture ratio, a wide range of continuous permittivity variation can be expected [32, 33]. Furthermore, if any solute is added to the liquid solvent, the permittivity will vary based on the concentration of the solute [34]. Thus, if a device can sense the permittivity change, a high volume of information on the fluids can be extracted. Many diverse microfluidic-enabled electromagnetic structures can be used to sense fluid permittivity changes, as will be shown.

Microwave resonators are very sensitive to permittivity changes and a resonance frequency shift is straightforward to observe with microwave measurement equipment. A split ring resonator (SRR) microfluidic-tunable filter is shown in Figure 10.10; the structure is inkjet-printed and it behaves as a bandstop filter. By loading the capacitive gap of the SRR with an array of fluids with different permittivities, the resonance frequency of the resonator can be tuned over a wide bandwidth. Utilizing only 6 μL of fluid, which is approximately one-twentieth of a drop of water, a 30% or $0.4\%/\epsilon_r$ change in the resonance frequency can be achieved. Despite its low-cost implementation, this inkjet-printed microfluidic filter shows a high temperature stability; below 1% variance in resonance frequency is observed for operating temperatures ranging from 273 to 332 K.

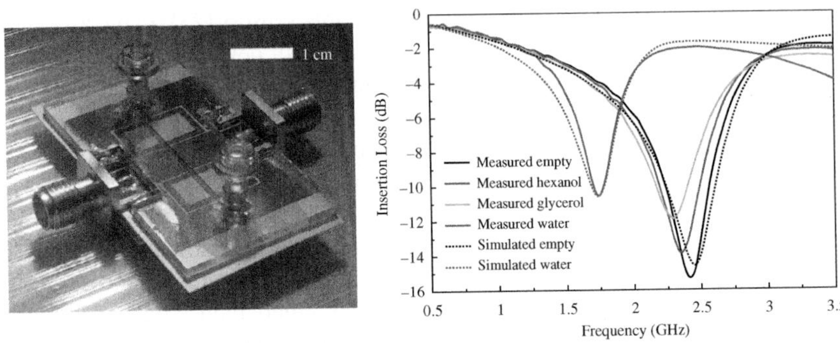

FIGURE 10.10 Left: A metamaterial-inspired inkjet-printed microfluidic-tunable bandstop filter. Right: Insertion loss of the microfluidic-tunable bandstop filter with different fluids in the channel in simulation and measurement. *Source:* Su et al. 2014 [35]. Reproduced with permission of IEEE.

With the potential of building wireless sensors, antennas are an appealing solution to be used as microfluidic-tunable elements. In Figure 10.11, an inkjet-printed microfluidic-tunable coplanar patch antenna is shown. Based on the fluid pumped into the sensor, the resonance frequency of the antenna is tuned due to variation of the fluid permittivity. A 13% frequency shift can be easily achieved with an operating frequency ranging from 3.8 GHz for an empty microfluidic channel to 3.3 GHz for a water-filled channel. The antenna features a return loss better than 30 dB in the whole tunable frequency range of 3.5 to 3.8 GHz. This low-profile inkjet-printed

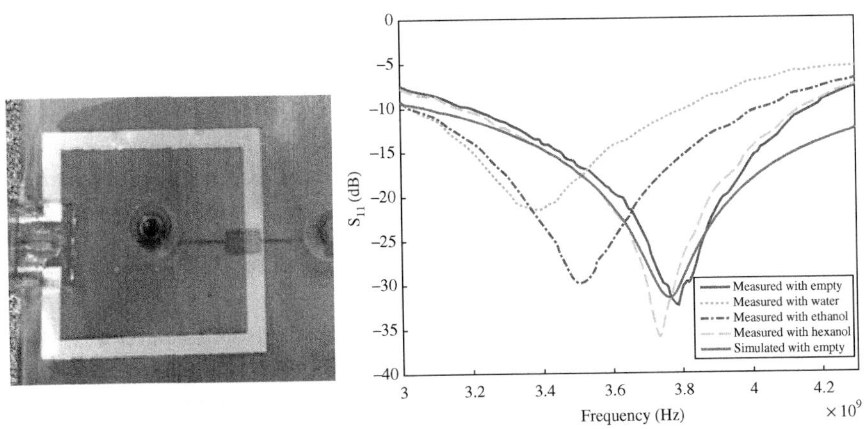

FIGURE 10.11 Left: An inkjet-printed microfluidic-tunable coplanar patch antenna. Right: The return loss of the microfluidic-tunable patch antenna with different fluids in the channel in simulation and measurement. *Source:* Su et al. 2014 [36]. Reproduced with permission of IEEE.

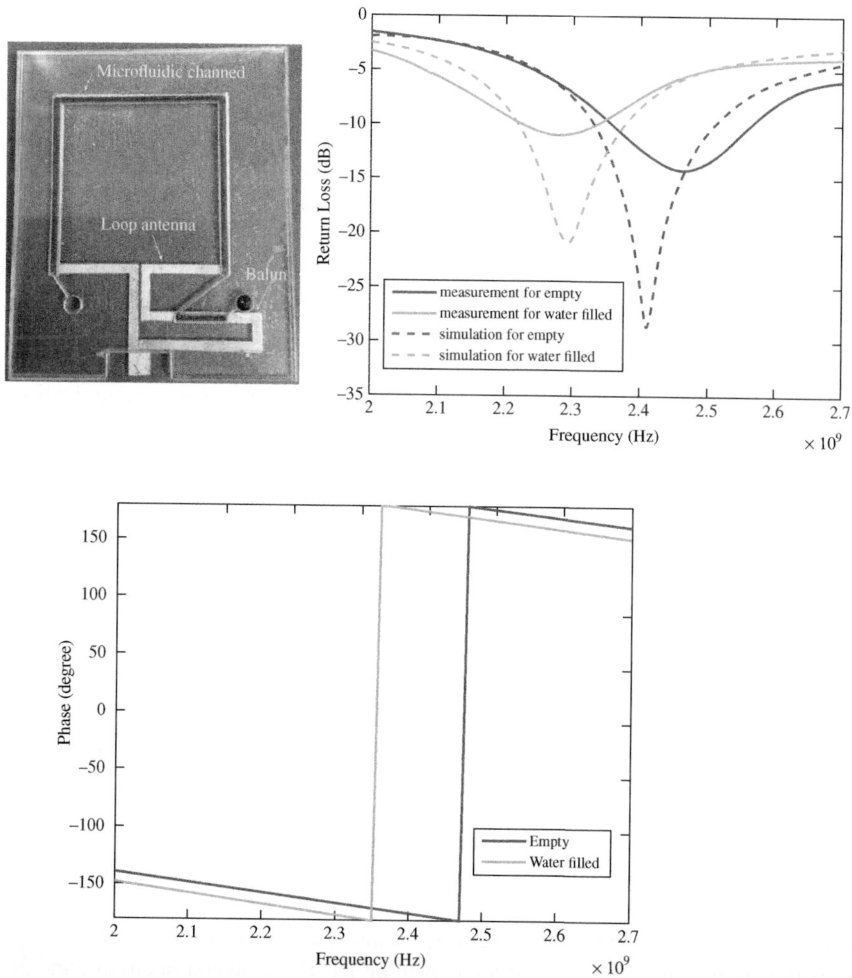

FIGURE 10.12 An inkjet-printed microfluidics-enabled loop antenna. Top-left: A photo of the fabricated loop antenna and balun. Top-right: The return loss of the microfluidics-enabled tunable antenna with empty channel and water-filled channel in simulations and measurements. Bottom: The phase difference of the microfluidics-enabled tunable balun with empty channel and water filled channel in the channel in simulation. *Source:* Su et al. 2015 [37]. Reproduced with permission of IEEE.

is suitable for a variety of applications, such as water quality monitoring, wearable liquid monitoring, bioliquids sensing, and identification.

Several elements of the antenna can utilize microfluidics, such as the balun of a differential antenna. A low-cost and disposable loop antenna is shown in Figure 10.12, which consists of a microfluidics-enabled tunable square loop and a microfluidics-enabled tunable balun to guarantee a similar radiation pattern over a reconfigurable "liquid-controlled" frequency range. By inserting different fluids over partial sections of the loop and the balun, the proposed antenna can achieve a frequency range from 2.28 GHz for a water-filled channel to 2.45 GHz for an empty channel. This antenna can be used as a fluid sensor for biomedical or chemical assay, or as a tunable antenna for communication applications. The microfluidics-enabled balun can be applied independently as a wideband balun for diverse communication, sensing, and biomonitoring applications that require tuning among different frequencies.

10.8 WIRELESS ENERGY HARVESTING FOR WEARABLES

In the last couple of decades, energy harvesting from ambient RF signals has attracted the interest following the explosive increase of use of RF waves for communication and broadcasting and the improvement of low-power consumption electronics. There are plenty of different kinds of RF sources, such as Very High Frequencies (VHF) radio, Ultra High Frequencies (UHF) TV signals, or Wi-Fi signals, which can possibly be used as the power source of low-power devices and wearable sensors, especially in urban areas, for example [38], [39]. A typical characteristic of microwaves is their capability to penetrate opaque walls making them available in more locations than other ambient energy sources such as solar and vibration, although the power density of RF signals is usually lower than other sources [40]. However, there are some 'hotspots' where RF power density is quite high and can provide a sufficient amount of power for relatively high power consumption electronics. For example, a two-way talk radio is a commonly used device for short distance communication in emergency scenarios, vehicle-to-vehicle communication, etc. Since it does not use any base station and directly transmits the signal toward the other mobile devices, it generates relatively high RF power compared to other mobile communication electronics, especially in the near field. People usually use their hand to operate a handheld radio, and therefore, it is possible to harvest energy from the radio-emitted waves by placing a harvesting circuit on top of the operator's hand. Generally speaking, RF energy harvesting circuits are composed of a receiver, an RF-DC conversion circuit, and an energy storage that includes lumped circuit components. In order to make the energy harvesting circuits wearable (wrapped around a wrist or sticked on the hand), a flexible substrate is used and inkjet printing technology can be applied for the fabrication of the conductors. In this section, the design and fabrication processes of wearable flexible energy harvester for an off-the-shelf two-way talk radio are described.

10.9 MICROWAVE RECEIVER DESIGN

Since most people operate Walkie-Talkies with their hands, it can be assumed that there is a high possibility that a sufficiently high amount of power can be harvested out of the radio's near field by placing a harvester on the hand. As an example, an off-the-shelf handheld radio can be used as a power source. In order to avoid interfering with the movements of the hand, there are two possible places to position the harvester;

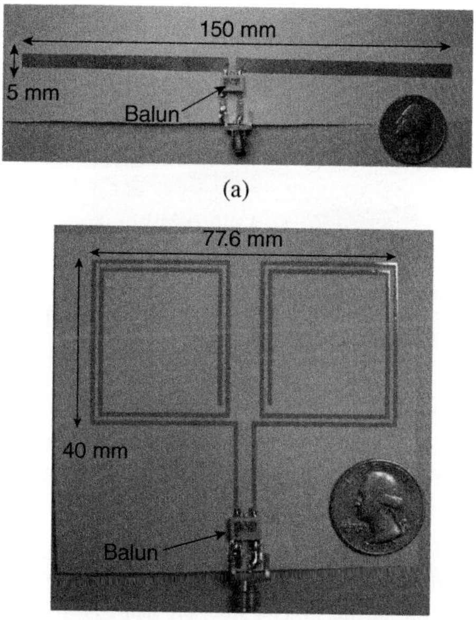

FIGURE 10.13 (a) Receiver for *E*-field. (b) Receiver for *H*-field. *Source:* Bito et al. 2015 [41]. Reproduced with permission of IEEE.

the wrist and the back of the hand (i.e., not the palm). Referring to the typical size of a wrist, a width of 5 cm, a length of 8 cm, and a circumference of 15 cm are adopted as the size constraints of the receiver. In order to be held by a typical-size palm, an 8 cm wide by 5 cm long rectangular is considered as the maximum size limitation for the receiver. Based on the near-field *E*- and *H*-field distribution and the above hand-dependent restrictions, the receiver circuits both for the wrist and the back of the hand are designed. The receiver on the wrist is aiming to harvest energy mainly from *E*-field, and the receiver on the back of the hand is mainly obtaining *H*-field energy. A dipole antenna for *E*-field and an open-type helical coil with four loops for *H*-field can be adopted for the wearable harvesting receiver [41]. Both can be fabricated by utilizing an inkjet printing masking process described below and they utilize a balun. The dimension of the *E*-and the *H*-field receivers are shown in Figure 10.13.

10.10 CIRCUIT FABRICATION WITH INKJET-PRINTED MASKING

Due to recent substantial improvements in fabrication process and performance up to the millimeter-wave frequency range, inkjet printing technology has become increasingly attractive for RF circuit fabrication. Inkjet printing technologies with silver nanoparticle-based inks have proven to be a very efficient solution for low-loss RF circuit patterning [42]. However, it is quite challenging to fabricate durable flexible

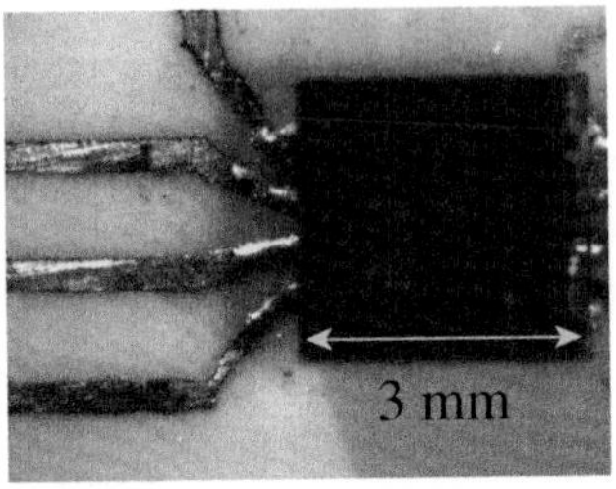

FIGURE 10.14 Circuit traces for a TDFN8 IC package fabricated with inkjet printing masking. *Source:* Bito et al. 2015 [41]. Reproduced with permission of IEEE.

circuits with printed traces and lumped circuit components because of the limited flexibility of the conductive epoxy, that is commonly utilized as the electrical interconnect between inkjet-printed conductive patterns. In order to create durable flexible circuits for wearable energy harvesting applications, the masking through the use of an inkjet-printed polymer is adopted. The polymer ink is made of 35 *w%* SU-8 polymer and is used as the mask on a copper-cladded LCP substrate. The thickness and dielectric constant of the substrate is 100 and 2.9 μm, respectively. Once the SU-8 ink is printed, the substrate is soft baked at 120° C for 10 minutes before the masking is exposed to 365 nm UV light for cross-linking. After the UV light exposure, the substrate is heated at 120° C for 5 minutes, yielding 4 to 6 μm thick SU-8 per layer [43]. Two layers of SU-8 are printed and then the uncovered copper metallization gets etched with $FeCl_3$. The etching time varies from 30 to 90 minutes depending on the temperature, size of substrate, thickness of metal layer, and freshness of $FeCl_3$ solution. The resolution of the conductive patterning is effectively equal to the resolution of inkjet printing, and it is possible to fabricate circuit trace width and spacing less than 400 μm, which are sufficiently good for many commonlyused packaged ICs. Figure 10.14 shows the printed conductive traces for a 3 mm by 3 mm TDFN8 package that have been fabricated with the above additive fabrication method.

10.11 INPUT POWER ESTIMATION AND RF-DC CONVERSION CIRCUIT DESIGN

In order to optimize the operation of the RF-DC conversion circuit, it is very important to estimate the level of the input power. However, since the transmitter and the receiver circuit are placed in the near-field, it is very difficult to estimate accurately how much power is actually transferred to the receiver through simulations. The requirement that the microwave receivers have to be wearable as well as the detrimental proximity effect of the human body further complicates this estimation. Therefore, in this case, the power transfer to the receiver port is computed from 2-port S-parameter measurements with a vector network analyzer (VNA). The energy harvesting system can be generally modeled as shown in Figure 10.15. In this figure, Z_S is the source impedance and Z_L is the load impedance. If input and output

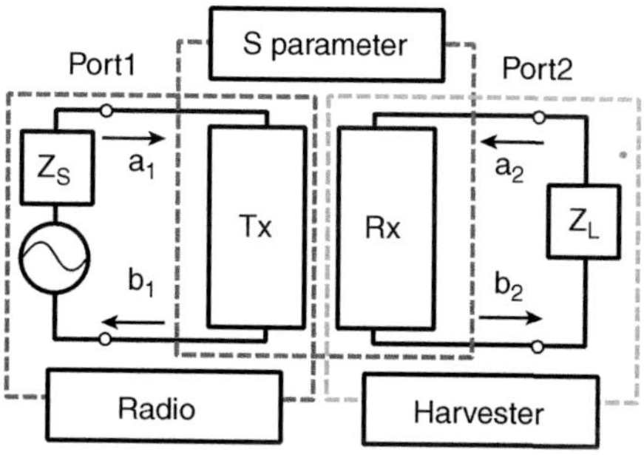

FIGURE 10.15 System configuration with equivalent S-parameter matrix. *Source:* Bito et al. 2015 [41]. Reproduced with permission of IEEE.

power at port1 and port2 are defined as a_1, b_1, a_2, and b_2, respectively, the topology of the transmitter antenna and of the harvesting receiver can be expressed as an S-parameter matrix. Once the power transferred to the load and the power from the source are defined as P_L and P_S, respectively, the power transfer efficiency from the source to the load can be expressed as:

$$\mu = \frac{P_L}{P_S} = \frac{b_2 b_2^* - a_2 a_2^*}{a_1 a_1^* - b_1 b_1^*} = \frac{(1 - |\Gamma_L|^2) S_{21}^2}{(1 - |\Gamma_{IN}|^2)|1 - \Gamma_L S_{22}|^2}. \tag{10.1}$$

If the reflection coefficient from source to port-1 (Γ_{IN}) in equation (10.1) is substituted with

$$\Gamma_{IN} = S_{11} + \frac{\Gamma_L S_{21} S_{12}}{1 - \Gamma_L S_{22}}, \tag{10.2}$$

where

$$Z_L = Z_0 \frac{1 + \Gamma_L}{1 - \Gamma_L}, \tag{10.3}$$

the efficiency can be expressed as a function of the reflection coefficient from port-2 to the load (Γ_L). Therefore, once the 2-port S-parameter matrix for the Tx-Rx propagation channel is obtained experimentally, the maximum power transfer efficiency can be analytically computed by sweeping the value of $\Gamma_L = Ae^{j\theta}$ in the range of $|A| = 1$ and $\theta = 0°$ to $360°$. At the same time, the load impedance value yielding the maximum efficiency can be computed from equation (10.2). For the measurement, it is possible to get a better idea of input power by simulating the real operation conditions. In order to mimic the two-way talk radio, a monopole antenna with similar properties to the one of the two-way radio can be placed at the bottom of a water bottle and used for the S-parameter measurements. The wearable harvesting receiver for *E*-field is wrapped at 7 cm from the bottom of the bottle and the receiver for *H*-field is placed at the bottom of the bottle for characterization. The measured S-parameters

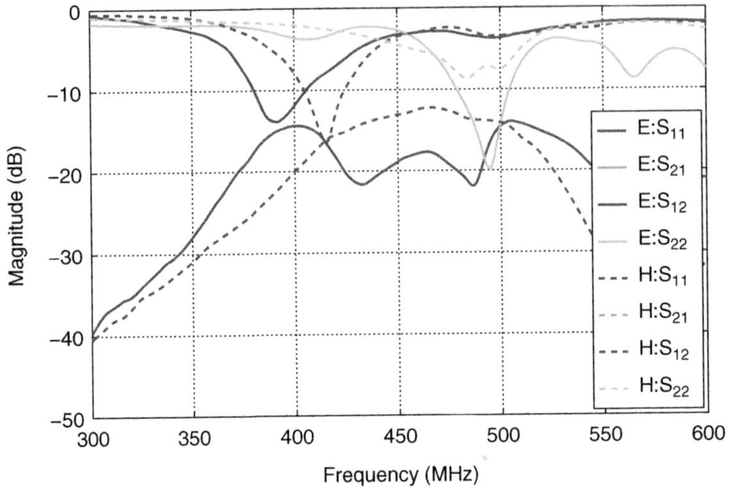

FIGURE 10.16 Measured S-parameters for *E*- and *H*-field receivers. *Source:* Bito et al. 2015 [41]. Reproduced with permission of IEEE.

for the transmitting monopole and for the *E*- and *H*-field receivers are plotted in Figure 10.16 for this specific configuration. Using these data in equations (10.1) and (10.2), the maximum potential power transfer efficiency of the *E*-field is 5.23% and of the *H*-field receiver is 16.5%, which are equivalently 52.3 mW and 165 mW of maximum transferred power since the output power of the two-way talk radio is 1 W.

Once the input power is estimated from the measurement, it is possible to design the RF-DC conversion circuit and the matching circuit. Since the operation frequency of the two-way talk radio is relatively low, around 464 MHz, lumped components are used for the matching circuit. In order to maximize the output voltage with the minimum possible circuit size, a single-stage Dickson voltage doubler with one Schottky diode chip are used as the rectifier. The circuits are initially designed and optimized with the help of microwave simulation tools and the matching circuit is further tweaked with the help of VNA measurements. The configuration of the rectifier prototypes are shown in Figure 10.17.

10.12 RF-DC CONVERSION EFFICIENCY MEASUREMENT AND PROTOTYPE OPERATION TESTS

The RF-DC conversion efficiency is one of the most important parameters which represents the quality of RF energy harvesting circuits. The output power can be computed with equation (10.4) where P_{in} is the input RF power, P_{out} is the output DC power, V_{out} is the output voltage, and R_{load} is the load resistance.

$$\eta = \frac{P_{out}}{P_{in}} \times 100[\%] \quad P_{out} = \frac{V_{out}^2}{R_{load}} \tag{10.4}$$

FIGURE 10.17 RF-DC conversion circuit topology for wearable (a) *E*-filed energy harvester (b) H-filed energy harvester. *Source:* Bito et al. 2015 [41]. Reproduced with permission of IEEE.

The output voltage measurements are initially conducted with the handheld radio by arranging the harvester and the radio on the side of a water bottle in a configuration similar to the one on the human arm holding the radio. The output voltage is measured by changing the load resistance of the circuit in the range of 100–6800 Ω. The measured RF-DC conversion efficiency curves for both *E*-field harvester and *H*-field harvester are shown in Figure 10.18, assuming the input power to be 52.3 mW for the E-filed harvester and 165 mW for the *H*-filed harvester, respectively. The maximum RF-DC conversion efficiency of 82.5% is achieved with the 1772 Ω load resistance and 89.2% is achieved with the 2996 Ω load resistance for *E*-field harvester and *H*-field harvester, respectively [41, 44]. The operation tests are conducted by replacing the load resistance with an LED and wrapping the prototype on the wrist or placing the

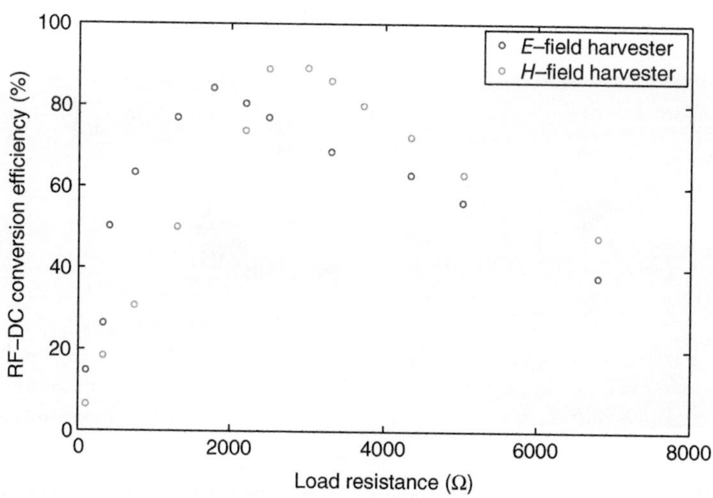

FIGURE 10.18 RF-DC conversion efficiency from ADS simulation and measurement with respect to load resistance. *Source:* Bito et al. 2015 [41]. Reproduced with permission of IEEE.

(a) (b)

FIGURE 10.19 (a) Open voltage measurement with an on-bottle setup. (b) Operation verification of the wearable harvester on a wrist. *Source:* Bito et al. 2015 [41]. Reproduced with permission of IEEE. For a color version of this figure, see the color plate section.

prototype on the back of the right hand. As a result, the LED is successfully turned on, utilizing only the harvested energy from the handheld radio under the typical radio operation condition with the E-field and H-field harvesters. This result implies that these wearable harvesters can provide power enough to run relatively high power consumption electronics with microcontrollers, sensors, and wireless communication functions. The measurement of the open-circuit voltage and the verified on-arm successful operation of the wearable E-field harvester and H-field harvester are shown respectively in Figures 10.19 and 10.20.

(a) (b)

FIGURE 10.20 (a) Open voltage measurement with an on-bottle setup. (b) Operation verification of the wearable harvester on a wrist. *Source:* Bito et al. 2015 [41]. Reproduced with permission of IEEE. For a color version of this figure, see the color plate section.

10.13 CONCLUSION

In this chapter, the design and fabrication process of inkjet-printed smart skins and wireless sensors has been discussed. Complex multilayer structures can be implemented by stacking printed layers of conductive and dielectric inks, and sensing modules can be fabricated using nanomaterials such as graphene oxide and CNTs. Microfluidic-tunable filters and antennas can be fabricated without the need of cleanroom or chemical treatment, significantly reducing the manufacturing cost of such devices. The feasibility of making fully flexible circuits and sensors has been clearly demonstrated by either printing conductors or inkjet-printing detailed low-cost masks followed by simple etching techniques. Finally, the easy realization of inkjet-printed energy harvesting circuits on flexible materials could potentially allow the full powering of microcontrollers and sensor circuitry, which would be necessary for the realization of truly autonomous wearable wireless systems for Internet of Things applications.

REFERENCES

[1] L. Yang, A. Rida, R. Vyas, and M. M. Tentzeris, "RFID tag and RF structures on a paper substrate using inkjet-printing technology," *IEEE Trans. Microw. Theory Techn.*, vol. 55, no. 12, pp. 2894–2901, December 2007.

[2] B. Cook and A. Shamim, "Inkjet printing of novel wideband and high gain antennas on low-cost paper substrate," *IEEE Trans. Antennas Propag.*, vol. 60, no. 9, pp. 4148–4156, September 2012.

[3] B. S. Cook, Y. Fang, S. Kim, T. Le, W. B. Goodwin, K. H. Sandhage, and M. M. Tentzeris, "Inkjet catalyst printing and electroless copper deposition for low-cost patterned microwave passive devices on paper," *Electron. Mater. Lett.*, vol. 9, no. 5, pp. 669–676, September 2013.

[4] B. S. Cook, J. R. Cooper, and M. M. Tentzeris, "Multi-layer RF capacitors on flexible substrates utilizing inkjet printed dielectric polymers," *IEEE Microw. Wireless Compon. Lett.*, vol. 23, no. 7, pp. 353–355, June 2013.

[5] S. H. Ko, J. Chung, H. Pan, C. P. Grigoropoulos, and D. Poulikakos, "Fabrication of multilayer passive and active electric components on polymer using inkjet printing and low temperature laser processing," *Sens. Actuat. A: Phys.*, vol. 134, no. 1, pp. 161–168, February 2007.

[6] B. J. Kang, C. K. Lee, and J. H. Oh, "All-inkjet-printed electrical components and circuit fabrication on a plastic substrate," *Microelectron. Eng.*, vol. 97, pp. 251–254, September 2012.

[7] Y. Li, R. Torah, S. Beeby, and J. Tudor, "An All-Inkjet Printed Flexible Capacitor for Wearable Applications," in 2012 Symposium on Design, Test, Integration and Packaging of MEMS/MOEMS (DTIP), Cannes, France, April 25–27, 2012, pp. 192–195.

[8] B. Tehrani, B. S. Cook, and M. M. Tentzeris, "Post-Process Fabrication of Multilayer mm-Wave On-Package Antennas with Inkjet Printing," in 2015 IEEE International Symposium on Antennas and Propagation (APSURSI), Vancouver, BC, Canada, July19–24, 2015, pp. 607–608.

[9] B. Tehrani, J. Bito, B. Cook, and M. Tentzeris, "Fully Inkjet-Printed Multilayer Microstrip and T-Resonator Structures for the RF Characterization of Printable Materials and Interconnects," in IEEE MTT-S International Microwave Symposium (IMS), Tampa, FL, June 1–6, 2014, pp. 1–4.

[10] A. Babakhani, X. Guan, A. Komijani, A. Natarajan, and A. Hajimiri, "A 77-GHz phased-array transceiver with on-chip antennas in silicon: Receiver and antennas," *IEEE J. Solid-State Circuits*, vol. 41, no. 12, pp. 2795–2806, December 2006.

[11] B. Cook, B. Tehrani, J. Cooper, and M. Tentzeris, "Multilayer inkjet printing of millimeter-wave proximity-fed patch arrays on flexible substrates," *IEEE Antennas Wireless Propag. Lett.*, vol. 12, pp. 1351–1354, October 2013.

[12] J. Jang, J. Ha, and J. Cho, "Fabrication of water-dispersible polyaniline-poly(4-styrenesulfonate) nanoparticles for inkjet-printed chemical-sensor applications," *Adv. Mater.*, vol. 19, no. 13, pp. 1772–1775, June 2007.

[13] H. Yoon, "Current trends in sensors based on conducting polymer nanomaterials," *Nanomaterials*, vol. 3, no. 3, pp. 524–549, August 2013.

[14] R. H. Baughman, A. A. Zakhidov, and W. A. de Heer, "Carbon nanotubes–the route toward applications," *Science*, vol. 297, no. 5582, pp. 787–792, August 2002.

[15] J. Kong, N. R. Franklin, C. Zhou, M. G. Chapline, S. Peng, K. Cho, and H. Dai, "Nanotube molecular wires as chemical sensors," *Science*, vol. 287, no. 5453, pp. 622–625, January 2000.

[16] E. S. Snow, F. K. Perkins, E. J. Houser, S. C. Badescu, and T. L. Reinecke, "Chemical detection with a single-walled carbon nanotube capacitor," *Science*, vol. 307, no. 5717, pp. 1942–1945, March 2005.

[17] Q. Cao and J. A. Rogers, "Ultrathin films of single-walled carbon nanotubes for electronics and sensors: A review of fundamental and applied aspects," *Adv. Mater.*, vol. 21, no. 1, pp. 29–53, June 2009.

[18] J. G. Hester, Y. Fang, and M. M. Tentzeris, "Inkjet-Printed, Flexible, High Performance, Carbon Nanomaterial Based Sensors for Ammonia and DMMP gas detection," in 45th European Microwave Conference (EuMC), Paris, France, September 7–10, 2015, pp. 857–860.

[19] R. Chen and S. Bangsaruntip, "Noncovalent Functionalization of Carbon Nanotubes for Highly Specific Electronic Biosensors," *in Proceedings of the National Academy of Sciences of the United States of America*, 2003.

[20] A. Pantelopoulos and N. G. Bourbakis, "A survey on wearable sensorbased systems for health monitoring and prognosis," *IEEE Trans. Syst., Man, Cybern. C*, vol. 40, no. 1, pp. 1–12, January 2010.

[21] Y. Hao and R. Foster, "Wireless body sensor networks for health-monitoring applications," *Physiol. Meas.*, vol. 29, no. 11, p. R27, October 2008.

[22] D. Diamond, S. Coyle, S. Scarmagnani, and J. Hayes, "Wireless sensor networks and chemo-/biosensing," *Chem. Rev.*, vol. 108, no. 2, pp. 652–679, February 2008.

[23] F. Benito-Lopez, S. Coyle, R. Byrne, and D. Diamond, "Sensing Sweat in Real-Time Using Wearable Microfluidics," in Proceedings of 7th International Workshop on Wearable and Implantable Body Sensor Networks, Singapore, June 26–28, 2010.

[24] G. M. Whitesides, "The origins and the future of microfluidics," *Nature*, vol. 442, no. 7101, pp. 368–373, July 2006.

[25] G. M. Whitesides and A. D. Stroock, "Flexible methods for microfluidics," *Phys. Today*, vol. 54, no. 6, pp. 42–48, June 2001.

[26] K. Abe, K. Suzuki, and D. Citterio, "Inkjet-printed microfluidic multianalyte chemical sensing paper," *Anal. Chem.*, vol. 80, no. 18, pp. 6928–6934, August 2008.

[27] B. S. Cook, J. R. Cooper, and M. M. Tentzeris, "An inkjet-printed microfluidic RFID-enabled platform for wireless lab-on-chip applications," *IEEE Trans. Microw. Theory Tech.*, vol. 61, no. 12, pp. 4714–4723, November 2013.

[28] A. Tidar, S. Shafiyoddin, S. Kamble, G. M. Dharne, S. S. Patil, P. W. Khirade, and S. Mehrotra, "Microwave Dielectric Relaxation Study of 1-hexanol with 1-propenol Mixture by Using Time Domain Reflectometry at 300 K," in Applied Electromagnetics Conference (AEMC), Kolkata, December 2009, pp. 1–4.

[29] R. Nigmatullin, M. A.-G. Jafar, N. Shinyashiki, S. Sudo, and S. Yagihara, "Recognition of a new permittivity function for glycerol by the use of the eigen-coordinates method," *J. Non-Cryst. Solids*, vol. 305, pp. 96–111, July 2002.

[30] K. Shibata, "Measurement of Complex Permittivity for Liquid Materials Using the Open-Ended Cut-Off Waveguide Reflection Method," in Proceedings of the 2011 China-Japan Joint Microwave Conference, Hangzhou, April 20–22, 2011, pp. 1–4.

[31] A. E. Lipton, M. K. Griffin, and A. G. Ling, "Microwave transfer model differences in remote sensing of cloud liquid water at low temperatures," *IEEE Trans. Geosci. Remote Sens.*, vol. 37, no. 1, pp. 620–623, January 1999.

[32] R. Behrends, K. Fuchs, U. Kaatze, Y. Hayashi, and Y. Feldman, "Dielectric properties of glycerol/water mixtures at temperatures between 10 and 50 degrees C," *J. Chem. Phys.*, vol. 124, no. 14, p. 144512, January 2006.

[33] A. H. Sihvola and J. A. Kong, "Effective permittivity of dielectric mixtures," *IEEE Trans. Geosci. Remote Sens.*, vol. 26, no. 4, pp. 420–429, July 1988.

[34] B. Hess, C. Holm, and N. van der Vegt, "Modeling multibody effects in ionic solutions with a concentration dependent dielectric permittivity," *Phys. Rev. Lett.*, vol. 96, no. 14, p. 147801, April 2006.

[35] W. Su, C. Mariotti, B. Cook, S. Lim, L. Roselli, and M. M. Tentzeris, "A Metamaterial-Inspired Temperature Stable Inkjet-Printed Microfluidic-Tunable Bandstop Filter," in 44th European Microwave Conference (EuMC), Rome, Italy, October 6–9, 2014, pp. 9–12.

[36] W. Su, B. Cook, M. M. Tentzeris, C. Mariotti, and L. Roselli, "A Novel Inkjet-Printed Microfluidic Tunable Coplanar Patch Antenna," in 2014 IEEE Int. Symp. on Antennas and Propagat. (APSURSI), Memphis, TN, July 2014, pp. 858–859.

[37] W. Su, B. Cook, and M. M. Tentzeris, "Low-Cost Microfluidics-Enabled Tunable Loop Antenna Using Inkjet-Printing Technologies," in 9th European Conference on Antennas and Propagation (EuCAP), Lisbon, Portugal, 12–17 April 2015, pp. 1–2.

[38] M. Pinuela, P. D. Mitcheson, and S. Lucyszyn, "Ambient RF energy harvesting in urban and semi-urban environments," *IEEE Trans. Microw. Theory Techn.*, vol. 61, no. 7, pp. 2715–2726, May 2013.

[39] R. J. Vyas, B. S. Cook, Y. Kawahara, and M. M. Tentzeris, "E-WEHP: A batteryless embedded sensor-platform wirelessly powered from ambient digital-TV signals," *IEEE Trans. Microw. Theory Tech.*, vol. 61, no. 6, pp. 2491–2505, May 2013.

[40] J. A. Paradiso and T. Starner, "Energy scavenging for mobile and wireless electronics," *IEEE Pervasive Comput.*, vol. 4, no. 1, pp. 18–27, March 2005.

[41] J. Bito, J. G. Hester, and M. M. Tentzeris, "Ambient Energy Harvesting From A Two-Way Talk Radio for Flexible Wearable Devices Utilizing Inkjet Printing Masking," in Microwave Symposium (IMS), 2014 IEEE MTT-S International, Phoenix, AZ, May 17–22, 2015, pp. 1–4.

[42] Y. Kawahara, S. Hodges, B. S. Cook, C. Zhang, and G. D. Abowd, "Instant Inkjet Circuits: Lab-Based Inkjet Printing To Support Rapid Prototyping of UbiComp Devices," in Proceedings of the 2013 ACM international joint Conference on Pervasive and Ubiquitous Computing, ACM, Zurich, Switzerland, September 2013, pp. 363–372.

[43] B. K. Tehrani, J. Bito, B. S. Cook, and M. M. Tentzeris, "Fully Inkjet-Printed Multilayer Microstrip and T-Resonator Structures for the RF Characterization of Printable Materials and Interconnects," in Microwave Symposium (IMS), 2014 IEEE MTT-S International, Tampa, FL, June 1–6, 2014, pp. 1–4.

[44] J. Bito and M. M. Tentzeris, "A Novel Flexible Wearable Magnetic Energy Harvester Utilizing Inkjet Masking Techniques," in Antennas and Propagation Society International Symposium (APSURSI), 2015 IEEE, Vancouver, BC, July 19–24, 2015, pp. 204–205.

11

CIRCUITS AND SYSTEMS FOR WIRELESS BODY AREA NETWORK

JOONSUNG BAE AND HOI-JUN YOO

Department of Electrical Engineering, Korea Advanced Institute of Science and Technology, Daejeon, Republic of Korea

11.1 INTRODUCTION

Recent achievements in information and communication technology (ICT) enable a new paradigm for the Internet of Things (IoT), in which everything is interconnected through the World Wide Web [1]. Among IoT applications, mobile healthcare is getting more and more attention due to the aging society, increasing medical expenses, and the necessity of managing chronic diseases which are the number one cause of death [2]. The key technologies for mobile healthcare or remote medical care are "Things"-oriented body area network (BAN) systems [3] connected to the Internet infrastructure via TCP/IP protocol (e.g., 6LowPAN) or access points (e.g., Wi-Fi), followed by "Internet"-oriented cloud computing [4] with big data analysis as shown in Figure 11.1.

We focus on the implementation of the "Things"-oriented MBAN (Medical Body Area Network) system which sheds new light on the short-range connectivity among various physiological sensors and wearable devices carried by human bodies for a continuous and ambulatory connection. Basically, the MBAN system in Figure 11.1 consists of a central hub and several sensor nodes, which are categorized into wearable (on-body) and implantable (in-body) nodes. The MBAN hub controls the nodes and gathers physiological data from sensors. Previously, these two types of sensor nodes were separately studied and developed as stand-alone wearable BAN or

Electromagnetics of Body-Area Networks: Antennas, Propagation, and RF Systems, First Edition.
Edited by Douglas H. Werner and Zhi Hao Jiang.

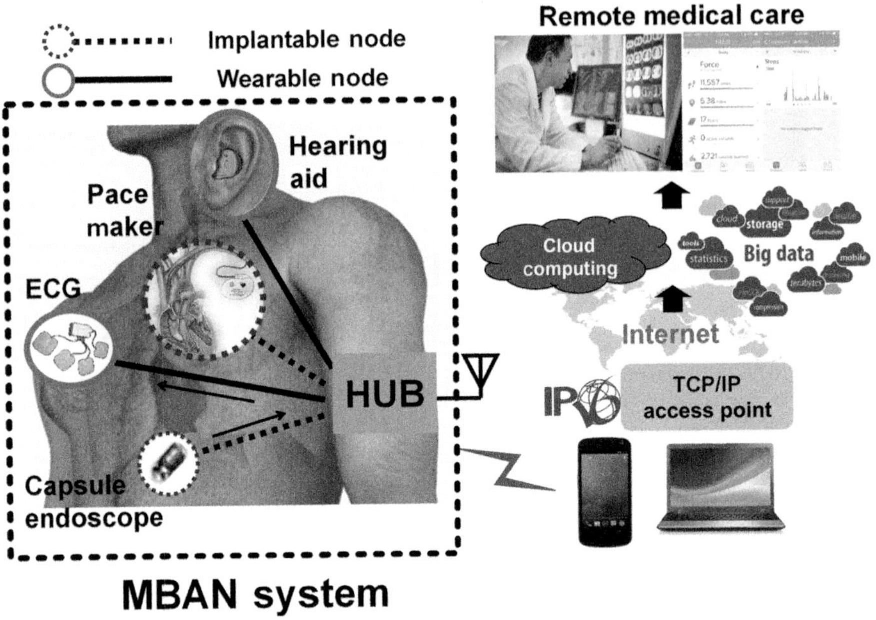

FIGURE 11.1 Medical body area network system for mobile healthcare. *Source:* Cho et al. 2009 [18]

implantable BAN devices [5–7]. Furthermore, most of the previous MBAN systems for both on- and in-body communication exploited the same frequency band for the wearable and implantable sensor nodes [8, 9]. However, since wave propagation on the body is quite different from propagation in the body [10], careful considerations for selecting the frequency band in the MBAN system are necessary for unobtrusive, inexpensive, wearable, and implantable sensor nodes.

Among the frequency bands for industry-science-medical (ISM), ultra-wide band (UWB), medical implant communication service (MICS), and human body communication (HBC), which have been utilized by short-range wireless standards, such as IEEE 802.15.1 (PHY/MAC of classic Bluetooth), IEEE 802.15.4 (PHY/MAC of Zigbee), and IEEE 802.15.6, the proposed MBAN system follows the MICS regulation established by the FCC for dedicated communication between implanted devices and devices outside of the body [11]. In addition, the 400 MHz MICS band shows less propagation loss in the body, compared to other high frequency ISM and UWB bands [12]. For on-body communications, we do not consider using the ISM and UWB bands for implementing energy-efficient sensor nodes to be feasible, since it has been found that wireless channels using frequencies over 1 GHz show significant path loss due to the body shadowing effect [13]. Recently, body channel communication (BCC), which uses the human body as a communication channel, has been optimized for wearable BAN applications [14, 15]. Since the high conductivity of the human body provides low path loss, and the communication frequency can be lowered without the need of a large-size antenna by using an electrode [16, 17], a customized BCC band has been adopted for on-body communication in order to substantially reduce the power consumption and the size of wearable nodes.

With regard to the MBAN sensor nodes, energy-efficient wireless technologies are a major design challenge that will extend the lifetime of the nodes under limited energy sources. Since energy consumption is power consumption multiplied by usage time, a low power transceiver with a fast turn-on time should be employed with tight control of the duty cycle between sleep and active modes. In addition, sensor node miniaturization and cost reduction are important considerations. Since the crystal oscillator (XO) is not only one of the most expensive ($1–$10), bulky (5 mm × 5 mm), power-hungry, and fragile components but also suffers from slow startup time of the order of milliseconds, developing a crystal-less transceiver in combination with an efficient duty cycle could further reduce the cost, size, and power consumption of the sensor nodes.

With these considerations in mind, this chapter presents an MBAN system following the implementation of a dual-band hub transceiver [18] with compact, crystal-less sensor transceivers that consume low energy and are low cost for mobile healthcare applications. The transceiver architectures and circuit techniques adopted in the LNA (Low Noise Amplifier) in the hub, the wearable node, and the implantable node make it possible to significantly decrease the power consumed by the transceivers. In particular, we focus on adopting frequency calibration techniques that do not require XOs in order to miniaturize the sensor node boards. As a result, a compact form factor is achieved for both an implantable temperature sensor board (15 mm × 15 mm), which utilizes system-in-package (SiP) technology, and for a wearable sensor board (12 mm × 20 mm) with electrocardiogram (ECG) data transmission for an MBAN system.

11.2 MBAN SYSTEM CONCEPT

11.2.1 System Overview

The proposed MBAN system is composed of wearable nodes and implantable nodes which are connected with a hub through the 30–70 MHz BCC and 402–405 MHz MICS bands. The MBAN hub manages several on-body and in-body sensor nodes to periodically obtain vital signals of the human body via a MICS antenna and a BCC electrode [19] as shown in Figure 11.2a. In such a system, the star network topology makes sense due to the limited energy capacity of the small batteries used for the sensor nodes. The hub is relatively free to consume more power and resources compared to the sensor nodes (especially implantable nodes), which should be miniaturized and energy efficient. Moreover, fast startup time is critical for the sensor nodes in order to achieve an efficient duty-cycled operation [20].

To reduce the form factor and energy consumption of the MBAN sensor nodes, the XO is replaced by an on-chip digitally controlled oscillator (DCO), aided by the reference frequency signal from the hub, which has an accurate frequency source such as

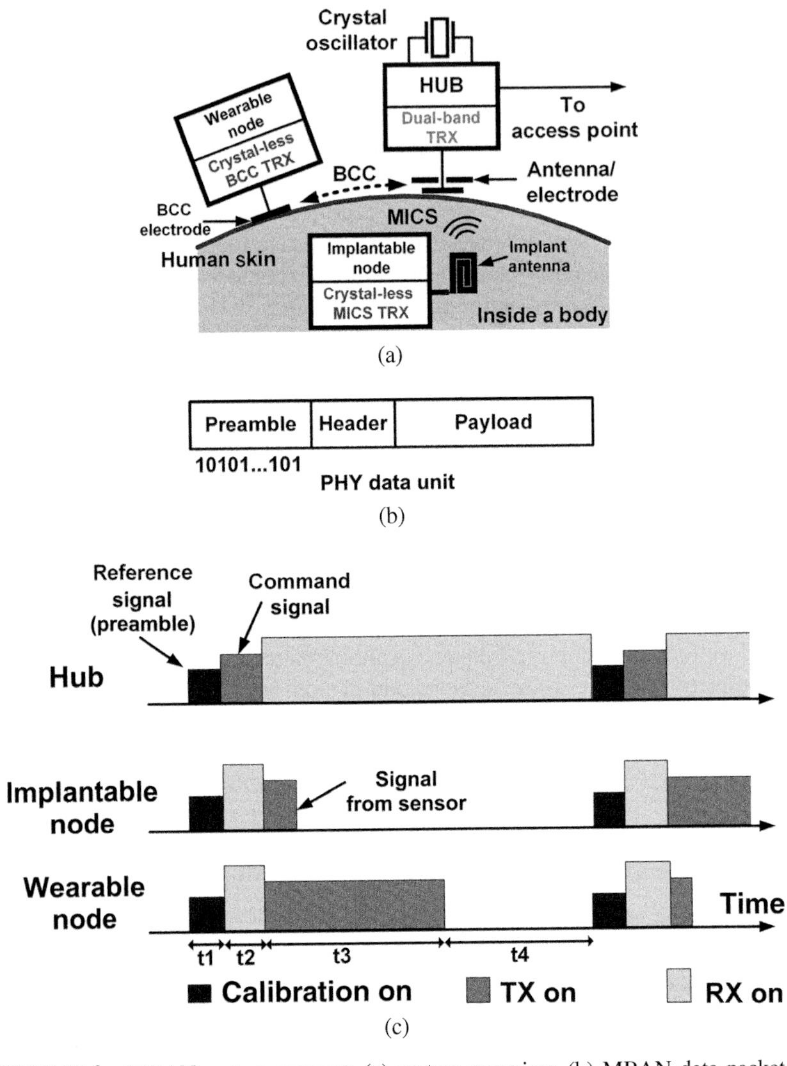

FIGURE 11.2 MBAN system concept: (a) system overview, (b) MBAN data packet, and (c) timing diagram with proposed frequency calibration concept. *Source:* Bae et al. 2012 [22]. Reproduced with permission of IEEE.

a XO. Although the integrated on-chip DCO is inexpensive ($0–$0.01 cost) and has a fast startup time (<1 μs), open-loop DCO has low frequency stability (>10,000 ppm) and suffers from frequency drift over temperature variation, since it is not phase locked to the stable frequency reference.

Given that the preamble signal in the data packet for bit synchronization consists of pre-determined bit sequences with a high transition density (Figure 11.2b), the hub wirelessly distributes an accurate frequency reference signal (preamble signal) periodically to the sensor nodes, and then the sensor nodes auto-calibrate the frequency of the DCO during the time period t_1 as depicted in Figure 11.2c. After calibration is completed by all sensor nodes, the hub configures each sensor, sending parameters such as the data rate, channel frequency, and percentage of the duty cycle by means of a command signal during time period t_2. Finally, each node transmits the sensed physiological data to the hub simultaneously by duty-cycled operation ($t_3/(t_3 + t_4)$). This operation is repeated every 100 ms ($t_1 + t_2 + t_3 + t_4$), based on the studies of the human gait [21], which guides the calibration period according to frequency drift on the human body. We also note that the energy overhead during the frequency calibration can be significantly reduced by shortening the time period t_1.

11.2.2 Crystal-Less Sensor Nodes

We can achieve a low cost, crystal-less sensor node implementation because of a frequency calibration technique based on the properties of an injection-locking oscillator (ILO) in the sensor node transceivers [22]. When the ILO is injection-locked by the proper injection signal, which has a frequency between $f_0 - f_L$ and $f_0 + f_L$, where f_0 is a free-running frequency of the oscillator and f_L is a locking range of the ILO, the output of the ILO follows the input signal. When the ILO is not injection-locked or injection-pulled by the injection signal out of the locking range, the phase difference between the output of the ILO and the injection signal is not constant but time varying [23]. From this property, Figure 11.3a explains the frequency calibration method using an injection-locking DCO (IL-DCO).

First, we assume that the output frequency of the DCO is linearly one-to-one matched to the control code of the DCO. To set the control code of the DCO, whose frequency is as close as possible to the reference frequency (f_{ref}), the reference signal from the hub is applied into the DCO. Then, the control codes of the DCO, CC_1 and CC_2, are found to correct the frequency of the DCO. By sweeping the N-bit control code of the DCO from 0 to $2^N - 1$, the values CC_1 and ($CC_1 + 1$) are identified, which make the DCO operate in the injection-pulling and injection-locking states, respectively. Similarly, ($CC_2 - 1$) and CC_2 are found where the DCO oscillates in the injection-locking and injection-pulling states, respectively, as shown in Figure 11.3a. Finally, the frequency of the DCO, coded with $[(CC_1 + CC_2)/2]$, is the closest frequency to the f_{ref} by the symmetric attribute of the locking range ($f_{ref} - f_L$ to $f_{ref} + f_L$).

With the proposed calibration method, detecting whether the IL-DCO is injection-locked or injection-pulled is a key technique in order to obtain the control codes CC_1 and CC_2. Since the output of the IL-DCO contains multi-tone frequency components when the IL-DCO is injection-pulling, an injection-locking detector (IL detector) with an envelope detector is employed to obtain the envelope of the IL-DCO output. The envelope is constant when the IL-DCO is injection-locked but amplitude modulated when the IL-DCO is injection-pulled. Figure 11.3b shows the IL detector, which is composed of the envelope detector, an AC-coupled amplifier, a Schmitt trigger, and a debouncing circuit. While applying the fixed reference signal (f_{ref}), the N-bit control code of the DCO is swept by the calibration controller. Then, the output of IL-DCO is monitored by the IL detector. The bottom of Figure 11.3b shows the

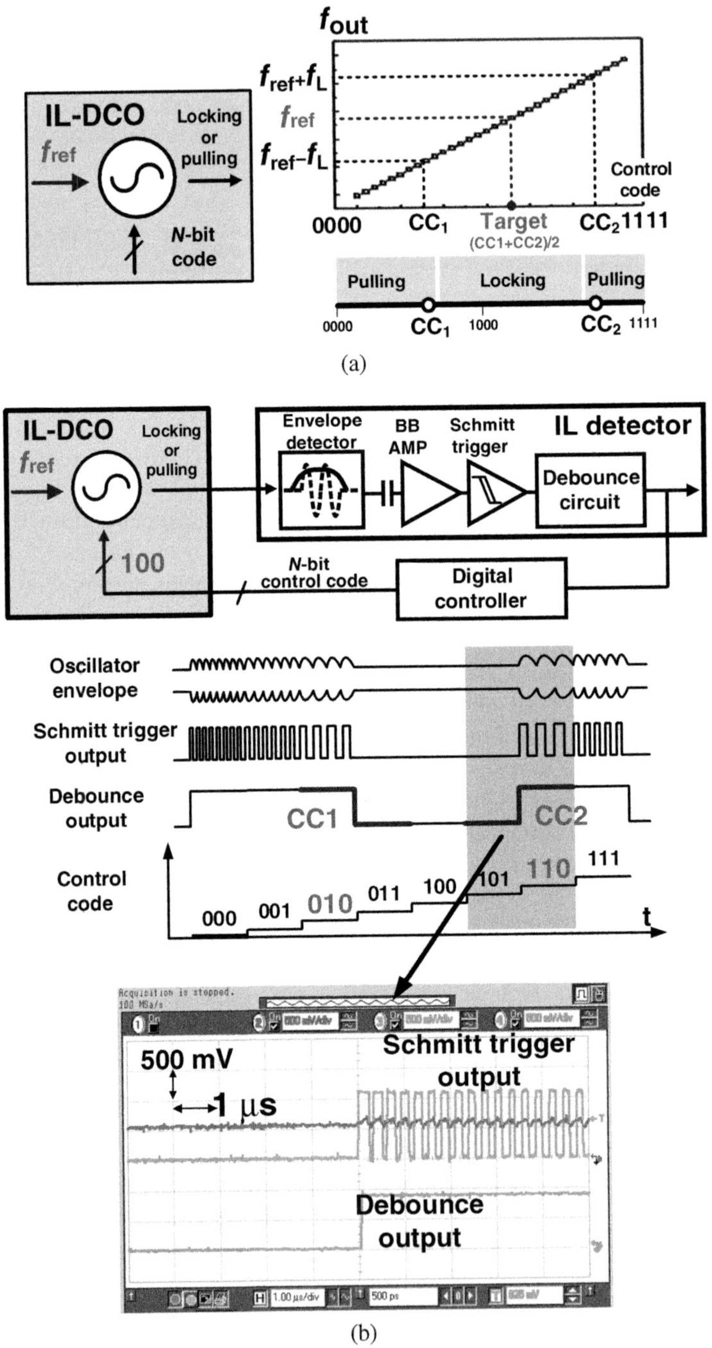

FIGURE 11.3 Frequency calibration technique: (a) calibration method, and (b) circuit implementation with operation waveform. *Source:* Bae et al. 2012 [22]. Reproduced with permission of IEEE.

operation example and measured waveform. While sweeping the control code from 000 to 111, the calibration controller reads the output of the IL detector and determines the control codes CC_1 and CC_2 closest to boundaries between injection locking and pulling. Finally, CC_1 and CC_2 are averaged by adding and shifting operations in order to correct the control code of DCO to the desired reference frequency. Furthermore, in order to quickly detect CC_1 and CC_2, a binary search algorithm could be exploited by the successive approximation register (SAR) digital controller. In brief, by taking advantage of IL-DCO and the reference signal from the hub, the sensor nodes need not employ a bulky and expensive XO, and the frequency drift of the DCO can be adjusted by finding the edges of the locking range.

11.3 ENERGY-EFFICIENT MBAN HUB DESIGN

11.3.1 Concurrent Dual-Band Operation

The MICS regulation established by the FCC partitions the spectrum from 402 to 405 MHz into 300 kHz channels [11]. These 10 channels are used under the listen-before-talk (LBT) protocol to avoid possible interference with primary users, such as aeronautical radio navigation and earth exploring satellites. The effective isotropic radiated power (EIRP) of a MICS transmitter must not exceed 25 μW or −16 dBm. On the other hand, the 30–70 MHz band is utilized for BCC in order to guarantee 10^{-6} BER (Bit Error Rate) based on the body channel analysis [16, 17]. FSK (Frequency Shift Keying) modulation is employed in both the BCC and MICS channels, since it has better spectral efficiency than ASK (Amplitude Shift Keying) and requires less hardware complexity than PSK (Phase Shift Keying). The MICS regulation stipulates the LBT protocol as shown in Figure 11.4a. Initially, the 10 channels in the MICS band are scanned, and after the most undisturbed two channels among them are determined for communication, the data transmission can be started. Therefore, in the conventional mode switching between MICS and BCC, the total energy consumption of the transceiver is the summation of the energy for both BCC and MICS following the LBT protocol. In the proposed system, the BCC and MICS dual-band signals are received concurrently, and the energy consumed for the LBT protocol and the MICS data reception is recycled by sharing the front-end circuits between the MICS and BCC receivers. Provided that the data reception periods for BCC and MICS are matched, a 30% energy reduction can be achieved by concurrent operation.

The on-body antenna needs special consideration in its design to support dual-band concurrent operation [19]. As an electrode for BCC, the radiating body of the antenna should be attached on the human skin securely to minimize the contact impedance at the signal interface. Due to the 0.1–1 S/m conductivity (σ) and 40–60 relative permittivity (ε_r) of skin, the resonant frequency of the antenna is shifted to a lower value and its radiation efficiency could be degraded on the skin as compared to when the antenna is in the air. The antenna is designed and optimized through electromagnetic (EM) field simulations as shown in Figure 11.4b. The radiating body, which

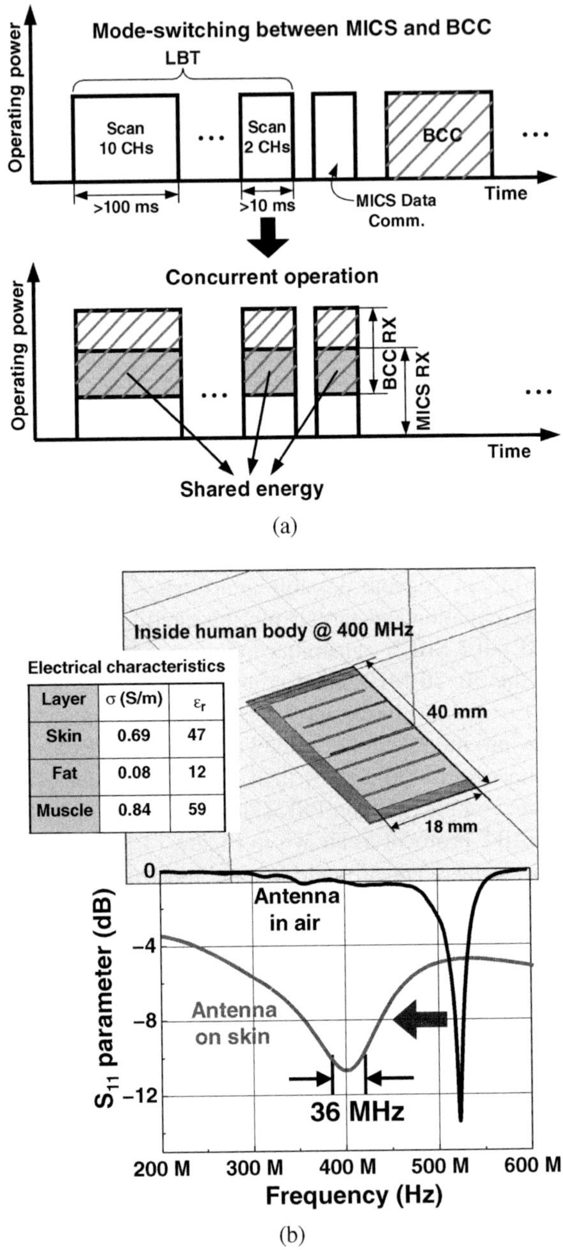

FIGURE 11.4 Concurrent dual-band operation: (a) energy saving profile, (b) antenna–electrode interface and measured return loss on human body and in air. *Source:* Cho et al. 2009 [18]. Reproduced with permission of IEEE.

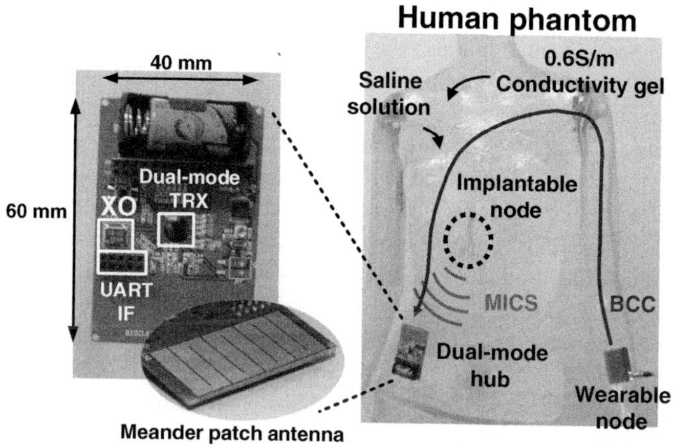

(a)

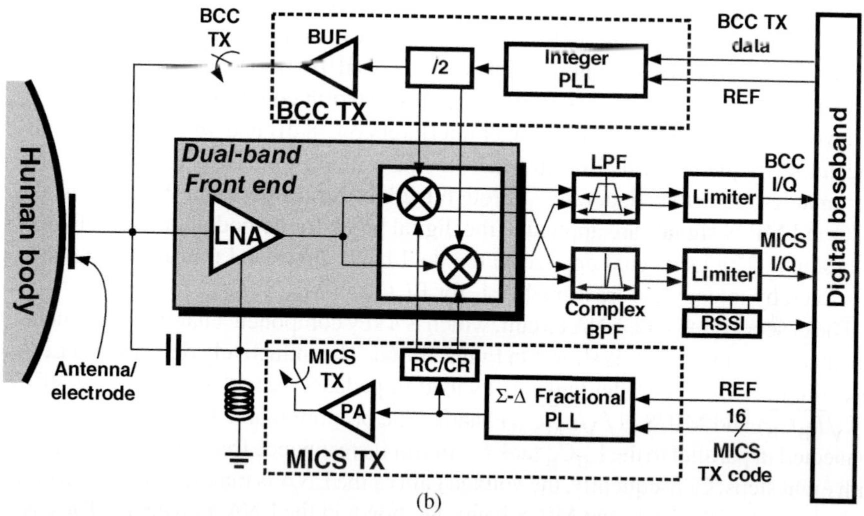

(b)

FIGURE 11.5 Dual-band hub: (a) board implementation, (b) transceiver architecture. *Source:* Cho et al. 2009 [18]. Reproduced with permission of IEEE.

makes direct contact with the skin, is covered with a 100 μm photo solder resist (PSR) layer to prevent severe degradation of the antenna matching performance. The measured return loss of the antenna in the air and on the skin is shown at the bottom of Figure 11.4b.

11.3.2 Dual-Band Hub Transceiver

Figure 11.5a shows the MBAN system concept with the dual-band hub board (40 mm × 60 mm) consisting of a dual-band transceiver chip, XO, and a UART (Universal Asynchronous Receiver/Transmitter) interface to transfer the received signal to other devices, such as a mobile smart device or laptop PC, which could be connected to the Internet infrastructure. The human phantom is prepared to mimic not only the in-body channel but also the on-body channel and verify the concurrent operation of BCC and MICS. The human phantom is filled with saline solution, which is often used to simulate the in-body environment, while its outer surface is covered with conducting gel that gives it the conductivity of human tissue, which is around 0.6 S/m. The wearable node is placed on the phantom surface, and the implantable node is coated with silicone for waterproofing and placed in the saline solution filling the phantom. The BCC signal is electrically coupled to the electrode of the hub transceiver through the conductive layer, whereas the MICS signal in the form of an electromagnetic wave propagates through the saline solution to the antenna on the hub transceiver. The antenna combines with the electrode interface on the hub for concurrent reception of both BCC and MICS signals.

Figure 11.5b shows a block diagram of the MBAN hub transceiver, which consists of two distinct chains for dual-band operation. The front-end circuits, which consume most of the power, are shared between BCC and MICS. The MICS band antenna attached to the human body also functions as an electrode that couples the BCC signal. Following the dual-band front-ends, the BCC and MICS signals are down-converted to zero and a low intermediate frequency (IF) of 450 kHz concurrently. The baseband filter for the MICS is a fifth-order Butterworth complex band-pass filter to remove the image and adjacent channel interferers by 38 dB and 48 dB, respectively. The cutoff frequency of the fourth-order low-pass filter in the BCC baseband is programmable for various data rates. Through the baseband filters and the limiters, the BCC and MICS signals are applied to the digital block for demodulation and channel assessment. The LO signal applied to the dual-band mixer and transmitter signal is generated by integer and fractional PLL for BCC and MICS, respectively.

The cascoded LC tank LNA circuit, which is a key component enabling the concurrent dual-band operation, is shown in Figure 11.6a. The inductively degenerated LNA cascoding the 2-LC parallel tanks as its load impedance provides gains in the BCC ($1/\sqrt{L_B C_B}$) and MICS ($1/\sqrt{L_M C_M}$) bands. The programmable R_{array} and C_{array} are connected in parallel to the $L_B C_B$ tank to control its resonance frequency and Q-factor with eight steps. Consequently, the voltage gain of the LNA is maintained at 23 dB for each channel of the BCC and MICS bands as shown in the LNA gain curve. The cascoded LC tanks also have an additional series resonance ($1/\sqrt{L_M || L_B(C_B + C_M)}$), which appears between the BCC and MICS bands around 200 MHz. This series resonance attenuates possible interferences at the dual-band receiver, such as FM radio signals, relaxing the linearity requirement of the following circuits.

The BCC and MICS signals amplified by the LNA need to be applied to separate down-conversion mixers driven by different LOs. Figure 11.6b shows the circuit topology which makes the two mixers operate concurrently while consuming no extra current. The single-balanced active mixer shifts the frequency band of the MICS signal to 450 kHz with 10 dB gain. Abrupt switching of M_2 and M_3 is important for higher conversion gain and less noise injection in the output. Therefore, it is common to provide half of the bias current I_B of M_1 externally to reduce the current commutated by the switching transistors. In this work, the extra current is recycled to activate a single-to-differential buffer driving the passive BCC mixer. The $R_C C_C$ low-pass filter decouples the two mixers by attenuating the >200 mV signal fluctuation on the

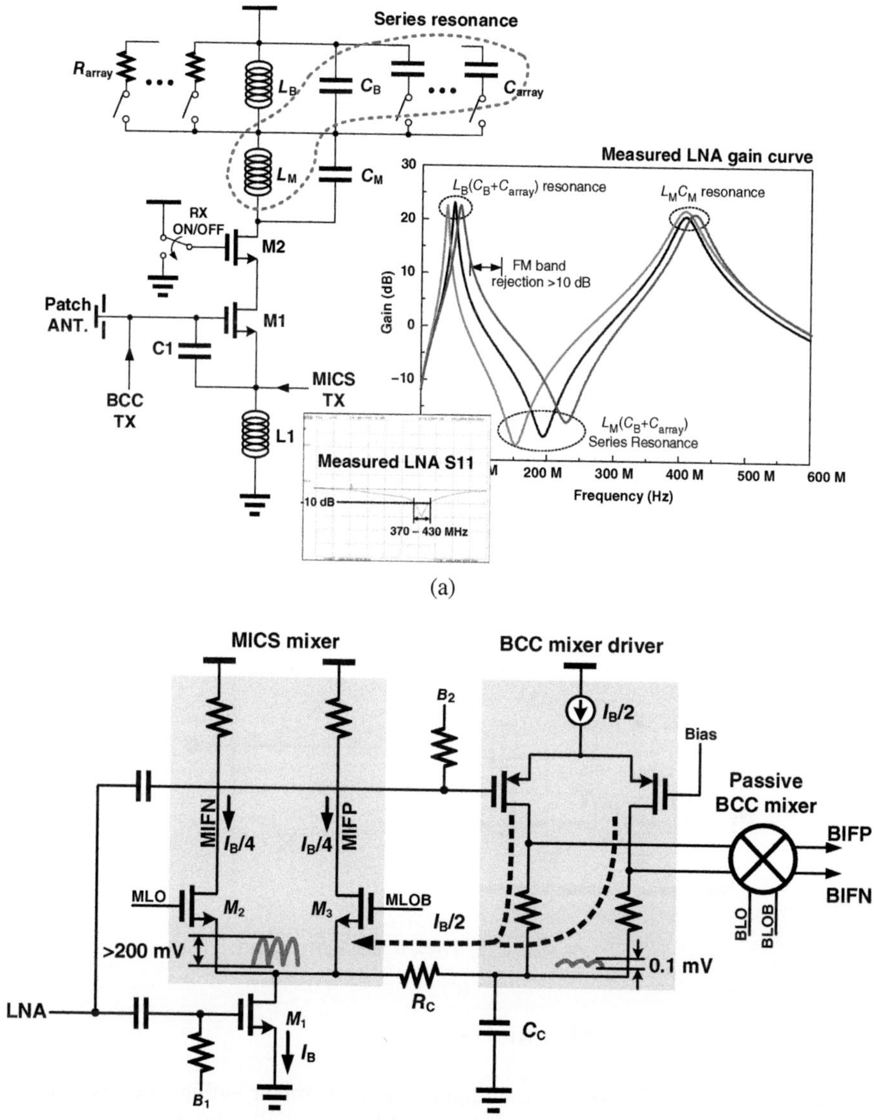

FIGURE 11.6 Building blocks for dual-band hub transceiver: (a) cascoded LC tank LNA, (b) current-recycling dual-band mixer, (c) unit stage of the fifth-order complex band-pass baseband filter for MICS, (d) biquad of the fourth-order low-pass baseband filter for BCC, and (e) I/Q LO generation circuit. *Source:* Cho et al. 2009 [18]. Reproduced with permission of IEEE.

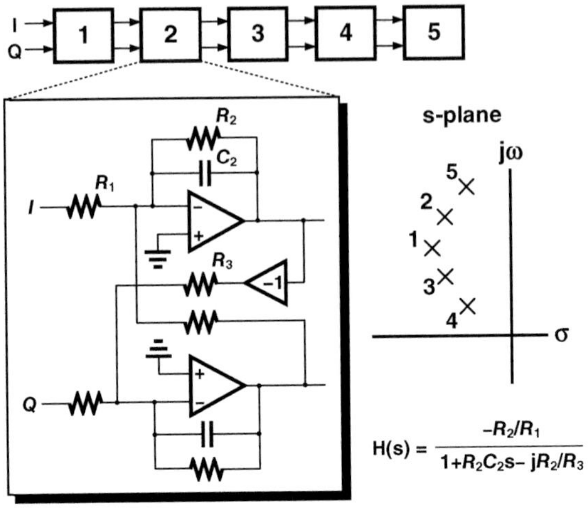

(c)

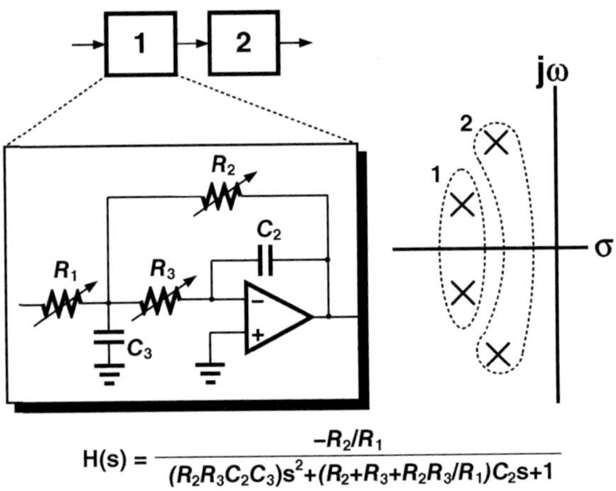

(d)

FIGURE 11.6 *(Continued)*

M_1 drain to less than 0.1 mV at the output of the BCC mixer buffer. The resistance R_C should be much higher than $1/g_m$ of M_2 and M_3 to minimize the amount of signal current lost through R_C after amplification by M_1. However, R_C cannot be arbitrarily large because the output swing level of the BCC buffer is significantly limited owing to the voltage drop of $R_C I_B/2$. Due to this tradeoff, the resistance of R_C is chosen to be 2.5 kΩ, which is seven times larger than $1/g_m$ of M_2. The BCC buffer with 15 dB gain compensates for the loss incurred by passive mixing. In addition, the passive mixer is applicable to the direct-conversion RX for BCC since it introduces little flicker noise. According to simulations using two tones from the BCC and MICS bands, the concurrent operation has no effect on the input-referred noise of each mixer, and <2 dB degradation of P_{1dB} is observed.

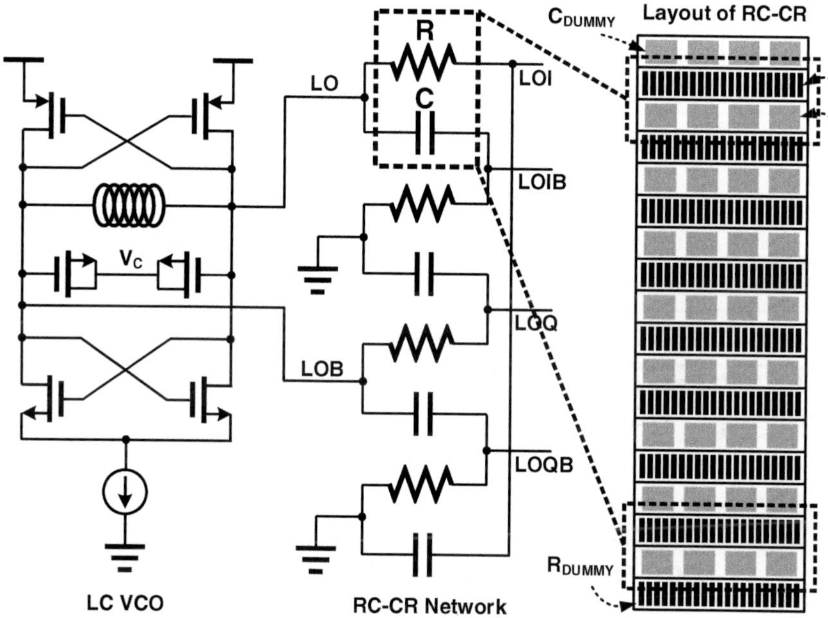

(e)

FIGURE 11.6 (*Continued*)

After the LNA/mixer cascade, the BCC and MICS baseband signals go through a fourth-order low-pass filter and a fifth-order complex band-pass filter, respectively, for channel selection. Schematics of the unit stages constituting these two baseband filters are shown in Figures 11.6c and 11.6d. For simplicity, all circuits are represented in single-ended form, though their actual implementations are fully differential. The complex integrator of Figure 11.6c generates a complex pole at $-1/R_2C_2 + j/R_3C_2$. Because there is no conjugate pole in the third quadrant of the *s*-plane, the image signal, which has negative frequency after down-conversion, can be filtered out. The poles of the cascaded integrators are selected to get a Butterworth filter response. The filter gain of 9 dB is distributed to the first two stages for minimum input-referred noise. The BCC signal down-converted to zero-IF does not suffer from the image issue, but the signal bandwidth can be adjusted for the variable data rates. The biquad of Figure 11.6d has resistor arrays programmable with 4 bits to move the conjugate poles and, hence, control the cutoff frequency of the filter. The unit gain frequency of the op-amp is set to 400 MHz to guarantee a >70° phase margin with all configurations of the resistor arrays.

The image rejection of a complex filter is very sensitive to gain and phase errors of the I/Q paths. The image rejection of 70 dB with perfect I/Q matching is reduced to 30 dB with just a 3° phase or a 1 dB gain imbalance. Figure 11.6e shows the I/Q LO generation circuits utilized in the transceiver. In contrast to using a divide-by-2 circuit or a quadrature VCO, adding an RC–CR network after the LC VCO requires

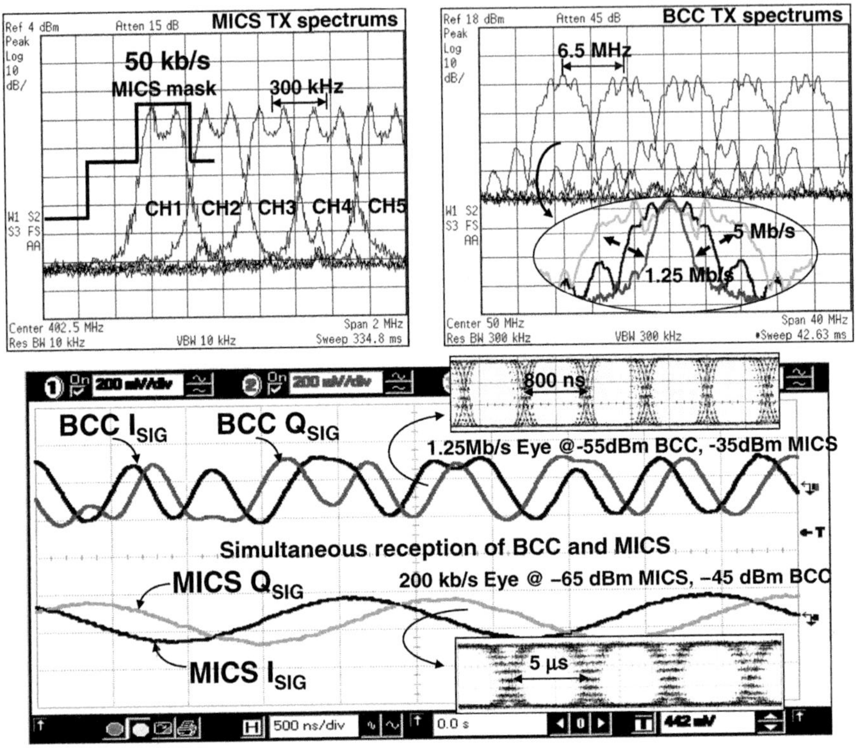

FIGURE 11.7 Measurement results for the dual-band hub transceiver. *Source:* Cho et al. 2009 [18]. Reproduced with permission of IEEE.

much less operating power if the Q-factor of the LC tank is sufficiently high. The I/Q accuracy of the circuit depends on the harmonic components of the LO signal generated by the oscillator and on the matching among the resistors and capacitors employed. The off-chip inductor with high Q (~15) is driven by both the nMOS and pMOS cross-coupled latches to generate a sinusoid with low distortion. As a result, the output of the LC VCO has the second and third harmonics below 33 and 35 dBc, which contribute to a one degree I/Q phase error. All resistors and capacitors constituting the RC–CR network are split and placed to compensate for the linear gradient.

Figure 11.7 shows the measurement results of the dual-band transceiver. The TX spectra in the MICS band with a 50 kb/s data rate show 300 kHz separation among their center frequencies and satisfy the FCC MICS regulation mask, which must be less than the −16 dBm EIRP limit. The spectra of the BCC signals modulated by 2.5 Mb/s data within the 30–70 MHz band are also shown with various data rates from 1.25 to 5 Mb/s. The time domain waveforms show the BCC and MICS baseband signal down-converted to zero-IF and low IF simultaneously. The eye diagrams of the demodulated data of 1.25 Mb/s in BCC and 200 kb/s in MICS show that the BCC signal with 20 dB larger MICS signal or vice versa can be amplified, down-converted, and filtered without distortion by the receiver chain. The dual-band hub transceiver chip is fabricated on a 0.18 μm CMOS process and dissipates 10.8 mW [18].

11.4 COMPACT SENSOR NODE DESIGNS

11.4.1 Implantable Sensor Node Transceiver

Since form factor is the most important consideration for implant devices, we focus on reducing the board size and cost for miniaturized implantable sensor nodes as shown in Figure 11.8a. First, we design the implantable sensor board assembled with the MICS transceiver chip, a micro controller unit (MCU) chip, and a commercial temperature sensor. After a functional verification, we further decrease the size of the implantable sensor by means of exploiting SiP technology. The chip array ball grid array (CABGA) type SiP module has a size of 10 mm × 10 mm, less than 2 mm thickness, and 100 balls containing 2 bare dies (the transceiver and MCU chips) and 1 temperature sensor package with 50 Ω line impedance. As a result, the size of the implantable node can be reduced from 20 mm × 20 mm to 15 mm × 15 mm without the use of a bulky XO, thanks to the IL-DCO frequency calibration technique.

The implantable MICS transceiver architecture shown in Figure 11.8b is composed of an RF front-end, an IF amplifier, a PLL-based demodulator, a direct-modulating transmitter, and an IL detector with a calibration controller. In the transmitter, FSK data is transmitted by directly modulating the DCO with the digital code from the MCU. During the calibration period, the IL-DCO replaces a frequency synthesizer, eliminating the need for a XO, which reduces the power, volume, cost, and transceiver startup time. In the receiver, a Q-enhancement LNA removes the interference at the RX input without using a polyphase filter for image rejection and therefore without the in-phase and quadrature-phase (I/Q) signal path in the receiver chain, mitigating the linearity/power requirements of the subsequent circuits. The sampling mixer down-converts the MICS signal to 2 MHz IF signal. Through the IF filter and limiter, the baseband signal is applied to the PLL-based demodulator to detect the received signal with only the single signal path without the I/Q path. The low power consumption of the transceiver is mainly caused by the Q-enhancement LNA, which eliminates the interferers to meet the required selectivity.

Figure 11.9a shows the Q-enhancement LNA with implantable antenna. As shown in the S-parameter (S11) curve, the L_sC_g tank is used for 50 Ω impedance matching at the input of the LNA and reused for the load impedance to drive the antenna at the output of the power amplifier (PA). The implantable antenna is designed through EM field simulations using the conductivity (σ = 1.2 S/m) and dielectric constant (ε_r = 80) of the saline solution. The antenna is covered with a PSR layer to prevent performance degradation. We measure the return loss of the silicone-coated antenna in the air and in the saline solution as shown at the bottom of Figure 11.9a. The L_DC_{array} tank controls its resonant frequency with a 7-bit coded capacitor bank, and its positive feedback loop enhances the Q-factor of the tank. The 6-bit DAC current bias determines the amount of feedback, and the measured Q-factor of the resonator in precise current control shows that the Q-factor is of the order of 1000. As a result,

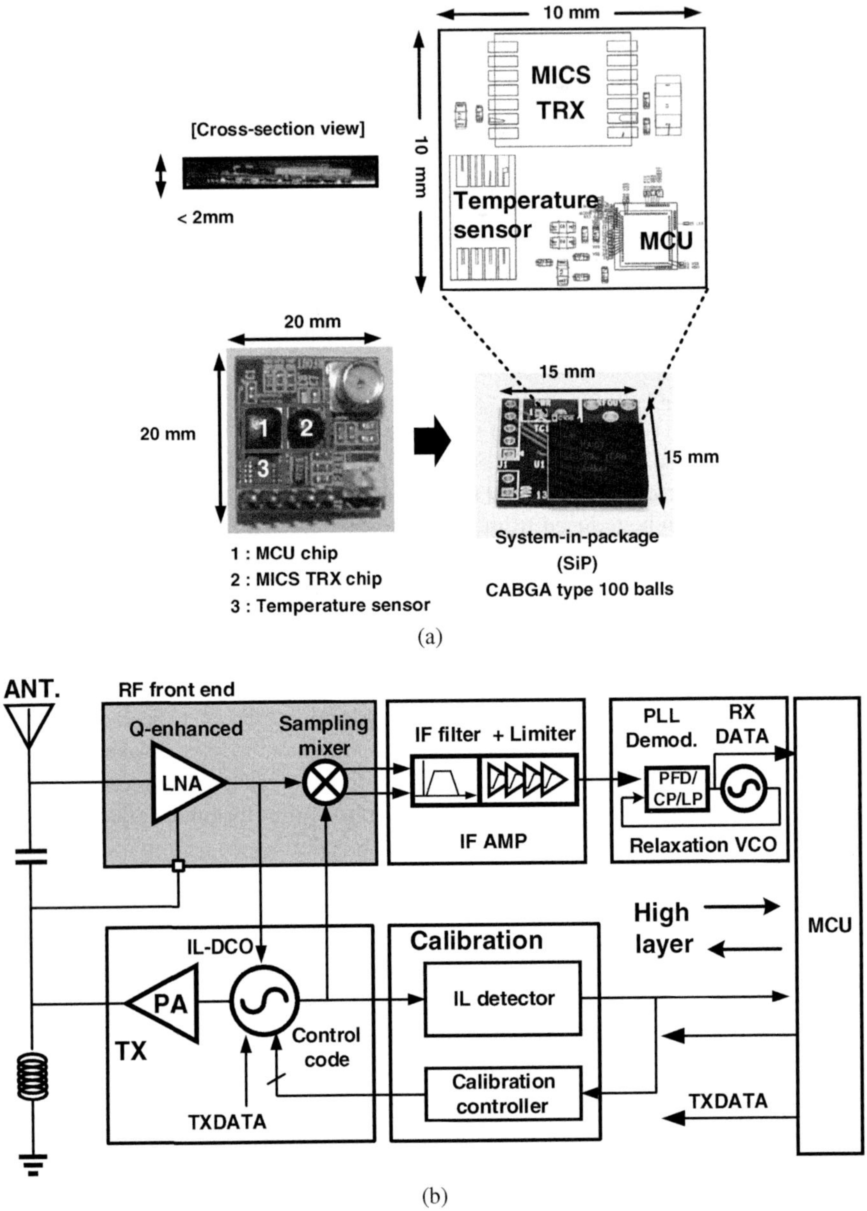

FIGURE 11.8 Implantable sensor node: (a) board implementation and (b) transceiver architecture.

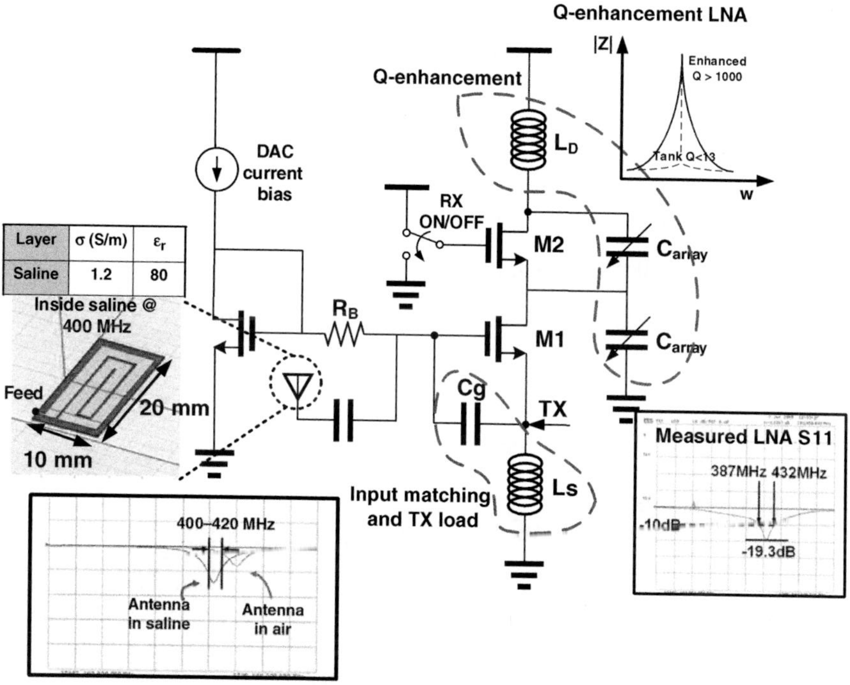

(a)

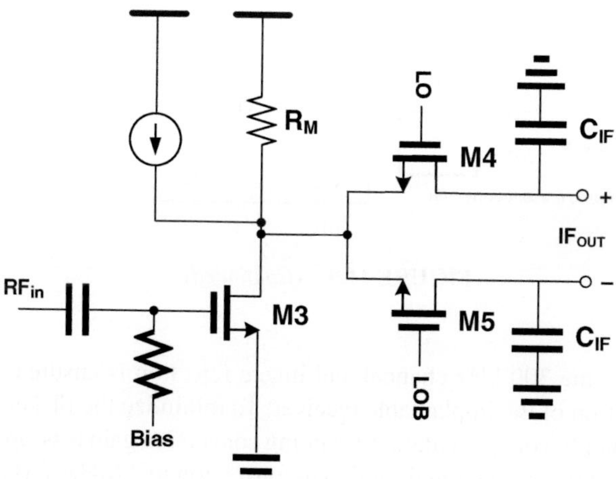

(b)

FIGURE 11.9 Building blocks for implantable node transceiver: (a) Q-enhancement LNA, (b) folded-sampling mixer, (c) IF amplifier, (d) relaxation oscillator in PLL demodulator, and (e) transmitter circuit. *Source:* Bae et al. 2009 [22]. Reproduced with permission of IEEE.

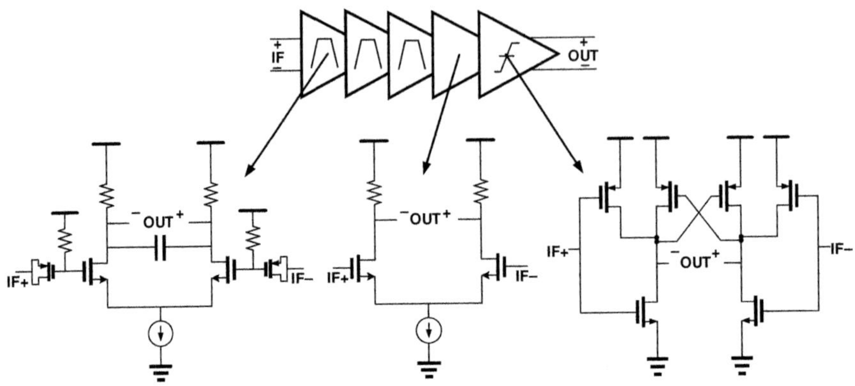

(c)

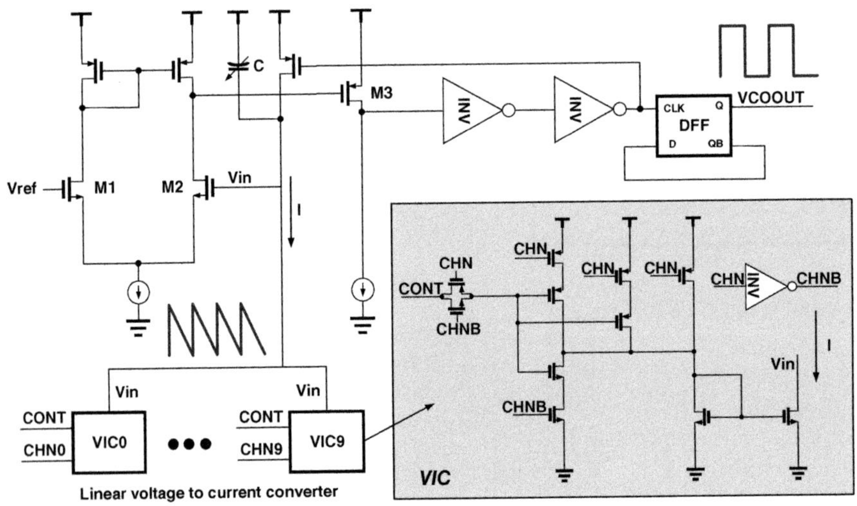

(d)

FIGURE 11.9 (*Continued*)

the selectivity of the 300 kHz channel and image rejection is ensured, relaxing the power consumption of the implantable receiver. To minimize the flicker noise and to operate in low supply voltage without sacrificing conversion gain loss, an AC-coupled folded sampling mixer is exploited for down-conversion to 2 MHz IF (Figure 11.9b).

The IF amplifier in Figure 11.9c provides gain over the IF bandwidth from 1 to 3 MHz and comprises a chain of four simple differential-pair gain stages and one level shifter. The first three stages are implemented with an AC-coupled band-pass topology to mitigate input common mode range for low supply voltage and to remove the high frequency signal from the mixer. The use of resistive loads simplifies biasing and provides reasonable gain with minimal voltage headroom. The last stage converts the analog IF signal to a digital baseband signal, which is applied to the PLL-like

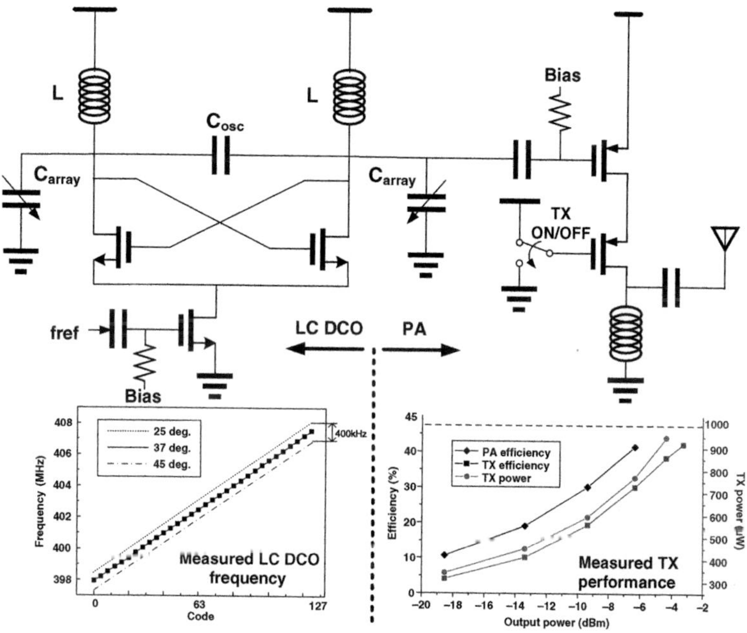

(e)

FIGURE 11.9 (*Continued*)

FSK demodulator. Since the frequency deviation of the received signal is relatively small, a relaxation oscillator (Figure 11.9d) is used to build the low gain VCO with low power consumption. To control the frequency of the oscillator, the current is linearly controlled to charge and discharge the capacitor *C*. The 10 voltage-to-current converters (VICs) are utilized to convert the voltage output of the loop filter (CONT) into a proportional current for each channel. After the VIC generates a linear current, the comparator converts the negative voltage ramp of the Vin node into a square wave. To operate in low supply voltage, transistors M_1, M_2, and M_3 are biased in the subthreshold region.

Figure 11.9e shows the transmitter, which consists of an LC DCO and a PA. The DCO employs a differential negative-G_m oscillator with a programmable capacitor array and an inductor. The PA uses a cascode transistor to improve the isolation between the antenna and the DCO and employs an LC tank to boost its load impedance. The DCO replaces the need for a frequency synthesizer with a wireless reference signal from the hub combined with the frequency calibration technique, eliminating the use of a bulky XO. The DCO frequency can be tuned to any desired frequency over the range from 398 to 408 MHz with 10 kHz accuracy by using a thermometer-coded capacitor bank which corrects for a drift of about 400 kHz over a 20 °C temperature range. The DCO phase noise is −118 dBc/Hz at a 1 MHz offset. The transmitter consumes 400 μW at an output power of −16 dBm. The output power level can be raised to −2 dBm by adjusting the bias current of the PA to compensate for the antenna loss or loss in the body.

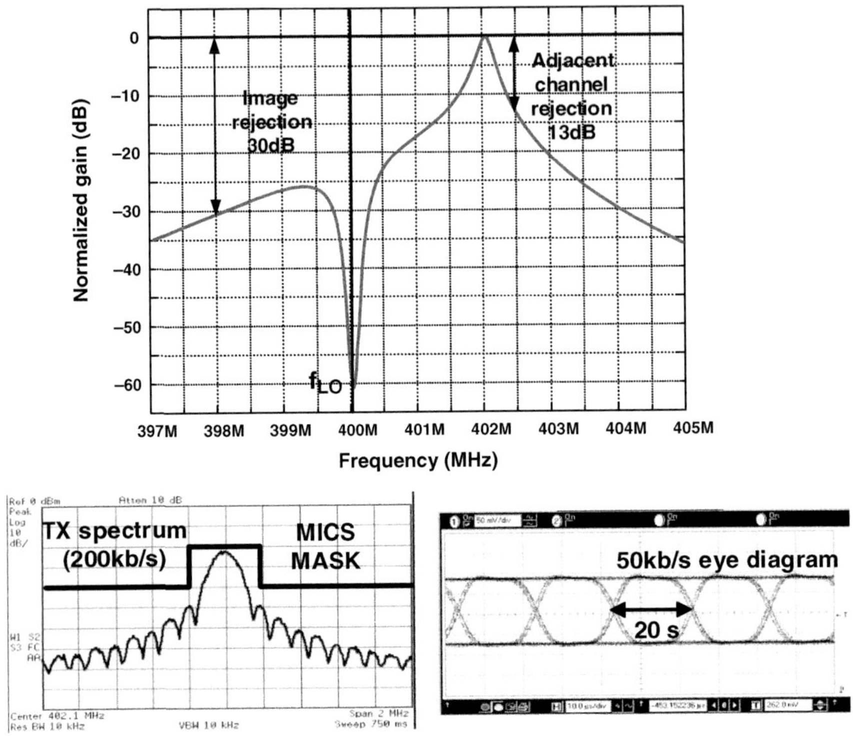

FIGURE 11.10 Measurement results for the implantable sensor node transceiver. *Source:* Bae et al. 2009 [22]. Reproduced with permission of IEEE.

The normalized gain of the receiver chain to baseband is shown in Figure 11.10, revealing that the Q-enhancement LNA provides an image and an adjacent channel rejection of more than 30 dB and 13 dB, respectively. This measurement is performed with a data rate of 50 kb/s for the down-link and 200 kb/s for the up-link using the human phantom. The data rates of the up-link and down-link can be increased up to 250 kb/s. According to the transmitter spectrum, the implantable node meets the MICS mask specification, and the output power level is below the −16 dBm EIRP limit. The MICS transceiver chip is fabricated on a 0.18 μm CMOS process and consumes 490 μW (RX) [24].

11.4.2 Wearable Sensor Node Transceiver

Figure 11.11a shows the compact wearable sensor node board (12 mm × 20 mm) with BCC transceiver chip. Compact sensor nodes can be achieved by using a highly integrated implementation of the transceiver chip. Moreover, the on-chip IL-DCO replaces the XO with the help of the reference signal from the hub for a crystal-less board design. Focusing on the low power consumption, the architecture of the wearable BCC node transceiver in Figure 11.11b is similar to the architecture of the BCC hub transceiver (Figure 11.5b), which employs the zero-IF receiver with passive

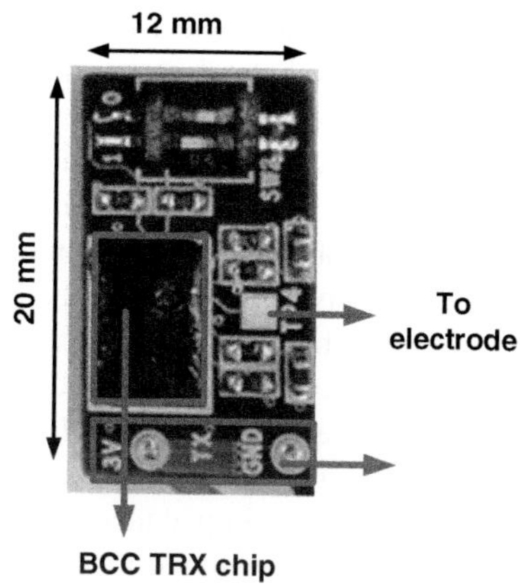

(a)

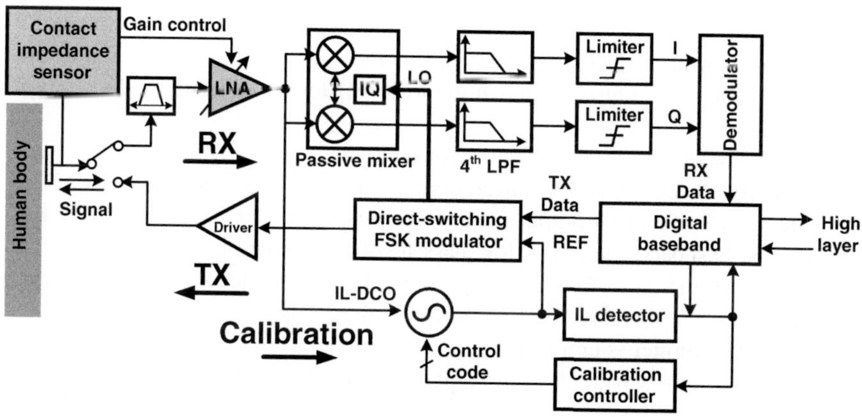

(b)

FIGURE 11.11 Wearable sensor node: (a) board implementation and (b) transceiver architecture.

I/Q mixer, fourth-order low-pass filters, and limiters. A single electrode (or vertical electrode) configuration rather than a two electrode (or horizontal electrode) configuration is adopted for both its simple interface and a high SNR at the input of the LNA [25]. In the transmitter, the binary data from the digital baseband is modulated by a direct switching modulator and applied to the human body through the single electrode. The calibration path with the IL-DCO, IL detector, and calibration controller provides the reference signal for the modulator in the transmitter as well as the LO signal in the direct conversion receiver. Furthermore, we employ a contact impedance sensor [15] to monitor the contact impedance between the electrode and the human body in order to improve the power efficiency of the LNA. Given that the contact impedance mainly determines the received signal quality into the LNA, we can optimize the power consumption of the LNA by adaptively changing its gain, noise figure, and linearity.

Figure 11.12a shows the schematic of the contact impedance aware LNA along with measurement results illustrating the channel variation in terms of the contact impedance variation using a circular electrode with a 1.5 cm diameter. The contact impedance between the electrode and the human body varies dynamically, resulting in up to 30 dB overall observed signal path loss variation when the electrode is in contact with or apart from the body. Such a dynamic variation makes the receiver consume significant power because the receiver should simultaneously satisfy not only the high linearity requirement in Case 1 but also the high sensitivity requirement in Case 2, as indicated in Figure 11.12a. To compensate for body channel quality degradation and to mitigate the additional linearity and sensitivity requirement caused by the contact impedance variation, we employ a contact impedance sensor to adjust the LNA gain control. By means of detecting the impedance variation, the control code for the operation mode of the LNA can be determined for better power efficiency. As a result, the receiver need not satisfy the high linearity and high noise performance requirements simultaneously, which significantly reduces the circuit power dissipated by the receiver. The LNA is composed of a parallel combination of common source (CS) and common gate (CG) stages to convert the single-ended signal to a differential signal. The gain control switch at the load resistors is adjusted by the contact impedance sensor.

The filtered baseband signal is digitized through the limiter (Figure 11.12b), which is a direct-coupled cascade of five differential amplifiers. Due to the high gain of the limiter, the duty cycle of its output is very sensitive to the DC offset added by the mixer and the baseband filter. The low-pass feedback formed by the RC filter can reject up to a 400 mV DC offset, maintaining a 48% duty cycle. The high-pass filter characteristics of the signal path also remove the flicker noise accompanied with the input signal. The limiter is employed as a received signal strength indicator (RSSI) with five rectifiers for a wide dynamic range obtained by adding the different input power ranges from the rectifiers. Along with the contact impedance sensor, the RSSI is useful for the LNA gain control.

Figure 11.12c shows the direct-switching modulator with two frequency synthesizers. The internal PLLs generate two consecutive frequencies, $FS1_{OUT}$ and $FS2_{OUT}$, representing binary data, and one of them is switched out to FS_{OUT}. The modulator can change the frequency without disturbing the VCO control voltage or the division ratio of the PLL that alters its loop dynamics. Hence, the loop bandwidth can be as large as possible, minimizing the frequency switching time without degrading the modulation speed. In addition, there is no need for DC-free coding of the data modulation and calibration of the VCO. Since FSK is performed by direct switching with <500 ps delay, the target data rate of 5 Mb/s can be easily obtained. The divide-by-4 unit following the switch quadruples the internal frequency separation and also the reference frequency separation. The 4× reference signal from the calibrated IL-DCO reduces the PLL's band switching time by a factor of 4. The frequency division also compensates for the phase noise and the reference spur increment due to the large loop bandwidth. Since the required frequency is under 100 MHz, the power overhead due to the two PLLs can be controlled under 1 mW. The output phase of the two integer-N PLLs using a common reference (REF CLK) are aligned at every rising

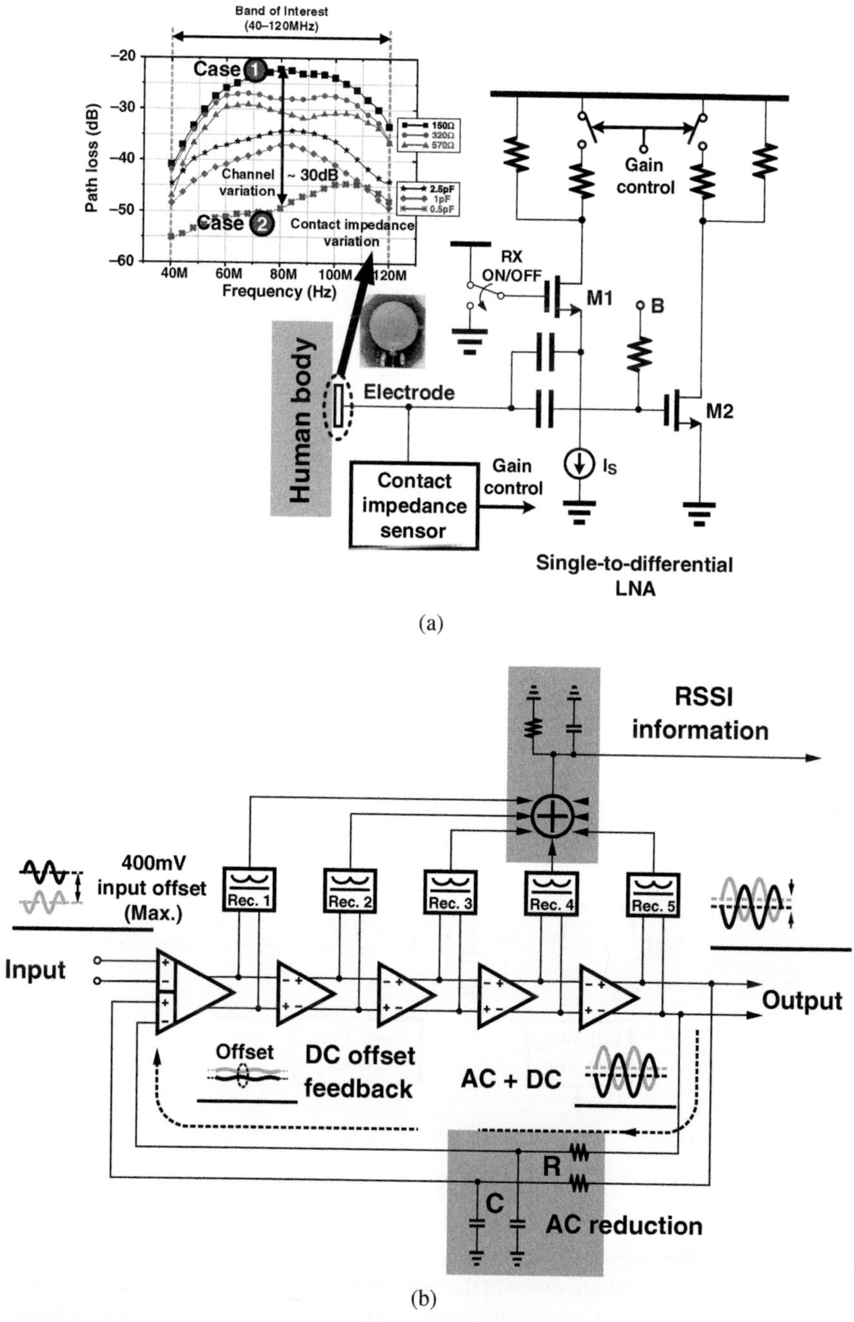

FIGURE 11.12 Building blocks for wearable node transceiver: (a) contact impedance aware LNA, (b) limiter and RSSI, (c) direct-switching FSK modulator, and (d) low power integer-N PLL. *Source:* Cho et al. 2009 [14]. Reproduced with permission of IEEE.

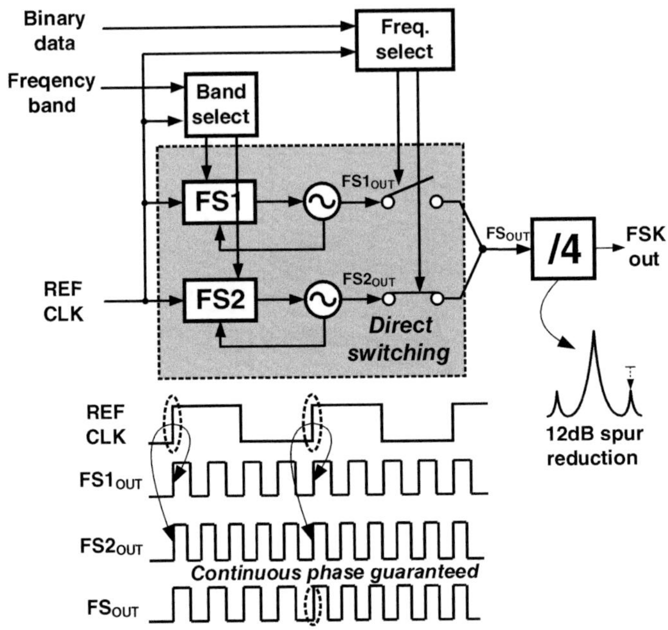

(c)

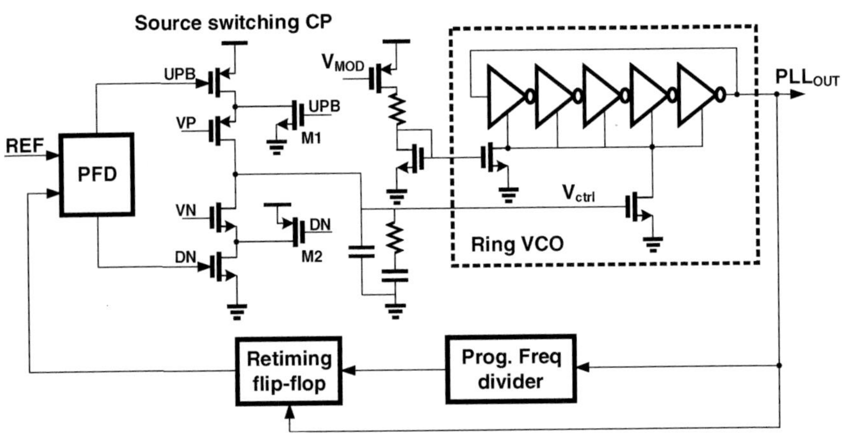

(d)

FIGURE 11.12 (*Continued*)

edge of the reference signal as shown in the waveform of Figure 11.12c. If these two signals are switched simultaneously at the synchronization edge, the phase continuity of the resulting FSK signal can be guaranteed. However, in the real implementation of the direct-switching modulator, a phase discontinuity may occur at the instant of frequency change because of the timing mismatch between the two PLL outputs. The timing mismatch mainly comes from the current mismatch in the charge pumps and the logic delay of the programmable dividers. In order to compensate for these, a fast-switching charge pump and a retiming flip-flop are designed in the frequency synthesizer as shown in Figure 11.12d. Cascode biasing is used in the charge pump to reduce the mismatch between the up and down currents when V_{ctrl} changes. Transistors M_1 and M_2 are added to improve the switching time by providing a discharge path for the pumping currents. The flip-flop in the feedback path cancels the delay of the programmable divider by retiming the divided signal with PLL_{OUT}. SPICE simulation confirms that the timing mismatch between two PLLs is reduced to less than 50 ps under process variations. The ring-type VCO is chosen for its wide tuning range. To increase the tuning range further without degrading noise sensitivity, switching capacitors are placed on the intermediate nodes of the delay chain.

The wearable BCC node transceiver is designed and fabricated on a 0.18 μm CMOS process with 2.4 mW power consumption. The receiver eye diagram at a data rate of 5 Mb/s and the transmitter spectrum at 1.25 Mb/s are shown in Figure 11.13. The noise figure (sensitivity performance) and 1 dB compression power (linearity

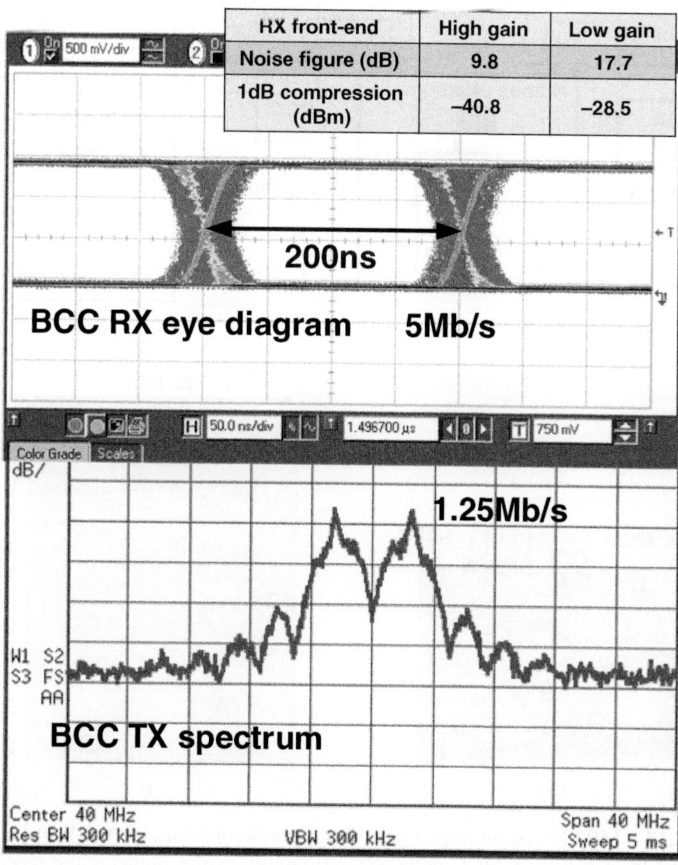

FIGURE 11.13 Measurement results for the wearable sensor node transceiver.

performance) of the receiver chain before the limiter circuit can be varied from 9.8 to 17.7 dB and from −40.8 to −28.5 dBm, respectively. Thanks to the contact impedance monitoring and the LNA gain control, the LNA does not have to fulfill both the high linearity and high noise performance simultaneously.

11.5 SYSTEM IMPLEMENTATION

Figure 11.14a shows the proposed MBAN system with the implemented crystal-less compact wearable and implantable nodes and the dual-band hub node. Although the MBAN system using on-body communication is feasible to field test, the *in vivo* configurations for in-body communication is beyond the bounds of this work. Therefore, the operation of the MBAN system is evaluated using a human phantom filled with saline solution to simulate the in-body environment as shown in Figure 11.5a.

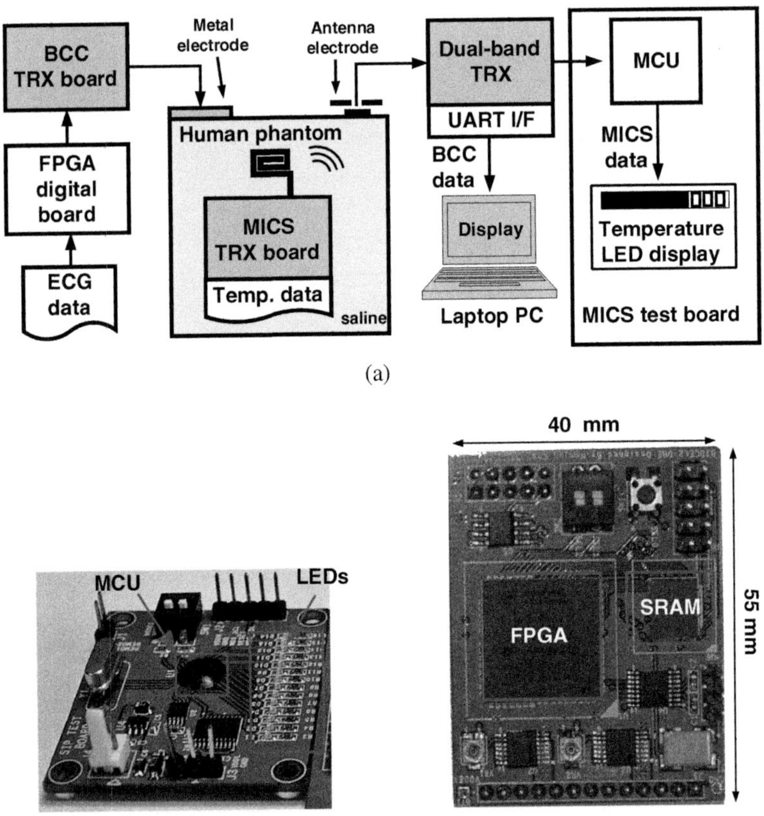

(b) (c)

FIGURE 11.14 MBAN system implementation: (a) system setup, (b) test board for MICS node, (c) test FPGA board for BCC node, and (d) laptop display showing the received raw ECG data.

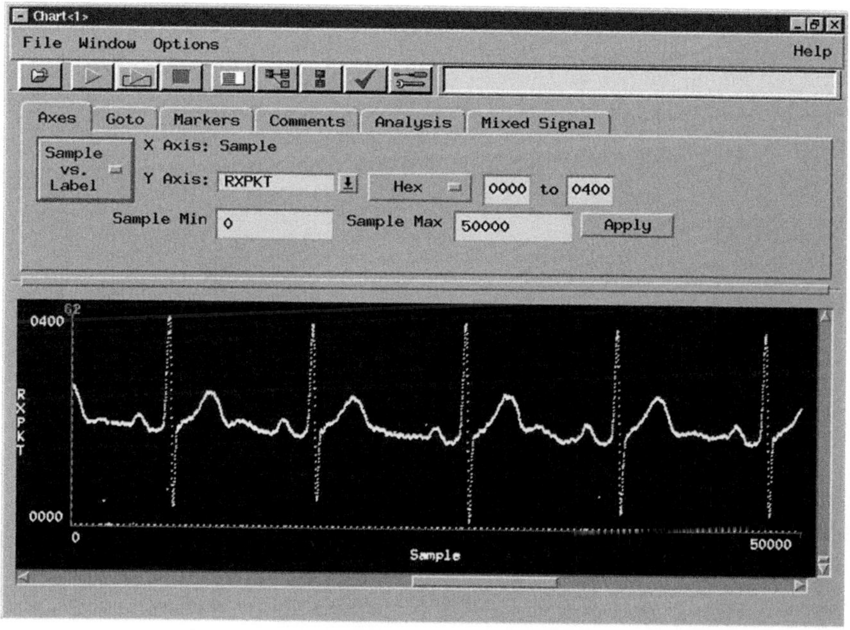

FIGURE 11.14 (*Continued*)

The dual-band hub board is powered by a 1000 mAh capacity, 1/2-AA-sized battery, whereas the power for the wearable and implantable node boards is supplied by 130 mAh capacity, 16 mm sized coin batteries. The implant node with temperature sensor sends its data through the implantable antenna, and the dual-band transceiver recovers the raw data from the temperature sensor. To convert the raw data into a meaningful temperature value with a visual display, we designed an MICS test board with an MCU chip and LED gage as shown in Figure 11.14b.

Figure 11.14c is the FPGA board with SRAM to provide the raw ECG data to the BCC transceiver board. Although this version of an ECG sensor is implemented as a separate FPGA board, future versions will integrate the ECG sensor on the wearable node with the transceiver chip. The node transmitter generates the modulated data packet after encoding the raw ECG data. The packet data is transmitted through the channel and is received by the dual-band hub transceiver. Then, using a UART interface, the received raw ECG data is displayed on the monitor of a laptop PC as shown in Figure 11.14d.

11.6 CONCLUSION

In this chapter, we have described an energy-efficient MBAN system for mobile healthcare applications that has been implemented with crystal-less compact wearable and implantable sensor nodes. The frequency calibration method utilized effectively eliminates the XOs in the sensor nodes with the help of the reference signal from the hub. The low power circuit techniques for the sensors and hub transceiver were introduced in our previous research. The prototype MBAN system was implemented to verify the proposed system, and the successful measurement results were provided.

REFERENCES

[1] L. Atzori, A. Iera, and G. Morabito, "The internet of things: a survey," *Comput. Netw.*, vol. 54, no. 15, pp. 2787–2805, October 2010.

[2] A. Darwish and A. Hassanien, "Wearable and implantable wireless sensor network solutions for healthcare monitoring," *Sensors*, vol. 11, no. 6, pp. 5561–5595, June 2011.

[3] B. Latre, B. Braem, I. Moerman, C. Blondia, and P. Demeester, "A survey on wireless body area networks," *Wireless Netw.*, vol. 17, no. 1, pp. 118, January 2011.

[4] C. Doukas, T. Pliakas, and I. Maglogiannis, "Mobile Healthcare Information Management Utilizing Cloud Computing and Android OS," in Proceedings of the 32nd IEEE Engineering in Medicine and Biology Society Conference, Buenos Aires, September 2010, pp. 1037–1040.

[5] P. D. Bradley, "An Ultra-Low-Power High-Performance Medical Implant Communication System Transceiver for Implantable Devices," in Proceedings of the 2006 IEEE Biomedical Circuits and Systems Conference, London, December 2006, pp. 158–162.

[6] B. Zhen, H.B. Li, and R. Kohno. "IEEE body area networks and medical implant communications," in Proceedings of the ICST 3rd international conference on Body area networks (p. 26). ICST (Institute for Computer Sciences, Social-Informatics and Telecommunications Engineering), March 2008.

[7] J. Yoo, L. Yan, S. Lee, Y. Kim, H. Kim, B. Kim, and H.-J. Yoo, "A 5.2 mW self-configured wearable body sensor network controller and a 12 μW wirelessly powered sensor for a continuous health monitoring system," *IEEE J. Solid-State Circuits*, vol. 45, no. 1, pp. 178–188, January 2010.

[8] M. Yuce, "Implementation of wireless body area networks for healthcare systems," *Sens. Actuators A: Phys.*, vol. 162, no. 1, pp. 116–129, July 2010.

[9] R. F. Yazicioglu, T. Torfs, P. Merken, J. Penders, V. Leonov, R. Puers, B. Gyselinckx, and C. Van Hoof, "Ultra-low-power biopotential interfaces and their applications in wearable and implantable systems," *Microelectronics Journal*, vol. 40, no. 9, pp. 1313–1321, 2009.

[10] K. Y. Yazdandoost and K. Sayrafian-Pour, "Channel models for body area networks," IEEE P802.15 Working Group for Wireless Personal Area Networks, IEEE P802.15-08-0780-02-0006, November 2008.

[11] FCC Rules and Regulations, MICS Band Plan, Part 95, 2003.

[12] J. Gemio, J. Parron, and J. Soler, "Human body effects on implantable antennas for ISM bands applications: models comparison and propagation losses study," *Prog. Electromagn. Res.*, vol. 110, pp. 437–452, 2010.

[13] A. Fort, J. Ryckaert, C. Desset, P. D. Doncker, P. Wambacq, and L. V. Biesen, "Ultra-wideband channel model for communication around the human body," *IEEE J. Sel. Areas Commun.*, vol. 24, no. 4, pp. 927–933, April 2006.

[14] N. Cho, L. Yan, J. Bae, and H.-J. Yoo, "A 60 kb/s-10 Mb/s adaptive frequency hopping transceiver for interference-resilient body channel communication," *IEEE J. Solid-State Circuits*, vol. 44, no. 3, pp. 708–717, March 2009.

[15] J. Bae, K. Song, H. Lee, H. Cho, and H.-J. Yoo, "A 0.24 nJ/b wireless body-area-network transceiver with scalable double-FSK modulation," *IEEE J. Solid-State Circuits*, vol. 47, no. 1, pp. 310–322, January 2012.

[16] N. Cho, J. Yoo, S. Song, J. Lee, S. Jeon, and H.-J. Yoo, "Human body characteristics as a signal transmission medium for intrabody communication," *IEEE Trans. Microw. Theory Tech.*, vol. 55, no. 5, pp. 1080–1086, May 2007.

[17] J. Bae, H. Cho, K. Song, H. Lee, and H.-J. Yoo, "The signal transmission mechanism on the surface of human body for body channel communication," *IEEE Trans. Microw. Theory Tech.*, vol. 60, no. 3, pp. 582–593, March 2012.

[18] N. Cho, J. Bae, and H.-J. Yoo, "A 10.8 mW body channel communication/MICS dual-band transceiver for a unified body sensor network controller," *IEEE J. Solid-State Circuits*, vol. 44, no. 12, pp. 3459–3468, December 2009.

[19] N. Cho, T. Roh, J. Bae, and H.-J. Yoo, "A planar MICS band antenna combined with a body channel communication electrode for body sensor network," *IEEE Trans. Microw. Theory Tech.*, vol. 57, no. 10, pp. 2515–2522, October 2009.

[20] J. Bae, L. Yan, and H.-J. Yoo, "A low energy injection-locked FSK transceiver with frequency-to-amplitude conversion for body sensor applications," *IEEE J. Solid-State Circuits*, vol. 46, no. 4, pp. 928–937, April 2011.

[21] C. Angeloni, P. O. Riley, and D. E. Krebs, "Frequency content of whole body gait kinematic data," *IEEE Trans. Rehabil. Eng.*, vol. 2, no. 1, pp. 40–46, March 1994.

[22] J. Bae, K. Song, H. Lee, H. Cho, and H.-J. Yoo, "A low-energy crystal-less double-FSK sensor node transceiver for wireless body-area network," *IEEE J. Solid-State Circuits*, vol. 47, no. 11, pp. 2678–2692, November 2012.

[23] B. Razavi, "A study of injection locking and pulling in oscillators," *IEEE J. Solid-State Circuits*, vol. 39, no. 9, pp. 1415–1424, September 2004.

[24] J. Bae, N. Cho, and H.-J. Yoo, "A 490 μW Fully MICS Compatible FSK Transceiver for Implantable Devices," in Symposium on VLSI Circuits Digest of Technical Papers, 2009, pp. 36–37.

[25] J. Bae and H.-J. Yoo, "The effects of electrode configuration on body channel communication based on analysis of vertical and horizontal electric dipoles," *IEEE Trans. Microw. Theory Tech.*, vol. 63, no. 4, pp. 1409–1420, April 2015.

12

ULTRA LOW-POWER MEMS-BASED RADIOS FOR WBAN

RAGHAVASIMHAN THIRUNARAYANAN,[1] ARAVIND PRASAD HERAGU,[2] AND CHRISTIAN ENZ[1]

[1] *Integrated Circuits Laboratory (ICLAB), IMT Institute, École Polytechnique Fédérale de Lausanne, Neuchâtel, Switzerland*
[2] *Semtech Neuchâtel Sàrl, Neuchâtel, Switzerland*

It is in your power to withdraw yourself whenever you desire. Perfect tranquility within consists in the good ordering of the mind, the realm of your own. Think of the universal power, of which thou has a very small portion; and of universal time, of which a short and indivisible interval has been assigned to thee; and of that which is fixed by destiny, and how small a part of it thou art.

Marcus Aurelius, *Meditations*

12.1 INTRODUCTION TO BODY AREA NETWORKS

Non communicable diseases (NCD) such as cardiovascular disease, diabetes, and cancer are the leading causes of global mortality today, accounting for around two-thirds of total deaths per year [1]. Further, data show that nearly 80% of the deaths by the NCDs occur in middle and low income countries which impose a large economic and social burden on the already under-staffed health-care systems [2]. This is further exacerbated by the fact that the population of people above the age of 60 is expected to rise to 1.2 billion or 15 % of the global population by the year 2025 [3]. Since the treatment of NCDs is expensive, prevention of these diseases in people with high risk factors by better and more accurate early diagnostics becomes important.

Electromagnetics of Body-Area Networks: Antennas, Propagation, and RF Systems, First Edition.
Edited by Douglas H. Werner and Zhi Hao Jiang.

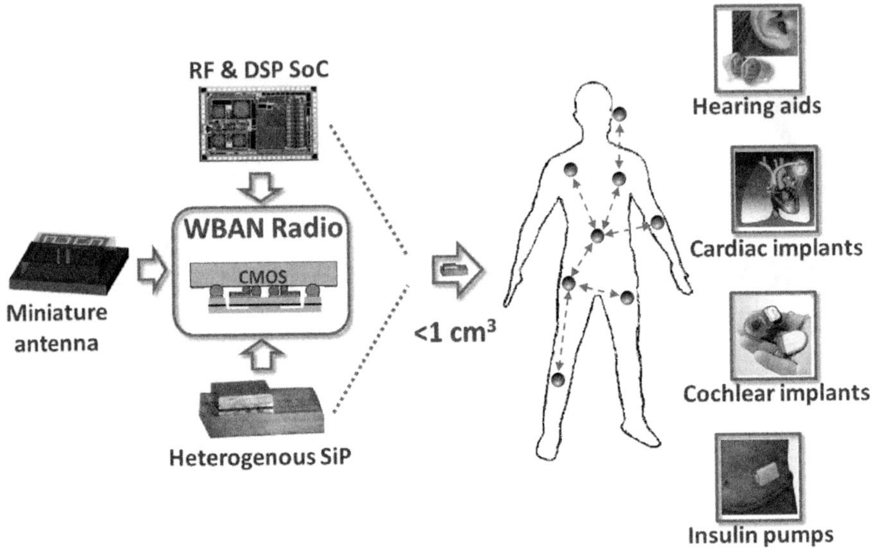

FIGURE 12.1 A typical WBAN system.

This requires a constant monitoring of various physiological parameters to enable an early detection of the onset of symptoms. Therefore, to alleviate the pressure on the health-care systems, there is a need for remote data collection from patients. Fortunately, wearable monitoring systems offer a low cost solution for this issue. These monitoring systems consist of an agglomeration of sensors that monitor various physiological parameters. In addition to this, these systems also usually utilize a wireless radio (since wired systems are expensive and cumbersome) to connect to the data collection hub for further analysis. Taken together, the sensors and the radio form what is known as a wireless body area network (WBAN). According to [4] *"A Wireless Body Area Network consists of small, intelligent devices attached on or implanted in the body which are capable of establishing a wireless communication link. These devices provide continuous health monitoring and real-time feedback to the user or medical personnel"* (Figure 12.1). There has been a rapid increase in the proliferation of these WBAN devices in the recent times with almost 1.7 million of these units being sold worldwide in 2012. This figure is projected to go up further and reach almost 18.2 million units by the year 2017 as shown in Figure 12.2 [5]. Due to this ever increasing presence, much research has been done in this field to improve and optimize the WBANs. Owing to their wearable/implantable nature, these BANs have a unique set of requirements unlike the normal wireless sensor networks (WSN). These requirements are detailed in the following section.

12.2 WBAN REQUIREMENTS

- The power consumption of the WBAN should be in the order of few tens of μW to ensure battery autonomy of such systems.

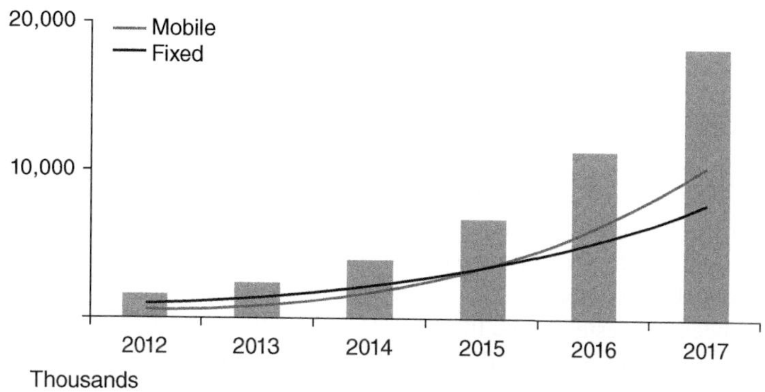

FIGURE 12.2 Projection of Global Health WBAN Shipments (2012–2017).

- WBAN devices should be very small in size of maximum 1 cm^3 in volume.
- They should have low output power to avoid any potential health concerns.
- The communication in some of these nodes which involve implantable devices occur through the human body which is a lossy medium that should be taken into consideration. Moreover, the fact that human body may be in motion should also be taken into account.
- Finally, strict security mechanisms are needed to ensure confidentiality of the patient's data.

12.3 LIMITATIONS OF CONVENTIONAL RADIOS FOR WBAN SYSTEMS

In the above list of WBAN requirements, the first point is of particular importance, since the size of the system is limited, which greatly constrains the energy sources. The main source of the power dissipation in the WBAN systems is the radio required to communicate with the data collection hub. To illustrate this point, let a hypothetical ultra low-powered radio that requires only 10 μW for nominal operation be utilized by the BAN. If such radio uses a CR2032 button cell battery (20 mm diameter and 3.2 mm thickness which amounts to around 1 cm^3) that provides about 225 mAh as an energy source, the battery life time can be calculated to be around 2.5 years. In case of further miniaturization of the battery, this lifetime will decrease even further, as in the case when a V335 battery providing 5 mAh is used for the same radio. This battery is expected to last only 20 days with the radio consuming 10 μW of power. Therefore, it is imperative to keep the power dissipation of radios used in WBANs as low as possible in order to extend the energy autonomy. Taking advantage of the fact that the average data that the WBAN needs to communicate are in the order of a few kb/s, the power reduction of the radios is accomplished by duty cycling. Duty cycling involves data transfer via radios capable of achieving high data rates, so that they communicate short bursts of data at this high rate and are asleep the rest of the time. Any increase in the data rate capability of these radios serves to reduce the time which they are active, thus leading to lower duty-cycle ratios and subsequently lower energy dissipation. But any such data rate increase poses many design challenges, especially in the frequency synthesizer part, due to the fact that the data rate is invariably linked to the synthesizer design. For instance, in the case of conventional phase locked loop (PLL)-based synthesizers, the data rate is directly proportional to the loop bandwidth. Therefore, any increase in data rate will also require an increase of the loop bandwidth which comes at a cost of increased quantization noise (QN) due to the Sigma-Delta modulator (SDM) appearing at the PLL output [6]. Many techniques have been suggested in the literature to reduce the impact of the SDM QN at the output of the PLL [7, 8]. Even though this reduction of the SDM QN would prima facie appear to help increasing the transmitter (TX) data rate, in reality it solves only a part of the problem. This is because of the fact that the PLL-based synthesizers accrue a constant energy overhead due to the frequency reference (which, in most cases consist of a crystal oscillator (XO). This is due to the reason that the XO has a long startup time, during which it has a significant power consumption. In addition, the settling time of the PLL also contributes significantly to the energy overhead of the system. This energy overhead therefore reduces the impact of any increase in data rate for reducing power consumption by increasing the rate of duty cycling. Therefore, it is necessary to look for an architecture that circumvents this energy overhead of the frequency synthesizer, thereby making a migration to higher peak data rates more effective. Apart from this overhead problem of the crystal oscillators, for applications requiring ultra miniaturized radio like cochlear implants, the quartz crystal remains one of the bulkiest part, preventing the introduction of a radio link in the ear canal. Therefore, there is a need to find an alternative to the crystal which not only can startup fast, but also is small so that it can be used in miniaturized systems. But before going into the potential solutions to the bulky crystal, there is a need to derive a figure-of-merit (FoM) that takes into account the energy overhead. This is due to the fact that the conventional FoM which calculates the energy spent per bit communicated (Energy/bit) does not address the energy overhead and is therefore of little use in Ultra Low Powered (ULP) duty-cycled systems.

12.4 COMPARISON METRICS FOR ULP RADIOS

The derivation of the energy metric here follows the work of [9] by assuming a ULP system which has to maintain a mean data rate of MDR (bit/s). The time taken for this system to communicate K bits of data is $T_d(s) = K/MDR$. If the system employs a radio capable of a peak data rate of PDR(bit/s), then it can be duty cycled with a ratio $DC = MDR/PDR$. The packet rate of the system is given by $R_p(packets/s) = MDR/L$ where L is the length of each packet.The duration of each packet is then $D_p(s) = L/PDR$. If this radio consumes a peak power of $P_p(W)$ while in operation, then the energy spent for communicating K bits of data is given by $E_c(J) = T_d \cdot R_p \cdot D_p \cdot P_p$ (communication energy). In addition, if the radio dissipates a power of $P_{st}(W)$ during wake-up, the overhead energy spent during each wake-up cycle will be $E_{oh}(J) = P_{st} \cdot T_w$, where $T_w(s)$ is the wake-up time of the radio (which is usually dominated by the frequency synthesizer wake-up time). In a PLL-based radio, the energy wasted during the settling time of the synthesizer also adds to this overhead ($E_{settle}(J) = P_p \cdot T_{settle}$). Subsequently, the energy wasted as overhead during the communication of K data bits is $E_{oh,tot}(J) = T_d \cdot R_p \cdot (E_{oh} + E_{settle})$. Then, the overall energy spent by the system for communicating K bits of data is $E_{p,tot}(J) = E_c + E_{oh,tot}$.

Useful Energy Threshold To illustrate the application of the above-derived FoM, let a WBAN communicating 10 kbit/s on average be assumed. Now let the radio used in the system utilize the state-of-the-art (SOTA) transmitters given by [10] and [11] which have a peak operational power dissipation (P_p) of 5.4 mW and are capable of a peak data rate (PDR) of 2 Mbit/s. To achieve the MDR of 10 kbit/s, the system can be duty cycled with a ratio of $DC = 0.5\%$. If the packet length is fixed to be $L = 32$ bytes, then the packet rate is calculated as $R_p = 40$ packets/s and the duration of each packet is $D_p = 125$ μs. If the data transmitted by the system is assumed to be $K = 10$ kb, then $T_d = 1$ s and the mean energy dissipated for communication is $E_c = 47$ μJ. If such a BAN employs a PLL-based radio for its radio, then the system will have a long wake-up time owing to the XO (typically $T_w = 1$ ms) and a high start-up power dissipation (P_{st}) of around 1 mW which worsens the $E_{p,tot}$ FoM. Coming to the energy overhead, due to its long startup time, the XO is usually left permanently on during which time it consumes 50–100 μW of power. Therefore the energy overhead for the transmission of 10 kb of data is ($E_{oh,tot} = 50$–100 μJ) which is higher than the energy spent to communicate. Even if a duty cycling of the XO along with the TX is pursued, based on the 1 ms startup time and 0.5 mW power, the energy overhead is 20 μJ (40 packets/s). In addition to this, assuming a best case PLL settling time of around 20 μs, the energy overhead due to the PLL will be around 4.3 μJ. This is shown in Figure 12.3 which graphically explains the aforementioned energy dissipation calculations. Thus, the total energy dissipated in the best case is then $E_{p,tot} = 51.3$ μJ. If the system uses a CR2032 button cell battery (225 mAh) as its power source, this energy dissipation translates to a battery lifetime of 200 days or about 4.4 days when a V335 (5 mAh) battery is used. This very small lifetime deters the use of button cell powered autonomous wearable/implantable nodes. An important conclusion that can be drawn from the figures presented above is that, the energy wasted as overhead during each transmit cycle is almost equal to the communication energy. As the data rate of the transmitter is increased further, the communication energy keeps decreasing and at a certain point, the overhead takes over as the dominant energy sink of the system. This crossover point (called useful energy threshold or UET) shown in Figure 12.4 represents an important point for duty-cycled applications. This is due to the fact that any data rate increase which brings E_c below this UET has a limited impact on the system efficiency due to the larger energy overhead. Therefore, in order to increase the UET, there is a need to reduce the energy overhead of the system by means of reducing its startup time.

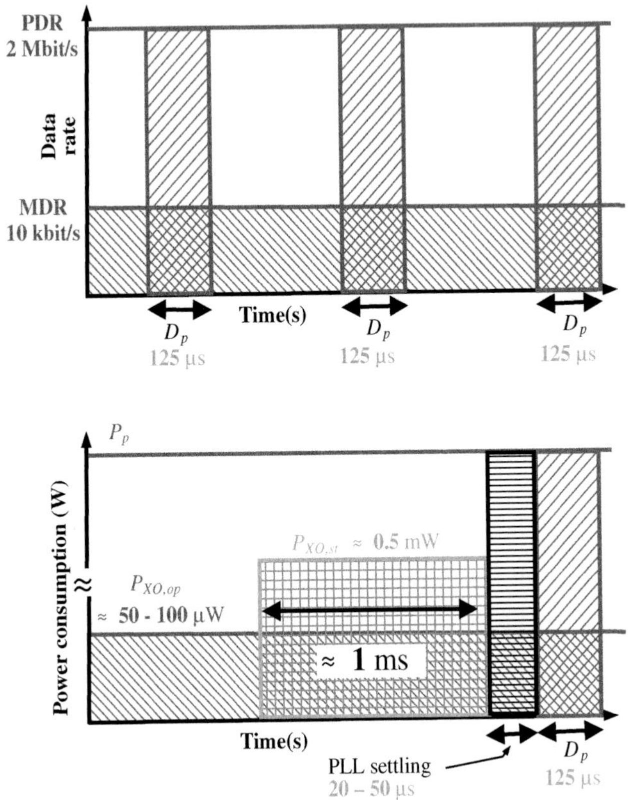

FIGURE 12.3 PLL-based transmitter duty cycling showing crystal oscillator startup overhead.

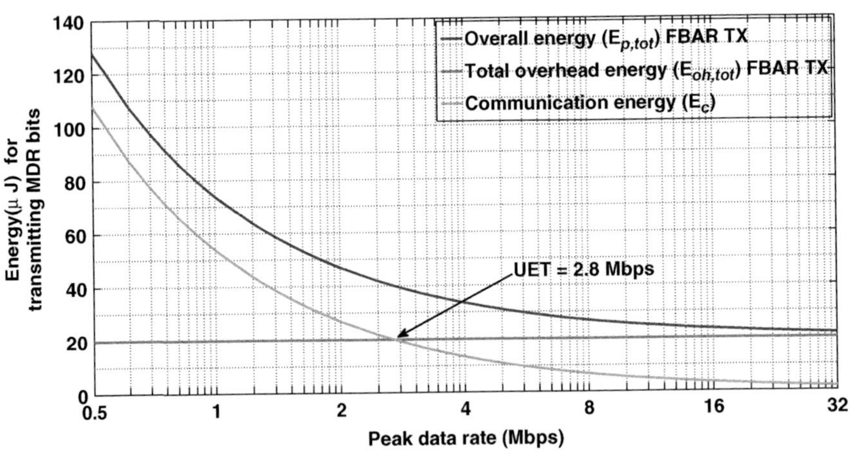

FIGURE 12.4 Energy dissipation break-up of a 10 kbit/s ULP system using the TX in [10]. *Source:* Wong et al. 2013 [10]. Reproduced with permission of IEEE.

12.5 MEMS RESONATORS—A SOLUTION TO BULKY CRYSTALS

MEMS resonators are an attractive alternative to the bulky quartz crystals due to their extremely small size with about 100 times smaller form factor. Among these micromachined resonators, a class of resonators called Bulk Acoustic Wave (BAW) resonators are of particular interest at RF, since they provide a stable frequency reference in the GHz range. Using such resonators, the loop-based synthesizer (PLL) employing crystal oscillator reference can be replaced by a loop- free, almost fully digital synthesizers which not only can startup in μs, but also avoid the latency due to the settling of the PLL. In addition, the high Q of the BAW resonators also makes them suitable for bandpass filters at RF with sharp roll off characteristics.

12.5.1 BAW Resonators

BAW resonators have found widespread use in duplexers due to their small size, high rejection, and low insertion loss [12]. They consist of a piezoelectric material (typically AlN) sandwiched between two electrodes as shown in Figure 12.5. When an electric field is applied between these electrodes, it causes mechanical deformation of the piezoelectric material. This results in an acoustic wave that travels in the direction of the thickness of the piezoelectric film for a particular orientation of electric field. The acoustic wave is reflected back at the film interface with the acoustic insulation layer due to impedance mismatch. When the thickness of the film equals an integer multiple of half wavelength, a standing wave is created by the forward travelling wave and the reflected wave. This acoustic wave in turn modifies the electric field distribution inside the piezoelectric film which changes the electrical impedance of the device. Thus the electrical impedance of the resonator changes with frequency [13]. Based on the nature of the acoustic insulator, the BAW resonators are classified into (a) Thin-Film Bulk Acoustic Resonators (FBAR) and (b) Solidly Mounted Resonators (SMR). While in an FBAR, the air interface serves as the acoustic insulator, the SMR employs Bragg mirrors made up of alternating layers of high and

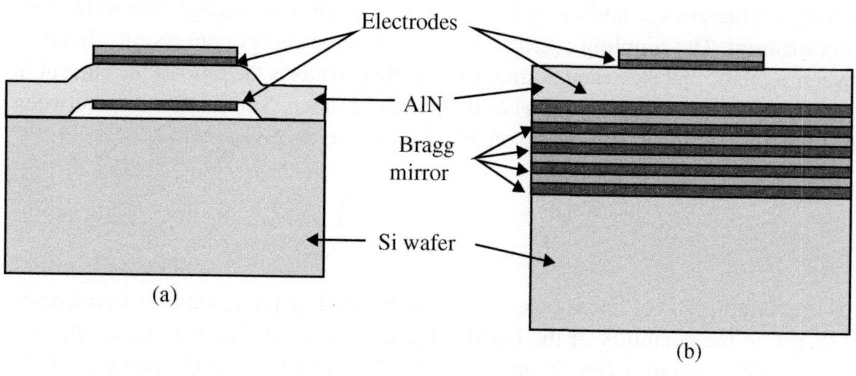

FIGURE 12.5 Cross section of BAW resonators: (a) FBAR, (b) SMR.

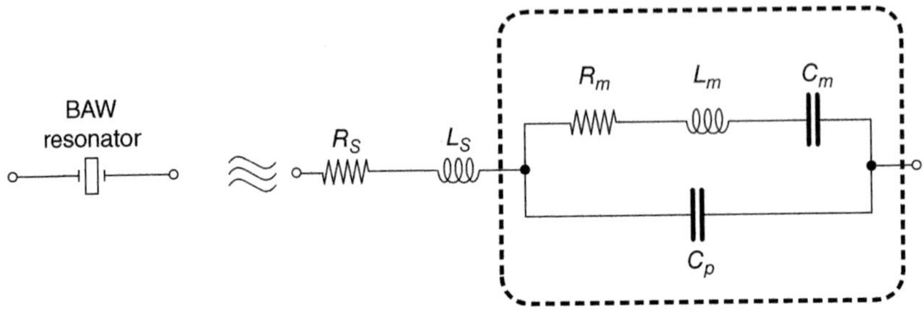

FIGURE 12.6 Butterworh-Van Dyke equivalent circuit of the BAW resonator.

low acoustic impedance. The thickness of the acoustic impedance layers must be designed with specified thickness such that there is a complete reflection of the acoustic waves into the device. The electrical equivalent of the BAW resonator is given by the Butterworth-Van Dyke model (Figure 12.6) and is similar to that of a quartz crystal [14]. It consists of a series Resistor-Inductor-Capacitor (RLC) network along with a parallel capacitance. In addition, there are parasitic resistances and inductances associated with the access connections. The intrinsic resonator thus has two resonance frequencies, one corresponding to the series RLC branch (series resonance) and the other corresponding to the total resonator itself (parallel or anti-resonance). The relation between these frequencies is given by

$$\omega_s = \frac{1}{\sqrt{L_m C_m}} \tag{12.1a}$$

$$\omega_p = \omega_s \sqrt{1 + \frac{C_m}{C_p}} \tag{12.1b}$$

The main performance parameters of the FBAR are the effective electromechanical coupling coefficient k_{eff} and the Q factor which specifies the energy loss in the resonator material. The coupling coefficient determines the energy conversion efficiency between acoustic and electrical domains or in other words is the ratio of the current in the motional branch (series branch) to the parallel branch. The relationship between this coupling coefficient and the resonance frequencies is given by [13]

$$k_{eff}^2 = \frac{\pi^2}{4} \left(\frac{\omega_p - \omega_s}{\omega_p} \right). \tag{12.2}$$

Thus, the coupling coefficient determines the interval of the resonance frequencies and therefore the tenability of the FBAR. Typical values of the coupling coefficient for an FBAR is about a few % and typical values of Q are in the range of 500–1000. Thus, the product $M = K \cdot Q_m$ (K is the ratio of the motional and the parallel capacitances i.e., C_m/C_p) is the figure-of-merit of a BAW resonator which needs to be maximized to improve the performance. Even though either of FBAR or SMR can be utilized in the design of the radio, the work presented in this chapter is based on FBAR and both the terms FBAR and BAW resonator will be used interchangeably henceforth in this chapter.

12.6 FBAR-BASED RADIOS

As mentioned in the previous sections, PLL-free radios based on FBARs have fast startup which aids in reducing the energy overhead in the case of WBANs. In addition, $Q-$ *boosted* FBARs serve as excellent band-pass filters at GHz frequencies. Both these properties of the FBAR have been taken advantage of in the radio presented in this chapter. Owing to different design methodologies, the design of the Transmitter (TX) and the Receiver (RX) have been dealt with separately.

12.7 FBAR-BASED TX ARCHITECTURE

In order to reduce the TX E_{oh}, an alternative method of frequency synthesis was proposed by Flatscher et al. [15] based on an FBAR. This method takes advantage of the fact that the FBAR has a very small startup time (a few μs) and hence less energy is wasted during each wake-up sequence. The modulation is performed by simply varying the digitally controlled oscillator (DCO) frequency (frequency shift keying - FSK). The main problem of the circuit presented in [15] is that it is confined to addressing a few channels owing to the limited frequency tuning range of the FBAR (a few MHz). A simple way to solve this problem is to divide down the FBAR Digitally Controlled Oscillator (DCO) local oscillator (LO) signal (f_{LO}) to generate an Intermediate Frequency (IF). The LO and the IF can then be up-converted ($f_{RF} = f_{LO} + f_{IF}$) to get the desired carrier frequency which is then power amplified and transmitted. Even this method of mixing does not enable this architecture to cover all the frequencies in a band like the 2.4 GHz ISM. Therefore, in order to circumvent the issue of limited channel addressing, the divider used to generate the IF has been implemented as a Phase Switching Divider (PSD) with a division step of 0.2 that further reduces the tuning range required on the FBAR to cover all channels in the band of interest.

Figure 12.7 shows the proposed PLL-free up-conversion transmitter architecture. It consists of an FBAR DCO generating the LO signal, which is then divided by a PSD to produce the desired IF signal, i.e., $f_{IF} = f_{LO}/N$. The LO and the IF are then fed to a mixer which up-converts these signals ($f_{RF} = f_{LO} + f_{IF}$) followed by a single-ended class-C PA doing the final amplification. The centre frequency of the FBAR is chosen such that the spurs due to IF harmonics fall outside the band of interest while also being able to address all the channels in the given band. For instance, with the band of interest in this case being f_{RF} = 2.36–2.5 GHz and the channel to be addressed is at 2.36 GHz, if f_{LO} is chosen to be greater than or equal to 2.22 GHz, the

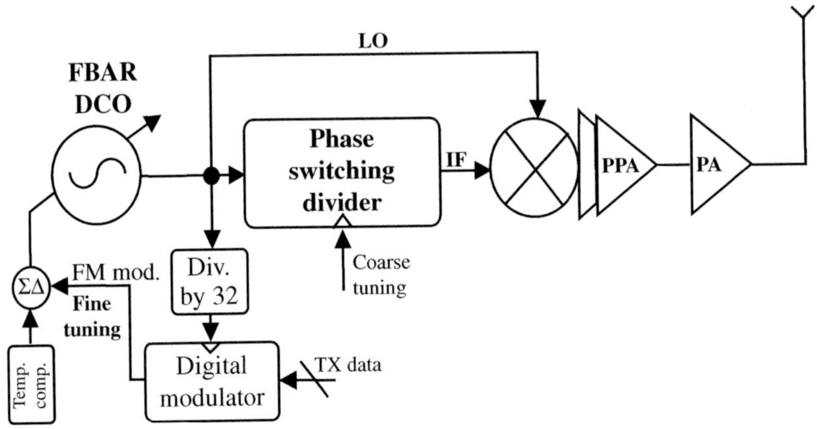

FIGURE 12.7 PLL-free up-conversion TX architecture.

second harmonic IF spur will be located at frequencies ≤ 2.5 GHz which is within the ISM band. Therefore the constraint on the LO is given by f_{LO} < 2.2 GHz. The other extreme is having an LO frequency far away from the wanted band. The drawback of this is as follows: greater the frequency difference between the LO and the wanted band, the higher the frequency of the IF signal will be. The IF is produced by the PSD which utilizes the IF itself as a clock (asynchronous feedback) for its state machines. Therefore, a high IF would result in more power dissipation in the PSD. Based on these constraints, a frequency of 2.2 GHz was chosen for the FBAR DCO. The DCO output is also divided by 32 to serve as a clock for a digital modulator that performs FSK by varying the DCO frequency.

12.7.1 FBAR DCO

Temperature compensated FBAR DCOs provide a highly stable reference at RF frequencies. These DCOs achieve an excellent phase noise performance along with low-power consumption [16] which is the main reason for choosing them for this TX architecture. The transistor level schematic of an FBAR DCO is shown in Figure 12.8. The DCO is implemented using a complementary structure in order to halve the power dissipation and reduce the flicker noise up-conversion. It consists of two cross-coupled pairs providing negative conductance to compensate for the energy loss in the FBAR. To avoid the latch-up of the circuit at startup, the cross-coupled pairs are AC coupled at their sources [17]. Furthermore, two common mode feedback (CMFB) transistors are implemented below the NMOS cross-coupled pair. To control the amplitude of the circuit while also ensuring a fast and reliable start up, an amplitude regulation loop (M_5–M_7) has been designed. This amplitude regulation loop is based on the concept of a proportional-to-absolute-temperature (PTAT) current [18]. A bank of 31 pairs of depletion/inversion MOS capacitors coarsely tune the DCO. Coarse tuning is also accomplished by changing the division ratio of the

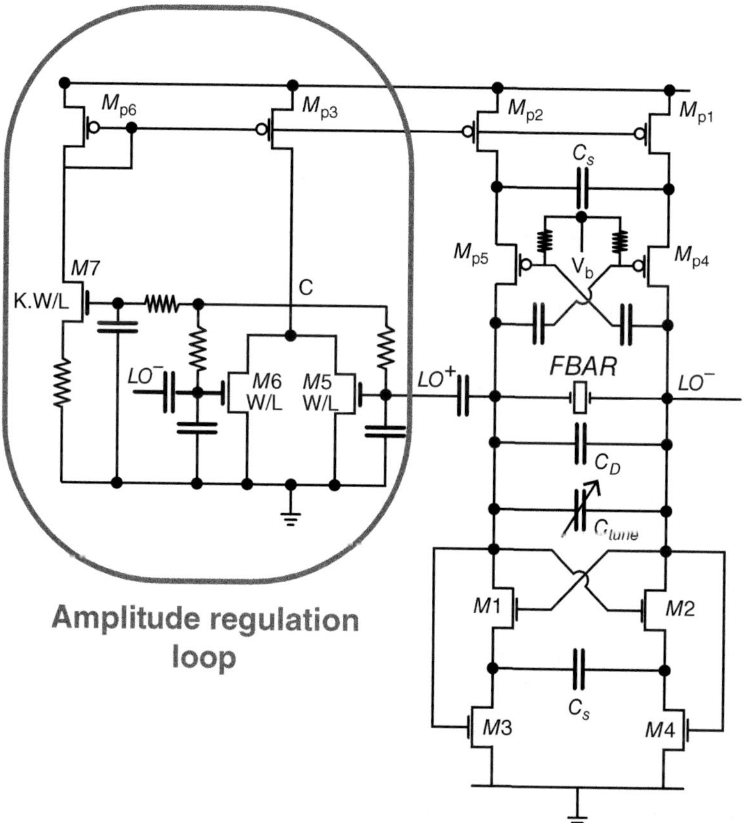

FIGURE 12.8 Schematic of the FBAR DCO.

PSD. The DCO also has three other MOS capacitances driven by the output of a 7-bit, second – order ΔΣ modulator enabling fine tuning. While the upper limit of the tuning range is limited by the resonator itself (a few MHz), the DCO has been designed to achieve a tuning resolution of around 0.9 ppm. Speaking of tuning, there exists a condition on the maximum DCO tuning range that must be satisfied for this architecture such that all the frequency channels within the particular band can be addressed. This condition is given as follows:

$$\Delta f_t = \frac{\Delta \cdot f_{LO}}{N_L^2 + (\Delta + 1)N_L + \Delta}, \quad (12.3)$$

where N_L is the lowest division ratio required for IF generation (which corresponds to the highest IF, provided low-side injection) and Δ is the division ratio step. With the f_{LO} being 2.2 GHz based on the power dissipation versus IF second harmonic tradeoff as mentioned in the previous section, the lowest division ratio N_L is 7.33 (for f_{RF} = 2.5 GHz and f_{IF} = 300 MHz). If a divider with a step size $\Delta = 1$ is used, the tuning range required to address all the channels according to equation (12.3) is $\Delta f_t = 31.6$ MHz or $\Delta f_t / f_{LO} = 14.4‰$. This is impossible to achieve for an FBAR, thus making this architecture unsuitable for multi-channel communication [15]. To circumvent this problem, the PSD with $\Delta = 0.2$ is used which decreases the tuning range needed to a more relaxed value of $\Delta f_t = 7$ MHz or $\Delta f_t / f_{LO} = 3.18‰$. This FBAR tuning along with the modulation index (MI) required on the transmit side for a successful demodulation by the receiver also sets the limit on the Maximum Achievable Data Rate (MADR) of the system (irrespective of single or multi-channel addressing). If the condition of all channel addressing is imposed on this, then if the maximum tuning range of the FBAR is given as *TR* (in‰) and the tuning required to compensate for the process and voltage variations is *PV* (in ‰) ($\simeq$ 1‰ [19]), the tuning remaining for modulation after accounting for Δf_t is

$$\Delta f_m = (TR - PV) * f_{LO} - \Delta f_t. \tag{12.4}$$

Then the MADR with All Channel Coverage (MADR-ACC) is given by

$$MADR - ACC = \frac{\Delta f_m}{MI}. \tag{12.5}$$

Any attempt to achieve data rates above this limit will see that the communication is restricted to a few channels.

12.7.2 Phase Switching Divider

The PSD implemented here was originally implemented to reduce the QN at the output of a fractional-N PLL [20]. The PSD block diagram is shown in Figure 12.9. It is made up of a five stage ring oscillator (ILRO) that is injection locked to the FBAR DCO. This ring oscillator produces five phases at the LO frequency that are spaced $0.2/f_{LO}$ apart. The PSD also consists of a finite state machine (FSM) that produces the select signals with each of them corresponding to a phase signal from the ILRO. The select signals are resynchronized using the edges of the phase signals following which the multiplication of the select signal and its phase occurs in a phase combiner. The resulting signals are then summed together using OR gates. This summed signal is then fed to an integer divider that is set to divide by the nearest integer of the division ratio required. The output of this integer divider is the wanted IF signal which also clocks the FSM. Further information about the implementation of this PSD is given in [20].

12.7.3 Mixer, Power Amplifier Stage, and Digital Baseband

The mixer is implemented as a single-balanced Gilbert cell with resonant load at 2.44 GHz followed by a push-pull preamplifier-buffer (PPA) which performs differential to single-ended conversion. The final power amplifier consists of a single-ended cascoded output stage and a resonant interface between the PPA and PA as shown in Figure 12.10. The integrated digital baseband contains an FSK modulator with

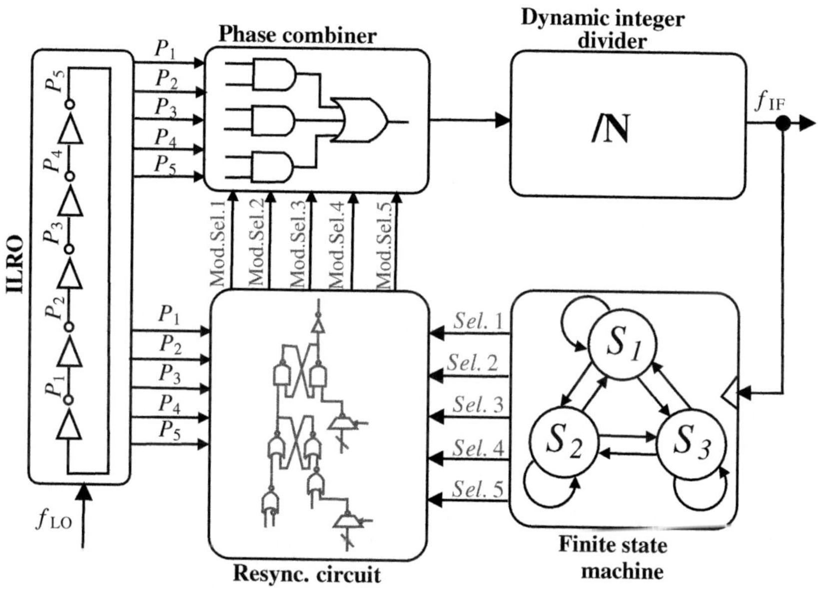

FIGURE 12.9 Phase switching divider.

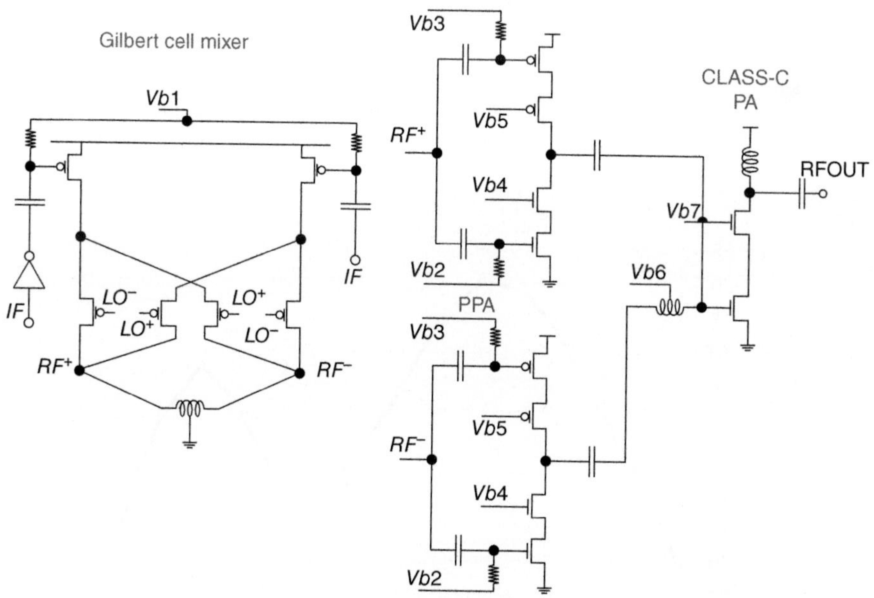

FIGURE 12.10 RF frontend mixer, PPA, and PA.

programmable pulse shape and modulation index, data rate selection, Manchester coding along with a high speed SPI interface. The input to the FSK modulator is the desired TX data pattern that is programmable via an SPI interface and can be set to a random sequence for the purpose of modulation analysis.

12.8 TRANSMITTER MEASUREMENT RESULTS

The structurally compensated (with SiO_2) FBAR used for this setup was measured to have a frequency sensitivity of −6.5 ppm/°C or 814 ppm over the temperature range −40 to 85°C. Further details about the SiO_2 compensation of the FBAR and its packaging can be found in [21]. To further improve the frequency stability, a three-points calibration scheme was employed [22] which reduced this large deviation to ±20 ppm over the entire range as shown in Figure 12.11. The maximum tuning of the FBAR DCO was measured to be 4.9‰. Accounting for FBAR PV variations (PV = 1‰) [19] and $\Delta f_t/f_{LO} = 3.18‰$ leaves $\Delta f_M/f_{LO} = 0.72‰$ for modulation. From equation (12.5), this translates to MADR-ACC of 3 Mbit/s at an MI of 0.5 in the range of 2.36—2.5 GHz. But if the focus is only on the 2.4—2.48 GHz ISM band or the other wing bands alone, then the f_{LO} frequency can be chosen to be 2.29 GHz so that the spur due to the IF second harmonic is outside the band. In this case, the tuning range required reduces to 0.99‰ resulting in an MADR-CC of 12 Mbit/s (2 FSK). Even though the MADR-CC is limited, specific channels can still be addressed at higher data rates. For this, the digital baseband of the implemented TX supports data rate up to 16 Mbit/s (4 FSK with 8 MS/s) and the eye diagrams corresponding to the different data rates are given in Figure 12.13, which show a trend of decreasing modulation accuracy with increasing data rates. Finally, the phase noise of the DCO shown in Figure 12.12 gives the value of −128 dBc/Hz at a frequency offset of 100 kHz and a

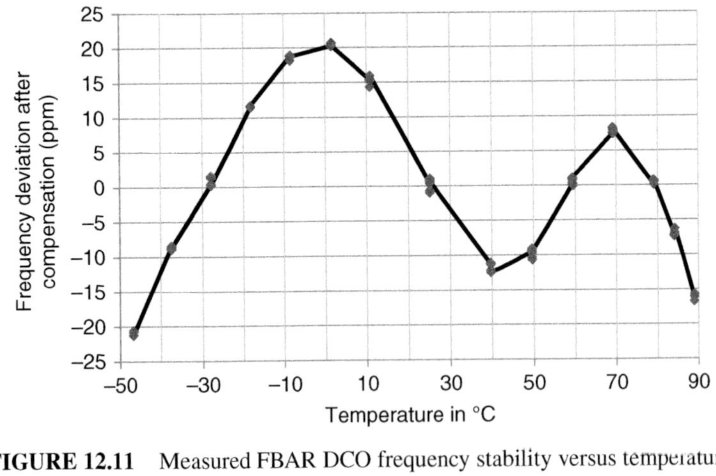

FIGURE 12.11 Measured FBAR DCO frequency stability versus temperature.

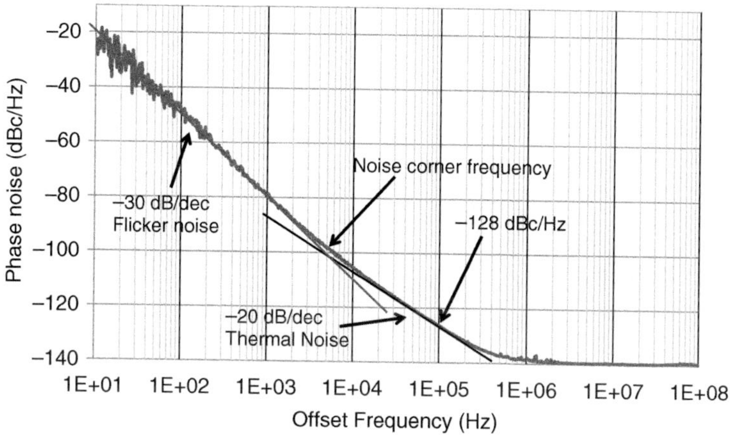

FIGURE 12.12 Measured phase noise of the FBAR DCO.

flicker to thermal noise corner frequency of 5 kHz. This value is also the phase noise of the carrier signal at low offset frequencies (due to injection locking).

The maximum output power of the Transmitter is −1 dBm which meets the Bluetooth-Smart requirements [23] (Figure 12.14). The spurs in the wideband output spectrum (Figure 12.15) mainly occur due to four mechanisms namely: (a) the fractional division step of the PSD (=0.2) causing spurs at multiples of $0.2 * f_{IF}$ offset, (b) clock feedthrough spurs located at multiples of $f_{LO}/32$ offset, (c) LO feedthrough,

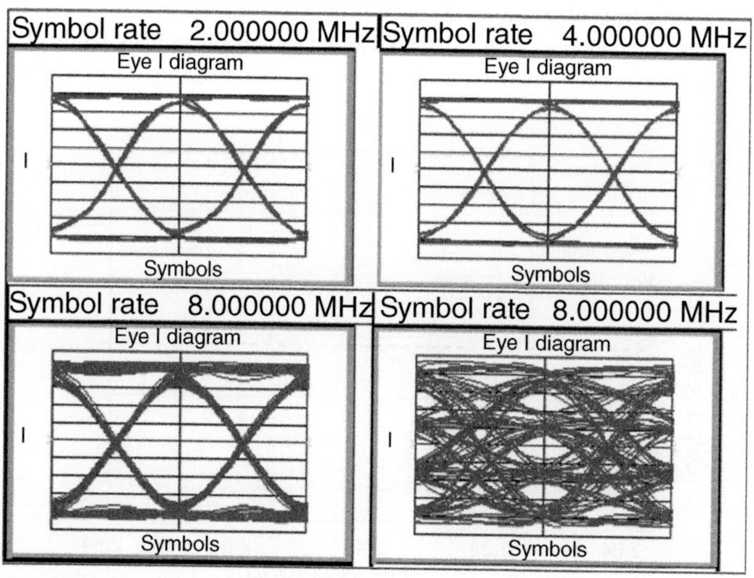

FIGURE 12.13 Eye diagrams for GFSK at various data rates *viz.* (a) 1 Mbit/s, (b) 2 Mbit/s, (c) 8 Mbit/s, (d) 16 Mbit/s, (8 MSps — 4FSK).

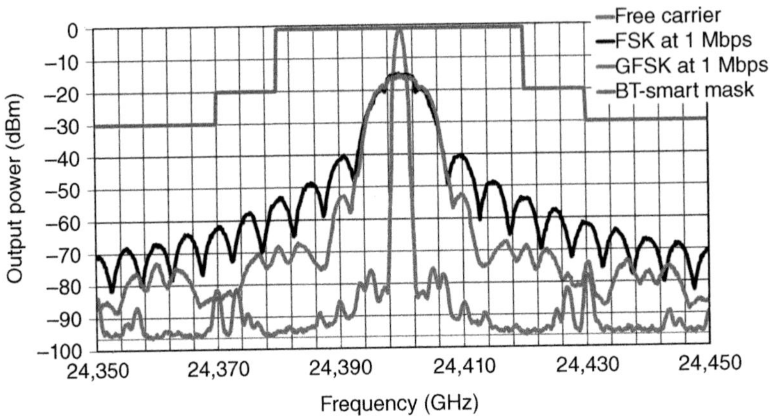

FIGURE 12.14 Close-in spectrum showing free carrier, FSK, and GFSK.

and (d) IF second harmonic. While the spurs due to both LO feedthrough and IF second harmonic are at −29 dBc, the fractional spurs due to the divider step are at least 47 dB below the carrier. The clock feedthrough spurs are at least −70 dBc as shown in Figure 12.15.

The dynamic operation of the TX is shown in Figure 12.16 which depicts the time domain TX output measured at 1 Mbit/s FSK with a modulation index of 0.5. The current profile shows that the TX takes only 5 μs to start transmission after wake-up. The wake-up times of the individual blocks are given as follows: 2 μs for the FBAR DCO, 2 μs for the PA, and 3 μs for the modulator. The PA and the modulator can be started at the same time after which the sample data pattern “0110 1010” can be seen on the frequency versus time plot. The TX can be turned off in 3 μs while channel switching also can be performed in just 3 μs. This frequency agility is one of the main

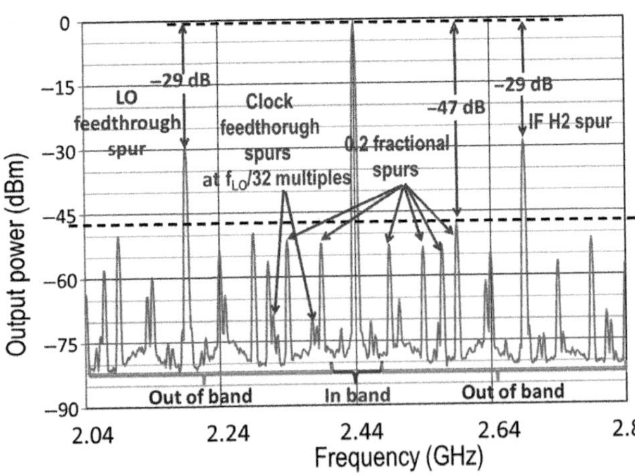

FIGURE 12.15 Wide-band spectrum showing the spurious characteristics.

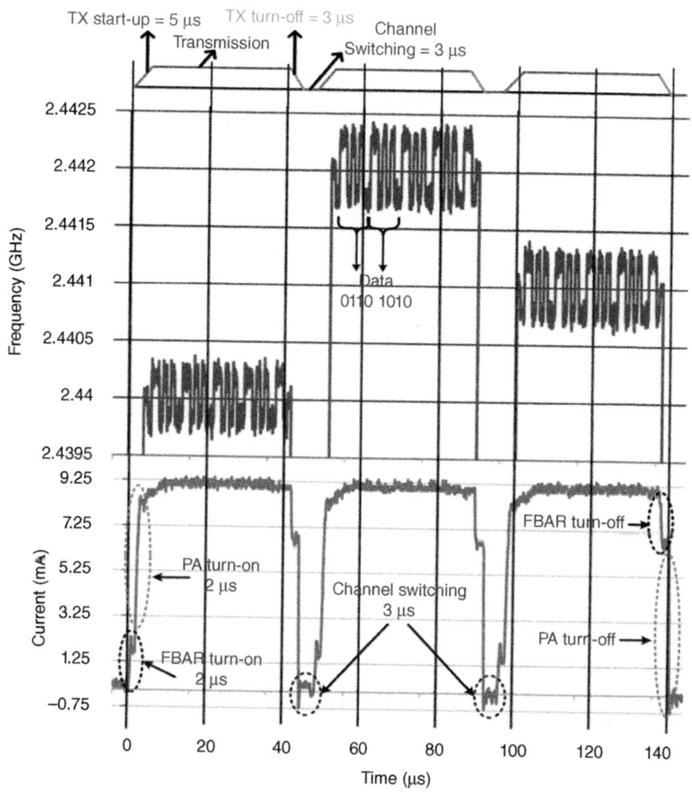

FIGURE 12.16 Frequency agility of the TX showing FSK data (top) and TX current profile (bottom) showing the startup/channel switching times.

advantages of this TX since it allows us to perform frequency hopping to any channel within the band in a span of 3 μs, while still being a narrow-band system.

The TX consumes 9.2 mA from a 1.2 V supply with the power breakdown shown in Figure 12.17. The power consumption during the 5 μs startup phase takes the following trajectory: Initially, the FBAR DCO starts during which time it consumes 2.5 mA, followed by a ramping up of the current due to the PA startup during the next 3 μs till the peak current is reached. To find an approximation for the startup energy overhead, the area under the current profile (Figure 12.16) was calculated and multiplied with the supply voltage to give the energy dissipation as E_{oh} = 18 nJ. At this juncture, it is imperative to note that this synthesizer does not suffer from the settling time that plagues the PLL-based synthesizer. Now, if this TX is implemented in the 10 kbit/s WBAN system described in Section 12.4 with L = 32 bytes and K = 10 kb, at a peak data rate of 16 Mbit/s this translates to the total startup energy dissipation $E_{oh,tot}$ = 0.72 μJ. The duration of each packet in this system is D_p = 16 μs and consequently, the energy dissipated in communication is now E_c = 7 μJ. Therefore, the total energy FoM is $E_{p,tot}$ = 7.72 μJ. A comparison of this TX with the

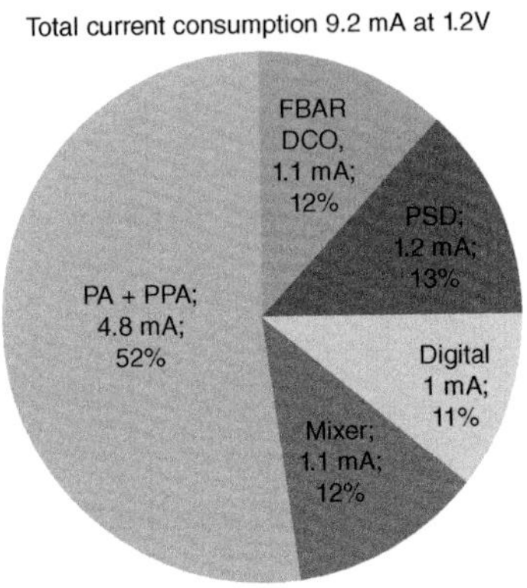

FIGURE 12.17 TX power consumption breakdown.

PLL-based SOTA TX at different peak data rates can be seen in Figure 12.18 from which the following conclusions can be drawn.

- The FBAR-based TX shows a 34 reduction in the startup energy overhead figures due to the very fast startup of the FBAR DCO and the absence of the synthesizer settling time.
- Due to this small startup overhead, the UET of the FBAR-based TX is 187.5 Mbit/s as opposed to 2.8 Mbit/s for PLL-based transmitter.
- The PLL-based TX performs slightly better at lower peak data rates due to the fact that the power spent for communication is much higher at these data rates. Since the PLL-based TX has lower peak power dissipation, this is an expected result.

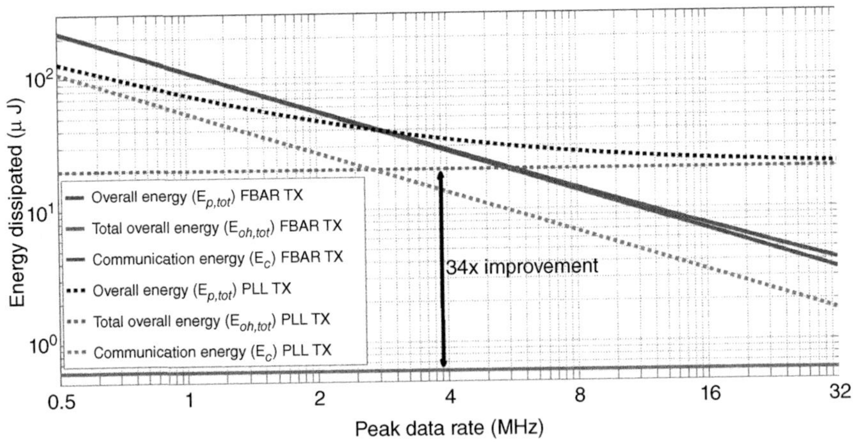

FIGURE 12.18 Variation of the energy dissipation with peak data rate in PLL-based SOTA TX and FBAR TX.

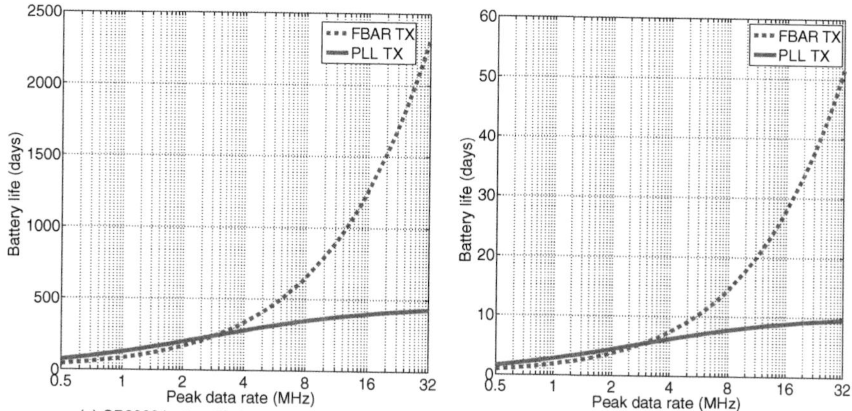

(a) CR2032 battery lifetime vs. peak data rate

(b) V335 battery lifetime vs. peak data rate

FIGURE 12.19 (a) Variation of lifetime of CR2032 battery with increasing data rate, (b) variation of lifetime of V335 battery with increasing data rate.

- At higher peak data rates, the longevity of the FBAR-based TX operating from a fixed power source such as a CR2032 battery is much more (Figures 12.19a and 12.19b) than that of the PLL-based TX due to its reduced overhead $E_{oh,tot}$. For instance, if under the assumption that the SOTA TX can be extended to operate at 16 Mbit/s peak data rate with the same power (5.4 μW), the FBAR-based TX outperforms the PLL-based SOTA TX by having a 3 times longer (1254 days vs. 400 days) battery life. But if the transmitters are operated at their maximum capable peak data rates (16 Mbit/s for FBAR TX and 2 Mbit/s for PLL TX), then the FBAR-based TX outperforms the PLL-based TX by a factor 7 in battery lifetime.

The performance comparison of this transmitter with other works of literature is shown in Table 12.1. The chip microphotograph given in Figure 12.20 shows the PLL-free, FBAR-based TX and the circuit of [20].

TABLE 12.1 Performance comparison with prior literature.

Parameter	[10]	[11]	[15]	This work
Peak data rate	2 Mbit/s	1 Mbit/s	50 kbit/s	16 Mbit/s
Modulation	FSK/PSK	FSK/PSK	FSK	FSK
I_{dc}(mA)	4.5	8.9	6.0	9.2
V_{DD}(V)	1.2	1.0	1.8	1.2
TX output power	0 dBm	0 dBm	1 dBm	−1 dBm
Startup time	40 μs	N/A	2 μs	5 μs
DCO PN (dBc/Hz)	−112 at 1 M	−87 at 130 k	−126 at 1 M	−121 at 100 k
Energy efficiency	2.7 nJ/b	6.5 nJ/b	144 nJ/b	0.7 nJ/b
$E_{st,tot}$ FoM	67 μJ	89 μJ	N/A	7.5 μJ
Technology	90 nm	130 nm	130 nm	65 nm

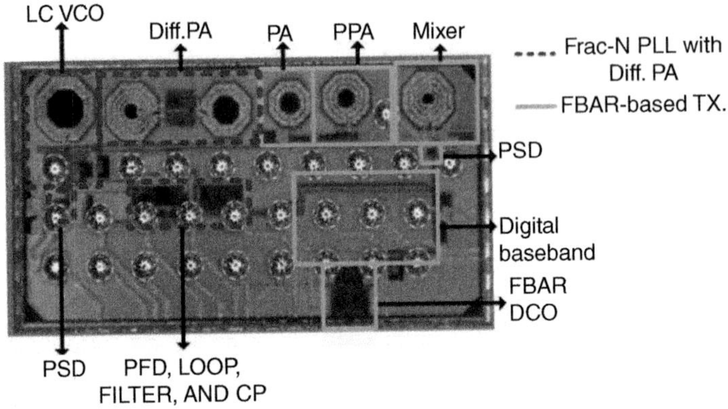

FIGURE 12.20 FBAR-based transmitter chip microphotograph.

12.9 SUMMARY OF THE FBAR-BASED TX

A PLL-free transmitter architecture based on an FBAR to reduce the startup energy overhead in duty-cycled low-power systems has been presented. Utilizing a PSD injection locked to the FBAR DCO to generate the IF thereby overcoming the limited frequency tuning of the FBAR, this TX can address all channels in the frequency range of 2.36–2.5 GHz. The FBAR also gives the TX the ability to start in just 5 μs, enabling this TX architecture to achieve a significant reduction in energy dissipation (a 7x reduction in the ULP example mentioned) at its MADR as compared to SOTA TX. In addition, the provision to switch channels in just 3 μs despite being a narrow-band system makes this transmitter an attractive proposition for ULP systems.

12.10 RECEIVER ARCHITECTURE

A receiver which makes the best use of the low phase noise of the FBAR DCO signal and eliminates the need for a PLL to address multiple channels is presented in this section. As shown in the block diagram of Figure 12.21, the receiver utilizes the intrinsic high Q of the FBAR in the RF front-end to provide in-band interferer rejection and in the frequency synthesis to provide clocks with high spectral purity. The PLL-free approach involving integer dividers, and the sub-sampling–based down-conversion, makes the receiver easily scalable for implementing it in an advanced deep sub-micron process.

The received RF signal from the antenna is pre-filtered and then fed differentially to the LNA. The LNA amplifies the RF signal and this amplified signal is fed to the input of an Amplifier-Mixer-Filter (AMF) cell which performs channel selection and filtering. The AMF cell can either be loaded with an FBAR full-lattice [24] or a FBAR pseudo-lattice [25] to provide narrow-band filtering at RF or at a super-high intermediate frequency (IF). This filtering can be used as an anti-aliasing filter to

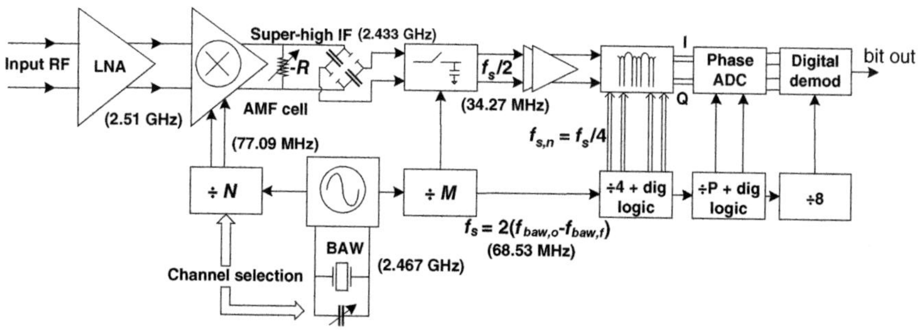

FIGURE 12.21 MEMS-based PLL-free sub-sampling receiver architecture. High Q FBARs are used in the channel selection filter and in the oscillator. The typical values of frequencies along the receiver chain are given in the parentheses.

support direct sampling of the input signal at RF. Further the use of a sub-sampling mixer provides discrete time filtering and this improves the overall filtering of the receiver chain. The signal converted to baseband in quadrature by the mixer is fed to a phase ADC and then to a digital demodulator to obtain the digital outputs. As it can be seen in Figure 12.21, the synthesis involves only a chain of integer dividers to provide the various clocks required in the chain.

12.10.1 Frequency Synthesis: The Integer-N Approach

Instead of running the FBAR DCO at a fixed frequency as in [26], if some frequency tuning is enabled by adding a switchable capacitor array across the resonator, then let us assume that the center frequency of each channel (which can be arbitrary) in the required band can be expressed as,

$$f_{rf,ch,n} = (f_{fbar,o} - \Delta f_n) \cdot \left(1 + \frac{1}{P_n}\right), \tag{12.6}$$

where $f_{fbar,o}$ is the intrinsic FBAR DCO frequency, Δf_n is the frequency tuning on the FBAR DCO.

$$P_n = N_l, (N_l + 1), (N_l + 2) \ldots N_h, \quad \text{where } N_l, N_h \text{ are integers.} \tag{12.7}$$

With no tuning ($\Delta f_n = 0$), the biggest step size (occurs for the lowest integer) is given by,

$$f_{step} = f_{fbar,o} \cdot \left(\frac{1}{N_l} - \frac{1}{N_l + 1}\right) = \frac{f_{fbar,o}}{N_l \cdot (N_l + 1).} \tag{12.8}$$

The step size reduces as N_l increases but the value of N_h that can be tolerated depends on image reject pre-filter characteristic. In a similar scenario (but the channel selection is done at IF) [26], N_l and N_h are chosen to be 15 and 30 with $f_{fbar,o}$ at 2.32 GHz.

To address any arbitrary frequency in the band of interest with integer division, there is a compulsive need for tuning the FBAR oscillator. The minimum tuning required depends on the biggest step size defined in equation (12.8). The minimum tuning required is decided by the relationship defined as,

$$(f_{fbar,o} - \Delta f_{min}) \cdot \left(1 + \frac{1}{N_l}\right) = f_{fbar,o} \cdot \left(1 + \frac{1}{N_l + 1}\right), \tag{12.9}$$

which when simplified results in,

$$\Delta f_{min} = \frac{f_{fbar,o}}{(N_l + 1)^2} = f_{step} \cdot \left(\frac{N_l}{N_l + 1}\right). \tag{12.10}$$

For nominal values of $f_{fbar,o}$ at 2.32 GHz and minimum value of 15 for N [26], the minimum tuning required is 9.06 MHz (0.4%). This is a reasonable requirement which can be achieved by adding a switched capacitor array across the FBAR in the DCO [27]. If N_h and N_l are the maximum and minimum integer division ratios respectively, then the range of frequencies that can be addressed by a single FBAR DCO and a set of integer dividers is given by,

$$(f_{fbar,o} - \Delta f_{min}) \cdot \left(1 + \frac{1}{N_h}\right) \leq f_{ch} \leq f_{fbar,o} \cdot \left(1 + \frac{1}{N_l}\right). \tag{12.11}$$

Similar to [26], N_h gets fixed by the pre-filter used for image rejection. For respective nominal values of N_l, N_h, $f_{fbar,o}$, and Δf_{min} at 15, 30, 2.32 GHz and 9.06 MHz, the range of frequencies that can be addressed is 86.7 MHz which is sufficient for covering the whole 2.4 GHz ISM band (80 MHz wide).

Sub-Sampling Clock Synthesis It is clear that with down-conversion by mixing in two steps, any arbitrary channel can be selected and down-converted to baseband. However, to preserve the frequency/phase-modulated information of the signal, there is a need for quadrature down-conversion to baseband. This requires generating quadrature either in the signal path or in the LO. By integer division, quadrature clocks can be generated only when the division ratio is even [28]. This additional constraint of even integer division results in the requirement of a higher tuning range on the FBAR DCO. To alleviate these issues, we choose a sub-sampling–based down-conversion approach for translating the filtered channel to baseband in quadrature. The critical anti-alias filtering needed to support this approach is readily provided by the FBAR channel selection filter.

If $f_{rf,ch,n}$ is the wanted channel frequency, it is selected by down-converting it to the peak frequency of the FBAR pseudo-lattice in the AMF cell. In other words,

$$f_{rf,ch,n} = f'_{fbar,f} + \frac{f'_{fbar,o}}{P_n}, \quad \text{with } P_n \text{ being an integer.} \tag{12.12}$$

In equation (12.12) $f'_{fbar,o}$ is the FBAR DCO frequency with frequency tuning and similarly $f'_{fbar,f}$ is the peak frequency of the AMF cell with tuning which are given by,

$$f'_{fbar,o} = f_{fbar,o} - \Delta f_o \quad \text{and} \quad f'_{fbar,f} = f_{fbar,f} - \Delta f_f. \tag{12.13}$$

If this filtered channel has to be down-converted (in two steps) to baseband in quadrature by sampling with differential clocks obtained by integer division of the FBAR DCO signal, then the initial sampling rate f_s can be shown to be given by [24],

$$f_s = \frac{f'_{fbar,o}}{m} = \frac{2f'_{fbar,f}}{2k-1}, \quad \text{with} \quad m, k = 1, 2, 3 \ldots. \tag{12.14}$$

It is clear from equation(12.14) that, the intrinsic anti-resonant frequencies of the FBARs used in the DCO ($f_{fbar,o}$) and in the filter ($f_{fbar,f}$) have to be different and the offset between these is at a minimum when $m = k$ in equation (12.14). Small offsets between the anti-resonant frequencies can be easily generated by depositing an oxide layer on the electrodes [29]. When the offset is minimum, the frequencies $f'_{fbar,o}$ and $f'_{fbar,f}$ are related as,

$$f'_{fbar,f} = \left(1 - \frac{1}{2m}\right) \cdot f'_{fbar,o}, \quad m = 1, 2, 3 \ldots. \tag{12.15}$$

Using equation (12.15) in equation (12.12), we get,

$$f_{rf,ch,n} = f'_{fbar,o} \cdot \left(1 + \frac{1}{P_n} - \frac{1}{2m}\right). \tag{12.16}$$

On the similar lines of the previous discussion, the minimum tuning required on the FBAR DCO ($\Delta f_{o,min}$) to address any arbitrary frequency in the band of interest is,

$$\Delta f_{o,min} = \frac{f_{fbar,o}}{(N_l + 1)^2 - \frac{N_l \cdot (N_l + 1)}{2m}}. \tag{12.17}$$

The tuning required (12.17) is similar to that in equation (12.10). For nominal values of $f_{fbar,o}$, N_l and m to be 2.4, 15 and 15 GHz respectively, the tuning required is 9.68 MHz (0.4%) which is a feasible requirement [27]. Similarly from equation (12.15) the tuning (Δf_f) required in the AMF cell is given by

$$\Delta f_f = \left(1 - \frac{1}{2m}\right) \cdot \Delta f_o, \quad m = 1, 2, 3 \ldots . \tag{12.18}$$

Further, with minimum difference between the intrinsic anti-resonant frequencies of the resonators used in the DCO and in the AMF cell, the sampling rate in equation (12.14) can be simplified as,

$$f_s = 2 \cdot (f'_{fbar,o} - f'_{fbar,f}). \tag{12.19}$$

Taking nominal values for the intrinsic anti-resonant frequencies $f_{fbar,o}$ and $f_{fbar,f}$ at 2.4 GHz and 2.32 GHz respectively, the sampling rate is 160 MHz with $m = 15$. The range of frequencies that can be addressed with this approach is

$$f'_{fbar,o} \cdot \left(1 + \frac{1}{N_h} - \frac{1}{2m}\right) \leq f_{ch} \leq f_{fbar,o} \cdot \left(1 + \frac{1}{N_l} - \frac{1}{2m}\right). \tag{12.20}$$

Taking the minimum and maximum integer division ratios for the channel selection LO to be 15 and 30 respectively, the range of frequencies that can be addressed with the sub-sampling approach is 89.68 -MHz which exceeds the bandwidth of the 2.4 GHz ISM band.

12.10.2 FBAR-Based RF Front-End

The RF front-end comprises of an LNA and an AMF cell as shown in Figure 12.21. A current reuse topology with g_m boosting is used for the LNA [25]. The AMF cell (see Figure 12.27) also involves current reuse to facilitate low-power consumption. Before going into the details of this cell, we shall first have a look at the load of this cell, which is an FBAR-based pseudo-lattice.

FBAR -Based Pseudo-Lattice Network The FBAR impedance has a peak at DC (capacitive at DC) and the DC offset or flicker noise arising from the transconductance stage get amplified and this causes a concern when the wanted signal is brought down to DC by sampling as these unwanted DC components are still present at DC (by sampling). This problem is partly alleviated in [24] by discrete time filtering which provides null at DC to attenuate the unwanted components before the signal down-conversion. Further, the absence of ideal bandpass filtering increases the effective noise bandwidth and which in-turn increases the amount of noise folding owing to sampling-based down-conversion [30, 31].

An FBAR pseudo-lattice structure as in [25] can be used (see Figure 12.22) to circumvent the issues in the previous paragraph. The term *pseudo* is used as the structure involves both FBARs and capacitors and differs from the FBAR lattice structures used in [32], [33], where only FBARs in the form of a lattice are used to provide wide band

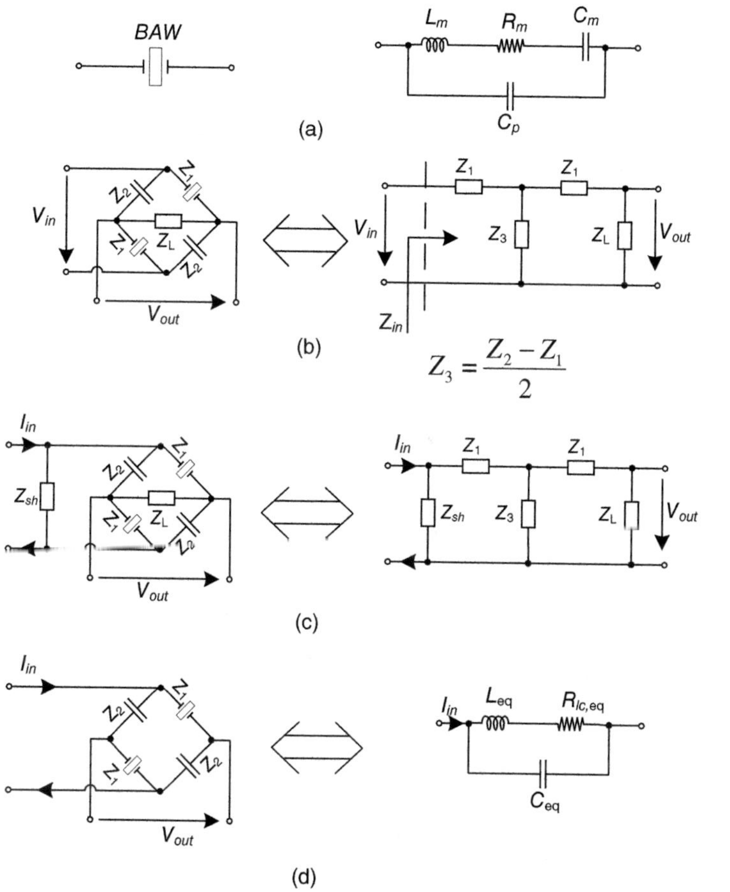

FIGURE 12.22 (a) FBAR equivalent model; FBAR-based lattice (b) with voltage input, (c) with current input. (d) The pseudo-lattice can be replaced by an LC equivalent with Q equal to that of the FBARs used in the lattice.

filtering. The pseudo-lattice comprises of two matched FBARs (Z_1) and two matched capacitances (Z_2) as shown.

The FBAR pseudo-lattice is not suitable for voltage inputs [25]. For a current input, the lattice provides a trans-impedance given by,

$$Z_{tran} = \frac{V_{out}}{I_{in}} = \frac{Z_L \cdot (Z_2 - Z_1)}{Z_1 + Z_2 + 2Z_L}, \text{ with } Z_{sh} \to \infty. \tag{12.21}$$

The magnitude plot of the trans-impedance as given in equation (12.21) is shown in Figure 12.23. As it can be seen it results in an ideal bandpass response similar to the one with voltage input. However, the peak of the trans-impedance is at the anti-resonant frequency of the matched FBARs used in the pseudo-lattice and the

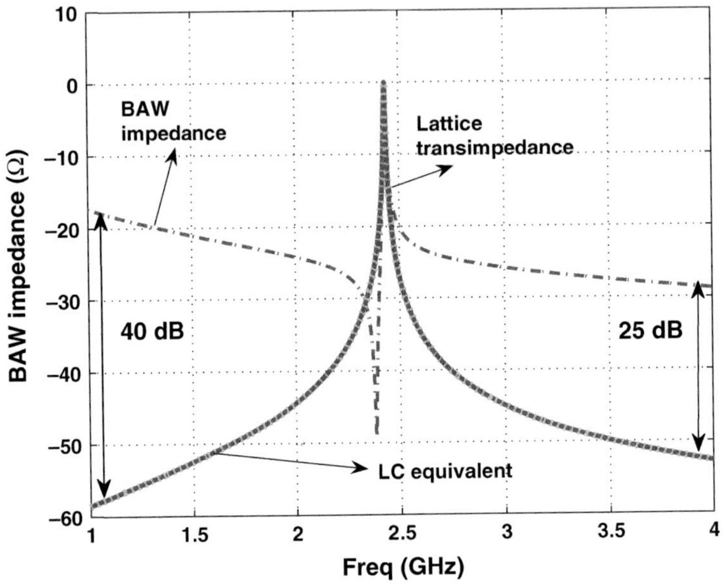

FIGURE 12.23 Normalized frequency plots of the FBAR impedance, trans-impedance of the FBAR pseudo-lattice. The pseudo-lattice significantly improves the rejection especially at far frequencies from the center frequency.

bandwidth of this filtering response is equal to that of the FBAR which depends on the Q of the resonator. It can be seen that the rejection provided by the FBAR pseudo-lattice is far superior compared to a single FBAR especially at frequencies far away from the center (peak) frequency. The FBAR pseudo-lattice can be modeled with an LC tank equivalent as shown in Figure 12.22 with the inductor L_{eq}, capacitor C_{eq}, and loss in the inductor $R_{lc,eq}$ given by

$$L_{eq} = L_m, \quad C_{eq} = \frac{1}{\omega_p^2 L_{eq}}, \quad \text{and} \quad R_{lc,eq} = R_m, \tag{12.22}$$

where ω_p is the anti-resonant frequency of the FBAR.

Bandwidth Tuning in the Lattice The intrinsic bandwidth of the lattice filter might not be sufficient for the targeted narrow-band standard. In [24], a solution is proposed to boost the Q of the FBAR by adding negative conductance across the resonator and without burning additional power. However in the lattice, two FBARs are used and compensation has to be provided to both the resonators to boost the Q and decrease the bandwidth. If a shunt impedance Z_{sh} is added in parallel with the current source as shown in Figure 12.22c, the trans-impedance is given by

$$Z_{tran} = \frac{Z_{sh} Z_L \cdot (Z_2 - Z_1)}{(Z_{sh} + Z_L) \cdot (Z_1 + Z_2) + 2 \cdot (Z_1 Z_2 + Z_{sh} Z_L)}. \tag{12.23}$$

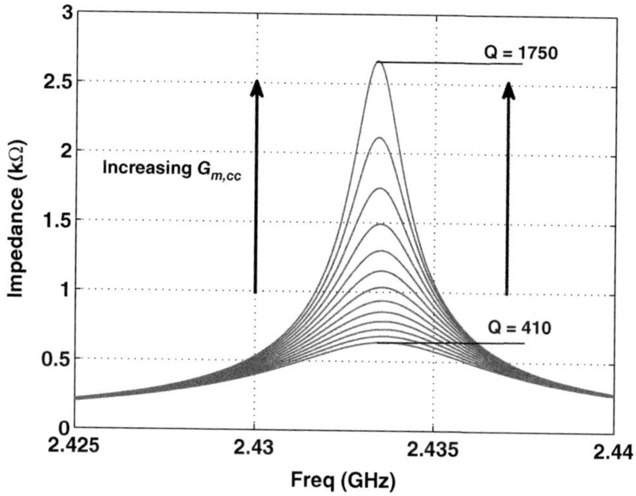

FIGURE 12.24 Q boosting of the FBAR pseudo-lattice by adding a negative conductance across it.

If Z_{sh} is a negative conductance $(-G_{m,cc}/2)$, then it can be shown that this negative conductance partially compensates for the losses in the resonators and thus boosts the Q of the filter and reduces its bandwidth. This phenomenon is similar to that in the Q-enhanced FBAR [24]. The bandwidth reduction and Q-boosting of the pseudo-lattice trans-impedance by adding a negative conductance is plotted in Figure 12.24. It is clear from the figure that the maximum impedance (impedance at the peak frequency) increases with the amount of compensation added. The critical conductance is the conductance which completely compensates for the losses in the resonators of the lattice and makes it to oscillate. This critical conductance is evaluated by equating the trans-admittance $(1/Z_{tran})$ to zero. In other words, equating the denominator of the expression in equation (12.23) to zero, we get

$$Z_{sh} = -\frac{Z_L \cdot (Z_1 + Z_2) + 2Z_1Z_2}{Z_1 + Z_2 + 2Z_L} = -Z_{in}. \tag{12.24}$$

Without any load, it can be shown that the Z_{in} is given by $(Z_1 + Z_2)/2$. The peak of the trans-impedance (defined in equation (12.23) is at the anti-resonant frequency of the FBARs used in the lattice. The impedance of these resonators at the anti-resonant frequency is resistive [34] and is given by

$$R_p = (KQ_m)^2 \cdot R_m. \tag{12.25}$$

So, Z_{in} at ω_p becomes

$$Z_{in} \cong \frac{R_p}{2} - \frac{j}{2 \cdot \omega_p \cdot C_{ltc}}. \tag{12.26}$$

Owing to intrinsically high Q of the FBARsand large capacitance C_{ltc}, the imaginary part of Z_{in} in equation (12.26) is negligible compared to R_p and the critical transconductance $G_{m,cc,max}$ can be shown to be given by

$$G_{m,cc,max} \cong \frac{4}{R_p}. \tag{12.27}$$

This critical conductance is twice that in the case of compensating a single FBAR [24]. This is justified because in the case of the lattice, two FBARs are used and the losses in both the resonators need to be compensated which requires more (negative) conductance.

The bandwidths of the filtering responses for various levels of compensation ($G_{m,cc}$) are evaluated numerically and variation of the bandwidth with the negative shunt conductance is plotted in Figure 12.25. In the plot the negative conductance is normalized to the maximum conductance as defined in equation (12.27) and the bandwidth is normalized to the bandwidth of the lattice filter without any compensation added across it. As seen in the plot, the bandwidth varies linearly with the increase in the negative conductance and can be expressed mathematically as

$$f_{bw,eq} = f_{bw,max} \cdot (1 - g_{m,cc}), \tag{12.28}$$

where $g_{m,cc}$ is the normalized conductance and is given by

$$g_{m,cc} = \frac{G_{m,cc}}{G_{m,cc,max}} \cong \frac{G_{m,cc}R_p}{4}. \tag{12.29}$$

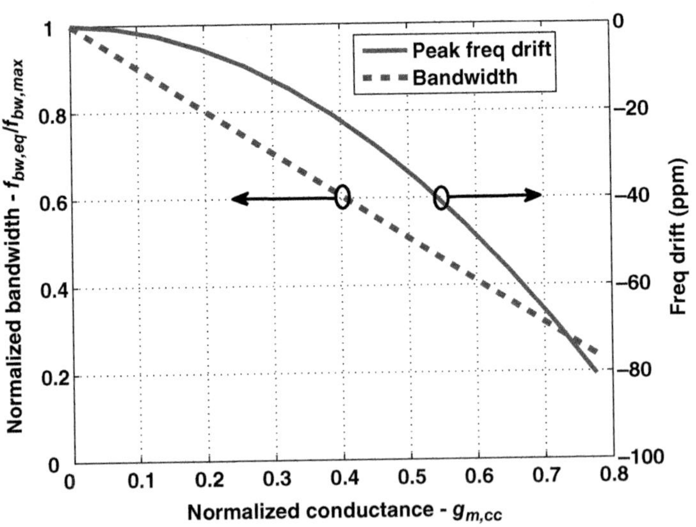

FIGURE 12.25 Bandwidth variation and the peak frequency drift with the addition of the negative conductance to boost the Q.

Adding negative conductance across the lattice not only reduces the bandwidth of the filter by boosting the Q, but also slightly drifts the peak frequency as seen in Figure 12.24. The variation of the peak frequency from the intrinsic anti-resonant frequency of the FBARs in the pseudo-lattice is plotted in Figure 12.25. The drift is very small compared to the anti-resonant frequency. For example, a bandwidth reduction by a factor of 2, the drift is around 50 ppm for an anti-resonant frequency of 2.433 GHz, which corresponds to a frequency drift of 120 kHz. This frequency drift owing to the shunt negative conductance becomes significant when the receiver is operated for very narrow-band standards where the channel bandwidth is a few hundred kHz.

Frequency Tuning in the Lattice The anti-resonant frequency of the FBAR can be tuned by adding a capacitance across it [24]. Similarly, adding a capacitance (Z_L) across the output terminals of the pseudo-lattice shifts the peak frequency of its trans-impedance. This can be observed from the expression for the trans-impedance in equation (12.23). By evaluating the peak frequency numerically (using the expression in equation (12.23) and taking $Z_L = 1/(j\omega C_L)$), the peak frequency tuning of the lattice is plotted in Figure 12.26. The peak frequency drift for a single FBAR is also shown in the plot. It can be seen that a certain amount of drift in the peak frequency of the pseudo-lattice requires approximately twice the capacitance required for drifting the peak frequency of a single FBAR by the same amount. The pseudo-lattice is a symmetrical structure, in that interchanging the FBARs with the capacitances and vice versa does not alter the performance of the lattice. So, addition of the load capacitance (Z_L) has similar impact on both the FBARs and on the capacitances. In other words, if the addition of capacitance C_L induces a capacitance ΔC in

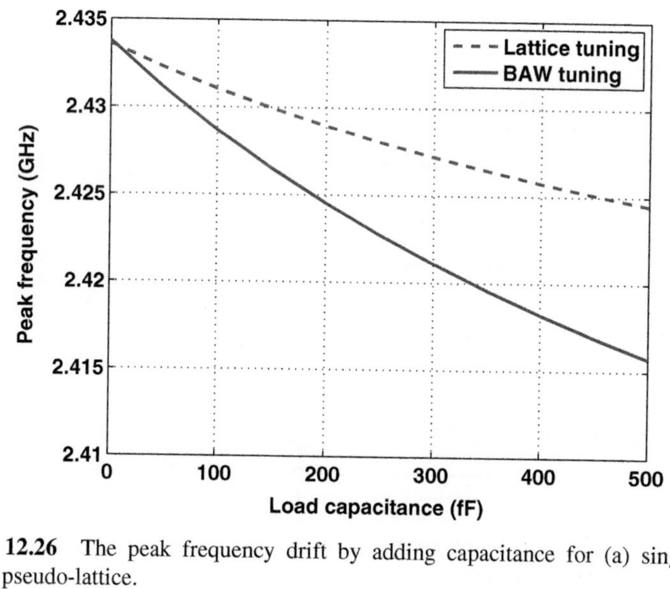

FIGURE 12.26 The peak frequency drift by adding capacitance for (a) single FBAR, (b) FBAR pseudo-lattice.

parallel with the FBARs, it induces a capacitance exactly equal to ΔC in parallel with the capacitors C_{ltc}. So at DC, the impedance of the FBARs and the capacitances in the pseudo-lattice is $(C_m + C_p + \Delta C)$ and thus results in an ideal bandpass response similar to that in an LC tank. As before, the inductance of the equivalent LC tank remains the same at L_m and the capacitance accounting for the frequency tuning can be empirically evaluated and can be expressed as

$$C_{eq,tune} \cong \frac{C_m \cdot (2C_p + C_L)}{2C_p + C_L + 2C_m}. \tag{12.30}$$

The AMF Cell The use of the FBARs in a pseudo-lattice structure significantly improves the selectivity compared to the use of a single resonator. However, this has to be complemented with a low-power transconductance stage which drives the pseudo-lattice so that the performance of the channel selection filter is highly efficient. In [24], a low-power solution called the AMF cell is proposed for the transconductance stage which employs current reuse to perform amplification, mixing, and Q-boosting of the FBAR used as the load. In this work too, the AMF cell is used but instead of a single FBAR, the FBAR pseudo-lattice is used. The AMF cell schematic is shown in Figure 12.27. It is a pseudo-differential structure with cascode transistors used for

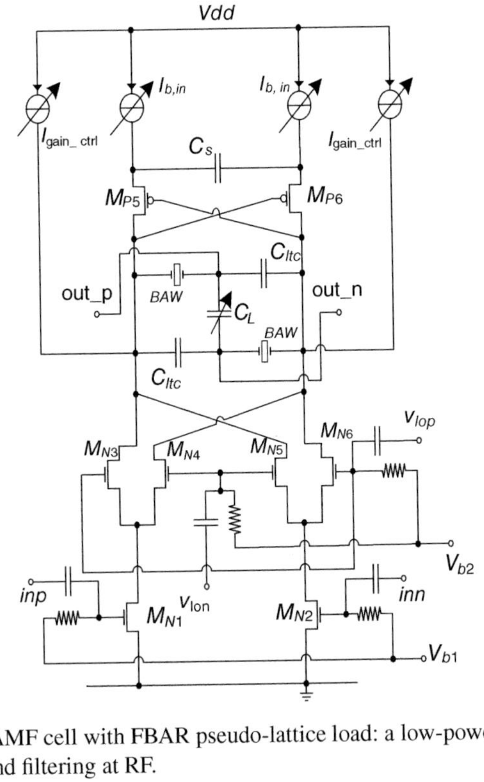

FIGURE 12.27 AMF cell with FBAR pseudo-lattice load: a low-power solution to perform channel selection and filtering at RF.

mixing. A CMFB structure (not shown) is used to set the DC voltage at the drains of the mixing transistors.

The input RF signal amplified by the LNA is fed to the AMF cell which performs channel selection and filtering. The input transistors $M_{N,1}$ and $M_{N,2}$ convert this to current (with transconductance $G_{m,in}$) which is then mixed with a low-frequency LO by the transistors $M_{N,3}$–$M_{N,6}$. This is then filtered by the FBAR pseudo-lattice placed at the drains of these mixing transistors. The Q-boosting of the lattice is provided by the PMOS cross-coupled pairs $M_{P,5}$ and $M_{P,6}$, with $G_{m,cc}$ being the transconductance of each of them. A switched capacitor array (C_L) which is digitally controlled is used as a load in the lattice to provide frequency tuning to the filter as shown in Figure 12.27. Capacitor C_s is added to ensure that the sources of the cross-coupled pair are at AC ground. To decouple the gain and the bandwidth settings, two parallel current sources (I_{gain_ctrl}) are added as in [24]. The mixing transistors are hard switched and the conversion gain of the AMF cell is given by

$$A_{conv} = \frac{2}{\pi} \cdot G_{m,in} \cdot |Z_{tran}|, \tag{12.31}$$

with $Z_{sh} = -G_{m,cc}/2$ in Z_{tran} as defined in equation (12.23).

Significance of the Capacitors C_{ltc} in the Lattice So far, we have seen that choosing two matched capacitances with a value of $C_{ltc} = C_m + C_p$, with C_m and C_p determined by the matched FBARs used in the pseudo-lattice, an ideal bandpass response similar to that of a high-Q LC tank is obtained. The term $(Z_2 - Z_1)$ in the numerator of the expression for the trans-impedance is the reason for a zero at the frequency where impedance of the FBAR Z_1 is equal to that of the capacitor $C_{ltc}(Z_2)$. At DC, these impedances are purely capacitive and equal to $(C_m + C_p)$. This is the reason for choosing a value of $(C_m + C_p)$ for the capacitances used in the pseudo-lattice and this results in a response similar to that in an LC tank with null at DC. If the capacitors C_{ltc} are tuned, then based on the impedance of these capacitors and the FBAR, a null appears at a particular frequency where all the impedances in the pseudo-lattice are exactly equal.

To better understand this phenomenon, consider the lattice network shown in Figure 12.28. When matched FBARs and matched capacitors are used, and without a load the input current divides into equal parts and flows through branches of the lattice as shown. So, the differential output voltage across the output terminals of the lattice is given by

$$V_{out} = \left(\frac{I_{in}}{2} \cdot Z_1 + V_X\right) - \left(\frac{I_{in}}{2} \cdot Z_2 + V_X\right) = \frac{I_{in}}{2} \cdot (Z_1 - Z_2). \tag{12.32}$$

It is evident from equation (12.32) that at the frequency where $Z_1 = Z_2$, the differential voltage is zero. The presence of a load does not affect this zero as already validated by the expression in equation (12.23). This property is attractive as it helps in rejecting out any strong interferer at any frequency which affects the performance

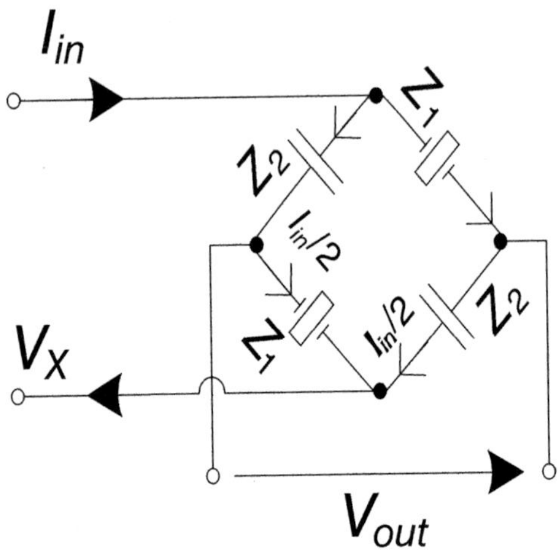

FIGURE 12.28 Significance of the capacitors C_{ltc} in the pseudo-lattice. By tuning these capacitors, the null can be positioned to reject any strong interferer.

of the receiver. This property can also be exploited to reduce the LO feedthrough arising from the mismatches in the mixing transistors. The flicker noise/DC offsets from the input pair and from the LNA preceding the AMF cell get up-converted to the frequency of the channel selection LO and can be rejected out by tuning appropriately the capacitance C_{ltc}. The simulated AMF cell response with different values of C_{ltc} is shown in Figure 12.29. It can be seen from this figure that, the peak of the

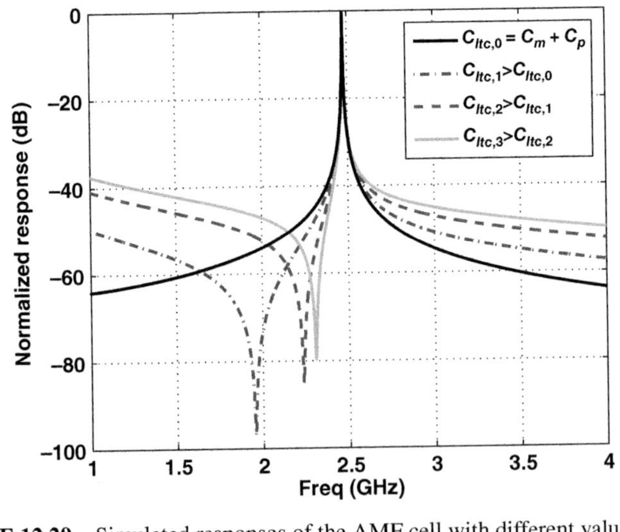

FIGURE 12.29 Simulated responses of the AMF cell with different values of C_{ltc}.

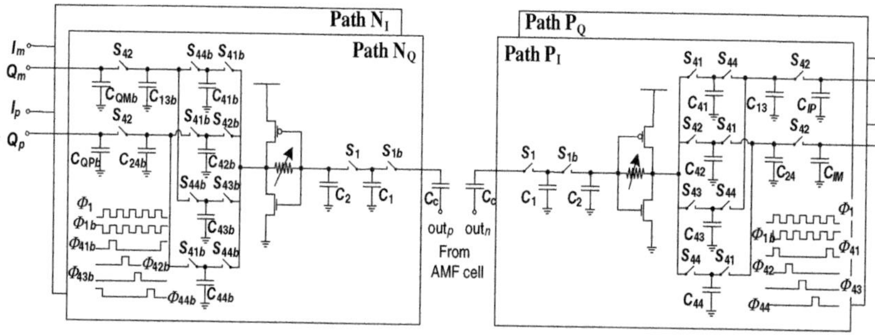

FIGURE 12.30 Quadrature two-stage sub-sampling down-conversion mixer. The second stage also provides discrete time filtering and decimates the sampling rate.

frequency response remains the same irrespective of the tuning of C_{ltc} and only the position of the null varies. This is an attractive property which allows rejecting out an strong unwanted signal without affecting the wanted signal.

12.10.3 Quadrature Sub-Sampling Mixer

The selected and filtered channel at the output of the AMF cell has to be down-converted to baseband in quadrature. Sampling the filtered signal at $f'_{fbar,f}$ by a clock with frequency of f_s as defined in equation (12.14) translates it to an intermediate frequency of $f_s/2$. This has to be further down-converted to baseband which requires the second stage of sampling. The two-stage sub-sampling mixer is shown in Figure 12.30. As seen, the differential outputs from the AMF cell are AC coupled to the sampling mixer. Four identical mixing circuits are placed, two on each side of the differential outputs of the AMF cell. This is done to make sure that the pseudo-lattice always sees a constant load on both the sides when either of the phases of the sampling clock ϕ_1 or ϕ_{1b} are high. To improve the overall conversion gain of the receiver, self-biased inverter-based amplifiers are placed between the two stages of the sub-sampling mixer as shown in the Figure 12.30.

The first stage of the sampling mixer has initial sampling switch clocked at a rate of f_s followed by another which removes the track component. The samples at $f_s/2$ are fed to the inverter-based amplifier and the small signal current is integrated on to the capacitors $C_{41(b)} - C_{44(b)}$ when the corresponding switches $S_{41(b)} - S_{44(b)}$ are turned on as shown in Figure 12.31. It is clear from Figure 12.30 that this integration is done periodically for a duration of $T_s/2$ which results in a windowed integration response (taking the bias resistance R_b to be large compared to the impedance of the capacitor C_{4x}) given by [35]

$$|H_{WI}(f)| = \frac{G_m}{2f_s C_{4x}} \cdot \left| \frac{sin(\pi f/2f_s)}{(\pi f/2f_s)} \right|. \tag{12.33}$$

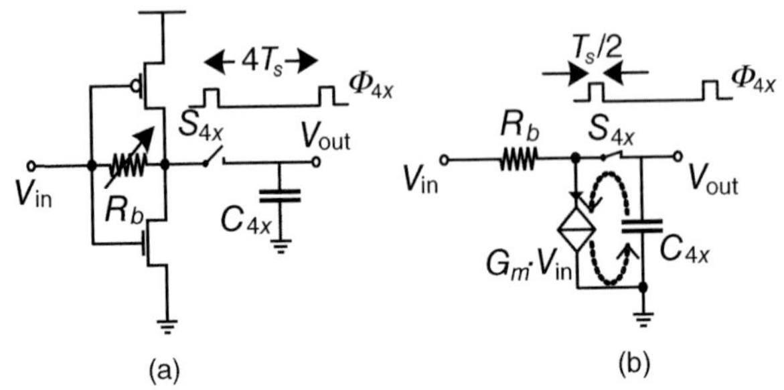

FIGURE 12.31 (a) Inverter-based IF amplifier, (b) small signal model – windowed integration.

The windowed integration results in a *sinc* response as shown in Figure 12.32. The response has nulls at all even multiples of f_s (resulting from the integration for a duration of $T_s/2$). This helps in reducing the amount of noise folding resulting from the second stage of sampling as in [35]. It is clear from the various phases of the clocks used, that the first sample from the initial sampling mixer is read out to the capacitor $C_{41(b)}$, second to $C_{42(b)}$, third to $C_{43(b)}$, fourth to $C_{44(b)}$, fifth again to capacitor $C_{41(b)}$, and so on. This periodically repeats and at the end of every cycle, the charges on capacitors $C_{41(b)}$, $C_{43(b)}$ are added and total charge is read out to capacitor $C_{13(b)}$. Similarly, capacitors $C_{42(b)}$ and $C_{44(b)}$ read out to capacitor $C_{24(b)}$. The charges from these capacitors are read out differentially as shown. To synchronize the reading out of the samples, a pair of switched capacitors driven by clock ϕ_{42} are added as shown. This way, out of 4 samples, a single sample is produced which results in re-sampling at a rate of $f_s/4$. So, this translates the channel at $f_s/2$ to baseband in quadrature. In this

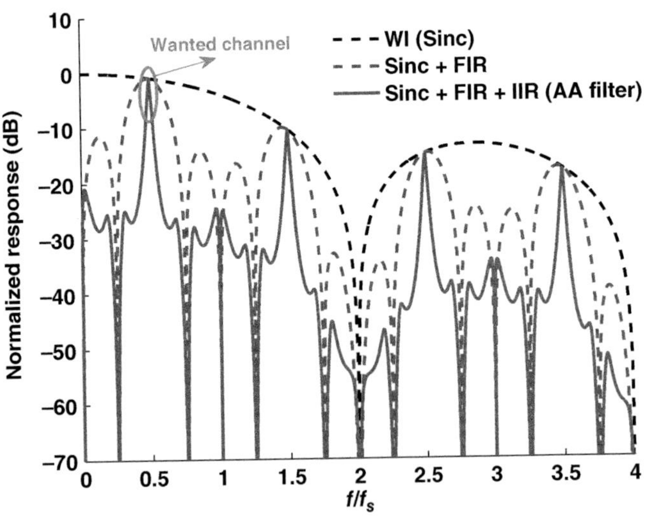

FIGURE 12.32 Filtering in the sub-sampling mixer.

process of the second down-conversion, the sampling mixer decimates the sampling rate which reduces the power consumed by the ADC. The adding of alternate samples and taking out a differential output results in an FIR response which is given by

$$H_{FIR}(z_1) = (1 + z_1^{-2}) - (z_1^{-1} + z_1^{-3}), \quad \text{with} \quad z_1 = e^{j2\pi f T_s}. \tag{12.34}$$

The capacitors in both the stages of the sampling mixer are not reset and the charge redistribution among the various capacitors in the mixer results in an IIR effect. It can be shown that the total IIR response is given by

$$|H_{iir}| = |H_{iir,1}||H_{iir,2}|, \tag{12.35}$$

$$\text{where,} \quad |H_{iir,1}(z_1)| = \frac{C_1}{C_2} \cdot \left| \frac{1}{1 - \alpha_1 \cdot z_1^{-1}} \right| \tag{12.36}$$

$$\text{and} \quad |H_{iir,2}(z)| = \frac{\alpha_2 C_a}{C_o} \left| \frac{1}{X(z)} \right| \quad \text{with,} \tag{12.37}$$

$$X(z) = M(z) \cdot (1 - \alpha_2\alpha_3 z^{-1} + \alpha_2^2 z^{-1} - z^{-1}) + \\ \alpha_2\alpha_3 \cdot ((M(z) + 1) \cdot (1 - \alpha_2) \cdot z^{-2} - z^{-1}), \tag{12.38}$$

$$M(z) = \frac{1 - (1 - \alpha_3) \cdot z^{-1}}{1 - \alpha_3} \quad \text{and} \quad z = e^{j8\pi f T_s}. \tag{12.39}$$

The capacitor ratios are given by,

$$\alpha_1 = \frac{C_2}{C_1 + C_2}, \quad \alpha_2 = \frac{C_{13(b)}}{2C_{41(b)} + C_{13(b)}} \quad \text{and} \\ \alpha_3 = \frac{C_{13(b)}}{C_{13(b)} + C_{IP(QMb)}} = \frac{C_{24(b)}}{C_{24(b)} + C_{IM(QPb)}}. \tag{12.40}$$

The second stage of the sampling mixer also requires an anti-alias filter as it re-samples at a rate of $f_{s,n} = f_s/4$. However, an interesting feature of the sampling mixer is that the second stage of the mixer has inherent anti-alias filtering functionality which prevents the unwanted signals from aliasing into the signal of interest. The windowed integration, the FIR filtering, and the IIR filtering combined together form the built-in anti-alias filtering for the second stage of the mixer, whose response is shown in Figure 12.32. As seen, the anti-alias filtering has nulls at every multiple of the new sampling rate $f_{s,n}$ except at those which are odd multiples of $f_s/2$ where the wanted signal is present. These nulls arising from the FIR filtering prevent the unwanted signals from aliasing into the signal of interest in the second stage of

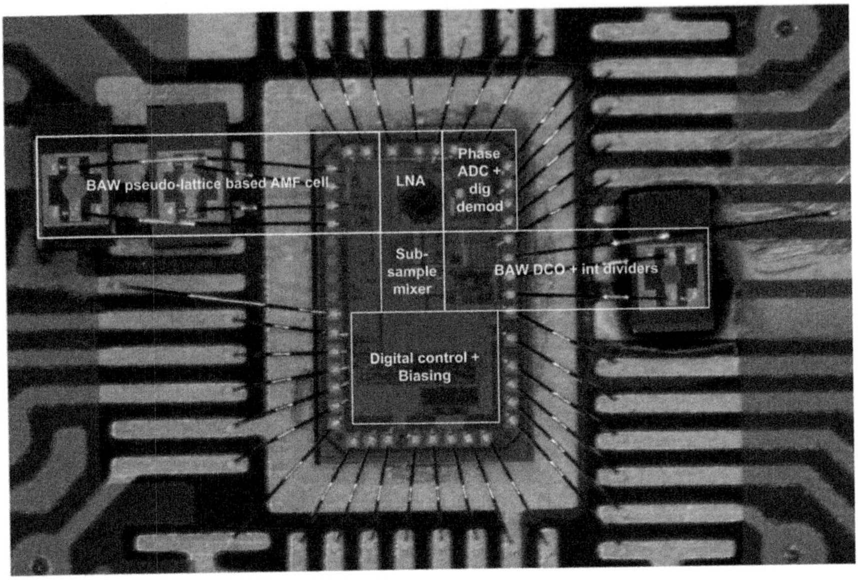

FIGURE 12.33 Chip microphotograph of the receiver. For a color version of this figure, see the color plate section.

the sampling mixer. However, it should be noted that these nulls are as a result of the differential output in the mixer and so reject only the signals at these frequencies and do not attenuate the noise.

This way the two-stage sub-sampling mixer down-converts the filtered channel to baseband in quadrature. The second stage provides discrete time filtering and feeds the down-converted samples to a phase Analog to Digital Converter (ADC) [36] at a decimated rate to reduce the power consumption of the ADC. A digital demodulator for Binary Frequency Shift Keying (BFSK) modulation scheme provides the demodulated bit out.

12.10.4 Receiver Measurement Results

The chip microphotograph of the multi-channel receiver is shown in Figure 12.33. The FBARs used for the channel filter (pseudo-lattice) and for the oscillator are wire bonded to the test chip. The complete receiver including the frequency synthesis consumes 5.94 mA current from a 1.8 V supply. The current breakdown is given in Table 12.2. The FBAR DCO without any additional capacitance (for tuning) works at 2.467 GHz and consumes 1.21 mA. A maximum tuning of 8.95 MHz is observed on the DCO and the current consumption for this maximum tuning is 1.74 mA. The FBAR DCO current consumption can be reduced when a complementary current reuse structure is used as in [37]. The AMF cell with the FBAR pseudo-lattice load peaks at 2.433 GHz with no additional tuning capacitance. The intrinsic Q of the FBARs used is around 400. This Q is boosted in the AMF cell to provide bandwidths down to 1 MHz ($Q = 2430$). The AMF cell consumes 1.08 mA with no

TABLE 12.2 Power consumption, V_{DD} = 1.8 V.

Block	Current (mA)
LNA	1.5
FBAR DCO	1.21
AMF cell	1.08
IF amplifier	4 × 0.1
Dividers, buffers, ADC, demodulator	1.75
Total	5.94

additional capacitance and for the maximum tuning it increases by 0.35 mA. The initial sampling rate is 68.5 MHz which is accordance with equation (12.19), while the decimated sampling rate is at 17.12 MHz.

Filtering in the Receiver The filtering responses (simulated and measured) of the AMF cell for various frequency tuning settings are shown in Figure 12.34. The maximum conversion gain of the RF front-end including the LNA and the AMF cell is 30.4 dB. The overall filtering response of the receiver including the AMF cell and the discrete time filtering is shown in Figure 12.35. A rejection of around 50 dB is measured at an offset of 15 MHz from the center frequency. The overall conversion gain of the receiver from RF to baseband is measured to be 46.2 dB.

Owing to the mismatches between the mixing transistors in the AMF cell, the large-amplitude, low-frequency channel selection clock feeds through to the output

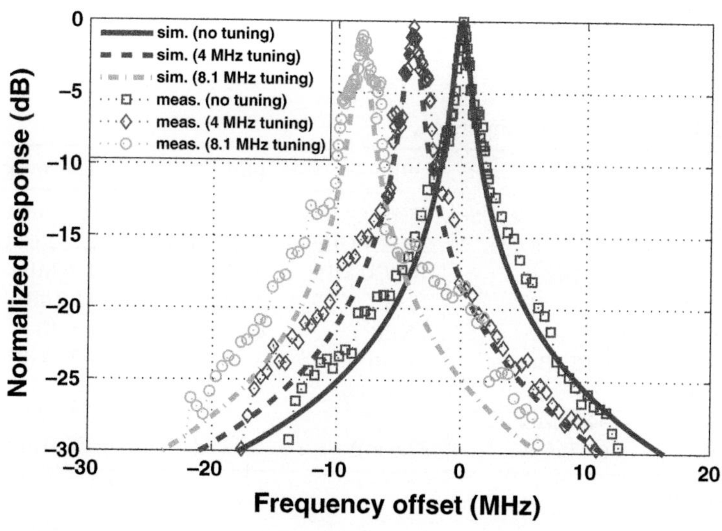

FIGURE 12.34 FBAR pseudo-lattice–based AMF cell filtering response with different tuning settings. A maximum tuning of 8.1 MHz is observed for the AMF cell. The plot is normalized to 30.4 dB which corresponds to the maximum gain with no tuning.

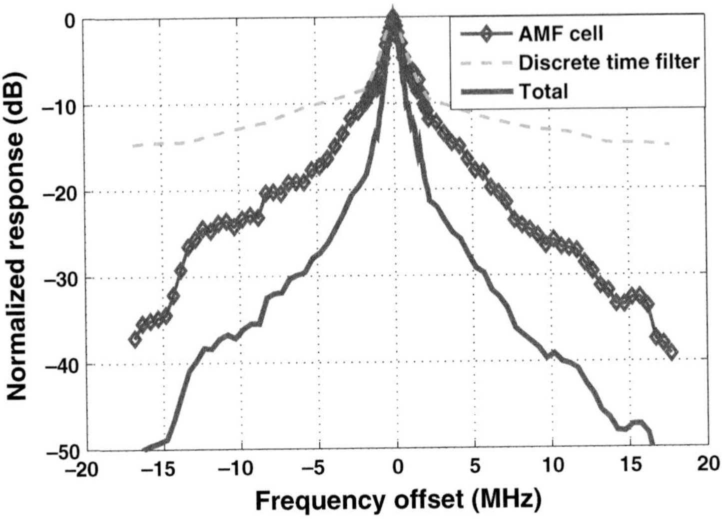

FIGURE 12.35 Overall measured response (normalized to 46.2 dB) of the receiver.

and appears close to the wanted signal after down-conversion by sub-sampling. It should be noted here that the DC offset or flicker noise from the LNA and from the input transistors of the AMF cell also get up-converted to the same frequency of the LO used for channel selection. As mentioned previously, this issue is solved by tuning the pseudo-lattice by the capacitances C_{ltc} so as to place the null at the required frequency. This attenuates the unwanted signal arising from LO feedthrough by 20 dB.

TABLE 12.3 Performance comparison.

Parameter	[38]	[39]	[40]	This work
Technology	130 nm CMOS	65 nm CMOS	65 nm CMOS	0.18 μm CMOS
Input frequency	2.4–2.48 GHz	2.4–2.7 GHz	1.8–2 GHz	2.5–2.63 GHz
Filter bandwidth	1 MHz	5/10/20 MHz	4 MHz	1–6 MHz
Maximum gain	—	30 dB	55.8 dB	46.2 dB
Noise figure	—	4.8 dB	2.8 dB	8.6 dB
Third Order Intercept (IIP3)	−15 dBm	−27 dBm	−8.4 dBm[a]	−38.7 dBm[b]
Sensitivity	−83 dBm[c]	—	—	−78 dBm[d]
Power Cons.	61.5 mW	77.6 mW	25.2 mW[e]	10.69 mW

[a]10 MHz, 20 MHz.
[b]±500 kHz.
[c]A bluetooth receiver, so the data rate is taken to be 1 Mbit/s.
[d]Data rate of 268 kbit/s.
[e]Does not include ADC and DBB.

12.11 SUMMARY OF THE FBAR-BASED RX

A novel way to perform channel selection and filtering at RF using FBARs has been presented herein. The FBAR pseudo-lattice filter limits the noise bandwidth which prevents noise folding, a major drawback in sampling receivers. The sub-sampling mixer provides discrete time filtering which complements the channel filter and improves the overall performance of the receiver. Employing current reuse in the LNA and in the AMF cell helps in achieving a low-power solution. A PLL-free approach for addressing multiple channels (arbitrary frequency) in the band of interest is presented. The use of integer dividers and sampling-based down-conversion helps in easy scalability to deep sub-micron processes.

12.12 CONCLUSION

MEM resonators (FBAR) prove superior to quartz crystals for building ULP WBAN systems owing to their fast wake-up which serves to reduce the energy overhead in the frequency synthesizer. In addition, they provide $high - Q$ band-pass filtering at RF which aids in channel selection. Combined with their very small size (form factor) as opposed to the bulky quartz crystals, they prove to be an ideal fit for implantable/wearable wireless sensor nodes.

REFERENCES

[1] A. Alwan, *Global Status Report on Noncommunicable Diseases 2010.* World Health Organization, 2011.

[2] S. Movassaghi, M. Abolhasan, J. Lipman, D. Smith, and A. Jamalipour, "Wireless body area networks: A survey," *IEEE Commun. Surveys Tutorials*, vol. 16, no. 3, pp. 1658–1686, 2014.

[3] M. Burrows, *Global Trends 2030: Alternative Worlds. US National Intelligence Council*, 2012.

[4] B. Latré, B. Braem, I. Moerman, C. Blondia, and P. Demeester, "A survey on wireless body area networks," *Wireless Netw.*, vol. 17, no. 1, pp. 1–18, 2011.

[5] ON world, "Mobile Healthnews: Prediction: Wearables to lead the 515 million sensors to ship in 2017," June 2013. Available at http://mobihealthnews.com/22752

[6] J. W. Rogers, C. Plett, and F. Dai, *Integrated Circuit Design for High-Speed Frequency Synthesis*. Artech House, Boston, London, 2006.

[7] J. Jin, X. Liu, T. Mo, and J. Zhou, "Quantization noise suppression in fractional-*N*PLLs utilizing glitch-free phase switching multi-modulus frequency divider," *IEEE Trans. Circuits Syst. I: Regular Pap.*, vol. 59, no. 5, pp. 926–937, April 2012.

[8] S. Pamarti, L. Jansson, and I. Galton, "A wideband 2.4-GHz delta-sigma fractional-NPLL with 1-Mb/s in-loop modulation," *IEEE J. Solid-State Circuits*, vol. 39, no. 1, pp. 49–62, January 2004.

[9] R. Dutta, R. A. van der Zee, M. J. Bentum, and A. B. Kokkeler, "Choosing Optimum Noise Figure and Data Rate in Wireless Sensor Network Radio Transceivers," in 2011 IEEE International Conference on Communications (ICC), Kyoto, June 5–9, 2011, pp. 1–5.

[10] Y.-H. Liu, X. Huang, M. Vidojkovic, A. Ba, P. Harpe, G. Dolmans, and H. de Groot, "A 1.9 nJ/b 2.4 GHz Multistandard (Bluetooth Low Energy/Zigbee/IEEE802. 15.6) Transceiver for Personal/Body-Area Networks," in 2013 IEEE International Solid-State Circuits Conference Digest of Technical Papers (ISSCC), San Francisco, CA, February 17–21, 2013, pp. 446–447.

[11] A. C. W. Wong, M. Dawkins, G. Devita, N. Kasparidis, A. Katsiamis, O. King, F. Lauria, J. Schiff, and A. J. Burdett, "A 1 V 5 mA Multimode IEEE 802.15. 6/Bluetooth Low-Energy WBAN Transceiver for Biotelemetry Applications," *IEEE J. Solid-State Circuits*, vol. 48, no. 1, pp. 186–198, January 2013.

[12] Y. Zhang and D. Chen, *Multilayer Integrated Film Bulk Acoustic Resonators*. Springer Science & Business Media, 2012.

[13] E. A. Vittoz, *Low-Power Crystal and MEMS Oscillators: The Experience of Watch Developments*. Springer Science & Business Media, 2010.

[14] K. Wang, J. Koo, R. Ruby, and B. Otis, "A 1.8 mW PLL-Free Channelized 2.4 GHz ZigBee Receiver Utilizing Fixed-LO Temperature-Compensated FBAR Resonator," in 2014 IEEE International Solid-State Circuits Conference Digest of Technical Papers (ISSCC), San Francisco, CA, February 9–13, 2014, pp. 372–373.

[15] M. Flatscher, M. Dielacher, T. Herndl, T. Lentsch, R. Matischek, J. Prainsack, W. Pribyl, H. Theuss, and W. Weber, "A Robust Wireless Sensor Node for In-Tire-Pressure Monitoring," in IEEE International Solid-State Circuits Conference-Digest of Technical Papers, 2009. ISSCC 2009, San Francisco, CA, February 8–12, 2009, pp. 286–287.

[16] J. Shi and B. P. Otis, "A sub-100 μW 2GHz differential colpitts CMOS/FBAR VCO," in 2011 IEEE Custom Integrated Circuits Conference (CICC), San Jose, CA, September 19–21, 2011, pp. 1–4.

[17] D. Ruffieux, "A High-Stability, Ultra-Low-Power Quartz Differential Oscillator Circuit for Demanding Radio Applications," in Proceedings of the 28th European Solid-State Circuits Conference, ESSCIRC 2002, Florence, Italy, September 24–26, 2002, pp. 85–88.

[18] E. Vittoz and J. Fellrath, "Cmos analog integrated circuits based on weak inversion operations," *IEEE J. Solid-State Circuits*, vol. 12, no. 3, pp. 224–231, June 1977.

[19] S. Gilbert, F. Zhang, R. Parker, M. Small, F. Bi, L. Callaghan, S. Ortiz, and R. Ruby, "Sub-10 fs Jitter S-band Oscillators and VCOs in a $1 \times 1 \times 0.23$ mm^3 Chip Scale Package," in 2013 Joint European Frequency and Time Forum & International Frequency Control Symposium (EFTF/IFC), Prague, July 21–25, 2013, pp. 171–174.

[20] R. Thirunarayanan, D. Ruffieux, F. Pengg, N. Scolari, P. Persechini, and C. Enz, "A 700 pJ/bit, 2.4 GHz, Narrowband, PLL-Free Burst Mode Transmitter Based on an FBAR with 5 μs Startup Time for Highly Duty-Cycled Systems," in 2014 IEEE Radio Frequency Integrated Circuits Symposium,Tampa, FL, June 1–3, 2014, pp. 25–28.

[21] C. C. Enz, J. Chabloz, J. Baborowski, C. Muller, and D. Ruffieux, "Building Blocks for an Ultra Low-Power MEMS-Based Radio," in IEEE International Workshop on Radio-Frequency Integration Technology, 2007, RFIT 007, Rasa Sentosa Resort, December 9–11, 2007, pp. 158–167.

[22] D. Ruffieux, F. Krummenacher, A. Pezous, and G. Spinola-Durante, "Silicon resonator based 3.2 W real time clock with 10 ppm frequency accuracy," *IEEE J. Solid-State Circuits*, vol. 45, no. 1, pp. 224–234, January 2010.

[23] "Bluetooth Core Specification 4.1," December 2013. Available at http://www.bluetooth.org/en-us/specification/adopted-specifications

[24] A. Heragu, D. Ruffieux, and C. Enz, "A low power BAW resonator based 2.4-GHz receiver with bandwidth tunable channel selection filter at RF," *IEEE J. Solid-State Circuits*, vol. 48, pp. 1343–1356, June 2013.

[25] A. Heragu, D. Ruffieux, and C. Enz, "A 2.4-GHz MEMS-based PLL-free multi-channel receiver with channel filtering at RF," *IEEE J. Solid-State Circuits*, vol. 48, no. 6, pp. 1343–1356, April 2013.

[26] D. Ruffieux, M. Contaldo, J. Chabloz, and C. Enz, "Ultra Low Power and Miniaturized MEMS-Based Radio for BAN and WSN Applications," in 2010 Proceedings of the ESSCIRC, Seville, September 14–16, 2010, pp. 71–80.

[27] J. Hu, R. Parkery, R. Ruby, and B. Otis, "A Wide-Tuning Digitally Controlled FBAR-Based Oscillator for Frequency Synthesis," in 2010 IEEE International Frequency Control Symposium (FCS), Newport Beach, CA, June 1–4, 2010, pp. 608–612.

[28] J. Chabloz, D. Ruffieux, and C. Enz, "A Low-Power Programmable Dynamic Frequency Divider," in 34th European Solid-State Circuits Conference, 2008, ESSCIRC 2008, Edinburgh, September 15–19, 2008, pp. 370–373.

[29] M.-A. Dubois, J.-F. Carpentier, P. Vincent, C. Billard, G. Parat, C. Muller, P. Ancey, and P. Conti, "Monolithic above-IC resonator technology for integrated architectures in mobile and wireless communication," *IEEE J. Solid-State Circuits*, vol. 41, pp. 7–16, January 2006.

[30] J. Fischer, "Noise sources and calculation techniques for switched capacitor filters," *IEEE J. Solid-State Circuits*, vol. 17, pp. 742–752, August 1982.

[31] B. Murmann, "Thermal noise in track-and-hold circuits: analysis and simulation techniques," *IEEE Solid-State Circuits Mag.*, vol. 4, pp. 46–54, June 2012.

[32] J. Chabloz, C. Muller, F. Pengg, A. Pezous, C. Enz, and M.-A. Dubois, "A Low-Power 2.4 GHz CMOS Receiver Front-end Using BAW Resonators," in IEEE International Solid-State Circuits Conference, 2006, ISSCC 2006, Digest of Technical Papers, San Francisco, CA, February 6–9, 2006, pp. 1244–1253.

[33] M. Contaldo, D. Ruffieux, and C. Enz, "A 5.4 dBm 42 mW 2.4 GHz CMOS BAW-Based Quasi-Direct Conversion Transmitter," in 2010 IEEE International Solid-State Circuits Conference Digest of Technical Papers (ISSCC), San Francisco, CA, February 7–11, 2010, pp. 498–499.

[34] C. Enz, J. Baborowski, J. Chabloz, M. Kucera, C. Muller, D. Ruffieux, and N. Scolari, "Ultra Low-Power MEMS-Based Radio for Wireless Sensor Networks," in 18th European Conference on Circuit Theory and Design, 2007, ECCTD 2007, Seville, August 27–30, 2007, pp. 320–331.

[35] A. Abidi, "The path to the software-defined radio receiver," *IEEE J. Solid-State Circuits*, vol. 42, pp. 954–966, May 2007.

[36] B. Banerjee, C. Enz, and E. Le Roux, "A 290 ua, 3.2 Mhz 4-bit Phase ADC for Constant Envelope, Ultra-Low Power Radio," in NORCHIP, 2010, Tampere, November 15–16, 2010, pp. 1–4.

[37] R. Thirunarayanan, A. Heragu, D. Ruffieux, and C. Enz, "Complementary BAW Oscillator for Ultra-Low Power Consumption and Low Phase Noise," in 2011 IEEE 9th International New Circuits and Systems Conference (NEWCAS), Bordeaux, June 26–29, 2011, pp. 97–100.

[38] R. Staszewski, K. Muhammad, D. Leipold, C.-M. Hung, Y.-C. Ho, J. Wallberg, C. Fernando, K. Maggio, R. Staszewski, T. Jung, J. Koh, S. John, I. Y. Deng, V. Sarda, O. Moreira-Tamayo, V. Mayega, R. Katz, O. Friedman, O. Eliezer, E. de Obaldia, and P. Balsara, "All-digital TX frequency synthesizer and discrete-time receiver for Bluetooth radio in 130-nm CMOS," *IEEE J. Solid-State Circuits*, vol. 39, pp. 2278–2291, December 2004.

[39] F. Montaudon, R. Mina, S. Le Tual, L. Joet, D. Saias, R. Hossain, F. Sibille, C. Corre, V. Carrat, E. Chataigner, J. Lajoinie, S. Dedieu, F. Paillardet, and E. Perea, "A Scalable 2.4-to-2.7 GHz Wi-Fi/WiMAX Discrete-Time Receiver in 65 nm CMOS," in IEEE International Solid-State Circuits Conference, 2008, ISSCC 2008, Digest of Technical Papers, San Francisco, CA, February 3–7, 2008, pp. 362–619.

[40] A. Mirzaei, H. Darabi, and D. Murphy, "A low-power process-scalable super-heterodyne receiver with integrated high- Q filters," *IEEE J. Solid-State Circuits*, vol. 46, pp. 2920–2932, December 2011.

13

EXPLORING PHYSIOLOGICAL FEATURES FROM ON-BODY RADIO CHANNELS

MAX O. MUNOZ AND YANG HAO

School of Electronic Engineering and Computer Science, Queen Mary University of London, London, UK

13.1 INTRODUCTION

The integration and connectivity of different electronic devices—such as smart phones, wrist-worn watches, and even home entertainment systems—have become a part of our daily life. The current devices offer not only human interaction, but also inter-operability between them, creating dedicated and independent wireless networks (wireless personal area networks (WPANs) and wireless body area networks (WBANs) [1, 2]). The new smart devices are capable of processing and controlling unique applications which are transmitted through seamless wireless channels operating at exclusive radio frequencies.

The continuous development of conformal and compact low-power electronics have enabled the miniaturization of hardware systems and, thus, led to a new generation of wearable sensors (Figure 13.1) [3, 4]. Some of these body-worn sensors are equipped with a type of piezoelectric transducer for harvesting energy from biomechanical and physiological processes—such as kinetic movements, external forces, vibrations, and thermal energy—into a source of power [4–6].

Most of the research efforts have been directed toward continuous monitoring and quantification of physiological body signals, as well as the development of personalized healthcare devices [7–9]. In recent years, the demand for non-invasive sensing

Electromagnetics of Body-Area Networks: Antennas, Propagation, and RF Systems, First Edition.
Edited by Douglas H. Werner and Zhi Hao Jiang.

FIGURE 13.1 A look inside the housing of a prototype of the electric potential integrated circuit (EPIC) electrocardiogram tracking device shows that it is actually an array of 16 sensors that together ensure accuracy and reliability. *Source:* Jones 2011 [3] and Gyselinckx et al. 2005 [4]. Reproduced with permission of IEEE.

solutions that can effectively replace existing physiological sensors have attracted great interest, not only from the academic research, but also from the industry. Continuous non-invasive remote sensing systems have remained an elusive technology; however, the development of novel and pioneering RF front-end architectures combined with advanced digital signal processing algorithms have demonstrated several potential benefits of non-contact sensing. These new contact-less sensing systems are able to detect and continuously monitor biological and biomechanical signals such as heartbeat variability, respiration rate, blood pressure, perspiration, and temperature [10].

Non-invasive monitoring systems that detect heart and respiration activity are particularly important in evaluating patient's pathologies, especially if the patient has conditions that can be perturbed or worsened by contact sensors, for example, neonates (infants at risk of sudden infant death syndrome) and burn victims [11–13]. Moreover, non-invasive sensing has also been evaluated in the search of survivors from different natural disasters (earthquakes, tsunamis) [14]. It is evident that this new trend, and the use of on-body wireless sensors within hospitals and home environments, can significantly reduce healthcare costs and also improve patient's daily lifestyle.

The chapter describes different non-invasive RF technologies that have evolved and shown the potential for remote monitoring of vital signs. Section 13.2 gives a brief overview of important physiological signals, with an emphasis on respiration rate and heartbeat. Section 13.3 presents distinct non-invasive sensing techniques and highlights some of the leading applications which can be adopted not only in clinical environments but also in home healthcare applications. Finally, concluding remarks of the chapter are presented in Section 13.4.

13.2 PHYSIOLOGICAL INFORMATION PARAMETERS

Extensive studies of the cardio-respiratory performance can be found in physiological literature which dwell in detail on the relationship between human activity and vital processes implicated in human cardiovascular and respiratory systems. Generally, the number of breaths and heartbeats per minute varies with the subject's age, gender, body mass index (BMI) (i.e., the ratio of body's weight and the square of height) and, of course, the subject's activity level. The pulse rate of a resting adult has an average frequency of 1–1.5 Hz, or 60–90 beats per minute (BPM). On the other hand, the resting respiration rate is about 6–12 breaths/min [15,16]. The biomechanical process of the cardiovascular and the respiration cycles produce abdominal displacements of 0.2–0.5 mm and 4–12 mm, respectively. During a normal walk, the normal pulse increases up to 95 BPM and the respiration rate varies between 12–18 breaths/min. The practice of any sport or fitness activities, such as jogging or swimming, causes the heartbeat rate to increase to almost twice that of a resting adult: 120–190 BPM (2–3.1 Hz, respectively) [16].

The cardiac recordings for a male test subject of 168 cm height and 80 kg weight is shown in Figure 13.2. The graphs depict time and spectral domain plots for two different scenarios: resting while standing and jogging at a constant speed of 5 km/h. The heart rate variability for both scenarios is listed in Table 13.1 considering a 5 minute time frame.

In the case of a resting scenario, the plot shows a main harmonic at 1.46 Hz describing a heart rate of 87.6 BPM. In contrast, a jogging scenario depicts two main frequency components: the first at 1.04 Hz and the second at 1.71 Hz. The former is mainly attributed to electrode movement due to constant human motion and the latter harmonic represents the cardiac signal, which outlines a heart rate of 102.6 BPM.

The transformation from time to frequency domain has introduced a heart rate (HR) deviation of 1.77% and 3.4% for resting and jogging scenarios, respectively. This deviation is mainly attributed to FFT coefficient rounding errors and floating point arithmetic quantization errors. It is evident that the ECG mainly records electrical variation of the heart over a period of time across the electrodes, whereas non-invasive EM-based physiological sensing systems detect the mechanical movement of the chest cavity [17,18].

13.3 METHODS FOR NON-INVASIVE PHYSIOLOGICAL DETECTION

In recent years, great attention has been given to the development and evaluation of non-invasive electromagnetic (EM) sensing systems. Nowadays, everyone is exposed to a complex mix of wireless communication sources such as Wi-Fi, GSM, GPS,

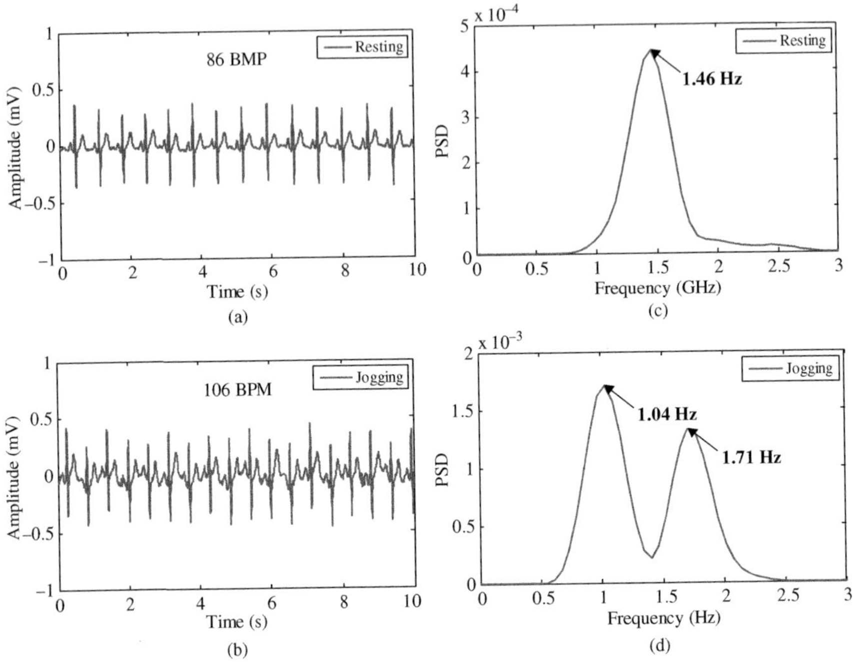

FIGURE 13.2 Heartbeat signal of the test subject while: (a) standing on the treadmill machine (resting stage), the average heartbeat was 86 BPM; (b) jogging at a constant speed of 5 km/h, the heartbeat rate was 106 BPM. Power spectral density (PSD) response using Welch's method for recorded ECG signals while the test subject is (c) standing on the treadmill machine (resting stage); (d) jogging at a constant speed of 5 km/h.

Bluetooth, and ZigBee which continuously radiate and propagate EM signals across our surrounding environment. In theory, the radio propagation information from each source can be used as a potential contact-less sensing method.

Several RF-based techniques have been proposed and tested at different microwave frequencies. The basic principle of this technology relies on the transmission of electromagnetic waves toward the sensing target. The reflected amplitude and phase are measured and recorded by other on/off-body receiving antenna architectures. The collected data embeds information not only from the radio channel link, but from the subject's bio-electromagnetic signature as well.

TABLE 13.1 Heart rate variability for resting and jogging scenarios calculated from a 5 min segment frame.

Scenario	HR mean	HR SD (σ)
Resting	86.07	4.83
Jogging at 5 km/h	106.3	4.29

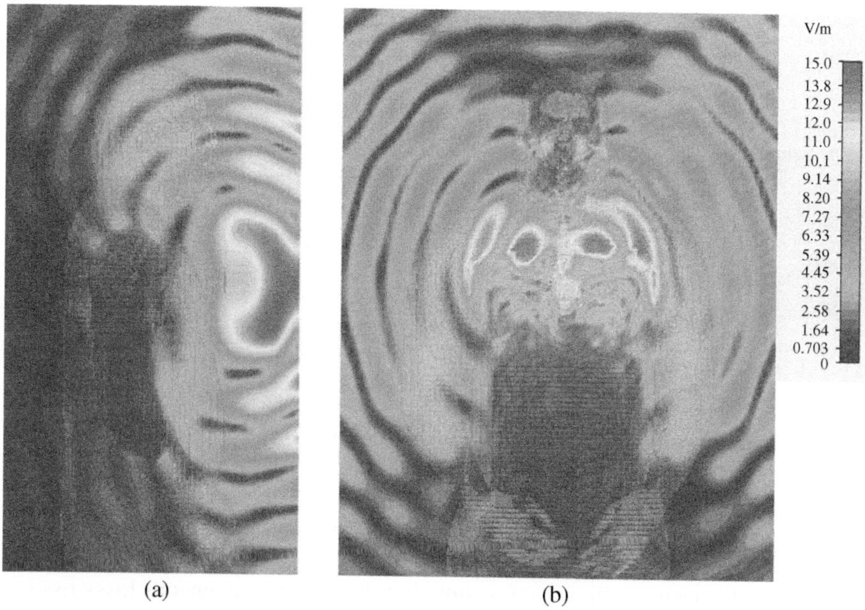

FIGURE 13.3 Simulation results of the EM wave propagation alongside the trunk section of the Hugo model implemented in CST Microwave Studio. The radiating element is a microstrip patch antenna. (a) Antenna on-body; (b) antenna off-body. For a color version of this figure, see the color plate section.

It is evident that the electromagnetic wave propagation of on/off-body antennas are highly influenced by the behavior of the transmission path. For instance, the antenna of an on-body wireless node placed at chest level (Figure 13.3a) radiates electromagnetic waves that travel in free space and alongside the body surface. In the case of an off-body antenna (Figure 13.3b), the travelling EM wave impinges on the surface of the body producing surface, scattered, and reflected waves (i.e., multipath, shadowing).

Most of the non-invasive EM sensing systems have been implemented within research laboratory environments adopting one of the following methods: received signal strength (RSS), continuous-wave (CW), and wide band pulsed (UWB) signals. In order to extract and classify physiological features, acquired data are post-processed using a wide range of advanced digital signal processing algorithms.

13.3.1 Received Signal Strength-Based Methods

The radio channel characteristics of different on/off-body wireless communication devices can be described using the RSS values (packet-based transmissions). The signal strength is dependent on a number of factors which include the output power of the transmitter, the sensitivity of the receiver, the gain of the antennas, and the attenuation of the signal as it travels through the air from the transmitter to the receiver (path loss). Wireless devices operating in close proximity to the human body considerably

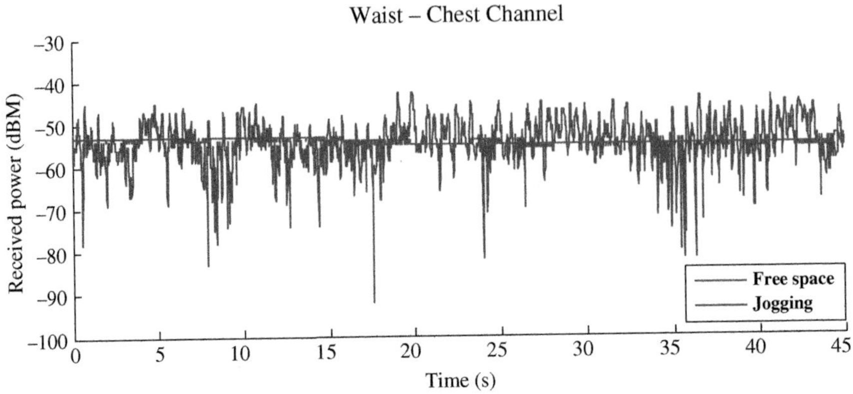

FIGURE 13.4 Recorded received signal strength (RSS) for a waist-to-chest channel recorded by custom-built wireless sensor nodes when the test subject is jogging and motionless in Queen Mary's Human Performance Laboratory.

attenuate the electromagnetic wave transmission due to inhomogeneous lossy tissue medium, and hence, drastically affect the received power signal. Moreover, external perturbations, such as human mobility and operation in cluttered environments, produce variant and complex EM wave propagation [19].

Empirical results of waist-to-chest radio channels have shown that dynamic scenarios create variations of ±15 dB from the average received signal; which is significantly greater than received signals of motionless users in free-space environments, where variations are ±3 dB (Figure 13.4) [18]. The practice of any sport activity produces a high level of fluctuations on the received signal, which are the consequence of the continuous movement of the human body.

Munoz et al. demonstrate the impact of motion on the radio propagation characteristics at 2.45 GHz from three different sport activities, namely, jogging, cycling, and rowing [17,18]. The packet-based RSS values are recorded for four on-body channels using custom-built wireless sensor nodes that are connected to a 2.15 dBi microstrip patch antenna (Figure 13.5b). The average power loss from on-body channels during dynamic tests have high variation amongst different links; however, back locations present the most correlated data between different actions (box plots, Figure 13.5c). The observed correlation is a consequence of the reduced movement on the back, in contrast to wrist and ankle sensor locations. In the case of the waist–ankle channel, data is highly correlated for two scenarios: jogging and cycling. This behavior is expected as a consequence of the constant speed of the user (5 km/h). The results suggest that the combination of unique radio propagation characteristics from two or three channels and implementing machine learning techniques is sufficient to determine the respective activity and even to classify the types of limb movements. The latter is accomplished using classification procedures such as support vector machine (SVM) and K-nearest neighbor methods, which are useful for rehabilitation procedures of individuals with functional limitations (kinesiotherapy applications) [20,21].

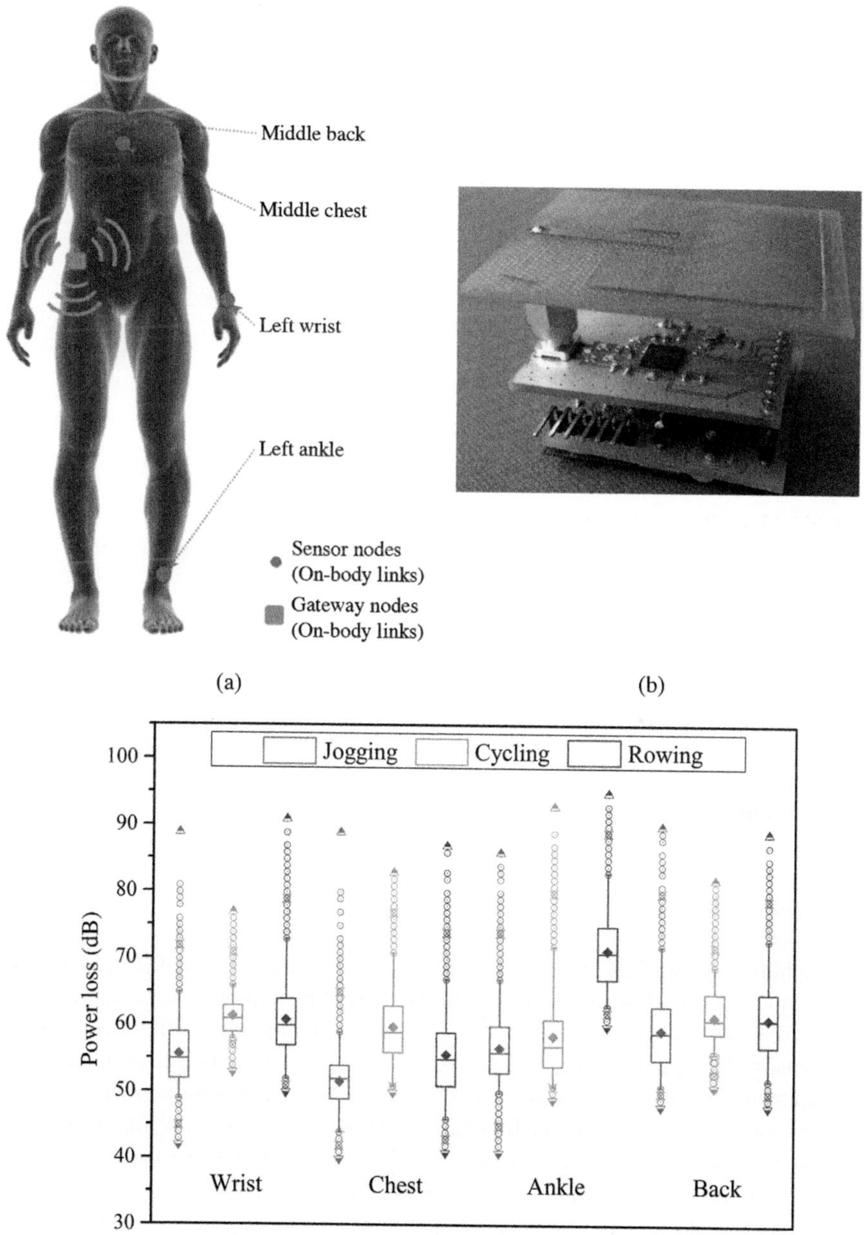

FIGURE 13.5 (a) Location of custom-built wireless sensors nodes used for dynamic WBAN radio channel characterization for each athletic activity; (b) custom-built wireless sensor nodes; (c) power loss for different on-body radio channels estimated from recorded RSS while test subject is under physical workout. The spacing between the different parts of the box indicates the degree of dispersion and skewness of the data.

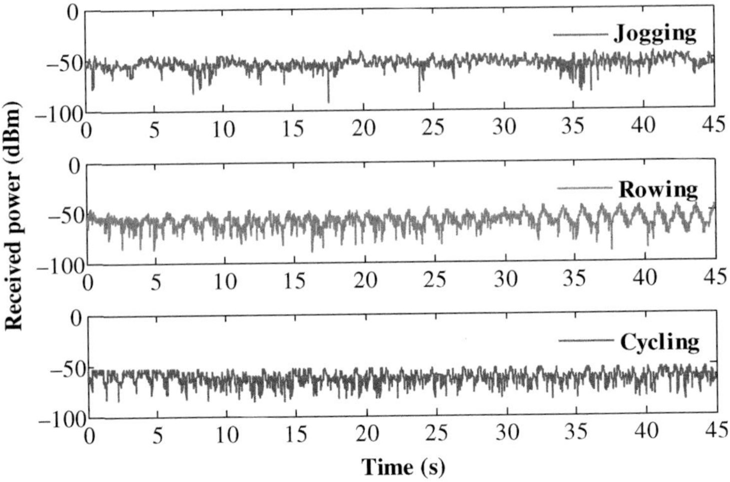

FIGURE 13.6 Received signal strength (RSS) for waist–chest channels recorded with custom-built wireless sensor nodes from three different activities: (a) jogging, (b) rowing, and (c) cycling. The measurements were performed in Queen Mary's Human Performance Laboratory.

Furthermore, Munoz et al. show that breathing rate can also be extracted from RSS measurements of waist-to-chest channels, even when the person is doing some exercises, but not the heart rate. The received signal of waist-to-chest channels for rowing, cycling, and jogging, with mean values of −50.19 dBm, −62.68 dBm, and −58.57 dBm, respectively, (Figure 13.6) is normalized and filtered using a 10th-order elliptic band-pass infinite impulse response (IIR) filter. In order to avoid frequency component confusion, two different band-pass filters are implemented: the first filter with cut-off frequencies of 0.1 Hz (lower frequency) and 0.8 Hz (higher frequency) for breathing analysis, and the second filter with cut-off frequencies of 0.8 Hz (lower frequency) and 3 Hz (higher frequency) for heart-beat movement.

The frequency domain analysis is based on the estimation of the power spectrum density (PSD) from filtered data. The PSD evaluation methods can be divided into parametric, for example, Autoregressive (AR) model, and non-parametric methods. The non-parametric methods, for example, Welch's periodogram, are in general, faster to compute and in the case of Welch's FFT method, it reduces the variance of the spectral density by averaging out uncorrelated noise.

This method divides the time series samples into overlapping sub-sequences. Each sub-sequence is windowed and then the estimated spectral density is averaged using a Hanning window (13.1) which smooths uncorrelated data located at the edges, diminishing aliasing of different rooted information and minimizing the amplitude dispersion into other harmonics (reduced spectral leakage). Additionally, the window function limits the extent of the sequence providing a more stationary spectral characteristic. The terms n and N in (13.1) represent the discrete-time index and the

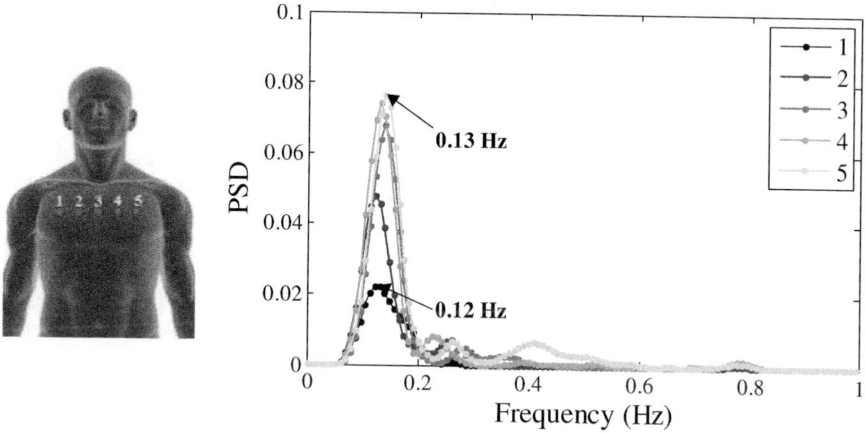

FIGURE 13.7 Power spectral response recorded by Wireless Sensor Networks (WSNs) for five different positions alongside the trunk section while the test subject is resting (standing on the treadmill machine).

length of the window, respectively. The window length was selected such that reasonable spectral performance is achieved. In the data analysis, N is limited to 4096 samples (without zero padding) with window sections of 1024 and window overlaps of 512.

$$w_n = 0.5\left[1 - \cos\left(2\pi\frac{n}{N}\right)\right] \quad \text{for } 0 \leq n \leq N \tag{13.1}$$

The spectral content of a resting scenario (test subject standing on the treadmill machine) is taken for five different locations, mainly along the upper chest section. The resulting plot (Figure 13.7) depicts the main component at these locations, which occurs between 0.12 Hz and 0.13 Hz; this would be equivalent to a breathing rate of 7.2–7.8 breaths/min.

In the case of dynamic scenarios, the spectral response of a jogging subject (Figure 13.8a) shows a sub-harmonic at 0.14 Hz (lower frequencies), which could be regarded as the thoracic displacement produced by the breathing process (approximately 8.5 breaths/min). The frequency domain plot of a rowing activity has two main components at lower frequencies (namely, 0.17 Hz and 0.71 Hz) (Figure 13.8b). The lower component, 0.17 Hz, is mainly attributed to the thoracic movement produced either by the human motion or the breathing process, which would relate to an average respiration rate of 10.2 breaths/min. The last spectrum plot corresponds to a cycling scenario (Figure 13.8c), where main components are observed at 0.16 Hz. The motion observed during cycling is mainly dominated by the leg movement, whereas the chest area remains fairly motionless, in contrast to the other activities. This type of body posture creates a cavity section for the electromagnetic propagation which maximizes the number of waves travelling along the body and hence the 0.16 Hz

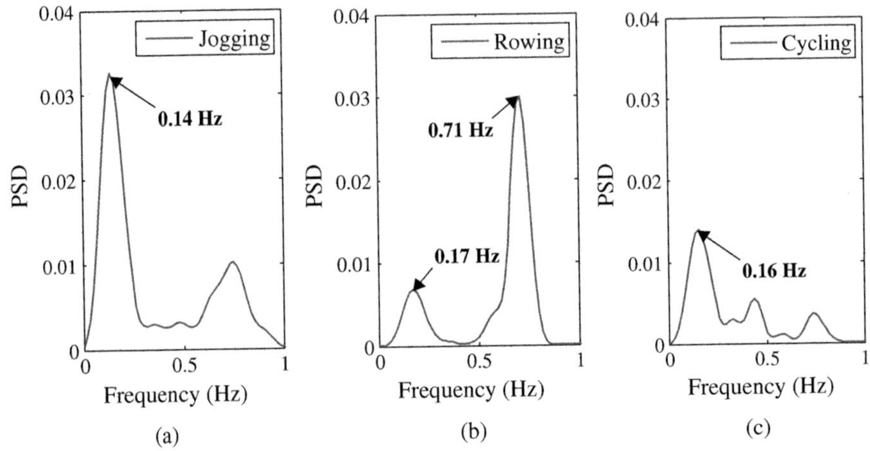

FIGURE 13.8 Power spectral density response using Welch's method for RSS signals while the test subject is (a) jogging; (b) rowing, and (c) cycling. All the activities are performed at a constant speed of 5 km/h.

harmonic is interpreted as a result of the respiration process which would be equivalent to 9.6 breaths/min.

It is shown that a wireless network topology has a better sensing performance than a single link. This non-invasive approach is applied by an application called BreathTaking which uses RSS values from a wireless sensor network of 20 MEMSIC TelosB following the IEEE 802.15.4 protocol [22] (Figure 13.9a).

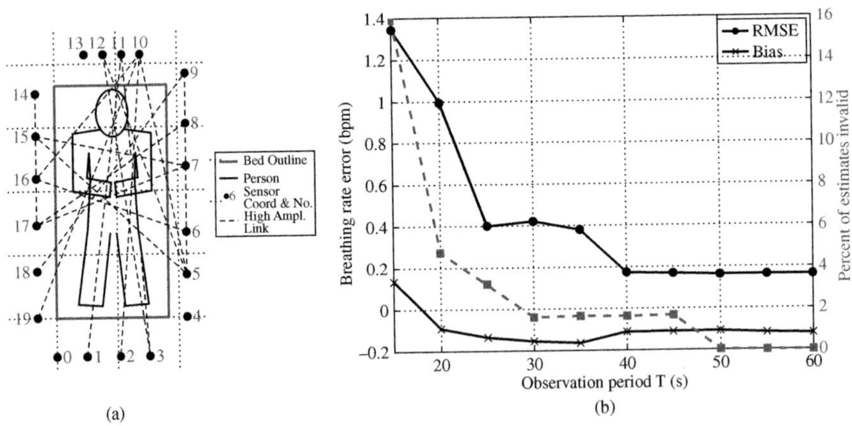

FIGURE 13.9 (a) Experimental layout showing sensors, bed, and person's approximate position; (b) RMSE and bias of valid BreathTaking frequency estimates for controlled breathing/patch antenna experiment (actual breathing rate of 15 breaths/min), and percent of rate estimates that are invalid (right y-axis), versus observation period T. *Source:* Patwari et al. 2014 [22]. Reproduced with permission of IEEE.

The application uses two different types of antennas: a directional patch antenna with an 8 dBi gain and an omnidirectional horizontal pattern dipole antenna with a 2.25 dBi gain. Although it remains possible to detect breathing using dipole antennas, it is clearly better, in terms of robustness to motion occurring outside of the test environment, to use directional antennas.

The non-invasive monitoring system reliably detects breathing and estimates breathing rate to within 0.3 breaths/min based on 30 seconds of data (Figure 13.9b). Moreover, the RSS-based sensing model is not only used to estimate respiration rate, but also to locate the breathing pattern of a person in home environments [23]. The extraction of the physiological parameter is based on a normalized sum of the squared amplitudes $|\widehat{A}|^2$ (maximum likelihood estimator (MLE)) over all measured radio links, which includes frequency, link amplitudes, and link phases. The method also removes motion interference produced by other movements, which generally cause large variations in RSS values, apart from a person's breathing. It is interesting to note that phase estimates of each link are affected by the breathing movement producing a bimodal distribution; in each case they are separated by 180°.

A further study is presented by Kaltiokallio et al. [24,25] who employ RSS measurements from a single pair of off-the-shelf wireless sensors which are based on Texas Instruments' CC2431. The stack protocol of each transceiver follows the IEEE 802.15.4 standard [26] for an operation at 2.4 GHz ISM band. The non-invasive wireless sensors use high performance directional flat patch antennas (8 dBi gain and 75° horizontal beam width).

The wireless sensors are put on opposite sides of a bed at a height of 0.85 m and 1.95 m apart from each other (Figure 13.10). The transmitter node transmits RSS packets over each of the 16 frequency channels with a sampling rate of 16 ms per channel. A high resolution RSS reference signal is also acquired by a real-time spectrum analyzer (RSA) that is connected to the receiver node through a directional coupler.

Motion interference is mitigated by using a hidden Markov model (HMM) to detect the time instances when breathing estimation is not possible, that is, presence of supplementary motion effects other than breathing. The suggested technique estimates breathing rate with an absolute mean error of 0.03 bmp, regardless of the person's posture. It is clear that the sensing performance improves with the channel diversity, the low-jitter periodic communication, and the oversampling.

13.3.2 Continuous-Wave (CW) Methods

Contrary to packet-based transmissions, continuous-wave (CW) methods radiate a continuous unmodulated EM signal at a specific frequency, thus maximizing the incident power on a target. The energy reflected off the target is detected by another receiving antenna module. The reflected wave is modulated in phase and is used to detect tiny physiological movements induced by breathing and heartbeats.

The early attempts of using CW microwave Doppler radar systems were implemented by Lin in the late 1970s [27, 28]. Lin's first experiment transmits a continuous-wave (CW) signal from 2.1 GHz to 2.5 GHz. The periodic displacement

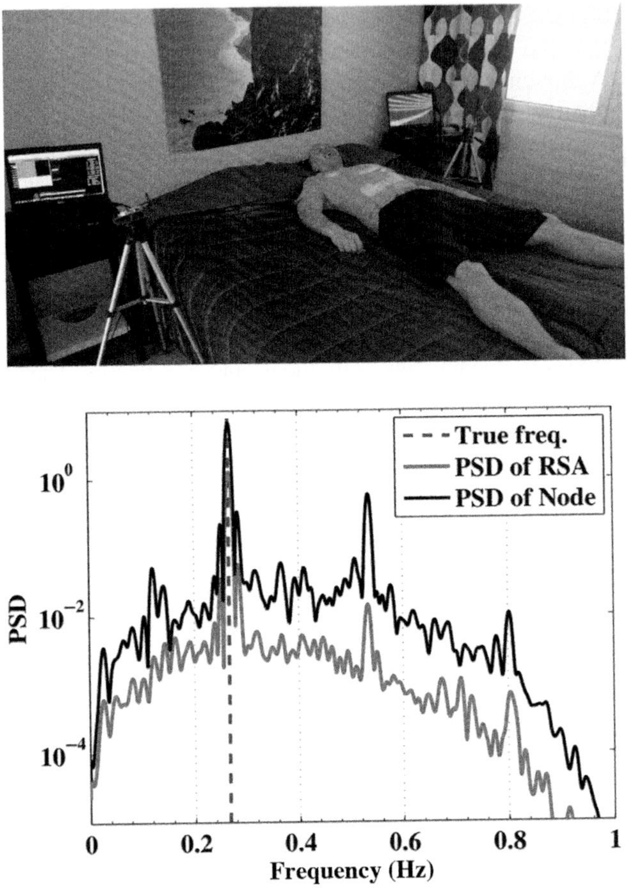

FIGURE 13.10 The experimental setup using one pair of TX-RX nodes and power spectrum density (PSD) of the estimated breathing rate of the test subject. *Source:* Kaltiokallio et al. 2013 [24] and Kaltiokallio et al. 2014 [25]. Reproduced with permission of IEEE.

of the thoracic cavity is recorded in the reflected wave, which is phase modulated (PM) by the time-varying position of the chest (concept related to the Doppler effect, Figure 13.11 [29]).

The phase of the reflected signal off the person's chest is directly proportional to the chest motion and is scaled by the wavelength of the signal. The relation between the chest displacement and the phase variation is modelled by

$$\theta(t) = {}^{4\pi x(t)}/_{\lambda} \tag{13.2}$$

where $\theta(t)$ is the phase of the reflected signal, $x(t)$ is the chest position, and λ is the wavelength of the transmitted EM waves. The information about respiration rate and heart rate is extracted from the recorded phase signal and travelling time.

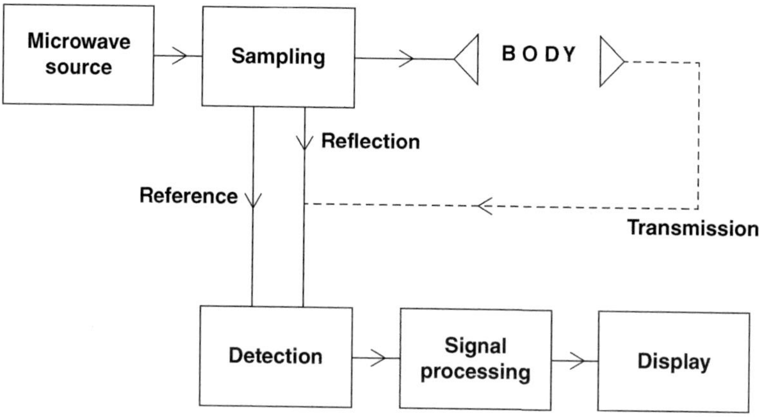

(a)

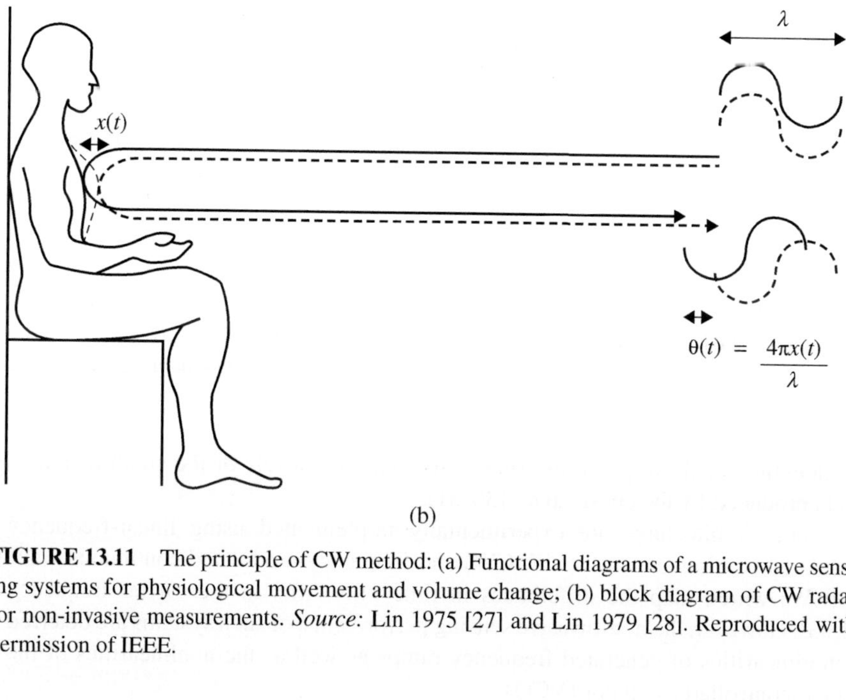

(b)

FIGURE 13.11 The principle of CW method: (a) Functional diagrams of a microwave sensing systems for physiological movement and volume change; (b) block diagram of CW radar for non-invasive measurements. *Source:* Lin 1975 [27] and Lin 1979 [28]. Reproduced with permission of IEEE.

Further advances in technology made possible the design of compact and lightweight systems, improving not only the detection accuracy, but also reducing the phase noise of the receiving signal. Doppler radar systems have been tested within academic research environments at 2.4, 5.8, 10, 16, and 60 GHz (Figure 13.12). The proposed system operates up to 1 m from the test subject. Signals are separated using Chebyshev filters and the heart rate variability is estimated by spectral analysis and peak detection methods. The results show that the use of higher frequencies (shorter

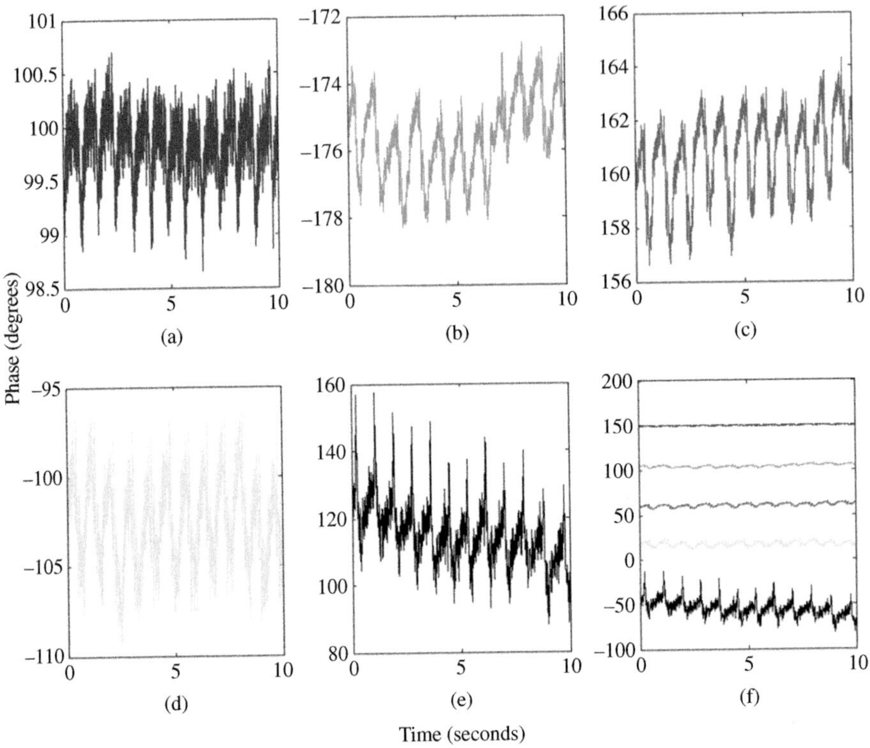

FIGURE 13.12 Contact-less sensing system operating at (a) 2.4 GHz, (b) 5.8 GHz, (c) 10 GHz, (d) 16 GHz, (e) 60 GHz, and (f) heartbeat signal detected from 2.4 GHz up to 60 GHz. *Source:* Obeid et al. 2008 [30] and Obeid et al. 2009 [31]. Reproduced with permission of IEEE.

wavelengths) vastly improve the sensitivity detection levels of the small displacements produced by the chest cavity [30, 31].

Several architectures are experimentally implemented using linear-frequency modulated continuous wave (LFMCW) and frequency-modulated continuous wave (FMCW) which map the differences in time into shifts in the carrier frequency [32,33]. However, the non-invasive sensing performance of such systems are deterred by nonlinearities of generated frequency ramps as well as the nonlinearities of the voltage-controlled oscillator (VCO).

The combination of such modulation techniques and multiport receiver architectures provide enhanced range resolutions and robustness against clutter and noise. A hybrid six-port radar system operating at 24 GHz CW signal is described in [33]. The radar prototype uses a rectangular horn antenna with a 20 dBi gain (Figure 13.13a). Different measurements are performed with the test subject lying and the six-port radar facing the chest region at a distance of approximately 1 m. The raw results for both physiological signals, breathing, and heartbeat are shown in Figures 13.13b–13.13c. A peak for the respiration rate is evident and the detected

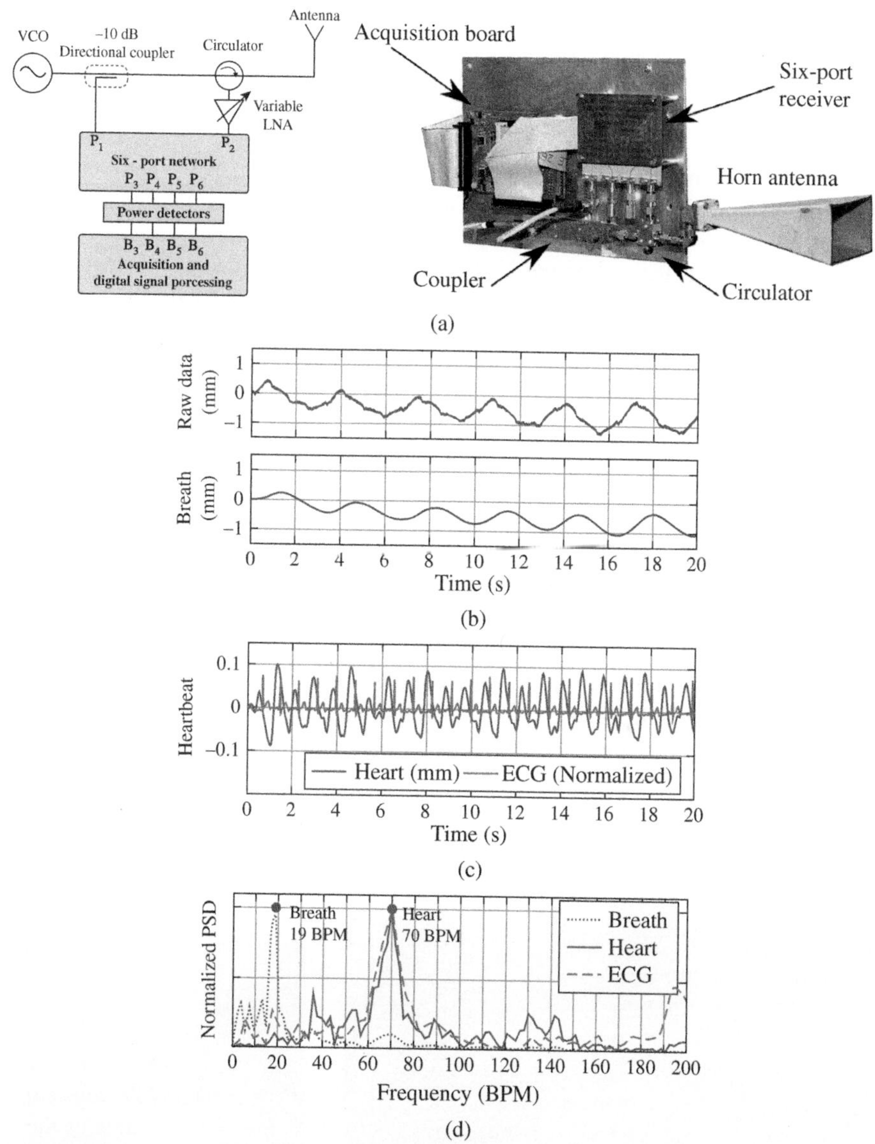

FIGURE 13.13 (a) Schematic of the proposed six-port-based monostatic radar and photograph of the hardware prototype. (b) Time domain signal of the measured raw data and low-pass-filtered breathing signal. The signal delay in the respiration signal is due to the response of the introduced filter; (c) measured heart beat signal in the time domain compared with a reference ECG; (d) Normalized PSD of the measured breath and heart rate signals compared with the PSD of a reference ECG. *Source:* Vinci et al. 2013 [33]. Reproduced with permission of IEEE.

heartbeat rate (70 BPM) is compared with a reference ECG from a medical ECG belt. The plots of normalized PSD for measured breath and heart rate signals are compared with the PSD of a reference ECG (Figure 13.13d).

While the concept of contact-less vital signs detection has been successfully demonstrated, the continuous development and evolution of RF front-end architectures—from heterodyne to homodyne and double-sideband architectures with direct IF sampling and self-injection locking—have enhanced the sensing sensitivity, reduced the interference, and maximized the detection range.

Many prototypes using homodyne double-sideband transmission systems and I/Q modulation (quadrature receivers) show significant reduction on the number of null points associated with single channel Doppler systems [35, 36]. Compact and lightweight CMOS/BiCMOS radar-on-chip solutions operating at 1.6 GHz and 2.4 GHz made use of such quadrature signals (Figure 13.14).

The manufactured chips are capable of detecting the movement produced by the heartbeat and breathing up to 50 cm of distance from the subject. The experimental results also reveal that by using quadrature receivers, the accuracy on heartbeat detection is approximately 80%, a significant improvement over single channel systems.

A high level system integration is shown by Kao's flip-chip CW radar sensor which embeds heterodyne architecture with quadrature demodulation [37]. The micro-radar operates at 60 GHz through a pair of microstrip patch antennas (Figure 13.15). The detection of heartbeat signal has been shown up to 0.3 m from the chest cavity. Although the results are compared to the human counting of the number of beats, complex signal demodulation (CSD) not only eliminates the null detection point problem, but clearly shows the movement of chest cavity due to heartbeat.

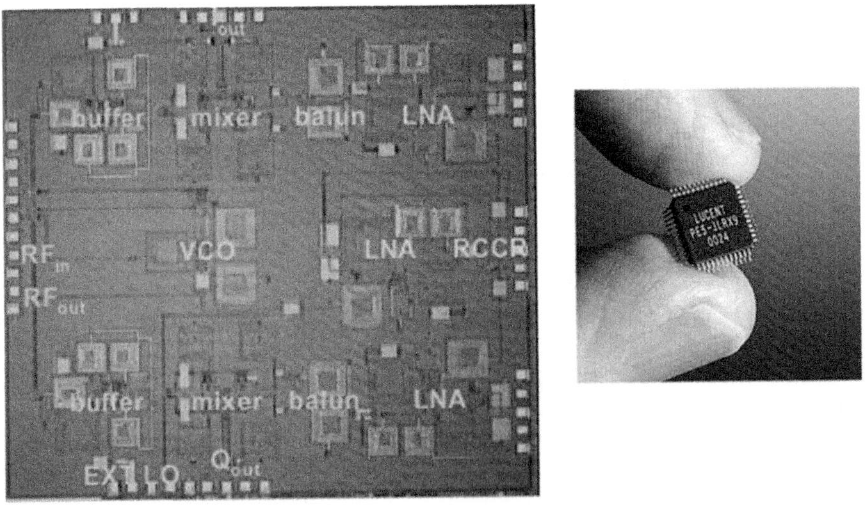

FIGURE 13.14 Photograph of the 0.25 μm silicon CMOS chip operating at 1.6 GHz and 2.4 GHz. *Source:* Droitcour et al. 2004 [34, 35]. Reproduced with permission of IEEE.

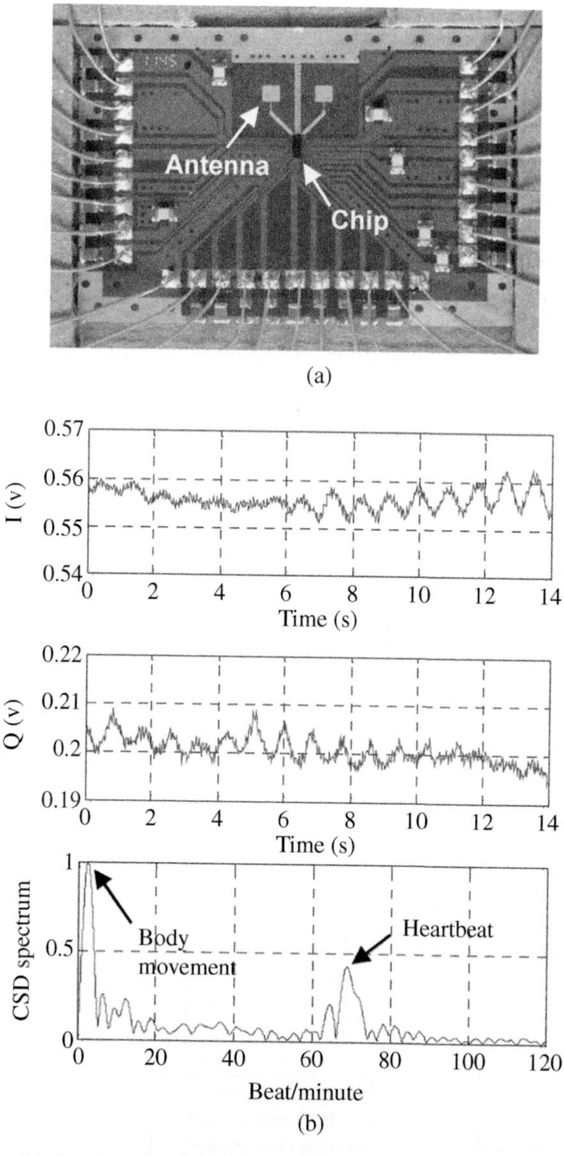

FIGURE 13.15 (a) Photograph of the integrated micro-radar operating at 60 GHz using a heterodyne architecture with quadrature demodulation; (b) baseband IQ outputs of heartbeat signal sampled and displayed by the oscilloscope and CSD spectrum showing heartbeat rate at 69 BPM. *Source:* Kao et al. 2012 [37]. Reproduced with permission of IEEE.

13.3.3 Wide Band Pulsed Radar Method

Ultra-wide band (UWB) pulse-based radio technologies transmit an impulse-like signal over a broad frequency range. When the fractional bandwidth of the transmitted signal is larger than 25% or the absolute bandwidth larger than 500 MHz, the radio signal is considered a UWB [38].

The Federal Communications Commission (FCC) allows the coexistence of UWB systems with existing communication services, such as GPS (global positioning system) and WLAN, but in order to minimize interference with them, UWB technologies must operate at a reduced mean effective isotropic radiated power (EIRP) density level of 41.3 dBm/MHz spread over several GHz of bandwidth [38]. The power spectral density (PSD) of the transmitters need to satisfy specific spectral masks which define the maximum radiated power levels, according to different regulatory bodies around the world (Figure 13.16).

The operation of UWB systems are categorized into two main transmission schemes: single band (impulse radio UWB, IR-UWB) and multiband (multi-carrier UWB). Single-band transmissions can directly modulate short duration pulses with low duty cycles (nanosecond base-band pulses), that can be either pulse amplitude modulation (PAM) or pulse position modulation (PPM), thus reducing the transceiver complexity due to the carrier-less communication.

In contrast to IR-UWB, multiband transmissions divide the UWB frequency band from 3.1 GHz to 10.6 GHz into several small bands, referred as sub-bands, where each sub-band has at least 500 MHz bandwidth (to comply with UWB regulations). The simultaneous transmission of multiple signals is achieved thanks to the digital

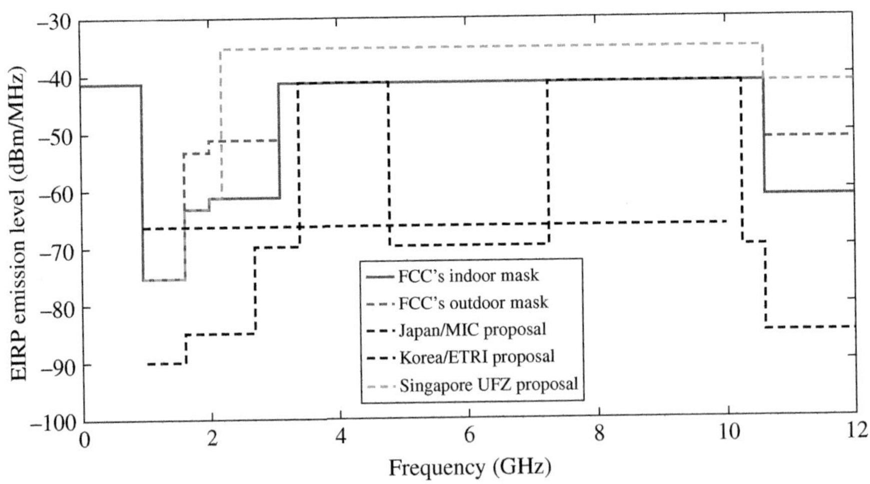

FIGURE 13.16 Proposed spectral masks of different regulatory bodies around the world. *Source:* Federal Communications Commission [38] (Public domain) and Y. Rahayu et al. 2008 [39]. (Reproduced with permission of IEEE).

encoding scheme and the orthogonal frequency division multiplexing (OFDM) (i.e., multi-carrier modulation) [39].

A wide range of UWB applications have been demonstrated and have shown several features that differentiate it from conventional narrowband systems. The high spatial resolution, the reduced inter-symbol interference (ISI), low PSD levels, and immunity to multipath are important characteristics that make UWB systems an alternative solution to CW radars and RSS-based methods for non-invasive physiological sensing.

A UWB transceiver front-end, operating from 3 GHz to 10 GHz, is implemented using standard 130 nm complementary metal oxide semiconductor technology and packaged in a high-frequency cavity. The core area of the transceiver circuit is 500 μm × 1100 μm and integrates a diplexer with a full UWB band carrier-based transmitter, a digital controlled oscillator (for pulse modulation) and a receiver front-end [40] (Figure 13.17a). The respiration patterns are detected from pulses that are

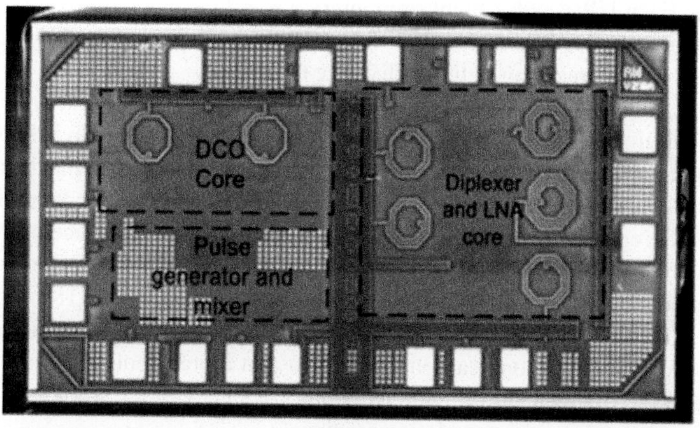

(a)

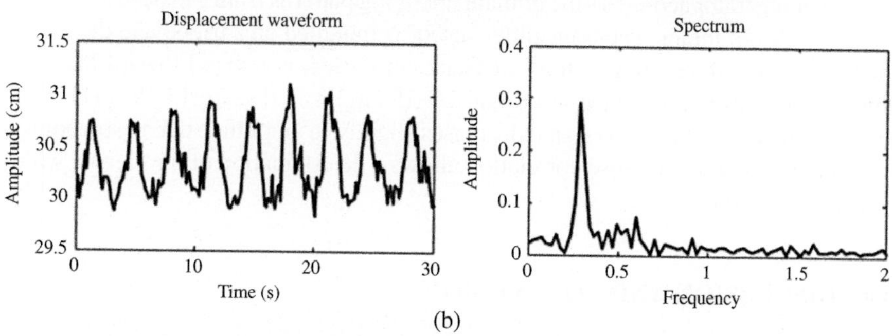

(b)

FIGURE 13.17 (a) Micrograph of the radar transceiver chip; (b) Measured breathing pattern in the time domain and the frequency domain for a time frame of 30 seconds and processed using FFT method. *Source:* Xubo et al. 2014 [40]. Reproduced with permission of IEEE.

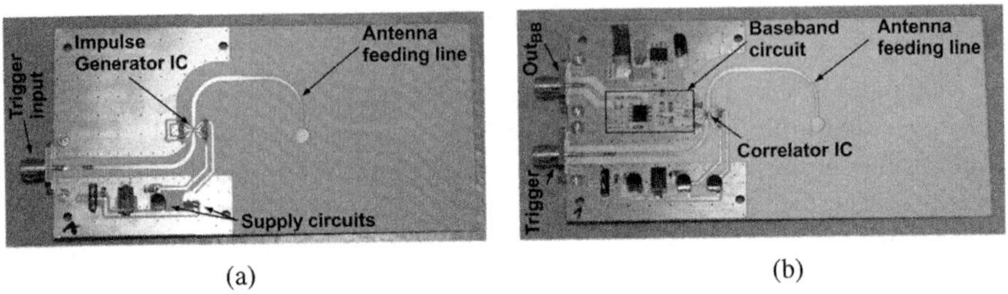

FIGURE 13.18 Photographs of the packaging of the monolithic microwave integrated circuits (MMIC) technology done by chip-on-board mounting on the antenna substrate: (a) compact impulse transmitter with antenna structure on backside; (b) correlation receiver IC and the BB circuit with the receiving the antenna structure on backside. *Source:* Schleicher et al. 2013 [41]. Reproduced with permission of IEEE.

transmitted every 0.1 μs toward the chest of a sitting test subject. The time interval for the transmitted and received signals determines the displacements of the chest movements. Each UWB pulse flight time interval is 4.2 ns which is converted to the distance of movement (Figure 13.19b). The implemented transceiver chip is capable of detecting human breathing within a distance of 50 cm.

Scheleicher et al. implemented an IR-UWB radar using monolithic microwave integrated circuits (MMICs) that are directly connected to 9.3 dBi Vivaldi antennas with an exponentially tapered slot line and a transition to a microstrip line by a Marchand balun. The transmitter structure houses an interpulse generator IC complying with the FCC regulations for indoor masks (Figure 13.18a). The receiver module embeds a correlator IC and a baseband circuit which provide a conversion from differential to single ended as well as the correlation stage (Figure 13.18b) [41]. The advantage of using a hardware correlation principle is that the impulses do not have to be sampled at a picosecond sampling rate, but at a more adequate sampling rate, depending on the frequency difference Δf (small frequency difference applied to the template impulse train in the correlation receiver).

The demonstrator acquires time domain breathing patterns from a male test subject and a 7-week-old infant. The transmitter sensor is mounted on a tripod and the main radiation beam, which emits a train of Gaussian pulses, is pointed toward the test subject's chest from a distance of around 20–25 cm away (Figure 13.19) [41]. For longer distances, the reflected signal decreases slightly in amplitude due to spreading loss, but the sweeping impulse correlation allows a determination of the signals with the same precision.

13.4 DISCUSSION AND CONCLUSION

The chapter summarized potential non-invasive technologies for future smart infrastructures (hospitals, military services, and care centers) supporting EM sensing technologies. The current solutions remain an essential and active part of patient/user awareness, not only to off-load demand of clinical services, but also to promote

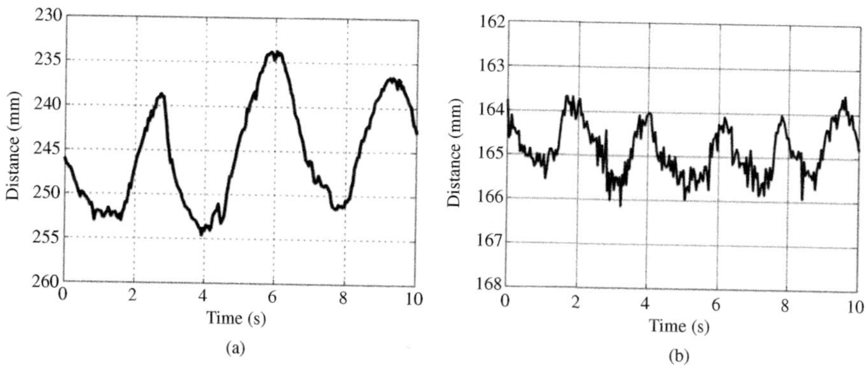

FIGURE 13.19 Time domain experimental results obtained by the IR-UWB radar sensor from (a) male test person breathing normally; (b) 7-week-old sleeping infant showing a sequence with a continuous breathing cycle. *Source:* Schleicher et al. 2013 [41]. Reproduced with permission of IEEE.

autonomous, unobtrusive, and flexible non-invasive monitoring solutions, especially for applications where the monitoring of vital signs are crucial.

Doppler radars adopting packet-based transmissions (RSS values), continuous wave (CW), and pulsed radar (UBW) have been recently introduced to a variety of applications, including home monitoring. Each of these non-invasive radio sensing technologies are described by highlighting the most promising ones. The EM wave travelling from on/off-body radio channels not only contain radio propagation characteristics, but also embed biomechanical information, such as motion pattern, breathing rate, and even heartbeat.

The received power signals, which are a combination of reflected, diffracted, and scattered waves, are highly non-stationary due to the constant change of frequency, amplitude, and phase with traveling time. For example, the continuous movement of the human body, trunk, and the limbs (dynamic scenarios) shadows the line-of-sight propagation, thus producing signal fluctuations of ± 15 dB from the average received signal.

Although received signals present high amplitude variance, different hardware and software methods are described to separate physiological signals. The latter solutions range from the simplest filtering and analysis of spectrum plots (FFT) to the most complex RF front-end architectures, such as homodyne with I/Q baseband modulation. In addition, high level system integration (top-end hardware) are presented from MMIC to CMOS/BiCMOS micro-radar-on-chips.

It is evident that greater sophistication is required in the presence of time-varying clutter and extraneous body movements. The use of powerful and robust digital signal processing techniques (e.g., template matching, blind source separation, and wavelet transforms) is undoubtedly advantageous to reduce the noise, and to extract vital signs successfully (for heart rate) and more accurately (for breathing rate).

There are several challenges that remain to be addressed, aside from making devices more compact. Directive narrow side lobe and higher gain antennas, like minimizing various noise sources (interference from external clutter), reduce the susceptibility to significant human motion artifacts. They can also achieve higher sensitivity levels with low levels of radiated power and ultimately, the detection of physiological signs on dynamic and outdoor scenarios.

We expect that next generation Doppler radars will integrate hybrid RF front-end architectures using background subtraction spectra algorithms and operating at UWB bands with antennas scanning at multiple frequencies, or even phased array systems, so that the angular coverage is extended up to 180°.

REFERENCES

[1] P. S. Hall and Y. Hao, in *Antennas and Propagation for Body-Centric Wireless Communications*. Artech House, 2006.

[2] Y. Hao and R. Foster, "Wireless body sensor networks for health-monitoring applications," *Physiol. Meas.*, vol. 29, p. R27, 2008.

[3] W. D. Jones, "New biosensor chip picks up heart signals remotely," *IEEE Spectrum, Inst. Electrical and Electron. Engrs.*, vol. 1, 2011.

[4] B. Gyselinckx, C. Van Hoof, J. Ryckaert, R. F. Yazicioglu, P. Fiorini, and V. Leonov, "Human++: Autonomous Wireless Sensors for Body Area Networks," in Proceedings of the IEEE Custom Integrated Circuits Conference, September 18–21, 2005, pp. 13–19.

[5] A. D. Joseph, "Energy harvesting projects," *Pervasive Comput., IEEE*, vol. 4, pp. 69–71, 2005.

[6] L. Xin and Y. Shuang-Hua, "Thermal Energy Harvesting for WSNs," in 2010 IEEE International Conference on Systems Man and Cybernetics (SMC), Istanbul, October 10–13, 2010, pp. 3045–3052.

[7] L. Schwiebert, S. K. S. Gupta, and J. Weinmann, "Research Challenges in Wireless Networks of Biomedical Sensors," in Proceedings of the 7th Annual International Conference on Mobile Computing and Networking, Rome, Italy, 2001.

[8] T. Arampatzis, J. Lygeros, and S. Manesis, "A Survey of Applications of Wireless Sensors and Wireless Sensor Networks," in Proceedings of the 2005 IEEE International Symposium on Intelligent Control, Mediterranean Conference on Control and Automation, Limassol, June 27—29, 2005, pp. 719–724.

[9] A. Pantelopoulos and N. G. Bourbakis, "A Survey on Wearable Sensor-Based Systems for Health Monitoring and Prognosis," *IEEE Transactions on Systems, Man, and Cybernetics, Part C (Applications and Reviews),* vol. 40, pp. 1–12, 2010.

[10] R. S. Hornung, R. F. Mahler, and E. B. Raftery, "Ambulatory blood pressure and heart rate in diabetic patients: an assessment of autonomic function," *Diabet. Med.*, vol. 6, pp. 579–585, October 1989.

[11] H. Forster, O. Ipsiroglu, R. Kerbl, and E. Paditz, "[Sudden infant death and pediatric sleep medicine]," *Wien Klin Wochenschr*, vol. 115, pp. 847–849, December 30 2003.

[12] N. Hafner, I. Mostafanezhad, V. M. Lubecke, O. Boric-Lubecke, and A. Host-Madsen, "Non-Contact Cardiopulmonary Sensing with a Baby Monitor," in 29th Annual International Conference of the IEEE Engineering in Medicine and Biology Society (EMBS), Lyon, August 22–26 2006, pp. 2300–2302.

[13] L. Yee Siong, P. N. Pathirana, and C. L. Steinfort, "Respiration Rate and Breathing Patterns from Doppler Radar Measurements," in 2014 IEEE Conference on Biomedical Engineering and Sciences (IECBES), Kuala Lumpur, December 8–10 2014, pp. 235–240.

[14] C. Kun-Mu, H. Yong, Z. Jianping, and A. Norman, "Microwave life-detection systems for searching human subjects under earthquake rubble or behind barrier," *IEEE Trans. Biomed. Eng.*, vol. 47, pp. 105–114, 2000.

[15] W. L. Kenney, J. Wilmore, and D. Costill, *Physiology of Sport and Exercise,* 6th ed. Human Kinetics Publishers, 2015.

[16] P. J. Maud and C. Foster, *Physiological Assessment of Human Fitness*. Human Kinetics Publishers, 2006.

[17] M. O. Munoz, R. Foster, and H. Yang, "Exploring physiological parameters in dynamic WBAN channels," *IEEE Trans. Antennas Propag.,* vol. 62, pp. 5268–5281, July 2014.

[18] M. Munoz Torrico, R. Foster, and Y. Hao, "Identifying Physiological Features from the Radio Propagation Signal of Low-Power Wireless Sensors," in Wireless Mobile Communication and Healthcare: Third International Conference, MobiHealth 2012, Paris, France, November 21–23, 2012, Revised Selected Papers, B. Godara and K. S. Nikita, Eds., 1 ed Berlin, Heidelberg: Springer Berlin Heidelberg, 2013, pp. 341–350.

[19] P. S. Hall and Y. Hao, *Antennas and Propagation for Body-Centric Wireless Communications*, 2nd ed. Artech House, 2012.

[20] A. R. Guraliuc, P. Barsocchi, X. F. Potorti, and P. Nepa, "Limb movements classification using wearable wireless transceivers," *IEEE Trans. Inf. Technol. Biomed.*, vol. 15, pp. 474–480, February 2011.

[21] A. R. Guraliuc, A. A. Serra, P. Nepa, G. Manara, and F. Potorti, "Detection and Classification of Human Arm Movements for Physical Rehabilitation," in IEEE Antennas and Propagation Society International Symposium (APSURSI), Toronto, July 11–17, 2010, pp. 1–4.

[22] N. Patwari, J. Wilson, S. Ananthanarayanan, S. K. Kasera, and D. R. Westenskow, "Monitoring breathing via signal strength in wireless networks," *IEEE Trans. Mobile Comput.,* vol. 13, pp. 1774–1786, 2014.

[23] N. Patwari, L. Brewer, Q. Tate, O. Kaltiokallio, and M. Bocca, "Breathfinding: a wireless network that monitors and locates breathing in a home," *IEEE J. Sel. Top. Signal Process.,* vol. 8, pp. 30–42, February 2014.

[24] O. Kaltiokallio, H. Yigitler, R. Jäntti, and N. Patwari, "Catch a Breath: Non-invasive Respiration Rate Monitoring via Wireless Communication," *arXiv preprint arXiv:1307.0084,* 2013.

[25] O. Kaltiokallio, H. Yigitler, R. Jantti, and N. Patwari, "Non-invasive Respiration Rate Monitoring Using a Single COTS TX-RX Pair," in IPSN-14 Proceedings of the 13th International Symposium on Information Processing in Sensor Networks, Berlin, April 15–17, 2014, pp. 59–69.

[26] IEEE 802.15.4 Standard. Wireless Medium Access Control (MAC) and Physical Layer (PHY) Specifications for Low Rate Wireless Personal Area Networks (LR-WPANs), 2006. Available at http://standards.ieee.org/getieee802/download/802.15.4-2006.pdf

[27] J. C. Lin, "Noninvasive Microwave Measurement of Respiration," in Proceedings of the IEEE, vol. 63, October 1975, pp. 1530–1530.

[28] J. C. Lin, J. Kiernicki, M. Kiernicki, and P. B. Wollschlaeger, "Microwave apexcardiography," *IEEE Trans. Microw. Theory Techn.*, vol. 27, pp. 618–620, June 1979.

[29] J. C. Lin, "Microwave sensing of physiological movement and volume change: a review," *Bioelectromagnetics*, vol. 13, pp. 557–65, 1992.

[30] D. Obeid, G. Issa, S. Sadek, G. Zaharia, and G. El Zein, "Low Power Microwave Systems for Heartbeat Rate Detection at 2.4, 5.8, 10 and 16 GHz," in First International Symposium on Applied Sciences on Biomedical and Communication Technologies, 2008. ISABEL '08, Aalborg, October 25–28, 2008, pp. 1–5.

[31] D. Obeid, S. Sadek, G. Zaharia, and G. El Zein, "Advances in Microwave Systems for Medical Applications," in International Conference on Advances in Computational Tools for Engineering Applications, 2009. ACTEA '09, Zouk Mosbeh, July 15–17, 2009, pp. 463–468.

[32] W. Guochao, J. M. Munoz-Ferreras, G. Changzhan, L. Changzhi, and R. Gomez-Garcia, "Application of linear-frequency-modulated continuous-wave (LFMCW) radars for tracking of vital signs," *IEEE Trans. Microw. Theory and Tech.*, vol. 62, pp. 1387–1399, 2014.

[33] G. Vinci, S. Lindner, F. Barbon, S. Mann, M. Hofmann, A. Duda, R. Weigel, and A. Koelpin, "Six-port radar sensor for remote respiration rate and heartbeat vital-sign monitoring," *IEEE Trans. Microw. Theory and Tech.*, vol. 61, pp. 2093–2100, 2013.

[34] A. D. Droitcour, O. Boric-Lubecke, V. M. Lubecke, and L. Jenshan, "0.25 /spl mu/m CMOS and BiCMOS Single-Chip Direct-Conversion Doppler Radars for Remote Sensing of Vital Signs," in 2002 IEEE International Solid-State Circuits Conference, ISSCC, San Francisco, CA, February 7–7, 2002, vol. 1, pp. 348–349.

[35] A. D. Droitcour, O. Boric-Lubecke, V. M. Lubecke, J. Lin, and G. T. A. Kovacs, "Range correlation and I/Q performance benefits in single-chip silicon Doppler radars for non-contact cardiopulmonary monitoring," *IEEE Trans. Microw. Theory and Tech.*, vol. 52, pp. 838–848, 2004.

[36] L. Changzhi, X. Yanming, and J. Lin, "Experiment and spectral analysis of a low-power K*a*-band heartbeat detector measuring from four sides of a human body," *IEEE Trans. Microw. Theory and Techn.*, vol. 54, pp. 4464–4471, 2006.

[37] T. Y. J. Kao, A. Y. K. Chen, Y. Yan, S. Tze-Min, and L. Jenshan, "A Flip-Chip-Packaged and Fully Integrated 60 GHz CMOS Micro-Radar Sensor for Heartbeat and Mechanical Vibration Detections," in 2012 IEEE Radio Frequency Integrated Circuits Symposium (RFIC), Montreal, QC, June 17–19, 2012, pp. 443–446.

[38] Federal Communications Commission (FCC) rules and regulations. Title 47: Telecommunication, Part 15: Radio Frequency Devices, Subpart F: Ultra Wideband Operation. (2011). Available at http://hraunfoss.fcc.gov/edocs_public/attachmatch/FCC-02-48A1.pdf

[39] Y. Rahayu, T. A. Rahman, R. Ngah, and P. S. Hall, "Ultra wideband technology and its applications," 2008 5th IFIP International Conference on Wireless and Optical Communications Networks (WOCN '08), Surabaya, 2008, pp. 1–5. doi: 10.1109/WOCN.2008.4542537

[40] W. Xubo, D. Anh, and D. Teng, "3–10 GHz ultra wideband front-end transceiver in 0.13 μm complementary metal oxide semiconductor for low-power biomedical radar," *IET Circuits, Devices & Syst.*, vol. 8, pp. 272–279, 2014.

[41] B. Schleicher, I. Nasr, A. Trasser, and H. Schumacher, "IR-UWB radar demonstrator for ultra-fine movement detection and vital-sign monitoring," *IEEE Trans. Microw. Theory Tech.*, vol. 61, pp. 2076–2085, March 2013.

14

POWER/DATA TELEMETRY TECHNIQUES FOR IMPLANTS OR WEARABLE SYSTEMS

ANIL K. RAMRAKHYANI AND GIANLUCA LAZZI

Department of Electrical and Computer Engineering, University of Utah, Salt Lake City, UT, USA

14.1 INTRODUCTION

With the advancement of microfabrication technology for low power and multi-functional devices (i.e., biosensors and neural stimulators), and their usage in the proximity of biological media, numerous medical devices and wearable electronics have been developed [1–5]. For each in-body (implant) and/or on-body (wearable) system, a stable power source and reliable communication data channel are key design requirements. While the design of the implanted device presents unique challenges such as small dimensions (1–2 cm), low power (1–100 mW), and high longevity (7–10 years), wearable electronics must be lightweight, and require multi-functionality and network connectivity with a standard wireless system (i.e., WiFi, Bluetooth). In the last decade, significant research has been devoted to create small, energy efficient implants that can be used to treat deafness (using cochlear implants), blindness (using retinal prostheses), pain management (using spinal cord stimulators), and Parkinson's disease (using deep brain stimulators). Wearable technologies have focused more on the choices that individuals can make in this digital age to access personal data (using biosensors) and information over the internet. With the vision of better information connectivity, multiple non-invasive devices such as Google Glass (for information

Electromagnetics of Body-Area Networks: Antennas, Propagation, and RF Systems, First Edition.
Edited by Douglas H. Werner and Zhi Hao Jiang.

access) [6], Apple Watch (for heart rate monitoring and information access) [3], and Fitbit (to create personal health log) [7] were developed.

For both implantable and wearable devices, specifications of the power and data telemetry need to be identified based on the underlying applications. However, few general methods and technologies have been developed to ease the constraints of these systems. For example, multiple wireless power transfer (WPT) methods (i.e., electromagnetic, acoustic) were proposed to create batteryless implants, achieving a small form factor. Similarly, to create a self-powered system, energy harvesting technologies (i.e., solar, biofuel, electromagnetic) have been developed. Some of these designs include life supporting devices (i.e., pacemaker, artificial heart), and therefore require a reliable data link for the regular monitoring of the device. Moreover, prosthetic devices such as retinal and cochlear implants require a stable (low error per packet) data link over which stimulation patterns can be transmitted in real time for the desired operation of the implant. For wearable electronics, the key requirement of the communication channel includes the use of standard protocol to achieve low power and a high data rate over a long (1–10 m) operating range.

In the following sections, different powering mechanisms and energy harvesting devices are discussed, including applications and their respective limitations. The data links for the wearable or implanted devices are categorized based on the use of low frequency propriety protocols versus the high frequency standard protocols. Moreover, two design examples are presented to discuss the implementation of a leading implanted retinal prosthesis and a wearable electronics based on body area network (BAN).

14.2 POWERING OF IMPLANTS AND WEARABLE SYSTEMS

In recent years, multiple powering techniques have been adopted to operate implanted and wearable devices. Even if traditional power storage units such as on-site battery remain the key source of power, multiple alternatives such as wireless power and energy harvesting devices were used to achieve long-term device operation of the miniaturized implanted systems. Similar approaches, which lead to larger discharging/replacement time of the battery, were taken for wearable electronics. Selection of specific powering mechanisms, such as an external power source (i.e., battery) versus the energy harvesting devices depend on the energy and size requirements of the underlying system. Figure 14.1 provides a detailed categorization of leading powering mechanisms. The following sections discuss different methods for powering implanted and wearable devices using non-regenerative and regenerative energy sources.

14.2.1 Non-Regenerative Power Sources

With the advancement of micro- and nano-electronics technologies, it is possible to accommodate complex and computationally intensive designs (i.e., microprocessor, image processor) while achieving the size constraints of the implantable and

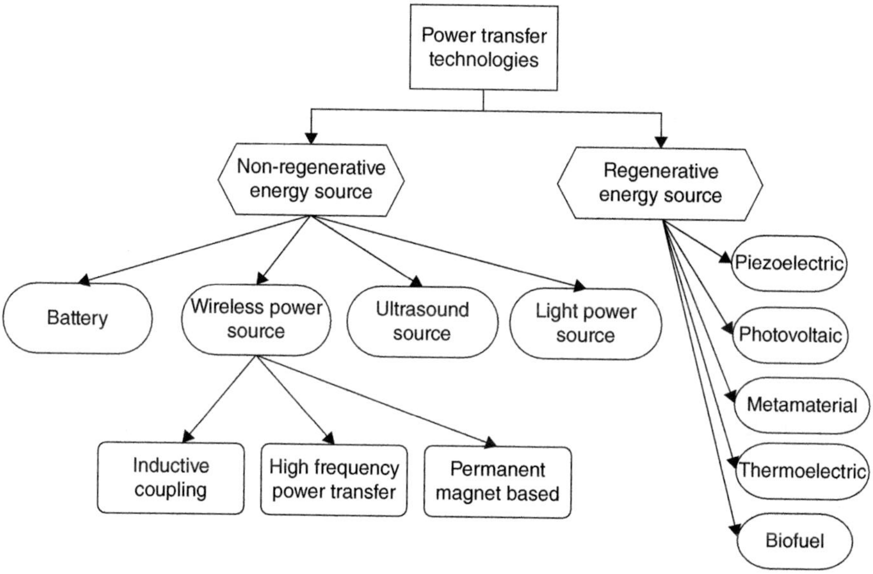

FIGURE 14.1 Different powering techniques for implanted and wearable devices. Multiple non-regenerative and regenerative methods can be chosen based on the underlying application's specifications.

wearable devices. Traditionally, dedicated power sources (non-regenerative) were used for systems with power requirements of the order of tens of milliwatts [8–10]. For many designs, especially for implanted devices, the moderate energy density of the implantable batteries restricts the device longevity. Therefore, many designs require alternate power transfer techniques from the dedicated external (outside the body) power source to the implanted electronics [11–13]. In the following sections, some of the dominant powering mechanisms are discussed for wearable as well as implanted systems.

14.2.1.1 On-Site Battery Operated Devices For wearable electronics, dedicated weight, volume, and form factor for the battery are significantly limited. With the growing demand for high energy density batteries and flexible form factors, the research community has utilized different materials [14] and developed new fabrication technologies [15]. Li-ion-based coin or polymer batteries dominate the consumer electronics market due to their wide availability in different form factors. Meanwhile, some research was conducted to achieve a Li-ion-based flexible battery [15]. To achieve paper-like flexibility, lightweight carbon nanotubes (CNT) were deposited on commercially available paper to create thin film current collectors (anode and cathode). This design achieved high energy density per unit mass (108 mAh/g) while maintaining its high flexibility and light weight, making it useful for powering wearable biometric gadgets used during sports activities. For most wearable electronics, the device lifetime is much longer than the battery discharge time. Therefore, rechargeable batteries with a large number of charge cycles (~1000–2000) and moderate discharge periods (~1–2 days) were used to achieve a battery lifespan of over 5 years.

For implantable devices, the energy capacity of the battery depends on the targeted lifetime and average power requirement of the implant. The battery is hermetically sealed inside a titanium case to meet the safety guidelines for medical devices. Spinal cord stimulators [16] or pacemakers [17] are examples of successful medical devices that use implanted batteries to achieve a device lifetime of 7–10 years. For some designs, rechargeable batteries, along with WPT capability, are used to reduce the dimensions of the implanted battery.

14.2.1.2 Wirelessly Powered Devices Using Electromagnetic Fields In the last decade, WPT using electromagnetic fields became the gold standard for contactless energy transfer between the external power source (i.e., battery) and the implanted electronics. For these systems, the efficiency of the power transfer depends on the operating frequency, operating distance, and the dimensions of the external and implant coils (or antennas). Similar to implanted devices, multiple wearable systems also integrate passive radio frequency identification (RFID) tags to achieve keyless entry and contactless payment options. In all these designs, the interrogator device (connected to a dedicated power source) generates the power and/or data signal and seeks for the transponder's (implant or wearable RFID) response. Depending on the power requirement of the transponder (implant or RFID tag) and the size constraint of the antenna or coil, different methods were considered. The following sections describe three methods based on the operating frequency and electromagnetic field source.

14.2.1.2.1 Inductive Coupling Inductive coupling-based WPT systems are typically preferred for designs in which the magnetic coil's dimensions and operating distance are much smaller than the operating wavelength. For most medical devices, operating distances between the external and implant coils range from 5 mm to 30 mm [1, 8]. Typically, the operating frequency of the design is kept below 20 MHz to maintain the absorbed electromagnetic energy in the tissue below the regulatory limits [20]. Depending on the number of coils used for the power transfer, the system can be characterized as either a two-coil [21, 18] or a multi-coil design [19, 22]. Figure 14.2 shows the block diagram of two-coil and multi-coil WPT systems. For both system types, the external coil generates the time-varying magnetic field, which couples with the implanted coil for the power transfer. Due to the evanescent nature of the magnetic field, the operating range of the inductive link is generally limited to only a few centimeters. Compared to the two-coil design, the multi-coil design is relatively new [19] and has shown significant advantages, achieving higher power transfer efficiency, tunable induced voltage at the implant, wider frequency bandwidth for the data communication, and lower susceptibility in system performance during the coils' misalignment [23]. Moreover, it was shown that using a multi-coil inductive link system, the absorbed energy in tissue [24] and the unintentional radiated field [25] can be significantly reduced compared to its two-coil alternative. Depending on

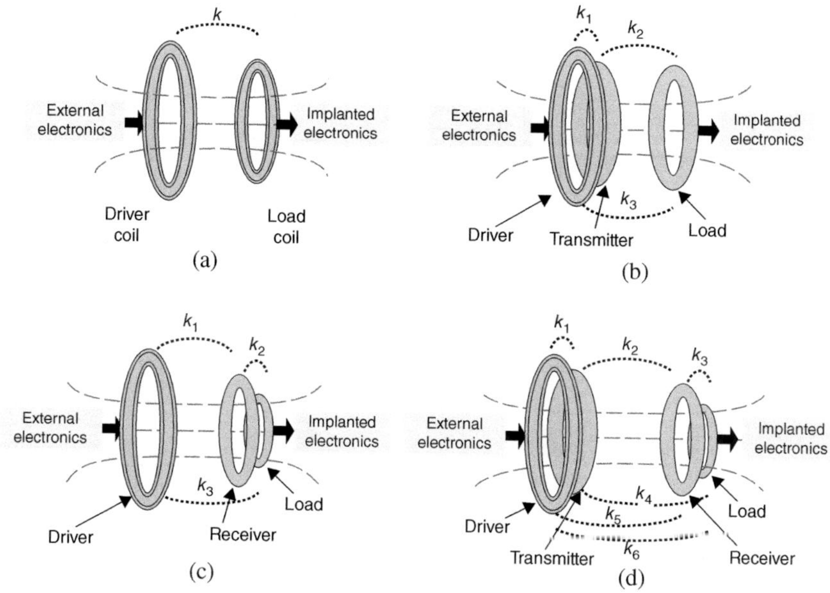

FIGURE 14.2 Two-coil and multi-coil wireless power transfer (WPT) system with different magnetic coupling between the coils. (a) Block diagram of the two-coil inductive link with the driver (external) and the load (implant) coil based on [18]. *Source:* Qusba et al. 2014 [18]. Reproduced with permission of IEEE. Block diagram of the three-coil WPT system with dual coils at (b) the external and (c) the implanted unit. (d) Four-coil WPT design with dual coils at the external and the implanted unit. *Source:* RamRakhyani et al. 2011 [22]. Reproduced with permission of IEEE. Compared to two-coil design, multi-coil WPT system achieves desired system's performance (efficiency, voltage gain, frequency bandwidth) by tuning vast design variables ($L_{1\text{–}4,}$ $k_{1\text{–}6}$).

the operating conditions, power transfer efficiency between 20% and 40% can be achieved over the inductive link. Therefore, this approach is quite suitable for high values of power transfer (100–300 mW) to the implant.

14.2.1.2.2 High Frequency Power Transfer High frequency WPT is a relatively new technique for operating low power (100–500 mW) and millimeter-size (2–4 mm in diameter) implants over mid-range (30–50 mm) coil separation [26, 27]. Due to low power transfer efficiency of the link (0.1–0.5%), this technique was primarily utilized for implantable sensors or low power neural stimulators [26]. Compared to an inductive link, high frequency WPT systems are operated over the frequency range of 1–2 GHz. The power generated by the external source is limited to 1–2 W to meet the specific absorption rate (SAR) and radiation guidelines of the federal regulatory agencies (Federal Communications Commission in the USA, International Electrotechnical Commission in the European Union). Figure 14.3 provides the block diagram of the high frequency wireless power link. Similar to an inductive link, the source generates an electromagnetic field near the implant site. Due to the

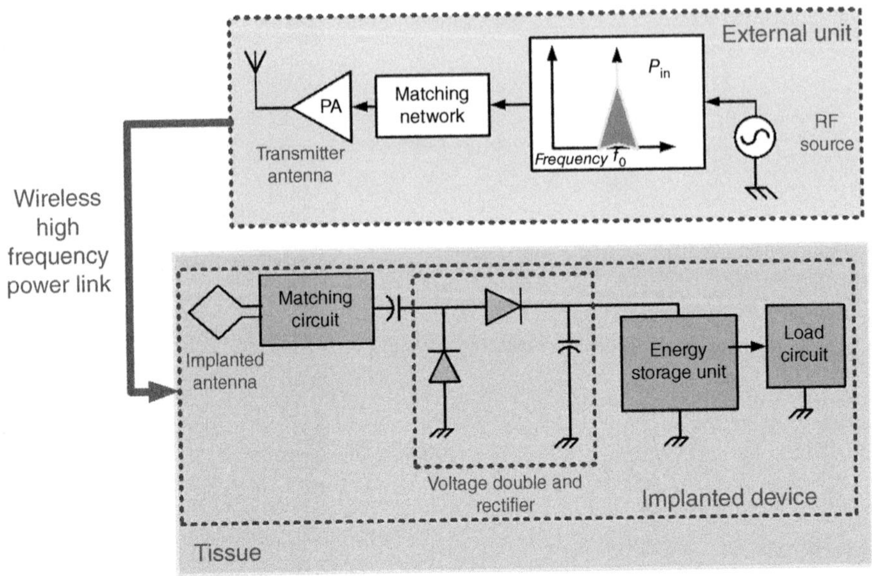

FIGURE 14.3 High frequency wireless power transfer system. The transmitter includes a high power radio frequency (RF) source with directional antenna. The implanted receiver unit includes a millimeter-size antenna with matching network and rectifier to convert the RF signal to DC voltage.

equivalent dimensions of the external coil with respect to the operating wavelength λ (speed of light in the medium/operating frequency), a significant amount of input power can be dissipated as radiated energy which does not couple with the implant coil. However, some designs have utilized directional antennas to focus the incident beam pattern at the implant site [28]. Due to its large operating distance and option of focused energy transfer, this technique was widely adopted for long range ultra-high frequency (UHF) RFID tags for keyless entry and resource management [29].

14.2.1.2.3 Permanent Magnet-Based Power Transfer Recently an alternative approach was proposed for generating a time-varying magnetic field using the rotation of permanent magnets [30]. In this approach, the energy was converted from stored chemical (battery) form to mechanical form, driving the motor. A high magnetic field intensity (75,000 A/m) permanent magnet, based on neodymium, iron, and boron (NdFeB) material, was connected to the motor. Figure 14.4 shows the block diagram of the system. With the rotation of the magnetic core, the magnetic field coupled to the receiver coil varied over time, resulting in an induced voltage at the implant. The operating frequency was limited to below 200 Hz due to the rotation speed of the motor (90 turns/second). It was argued that the low frequency operation led to significantly lower absorbed energy in the tissue, thus making this approach safer for biomedical applications. However, unintentional stimulation or modulation of the underlying neuronal network was not considered, which is critical

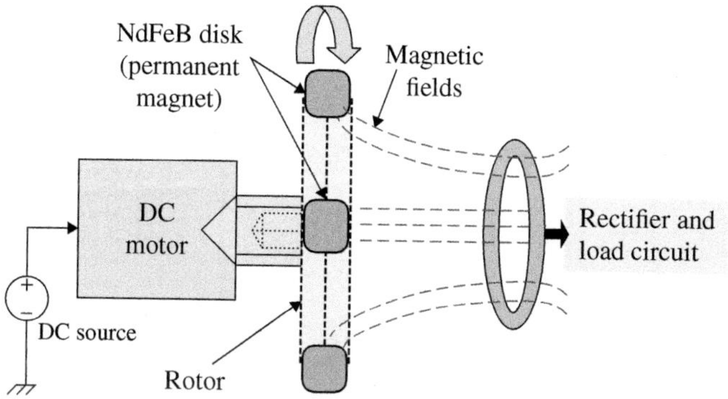

FIGURE 14.4 Permanent magnet-based wireless power transfer. The transmitter system includes high magnetic field intensity magnets positioned on the rotor. With the circular motion of the rotor, the coupled magnetic field in the receiver coils changes over time, which creates the voltage across the load circuit. Energy is converted from the electric form to mechanical form and back to electric form.

due to the low frequency temporal coding of neural signals. Much research is needed to address the safety and efficacy of this approach. For wearable electronics, this approach may not be useful due to the small operating range and total mass of the motor-based system.

14.2.1.3 Ultrasound Powered Source Ultrasound-based power transfer, from a dedicated energy source to a small-sized implanted or wearable system, is a novel technique. Compared to an electromagnetic field, acoustic waves travel many orders slower (velocity of acoustic wave/electromagnetic wave in air $\sim 10^{-6}$) inside the propagation mediums (i.e., air, tissue), which results in the acoustic signal having a small operating wavelength (1–2 mm) while operating at the MHz frequency range (typically called ultrasound). Instead of using direct or induced electric current for energy transfer, the ultrasound source creates pressure waves. These waves travel through the biological medium and are intercepted by the receiver which converts the acoustic energy back into electrical energy. Due to low losses of acoustic signals in biological tissue, implanted devices can be implanted deep inside the body (5–50 mm) [31,32]. Significant research has been dedicated to create focused ultrasound beams for the non-invasive imaging of soft tissues. Typically, piezoelectric transducers (PZT) were used to create acoustic signals [31]. However, due to high acoustic impedance of the PZT compared to air (acoustic impedance of PZT/acoustic impedance of air ~75,000), it requires direct contact between the transducer and the tissue to reduce the impedance mismatch between different propagating mediums. During the use of PZT, matching layers were applied to the device, to match the acoustic impedance of the PZT (Z_{PZT} = 31 MRayl) to the tissue impedance (Z_{tissue} = 1.5 MRayl). Figure 14.5 shows a typical block diagram of ultrasound-based power transfer to the implanted device. The tissue attenuation coefficient α increases with the operating frequency f ($\alpha = 0.7f^{1.2}$ dB/cm) and a frequency range of 200 kHz to 1.2 MHz was

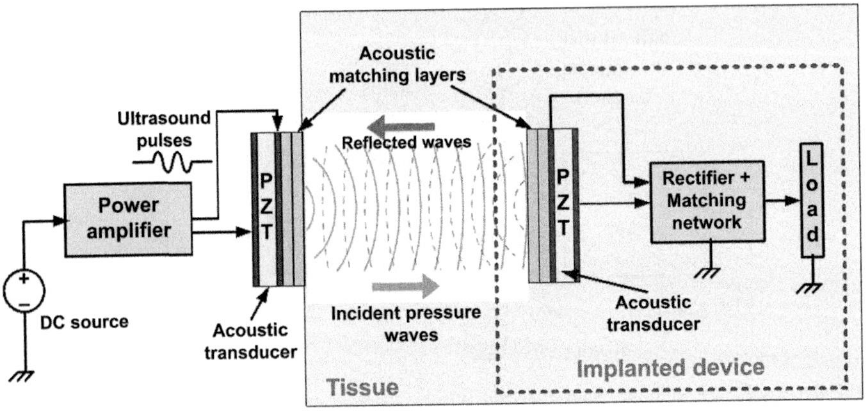

FIGURE 14.5 Ultrasound wireless power transfer system. The external source generates the ultrasound pressure waves which can travel through the tissue. The implanted acoustic transducer converts the acoustic energy to electric energy.

recommended for the ultrasound-based power transfer [31]. This design successfully transferred 70 mW of power to the deep implant (depth = 40 mm) with a power transfer efficiency of 27%.

14.2.1.4 Light Powered Source Recently, an incoherent light source was successfully used to transfer energy to an implant through tissue [33]. The design utilized a wearable near-infrared light emitting diode (LED)-based light source with a central wavelength of 850 nm (Figure 14.6). An implanted photo voltaic (PV) cell was used to convert the received photons to electric current. In the external unit, an LED source was used in combination with a parabolic reflector and a planoconvex lens to control the depth of focus inside the tissue. Due to the smaller penetration depth of the near-infrared signal compared to inductive or ultrasound sources, this method can only be used to charge or power shallow implants (depth of 5–10 mm). However, the abundance of incoherent light sources such as LEDs, makes this method a useful power transfer technique. For wearable devices, this technique can be exceptionally useful as there can be a direct path through the air between the light source and the receiver PV cell. As the near-infrared light propagates through the biological tissue, a significant amount of incident energy is absorbed, causing increased local tissue temperature. Therefore, a safety study of this technique needs to be considered before deciding on the input power of the light source.

14.2.2 Regenerative Energy Harvesting Devices

With the development of ultra-low power electronics such as micro- and nano-watt biosensors, it is now possible to power some of these systems directly through ambient energy. A significant amount of research was dedicated to convert abundant solar energy to useful electric energy. Similarly, some attempts were made at harvesting vibration and electromagnetic energy for device operation. Typically, these systems

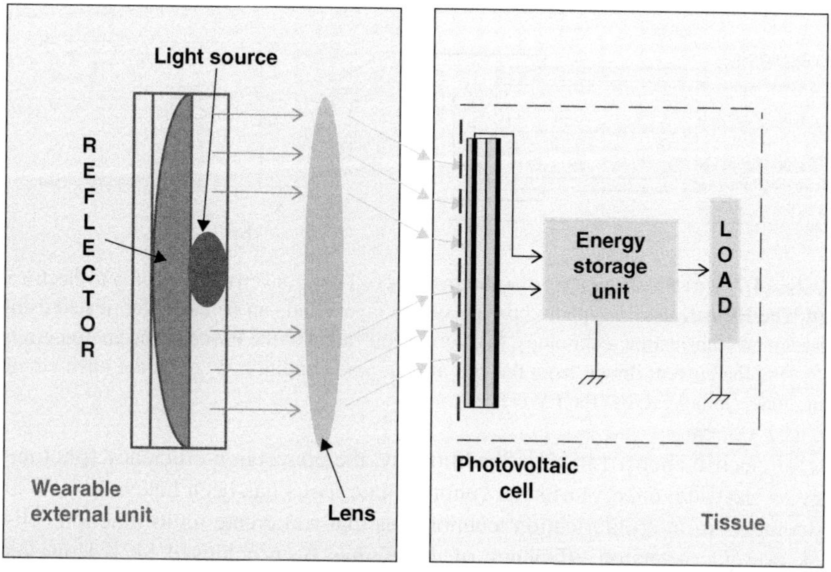

FIGURE 14.6 Light source-based wireless power transfer. The power source includes light source (LED), parabolic reflector, and convex lens to focus the light beam at the receive site. The implanted photovoltaic cell converts the light energy to electric form.

involve an energy harvester device which converts the ambient energy to electric current. In many cases, this energy is accumulated inside a storage unit (i.e., supercapacitor, battery) over the time span of a few seconds to hours, to meet the power requirement of the underlying application. In the following sections, different harvesting techniques are discussed.

14.2.2.1 Mechanical Energy Harvester Piezoelectric material is one of the leading devices that converts mechanical energy to electric form and vice versa. When an external pressure or force is applied across the piezoelectric material, the internal dipoles orient to create non-zero voltage across the terminals. In the presence of repetitive pressure at the resonance frequency of the piezoelectric structure, substantial mechanical energy can be converted to an electric signal. Depending on the vibration direction, different resonance modes (at separate frequencies) can be generated inside the material. Therefore, a single device can harvest energy from a wide frequency range of external vibrations [1, 8, 34–36]. In general, the human body generates multiple low frequency rhythmic signals (i.e., human heartbeats, respiratory action, and fluctuating blood pressure). Many voluntary actions such as walking and running can also create sufficient energy for low power wearable electronics [37]. So, some designs have proposed favorable energy harvesting locations (i.e., ankle, knee, hip, shoulder) for piezoelectric devices [36, 38, 39].

14.2.2.2 Photovoltaic Cell Since the discovery of the photoelectric effect, there has been a systematic effort to convert abundant solar energy (density ~1000 W/m^2)

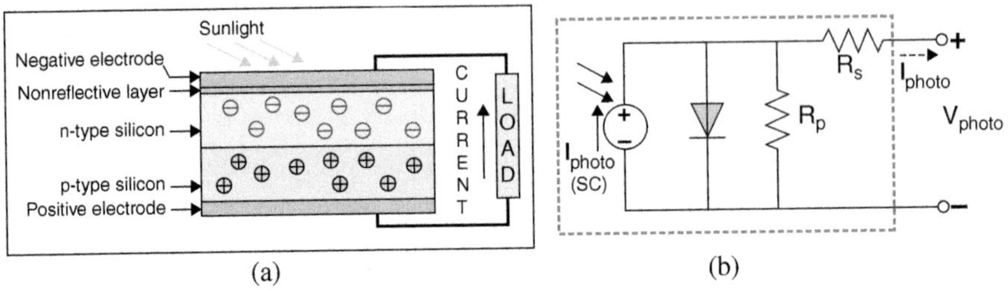

FIGURE 14.7 (a) Block diagram of a photovoltaic cell that converts solar energy to electrical energy. The PV cell uses n-type and p-type silicon layers and can be easily fabricated using standard micro-fabrication technology. (b) Norton equivalent of the PV cell. The photo-current ($I_{photo(SC)}$) is the current drawn from the cell at zero-load condition. V_{photo} is the open-circuit (infinite load) voltage across the PV cell.

to useful electric energy [40, 41]. Traditionally, the conversion efficiency (photonic energy to electrical energy) of solar (photovoltaic) cells has been below 10%. With advancements in microfabrication technologies that can create multi-junction cells, a peak power conversion efficiency of 44.7% has been achieved [42]. However, low cost and commercially available solar cells are still limited to efficiencies of 15–20% and can be useful to power only low or moderate power wearable devices. Figure 14.7 shows the block diagram of photovoltaic based energy harvesters. This technique can be very useful for powering wearable electronics such as health monitoring sensors. By including a tiny strip of solar cell (area = 1 cm^2), sufficient electric energy (~1–10 mW) can be generated and can be stored in the on-site batteries during daytime.

14.2.2.3 Electromagnetic Energy Harvester With the increasing connectivity of the wireless links (i.e., GSM, WiFi) to the mobile device, substantial electromagnetic energy can be present around us. Even if an individual wireless link is only short-lived and narrow band, the time-multiplexed operation of multiple devices results in a very low power broadband electromagnetic spectrum (~0.01–0.1 $\mu W/cm^2$). To demonstrate this, one study has created a prototype electromagnetic (EM) energy harvester system that captures EM signals over the global system for mobile (GSM) communication band (900 MHz, 1800 MHz) [43]. A similar approach was taken to harvest energy from 240 MHz to 710 MHz. The design used an active antenna based on a metamaterial split ring resonator (SRR) to achieve broadband impedance matching [11]. Typically, these designs require a broadband antenna, multistage voltage doubler, and high efficiency rectifier circuit to substantially enhance the input voltage across the antenna and rectify the signal as DC current to charge an energy storage unit (i.e., battery, super capacitor). Figure 14.8 shows the typical block diagram of an electromagnetic field-based energy transfer system.

14.2.2.4 Thermoelectric Substrate Thermoelectric generators (TEG) operate on the Seebeck effect, which states that in the presence of a temperature gradient in the TEG device, electric current can be generated and vice versa [37, 44]. In these devices, two thermocouples or thermojunctions are created by using two metals or alloys with

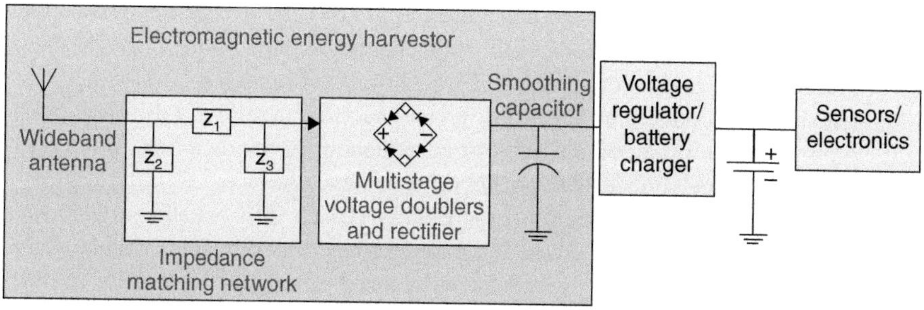

FIGURE 14.8 Block diagram of electromagnetic energy absorber. The module includes wide band antenna (i.e., spiral, log-periodic), matching network, multistage voltage doubler, and energy storage unit.

different Seebeck coefficients. When there is a temperature difference between these thermojunctions, an electric current flows in the device, which can be used to operate low power devices. Typically, p-doped and n-doped materials can be used as two materials to create a TEG device. In recent years, TEG devices have been found useful in powering implantable devices. The temperature of the human skin modulates with the ambient temperature. However, the internal temperature of the body is well regulated at 37°C. Therefore, a TEG device implanted near the skin surface experiences a temperature gradient between the two interfaces of the device. Depending on the thickness of the TEG device and its location, a substantial (1−2°C) temperature difference between the two interfaces (or thermojunctions) can be achieved. Such devices were successfully tested to power implanted electronics with power requirements of 10−100 μW [45,46]. Moreover, this technique can be useful to power wearable electronics as well. For example, recent research used a thermoelectric (TE) polymer coated over commercial fabric to create TE fabric suitable for harvesting body heat energy [47].

14.2.2.5 Biofuel Powered Devices In the field of bio-inspired electronics, using biofuel as an energy source has been long sought. The human body utilizes the binding energy of carbon-based molecules (i.e., sucrose, glucose) to perform bodily functions. Biofuel was considered in an attempt to mimic this process in self-powered implantable medical devices. The design converts chemical (or binding) energy of internal molecules, which are abundant and easily available through the food supply, to useful electrical energy. A functional glucose biofuel cell was proposed based on the wide variety of enzymes and redox mediators and was implanted inside a freely moving animal [48]. The design achieved peak power generation of 24.4 μW/mL which is sufficient to power low power sensors or neural stimulators (i.e., pacemaker).

14.3 DATA COMMUNICATION TO IMPLANTS AND WEARABLE SYSTEMS

Suitable operation of an implanted device depends on the individual patient and the device's operating conditions. Therefore, all implanted medical devices require a data communication channel between the external controller and the implanted device for remotely modifying the implant's parameters and/or to record internal biosignals. For any implanted device, specifications of the data channel depend on the required data rate, operating range, and its dimensions. Typically, each implanted device interacts with a dedicated external controller. Most device manufacturers use a different propriety communication protocol and operating frequency for the data link. However, in the age of wide data connectivity, recent designs have attempted to utilize industry standard protocols (i.e., Bluetooth Low Energy (BLE), MICS band) to achieve connection with the wireless network. In most designs, electromagnetic fields were created using a radio frequency source and were used effectively to transfer useful data between external and implanted systems. Some effort was dedicated to use an acoustic signal to achieve a low loss data link [31].

Wearable systems, in contrast, interact with a wide variety of devices. Therefore, these systems commonly adopted industry standard protocols (i.e., bluetooth, ZigBee, WiFi, ANT_+, NFC) while operating in the license free frequency bands. Selection of the individual standard depends on the required data rate, power budget for the communication, and the operating range. Depending on the underlying application and operating conditions, data communication channels are either low frequency (below 100 MHz) or high frequency (above 100 MHz). In the following sections, the differences between the different channels are discussed.

14.3.1 Low Frequency

Most implanted devices with a dedicated external controller operate over a short range and require few control instructions. For these devices, the dimensions of the implanted antenna needs to be small (below 1–2 cm) and the data communication channel needs to be reliable (low data packet loss). Due to low propagation loss (or attenuation) for low frequency (LF) signals inside the tissue, LF channels are commonly used for implanted medical devices (i.e., spinal cord stimulator, pacemaker, deep brain stimulator) [16, 17]. Using moderate size implant coils (1–2 cm diameter), a short operating range (1–5 cm) and a moderate data rate (100 kbps to 2 Mbps) was achieved for these designs [10].

Most implantable designs require both the wireless power and data link between the external and implanted system. In multiple designs, the power signal is modulated using the data signal, and control data from the external system is transferred over the power link. These designs are commonly referred to as single band power and data link. Typically, the absorbed energy (in the tissue) associated with the power carrier frequency is high and increases with the operating frequency. So, the associated data bandwidth is limited (500 kHz to 2 MHz) due to the use of a low power carrier frequency (below 10 MHz). To decouple this dependence, the dual band power and data link was proposed [10]. Using low frequencies (below 1 MHz) for the power transfer and moderate frequencies (10–50 MHz) for the data channel, data rates up to 1–2 Mbps were achieved without increasing the absorbed energy in the tissue [49]. Figure 14.9 shows the block diagram of single and dual band power and data transfer. To retrieve the status of the implant, an uplink channel (from the implant to the external) is required. Depending on the required data rate and implant complexity, the

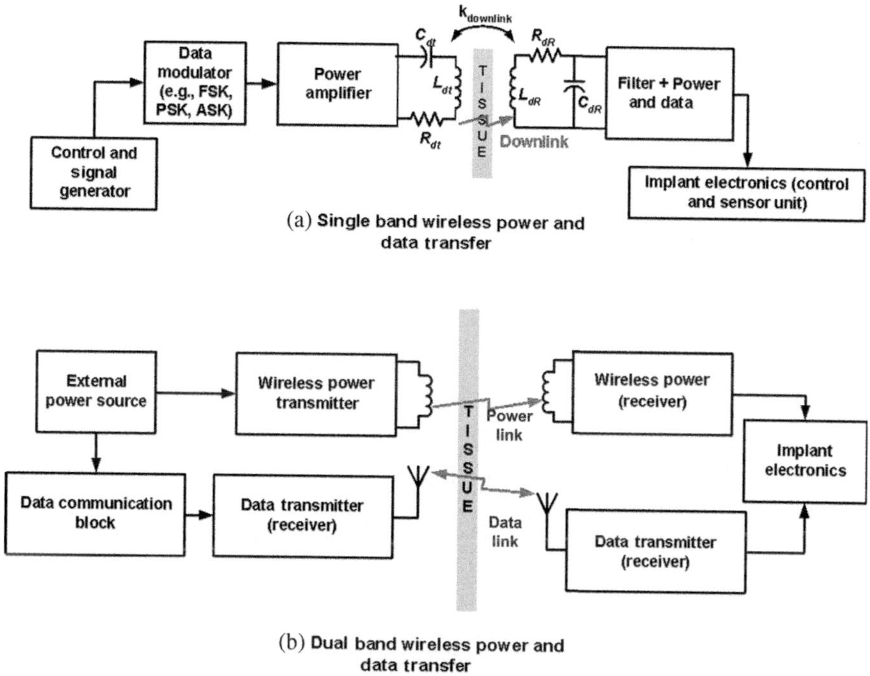

FIGURE 14.9 (a) Single band wireless power and data transfer. The data signal modulates the power signal in the form of amplitude, frequency, or phase. The achievable data rate for the single band system is limited by the carrier frequency of the power signal. (b) Dual band wireless and data transfer system. The power link uses a low frequency carrier to reduce energy absorption in tissue, while the data link utilizes the high frequency carrier to achieve a high data rate.

back telemetry system either uses a back scattering mechanism to deliver data over the power carrier signal [50] or uses a separate channel for the uplink communication.

For many wearable devices, the communication range between the reader and the wearable electronics is limited to few centimeters to achieve a secure data communication channel. For example, electronic wallets such as Google Wallet or Apply Pay use the in-built near field communication (NFC) system to achieve secure transactions. A similar approach was applied to RFID-based keyless entry or contactless access to health monitoring devices in a dense hospital environment.

14.3.2 High Frequency

High frequency data communication is commonly used for longer distances (1–10 m), high data rates (2–10 Mbps), and small implants. For these systems, the operating frequency is primarily limited to a few bands that can ensure no interference with commercial licensed bands. Depending on the region, some bands are kept license free for industrial, medical, and research usage. Table 14.1 provides

TABLE 14.1 Unlicensed frequency bands.

Frequency	Descriptions
433.05–434.79 MHz, 868–870 MHz	Europe unlicensed ISM bands
902–928 MHz	US unlicensed ISM band
2400–2483.5 MHz	Unlicensed ISM band (worldwide)
5150–5350 MHz, 5470–5825 MHz	ISM unlicensed (worldwide)
402–405 MHz	MICS band (worldwide)
401–406 MHz	MICS band extended (worldwide)
174–216 MHz, 470–608 MHz, 614–668 MHz	Wireless medical telemetry
608–614 MHz, 1395–1400 MHz, 1427–1432 MHz	WMTS bands (1.5 MHz/channel)
202.65–205.15 MHz	WMTS in New Zealand
420–429 MHz, 440–449 MHz	WMTS bands in Japan

a list of some dedicated bands operating in different regions of the world. Depending on the application, one or more frequency bands can be used. For example, the Industrial, Scientific, and Medical (ISM) band can be used for any medical and/or commercial applications. However, the Medical Implant Communications Service (MICS) band can only be used for communication with implanted devices. Since the MICS band shares the frequency spectrum with a licensed application, the transmitted power and data communication protocol is strictly regulated. The Wireless Medical Telemetry Service (WMTS) band can be used for any implantable and wearable medical devices. Due to worldwide license-free usage of the MICS (402−405 MHz) and ISM (2400−2483.5 MHz) bands, most medical, implanted, and wearable devices use antennas operating at these frequencies.

Typically, a communication network created by wearable or implanted devices is referred to as a BAN (Figure 14.10) [51,52]. Significant research was contributed to characterize the proportion channel between different sensor nodes [9, 53]. To achieve the large-scale, secured network connectivity for implanted and wearable devices, different communication protocols were developed. Bluetooth, Zigbee, and ANT_+ are some of the dominant communication protocols operating in the high frequency ISM band (2400−2483.5 MHz). For a wearable or implanted device, selection of individual protocols was determined based on the power budget, network size (number of nodes), data rate, and operating distance.

14.4 DISCUSSION

In the last decade, with the rapid advancements in micro- and nano-fabrication technology, low power, small dimension and multifunctional implants and wearable electronics were developed. While on-site batteries served as the common source of power for computationally intensive devices (i.e., smart watch), the new contactless power delivery methods (i.e., ultrasound, inductive) were employed on medical implants to achieve longevity and smaller device dimensions. For these designs, trade-offs were made between the dimensions, mass, and replacement/rechargeable period of the energy source. To achieve batteryless and life-long operation of the system, energy harvesting methods were proposed, which can convert ambient

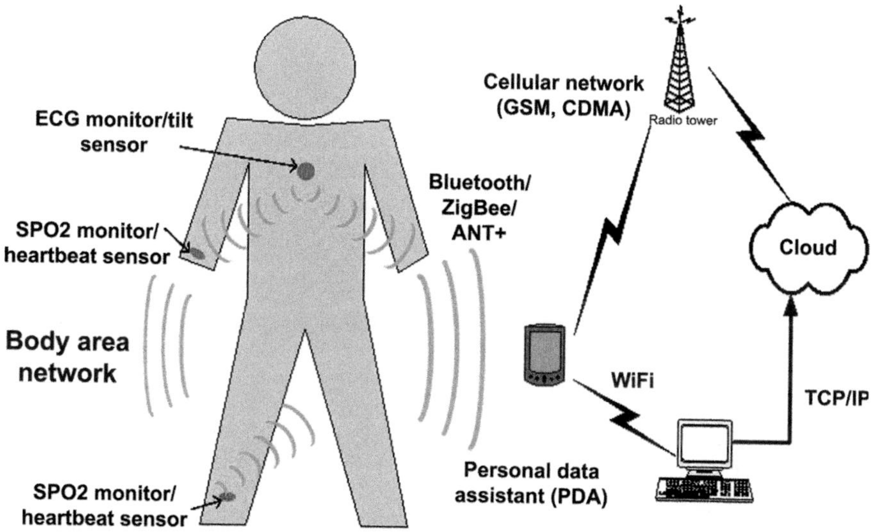

FIGURE 14.10 Block diagram of the typical body area network (BAN). The system includes the body-worn sensors/electronics with the capability of radio communication. The multi-node sensor network can communicate over the standard protocols such as Bluetooth, ZigBee and ANT_+. The range of the BAN can be extended by interfacing with the cellular or wired network.

mechanical, electromagnetic, solar, and/or chemical energy to useful electrical energy. Current energy harvesting technologies are limited to very low power (micro- and nano-watt) devices. However, further increases in the energy conversion efficiency (between ambient and electrical forms) will lead to self-powered and sustained energy sources for wearable and implanted electronics.

The specifications (frequency and data rate) of the communication channels vary across the different applications. Most implanted devices use application specific and proprietary communication links, which can be operated over a short range (less than 5 cm). However, with the desire for long-term implant monitoring and remote connectivity, new implants are equipped with high frequency radiofrequency (RF) transceivers. To communicate with the implanted system over large distances (2–10 m), the MICS band (402–405 MHz) was adopted, due to its worldwide usage and low attenuation compared to the ISM band (2400–2483.5 MHz). For on-body wearable electronics, the high frequency ISM band was utilized to create a multi-node network between the on-body sensors and the surrounding network. Table 14.2 provides the design selection and its limitations to effectively deliver power and data to implantable/wearable electronics.

14.5 DESIGN EXAMPLES

14.5.1 Wireless Power and Data for Retinal Prosthesis

The retinal prosthesis is one of the most successful medical devices that restores partial vision in patients with age-related macular degeneration (AMD) and retinitis

TABLE 14.2 Comparison between different power mechanism.

Non-regenerative power source	Mechanism	Power range	Primary application
On-site battery [3, 17]	Direct DC source	100mW–2 W	Wearable, implants
Wireless power [22]	Electromagnetic field coupling	1–50 mW	Implants (i.e., retinal prosthesis)
Ultrasound [31]	High frequency pressure wave	1–50 mW	Deep tissue implants
Optical [33]	Near-infrared light	1–20 mW	Wearable sensors, implants
Regenerative power source	Mechanism	Power range	Primary application
Mechanical [37]	Vibration, pressure	0.01–0.1 mW	Wearable sensors
Solar cell [40]	Photoelectric	0.1–10 mW	Wearable sensors
Electromagnetic field harvester [11]	High frequency (HF) field absorber	0.01–1 mW	Wearable sensors
Thermoelectric [46]	Seebeck effect	0.01–0.1 mW	Wearable sensors, shallow implants
Biofuel [48]	Chemical reaction	0.01–1 mW	Deep body implants

pigmentosa (RP) [4]. Figure 14.11 shows the block diagram of the retinal prosthesis system which uses a hybrid design of implanted and wearable electronics. The battery operated body-worn processing unit transfers energy and data to the implanted electronics to create visual perceptions. It uses epiretinal electrodes to bypass the defective photoreceptors and electrically stimulate retinal ganglion to generate neural activities. The visual perception of the patient depends on the spatial and temporal distribution of the stimulation pattern. Therefore, a body-worn processing unit is used to generate an optimal stimulating pattern based on the surroundings. In this design, an eyeglasses-mounted camera captures external images and a body-worn unit performs real-time processing on the images to generate stimulation patterns.

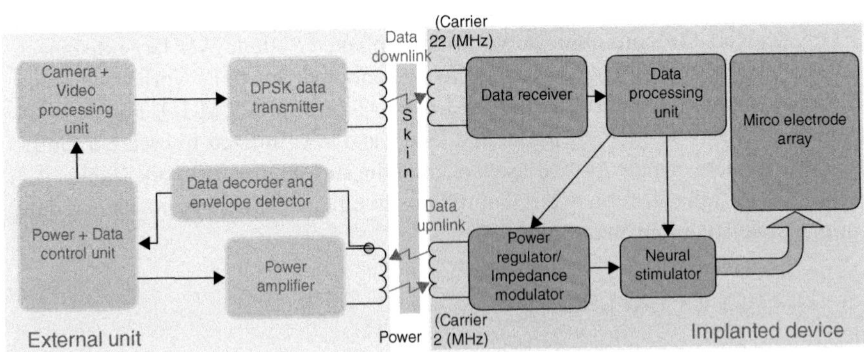

FIGURE 14.11 Block diagram of retinal prosthesis system based on the design proposed in [49]. *Source:* Chen et al. 2010 [49]. Reproduced with permission of IEEE.

Due to the complexity and energy requirement of the image processor, all the computation is performed outside the body while keeping the basic function of the implanted neural stimulator. Due to a large multi-electrode array being necessary for generating high resolution visual perception, the power requirement of the implant can vary largely from 50 mW to 250 mW [49]. In this design, an external power source (battery) is used and energy is transferred wirelessly (using low frequency inductive coupling) to the small-sized implant. Typically, such WPT systems achieve a total efficiency of 10–15%. Therefore, a small form factor rechargeable battery was used to achieve a time period of 4–6 hours between individual charging cycles [54]. To control the stimulation pattern, the design uses the forward telemetry link (downlink) modulated (amplitude modulation) over the power signal. During device operation, the eye can rotate, which may result in a temporary loss of power link. Therefore, a back telemetry link (uplink) is used to create a feedback loop that can result in constant output power at the implant.

14.5.2 Wireless Sensors for the Body Area Network

With the development of small and power-efficient health monitoring sensors, multiple applications have attempted to create a local and secure BAN around individuals [53]. Medtronic's data connectivity solutions is one of the leading patient management systems which helps with connecting an implanted cardiac device to a clinical system. Figure 14.12 provides the schematics of the patient management system. Some designs utilize sensor modules that can be scaled to create a network formation. For these designs, low weight, small form factor, and improved battery life are the key design requirements. To create a multi-node network, these devices use radio frequency (RF) transceivers while operating at standard communication protocols to achieve seamless compatibility with other devices. The high frequency

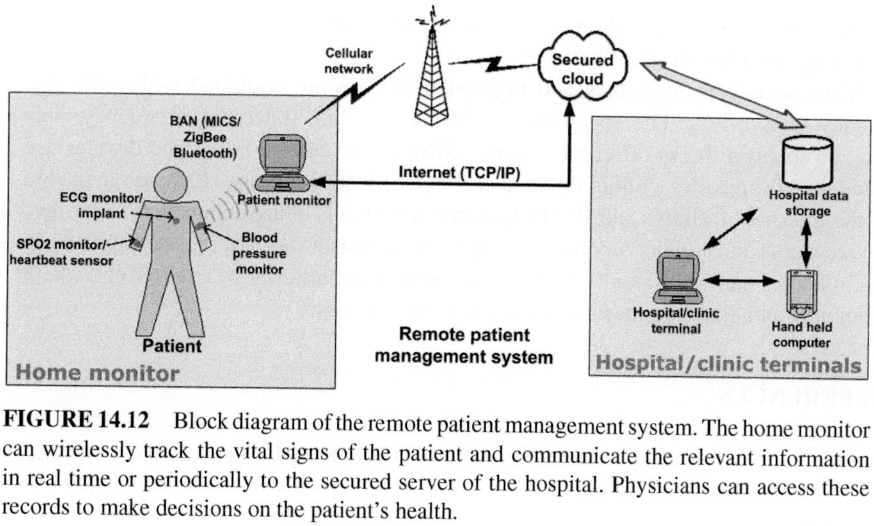

FIGURE 14.12 Block diagram of the remote patient management system. The home monitor can wirelessly track the vital signs of the patient and communicate the relevant information in real time or periodically to the secured server of the hospital. Physicians can access these records to make decisions on the patient's health.

ISM band (2400–2483.5 MHz) is a commonly used unlicensed band (worldwide) for the BAN. Due to wide usage of this spectrum, different form factors of antennas were proposed that can fit inside wearable device to create the desired radiation pattern. Patch antennas, inverted-F, monopoles, and meanders are commonly used antennas for wearable electronics. In some designs, novel antenna fabrication techniques were used in which antennas can be embedded inside the user's clothing or can be mounted as a flexible bandage. Independent of the antenna selection, multiple communication protocols such as WiFi, Bluetooth, BLE, ANT_+, ZigBee can be operated over the ISM spectrum. For different applications, selection among individual protocols is determined based on the required data rate, number of network nodes, power budget of the device and added license cost for the use of protocol. Bluetooth and WiFi can support high data rates (1–54 Mbps) at the expense of higher power requirements. Low energy Bluetooth devices transfer device status information in the form of small packets 8–27 bytes and require a fraction of the power compared to Bluetooth High Speed(HS). ANT_+ is a proprietary protocol which is used for sensor networks and remote control systems. Some applications include access to heart rate monitors, cycling power meter monitors, and control of music players.

14.6 CONCLUSION

The need for energy efficient and multi-functional implanted or wearable devices has driven research into novel power and data telemetry techniques. While batteries have remained the primary energy source for such devices, multiple efficient and safe WPT methods were developed to reduce the size of implants. To improve the longevity of the wearable and implanted devices, energy harvesting devices were proposed which can convert ambient energy (i.e., solar, vibration, pressure, electromagnetic, biofuel) into useful electric energy. Currently, such harvesting devices are limited to powering low power (0.1–10 mW) electronics. However, with the improvement of energy harvesting efficiency, such methods can be widely applied in conjunction with traditional powering mechanisms (i.e., battery).

Most wearable electronics and implanted devices are equipped with radio frequency transceivers. The specifications of the data link (operating frequency, data rate, protocol) differ in different systems. Most of the current implanted devices use a dedicated controller without the option of network connectivity. However, in the era of the Internet of Things and Big Data, some research is being undertaken to retrieve device status and capture a patient's long-term biosensor data over a secure network. Such studies can significantly benefit the medical community by improving system efficiency and allowing individuals to track their fitness.

REFERENCES

[1] A. Fowler, S. Moheimani, and S. Behrens, "A 3-DOF SOI MEMS ultrasonic energy harvester for implanted devices," *J. Phys.: Conf. Ser., PowerMEMS 2013*, vol. 476, pp. 1–5, 2013.

[2] W. Burleson, S. S. Clark, B. Ransford, and K. Fu, "Design Challenges for Secure Implantable Medical Devices," in Proceedings of the 49th Annual Design Automation Conference, ACM, 2012, pp. 12–17.

[3] D. Cooper, The Wearables Industry Needs to Brace Itself for the Apple Watch, 2015. Available at http://www.engadget.com/2015/03/09/wearables-industry-apple-watch/

[4] Second Sight, How is Argus II Designed to Produce Sight? 2014. Available at http://www.engadget.com/2015/03/09/wearables-industry-apple-watch/

[5] J. Paulo and P. Gaspar, "Review and Future Trend of Energy Harvesting Methods for Portable Medical Devices," in Proceedings of the World Congress on Engineering 2010 Vol II, June 30–July 2, London, 2010.

[6] Daniel Burrus, Google Glass is Transforming Wearable Technology, 2014. Available at http://www.huffingtonpost.com/daniel-burrus/google-glass-is-transform b 4949991 .html

[7] Wearable Technology Life, Wearable Technology Life, 2015. Available at http://www.wearabletechnologylife.com/fitbit/

[8] K. Cook-Chennault, N. Thambi, and A. Sastry, "Powering mems portable devices: a review of non-regenerative and regenerative power supply systems with special emphasis on piezoelectric energy harvesting systems," *Smart Mater. Struct.*, vol. 17, no. 4, pp. 043001–043033, 2008.

[9] J. Abouei, J. D. Brown, K. N. Plataniotis, and S. Pasupathy, "Energy efficiency and reliability in wireless biomedical implant systems," *IEEE Trans. Information Technol. Biomed.*, vol. 15, no. 3, pp. 456–466, 2011.

[10] G. Wang, W. Liu, M. Sivaprakasam, and G. A. Kendir, "Design and analysis of an adaptive transcutaneous power telemetry for biomedical implants," *IEEE Trans. Circuits Syst. I: Regular Papers*, vol. 52, no. 10, pp. 2109–2117, October 2005.

[11] G. Fu and S. Sonkusale, "Broadband energy harvesting using a metamaterial resonator embedded with non-foster impedance circuitry," *arXiv preprint arXiv*:1411.0662, 2014.

[12] E. Y. Chow, C.-L. Yang, Y. Ouyang, A. L. Chlebowski, P. P. Irazoqui, and W. J. Chappell, "Wireless powering and the study of RF propagation through ocular tissue for development of implantable sensors," *IEEE Trans. Antennas Propag.*, vol. 59, no. 6, pp. 2379–2387, April 2011.

[13] A. V. Nurmikko, J. P. Donoghue, L. R. Hochberg, W. R. Patterson, Y.-K. Song, C. W. Bull, D. A. Borton, F. Laiwalla, S. Park, Y. Ming, and J. Aceros, "Listening to brain microcircuits for interfacing with external world—progress in wireless implantable microelectronic neuroengineering devices," *Proc. IEEE*, vol. 98, no. 3, pp. 375–388, February 2010.

[14] M. Koo, K.-I. Park, S. H. Lee, M. Suh, D. Y. Jeon, J. W. Choi, K. Kang, and K. J. Lee, "Bendable inorganic thin-film battery for fully flexible electronic systems," *Nano Lett.*, vol. 12, no. 9, pp. 4810–4816, July 2012.

[15] L. Hu, H. Wu, F. La Mantia, Y. Yang, and Y. Cui, "Thin, flexible secondary Li-ion paper batteries," *ACS Nano.*, vol. 4, no. 10, pp. 5843–5848, October 2010.

[16] Boston Scientific Inc., Spinal Cord Stimulator Systems, 2015. Available at http://www.bostonscientific.com/en-US/products/spinal-cord-stimulator-systems.html

[17] Medtronic Inc., Bradycardia, 2015. Available at http://www.medtronic.com/patients/bradycardia/pacemaker/

[18] A. Qusba, A. RamRakhyani, J.-H. So, G. Hayes, M. Dickey, and G. Lazzi, "On the design of microfluidic implant coil for flexible telemetry system," *IEEE Sens. J.*, vol. 14, no. 4, pp. 1074–1080, April 2014.

[19] A. RamRakhyani, S. Mirabbasi, and M. Chiao, "Design and optimization of resonance-based efficient wireless power delivery systems for biomedical implants," *IEEE Trans. Biomed. Circuits Syst.*, vol. 5, no. 1, pp. 48–63, February 2011.

[20] V. Singh, A. Qusba, A. Roy, R. Castro, K. McClure, R. Dai, R. Greenberg, J. Weiland, M. Humayun, and G. Lazzi, "Specific absorption rate and current densities in the human eye and head induced by the telemetry link of an epiretinal prosthesis," *IEEE Trans. Antennas Propag.*, vol. 57, no. 10, pp. 3110–3118, October 2009.

[21] L. Wu, Z. Yang, E. Basham, and W. Liu, "An Efficient Wireless Power Link for High Voltage Retinal Implant," in 2008 IEEE Conference on Biomedical Circuits and Systems, BioCAS 2008, Baltimore, November 20–22, 2008, pp. 101–104.

[22] A. K. RamRakhyani and G. Lazzi, "On the design of efficient multi-coil telemetry system for biomedical implants," *IEEE Trans. Biomed. Circuits Syst.*, vol. 7, no. 1, pp. 11–23, 2013.

[23] A. RamRakhyani and G. Lazzi, "Use of Multi-Coil Telemetry System for High Tolerance Efficient Wireless Power System," in 2012 IEEE International Conference on Wireless Information Technology and Systems (ICWITS), Maui, HI, November11–16, 2012, pp. 1–4.

[24] A. RamRakhyani and G. Lazzi, "Multi-coil approach to reduce electromagnetic energy absorption for wireless powered implants," *IET Healthc. Technol. Lett.*, March 2014, pp. 21–25.

[25] A. RamRakhyani and G. Lazzi, "Interference-free wireless power transfer system for biomedical implants using multicoil approach," *Electron. Lett.*, vol. 50, no. 12, pp. 853–855, June 2014.

[26] D. Pivonka, A. Yakovlev, A. Poon, and T. Meng, "A mm-sized wirelessly powered and remotely controlled locomotive implant," *IEEE Trans. Biomed. Circuits Syst.*, vol. 6, no. 6, pp. 523–532, December 2012.

[27] A. Ma and A. Poon, "Midfield wireless power transfer for bioelectronics," *IEEE Circuits and Syst. Mag.*, vol. 15, no. 2, pp. 54–60, Secondquarter 2015.

[28] G. Yang, M. R. Islam, R. A. Dougal, and M. Ali, "A reconfigurable stacked patch antenna for wireless power transfer and data telemetry in sensors," *Prog. Electromagnetics Res.*, vol. 29, pp. 67–81, Secondquarter 2012.

[29] P. Nikitin, K. Rao, and S. Lazar, "An Overview of Near Field UHF RFID," in 2007 IEEE International Conference on RFID, Grapevine, TX, March 26–28, 2007, pp. 167–174.

[30] H. Jiang, J. Zhang, D. Lan, K. K. Chao, S. Liou, H. Shahnasser, R. Fechter, S. Hirose, M. Harrison, and S. Roy, "A low-frequency versatile wireless power transfer technology for biomedical implants," *IEEE Trans. Biomed. Circuits Syst.*, vol. 7, no. 4, pp. 526–535, 2013.

[31] S. Ozeri and D. Shmilovitz, "Ultrasonic transcutaneous energy transfer for powering implanted devices," *Ultrason.*, vol. 50, no. 6, pp. 556–566, 2010.

[32] H. Vihvelin, J. Leadbetter, M. Bance, J. A. Brown, and R. B. Adamson, "Compensating for tissue changes in an ultrasonic power link for implanted medical devices," *IEEE Trans. Biomed. Circuits Syst.*, vol. 10, no. 2, pp. 404–411, April 2016.

[33] Y. Song, Q. Hao, X. Kong, L. Hu, J. Cao, and T. Gao, "Simulation of the recharging method of implantable biosensors based on a wearable incoherent light source," *Sensors*, vol. 14, no. 11, pp. 20687–20701, 2014.

[34] S. R. Anton, "Multifunctional Piezoelectric Energy Harvesting Concepts," Ph.D. Dissertation, Virginia Polytechnic Institute and State University, 2011.

[35] A. Fowler, S. Moheimani, and S. Behrens, "An omnidirectional mems ultrasonic energy harvester for implanted devices," *J. Microelectromech. Syst.*, vol. 23, no. 6, pp. 1454–1462, December 2014.

[36] C. Dagdeviren, B. D. Yang, Y. Su, P. L. Tran, P. Joe, E. Anderson, J. Xia, V. Doraiswamy, B. Dehdashti, X. Feng, L. et al., "Conformal piezoelectric energy harvesting and storage from motions of the heart, lung, and diaphragm," *Proc. Nati. Acad. Sci.*, vol. 111, no. 5, pp. 1927–1932, 2014.

[37] A. Cadei, A. Dionisi, E. Sardini, and M. Serpelloni, "Kinetic and thermal energy harvesters for implantable medical devices and biomedical autonomous sensors," *Meas. Sci. Technol.*, vol. 25, no. 1, pp. 012003–0120016, 2014.

[38] E. Romero, R. O. Warrington, and M. R. Neuman, "Body motion for powering biomedical devices," in, *Conf. Proc. IEEE Eng. Med. Biol. Soc.,* pp. 2752–2755, 2009.

[39] T. Starner, "Human-powered wearable computing," *IBM syst. J.*, vol. 35, no. 3-4, pp. 618–629, 1996.

[40] J. P. Carmo, J. M. Gomes, L. M. Goncalves, and J. H. Correia, "A flexible thin-film for powering stand-alone electronic devices," *Measurement*, vol. 46, no. 10, pp. 4145–4151, 2013.

[41] Z. L. Wang and W. Wu, "Nanotechnology-enabled energy harvesting for self-powered micro-/nanosystems," *Angew. Chem. Int. Ed.*, vol. 51, no. 47, pp. 11700–11721, 2012.

[42] F. Dimroth, M. Grave, P. Beutel, U. Fiedeler, C. Karcher, T. N. D. Tibbits, E. Oliva, G. Siefer, M. Schachtner, A. Wekkeli, A. W. Bett, R. Krause, M. Piccin, N. Blanc, C. Drazek, E. Guiot, B. Ghyselen, T. Salvetat, A. Tauzin, T. Signamarcheix, A. Dobrich, T. Hannappel, and K. Schwarzburg, "Wafer bonded four-junction gainp/gaas/gainasp/gainas concentrator solar cells with 44.7% efficiency," *Prog. Photovoltaics: Res. Appl.*, vol. 22, no. 3, pp. 277–282, 2014.

[43] L. M. Borges, R. Chavez-Santiago, N. Barroca, F. J. Velez, and I. Balasingham, "Radiofrequency energy harvesting for wearable sensors," *IET Healthc. Technol. Lett.*, vol. 2, no. 1, pp. 22–27, 2015.

[44] H. Bottner, "Thermoelectric Micro Devices: Current State, Recent Developments and Future Aspects for Technological Progress and Applications," in Proceedings of the Twenty-First International Conference on Thermoelectrics, ICT '02, August 25–29, 2002, pp. 511–518.

[45] M. Wahbah, M. Alhawari, B. Mohammad, H. Saleh, and M. Ismail, "Characterization of human body-based thermal and vibration energy harvesting for wearable devices," *IEEE J. Emer. Sel. Top. Circuits Syst.*, vol. 4, no. 3, pp. 354–363, September 2014.

[46] Y. Yang, X.-J.Wei, and J. Liu, "Suitability of a thermoelectric power generator for implantable medical electronic devices," *J. Phys. D: Appl. Phys.*, vol. 40, no. 18, pp. 5790, 2007.

[47] Y. Du, K. Cai, S. Chen, H. Wang, S. Shen, R. Donelson, and T. Lin, "Thermoelectric fabrics: toward power generating clothing," *Sci. Rep.*, vol. 4, April 2014.

[48] P. Cinquin, C. Gondran, F. Giroud, S. Mazabrard, A. Pellissier, F. Boucher, J.-P. Alcaraz, K. Gorgy, F. Lenouvel, S. Mathé, P. Porcu, and S. Cosnier, "A glucose biofuel cell implanted in rats," *PLoS One*, vol. 5, no. 5, pp. e10476, May 2010.

[49] K. Chen, Z. Yang, L. Hoang, J.Weiland, M. Humayun, and W. Liu, "An integrated 256-channel epiretinal prosthesis," *IEEE J. Solid-State Circuits*, vol. 45, no. 9, pp. 1946–1956, September 2010.

[50] M. Suster, J. Guo, N. Chaimanonart, W. Ko, and D. Young, "A Wireless Strain Sensing Microsystem With External RF Power Source and Two-Channel Data Telemetry Capability," in 2007 IEEE International Digest of Solid-State Circuits Conference, San Francisco, CA, February 11–15, 2007, pp. 380–609.

[51] A. Bindra, "Medical info-communications signals an era of body area networking," *RF Design*, pp. 10–14, February 2008.

[52] S. K. Kumar, "Design of a Wireless Platform for Wearable and Home Automation Applications," Ph.D. Dissertation, University of Missouri–Kansas City, 2012.

[53] D. Raskovic, T. Martin, and E. Jovanov, "Medical monitoring applications for wearable computing," *Comput. J.*, vol. 47, no. 4, pp. 495–504, 2004.

[54] Second Sight, Argus II Retinal Prosthesis System, 2015. Available at http://www.fda.gov/ucm/groups/fdagov-public/@fdagov-afda-adcom/documents/document/ucm320423.pdf

INDEX

Electromagnetics of Body-Area Networks: Antennas, Propagation, and RF Systems, First Edition.
Edited by Douglas H. Werner and Zhi Hao Jiang.

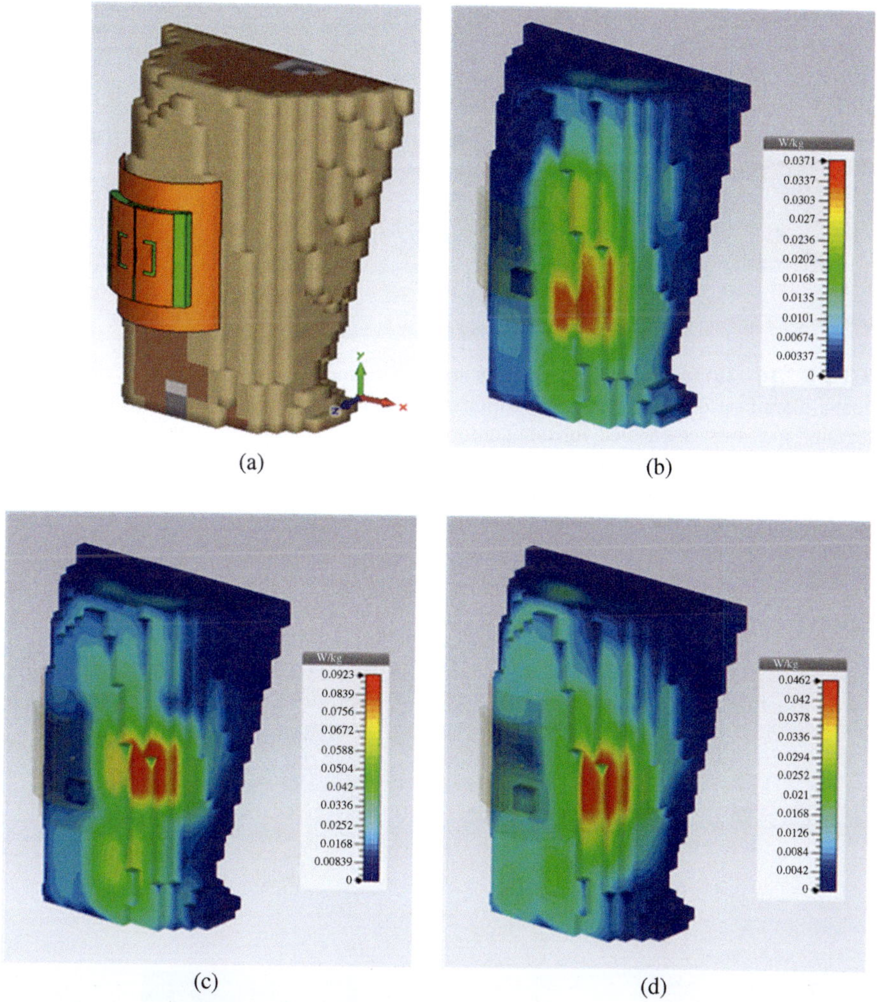

FIGURE 1.12 The SAR of a (a) bent dual-band antenna integrated with an artificial magnetic conductor (AMC) at (b) 2.45 GHz, (c) 5.2 GHz, and (d) 5.8 GHz evaluated on a voxel model. *Source:* Yan et al. 2015 [54]. Reproduced with permission of IEEE.

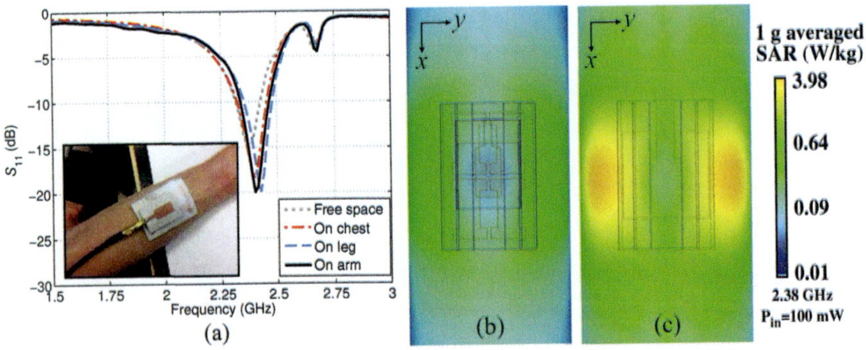

FIGURE 2.16 (a) Measured S_{11} of the structurally deformed integrated metasurface-enabled antenna placed on different parts of the human body. The inset shows a photograph of the integrated metasurface-enabled antenna conformal to a human arm. Simulated 1 g averaged SAR values for (b) the integrated metasurface-enabled antenna and (c) a conventional patch antenna with the same footprint at 1 mm away from the multilayer tissue model. *Source:* Jiang et al. 2014 [68]. Reproduced with permission of IEEE.

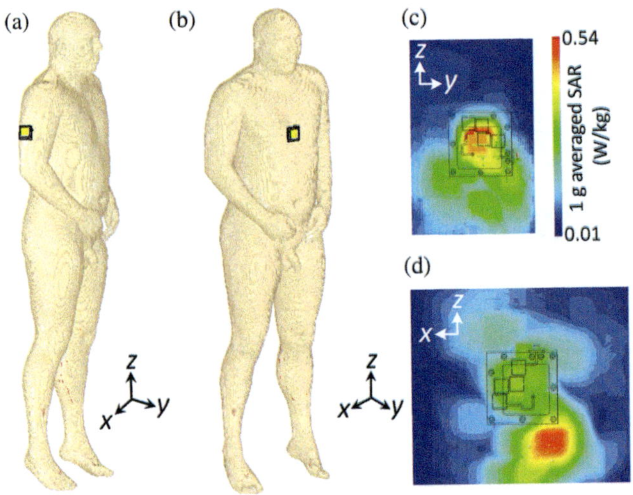

FIGURE 2.18 Configuration of the filtering antenna mounted on (a) the arm and (b) the chest of the HUGO human body model in CST MWS. Simulated 1g averaged SAR values for the filtering antenna when placed on (c) the arm and (d) the chest.

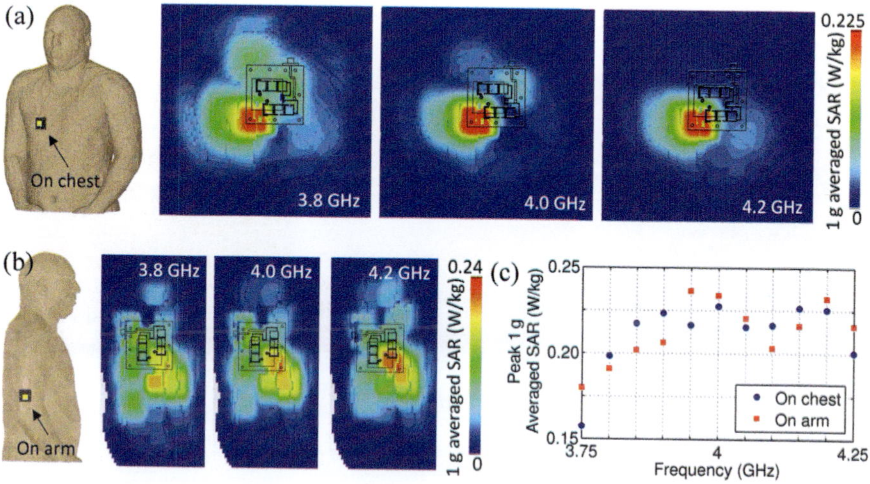

FIGURE 2.20 Simulated 1g averaged SAR values of the filtering antenna mounted on (a) the chest and (b) the arm of the HUGO human body model in CST MWS at 3.8, 4.0, and 4.2 GHz. (c) Simulated peak 1g averaged SAR values in the targeted band as a function of frequency. *Source:* Jiang and Werner 2015 [64]. Reproduced with permission of IEEE.

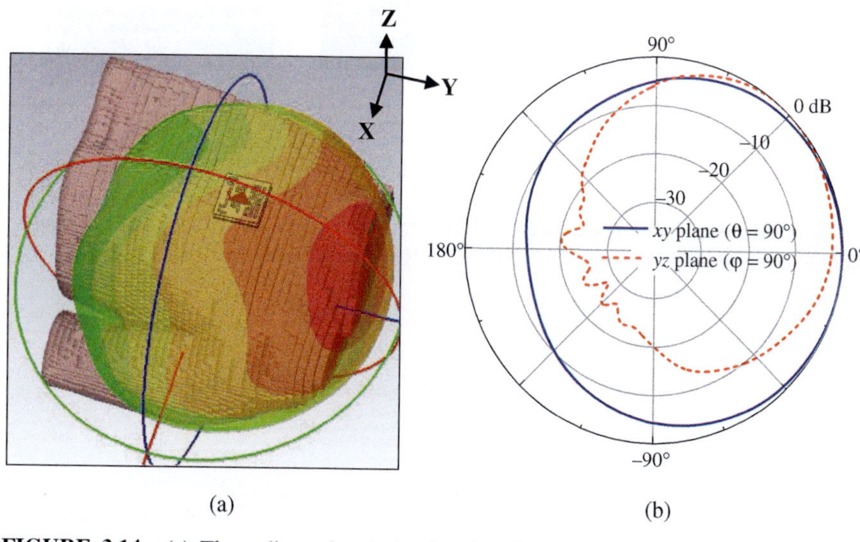

FIGURE 3.14 (a) Three-dimensional simulated radiation pattern of $YAGI_{D\text{-}AMC}$ antenna configuration at 2.45 GHz for on-voxel chest right location, and (b) normalized radiation polar plot for $\theta = 90°$ (xy plane) and $\varphi = 90°$ (yz plane). *Source:* Agarwal et al. 2016 [22]. Reproduced with permission of IEEE.

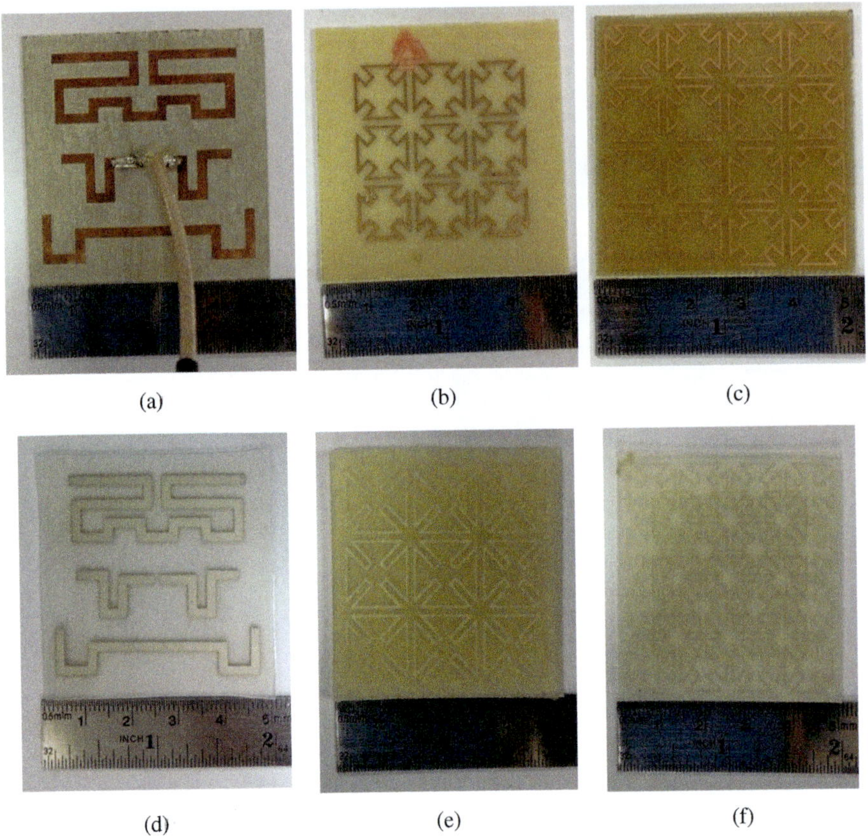

FIGURE 3.19 Antenna prototypes with AMC layers: on FR4 PCB (a) Yagi radiator; (b) double-layered top AMC layer of 3 × 3 units; (c) double-layered bottom AMC layer of 4 × 4 units. Also shown are latex-based AMC layers: (d) Yagi radiator; (e) single layer; and (f) double layer. *Source:* Agarwal et al. 2016 [22]. Reproduced with permission of IEEE.

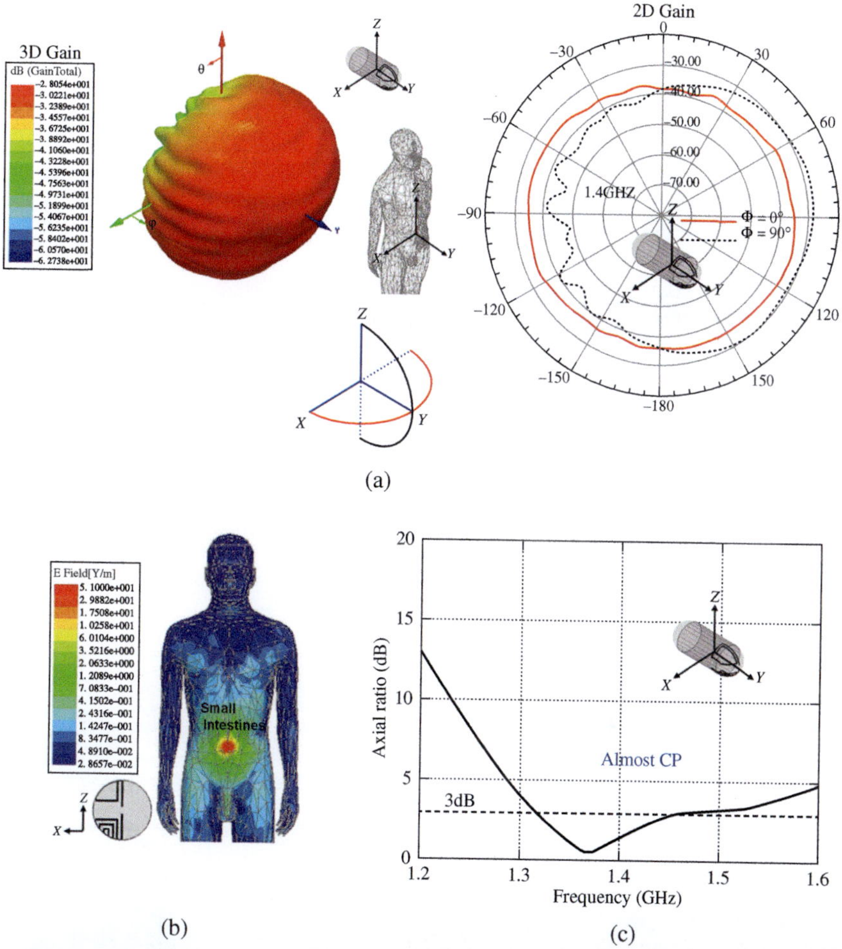

FIGURE 4.13 (a) Radiation pattern of the conformal antenna at 1.4 GHz when placed inside the human body. *Source:* Rajagopalan et al. 2011 [10]. Reproduced with permission of McGraw-Hill Education. (b) Electric field distribution of the antenna on the surface of human body model. *Source:* Rajagopalan and Rahmat-Samii 2010 [10]. Reproduced with permission of IEEE. (c) Axial ratio performance of the conformal antenna inside human body model. *Source:* Rajagopalan and Rahmat-Samii 2010 [10]. Reproduced with permission of IEEE.

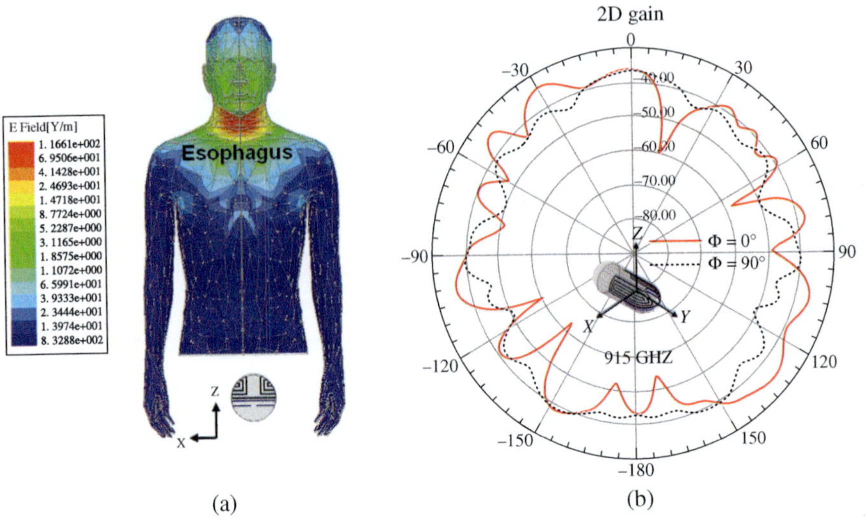

FIGURE 4.24 (a) Electric field distribution of the capsule tag antenna. (b) 2D radiation pattern of the tag.

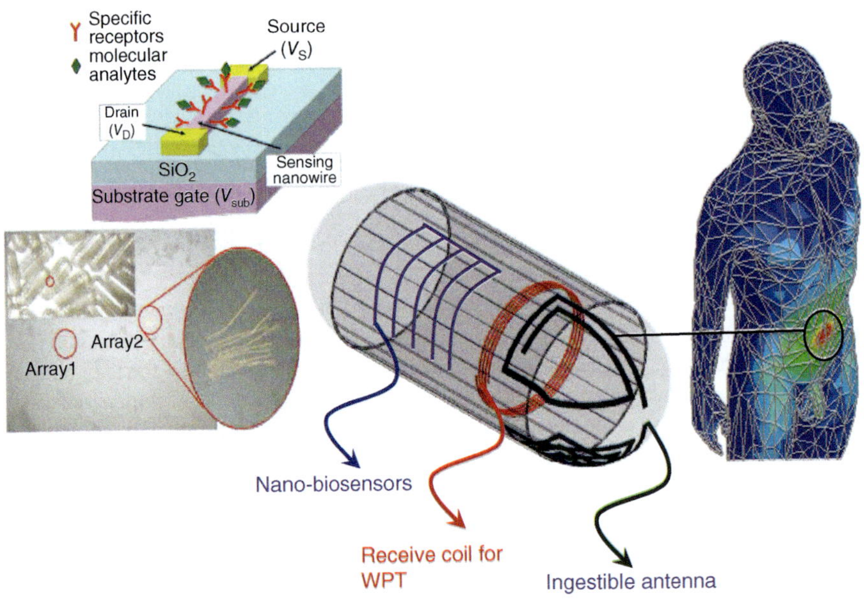

FIGURE 4.37 A diagnostic capsule system with nano-biosensors. *Source:* Rajagopalan et al. 2012 [75]. Reproduced with permission of IEEE.

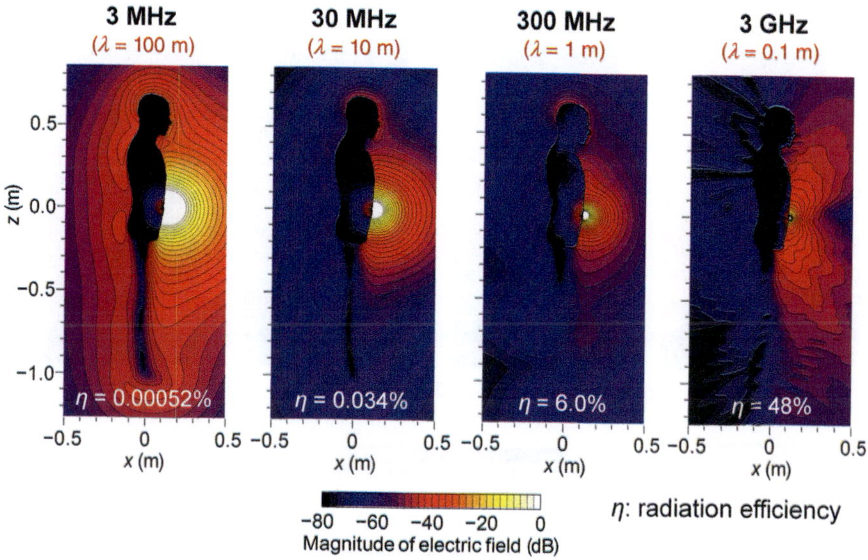

FIGURE 5.4 *E*-field distributions at different frequencies in free space. *Source:* Ito and Haga 2010 [16]. Reproduced with permission of IEEE.

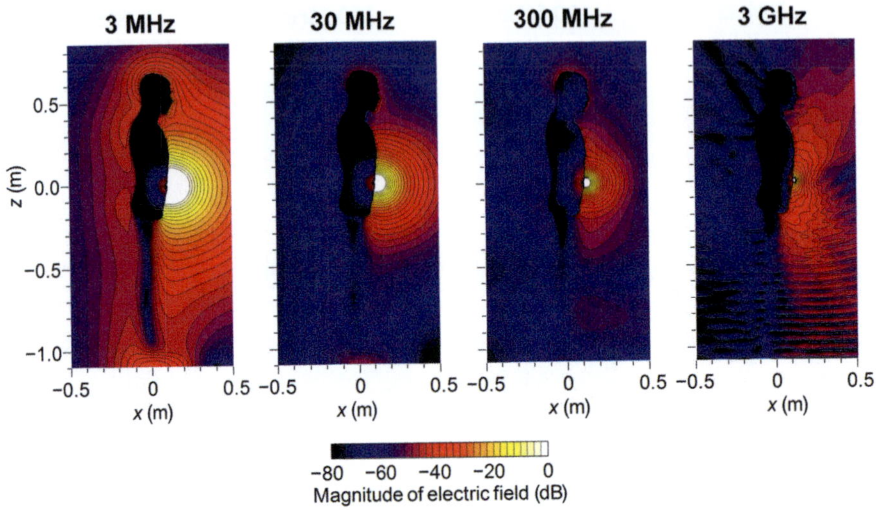

(a)

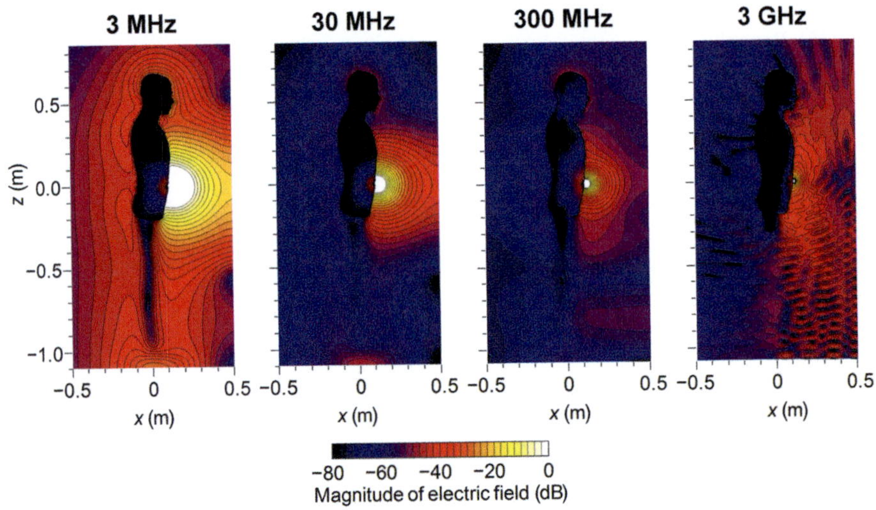

(b)

FIGURE 5.8 *E*-field distributions at different frequencies (a) only with floor and (b) with floor and wall. (Equal input power is assumed.)

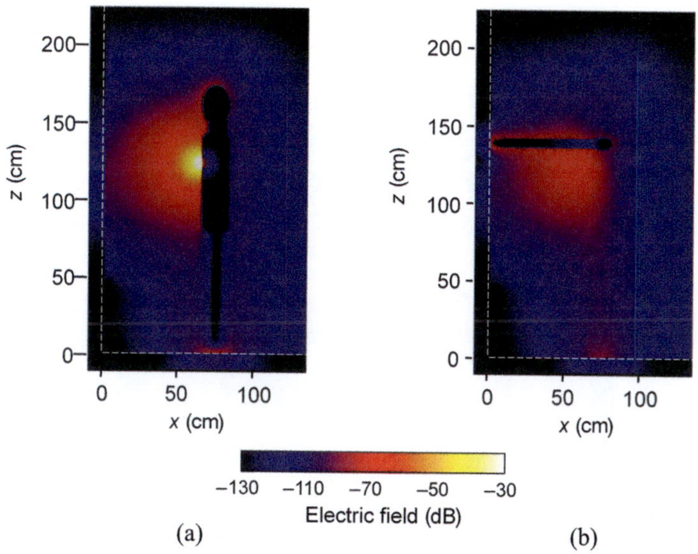

FIGURE 5.20 Electric field distributions at 3 MHz.

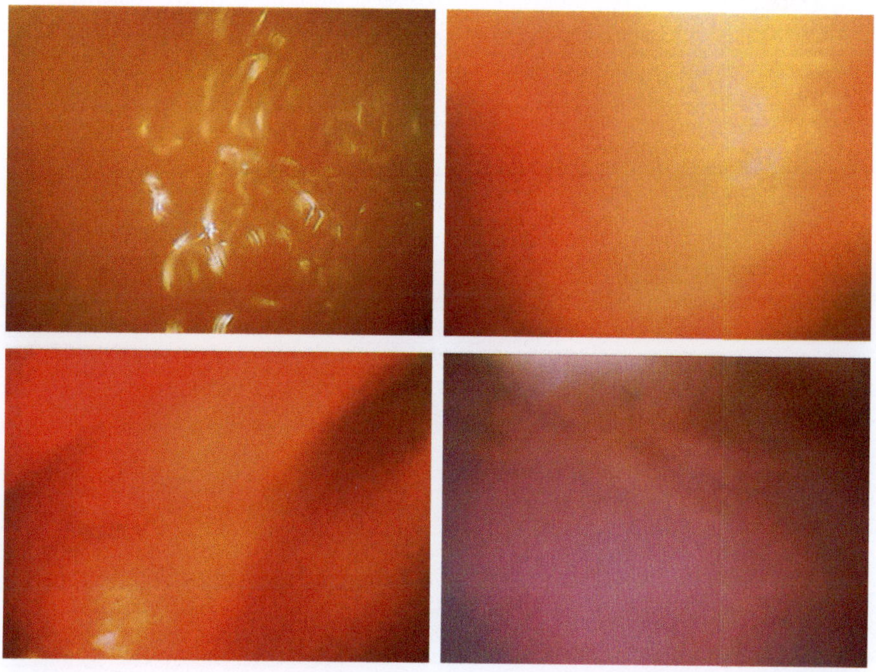

FIGURE 6.34 Photos of the images captured inside the gastrointestinal tract of the live pig.

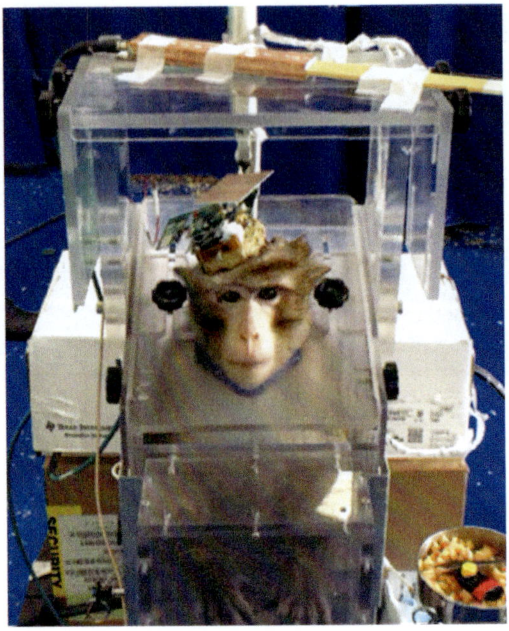

FIGURE 6.54 A wireless neuroprobe system demonstration on the primate.

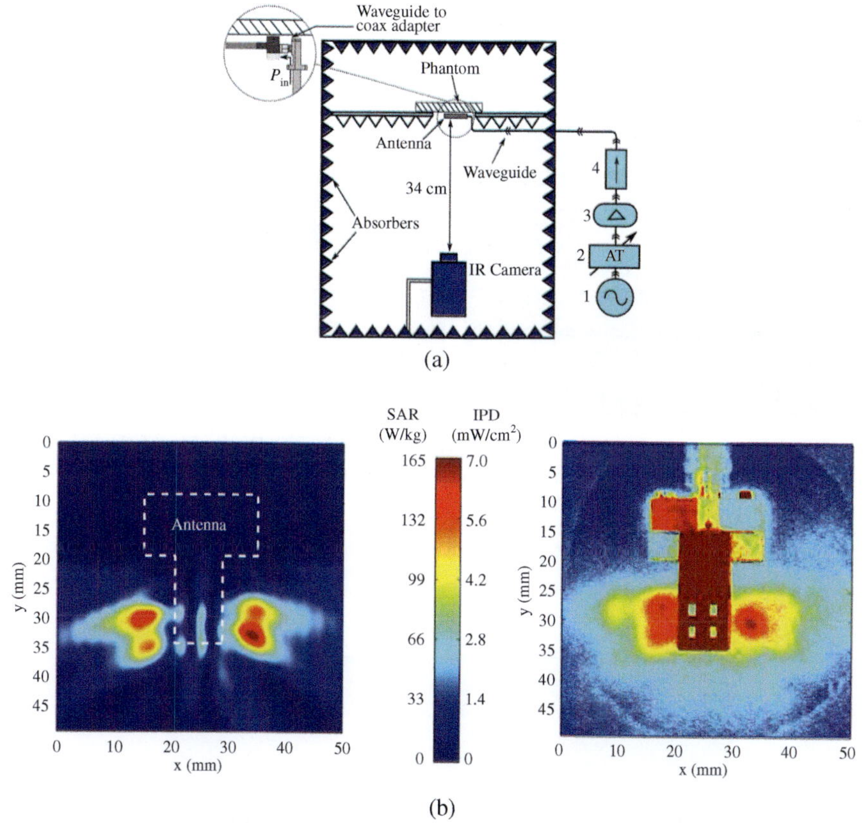

FIGURE 7.19 (a) SAR measurement setup. (b) SAR and IPD distributions at 60 GHz. (Left) Numerical results for the antenna on the skin. (Right) Measurements on the skin-equivalent phantom. P_{in} = 322 mW. *Source*: Chahat et al. 2012 [70]. Reproduced with permission of IEEE.

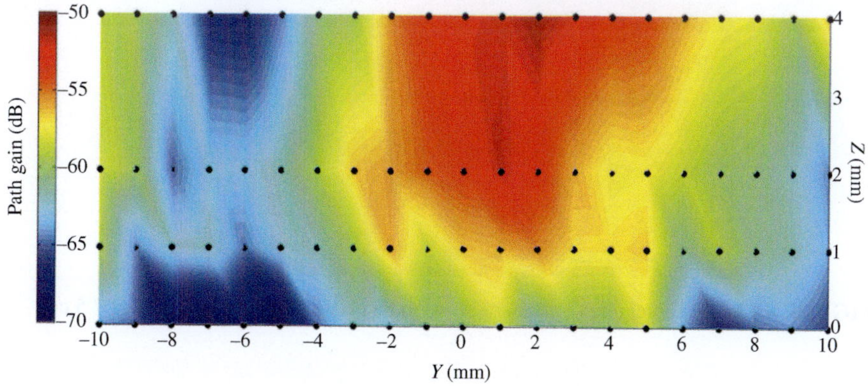

FIGURE 7.37 Measured path gain between end-fire antenna and waveguide in an off-body scenario. Distance between the hand and waveguide is $d = 5$ cm.

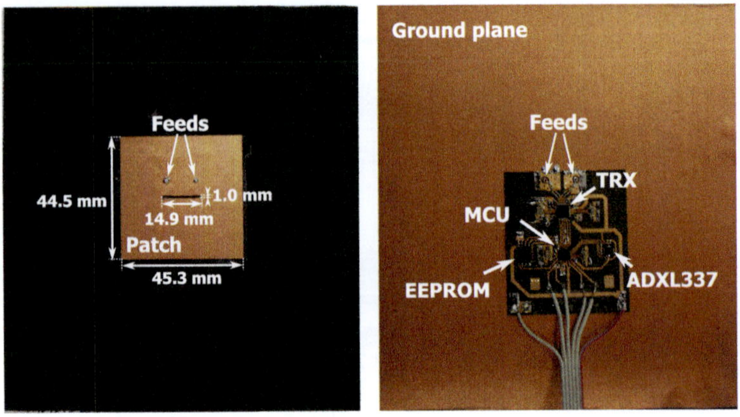

FIGURE 8.22 (a) Front side of the wireless sensor node. (b) Back side of the wireless sensor node with the electronic components mounted on the flexible substrate. *Source:* Vanveerdeghem et al. 2014 [42]. CC BY 4.0.

FIGURE 8.23 The integrated system bent around a plastic tube with a radius of 7.5 cm.

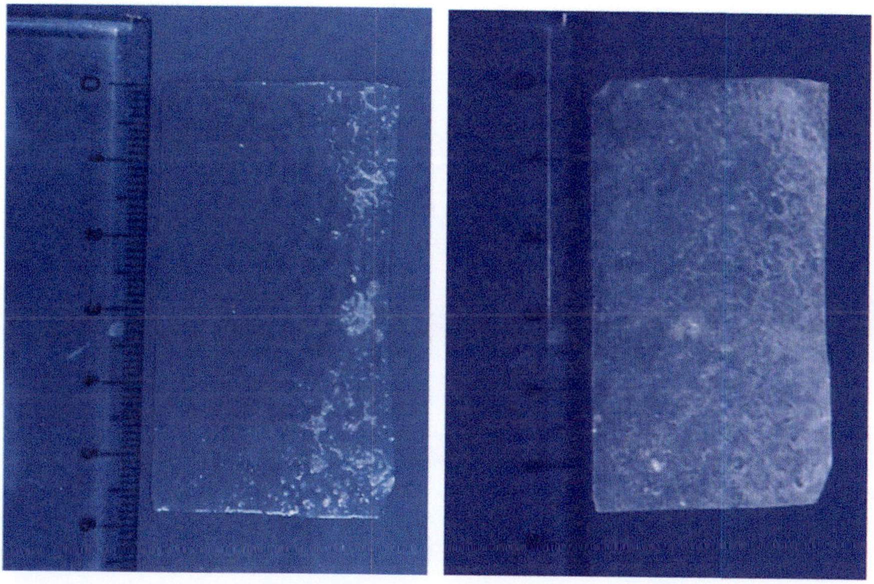

FIGURE 9.21 Samples of the fabricated hydrogel membrane. (Left) Swollen. (Right) Dried. *Source:* Occhiuzzi et al. 2015 [31]. Reproduced with permission of IEEE.

(a) (b)

FIGURE 10.19 (a) Open voltage measurement with an on-bottle setup. (b) Operation verification of the wearable harvester on a wrist. *Source:* Bito et al. 2015 [41]. Reproduced with permission of IEEE.

(a) (b)

FIGURE 10.20 (a) Open voltage measurement with an on-bottle setup. (b) Operation verification of the wearable harvester on a wrist. *Source:* Bito et al. 2015 [41]. Reproduced with permission of IEEE.

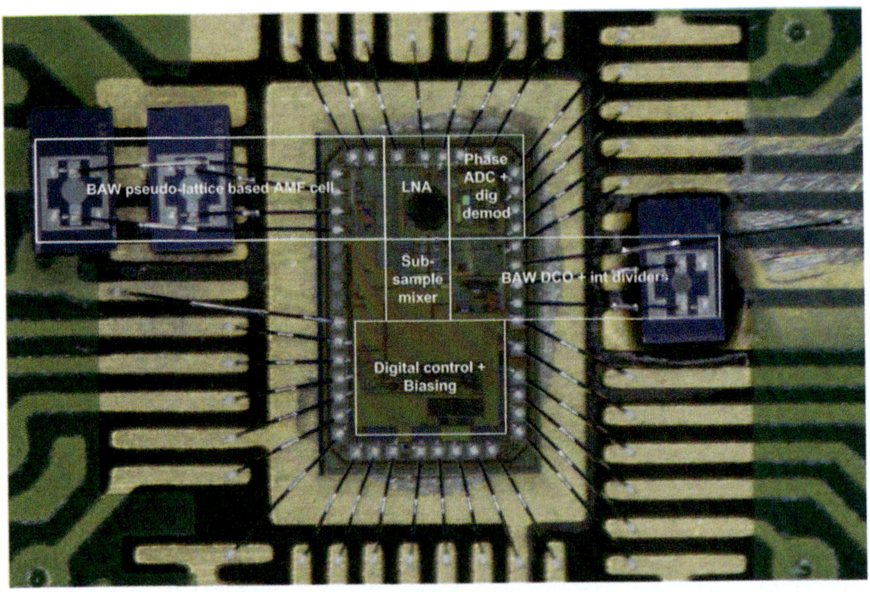

FIGURE 12.33 Chip microphotograph of the receiver.

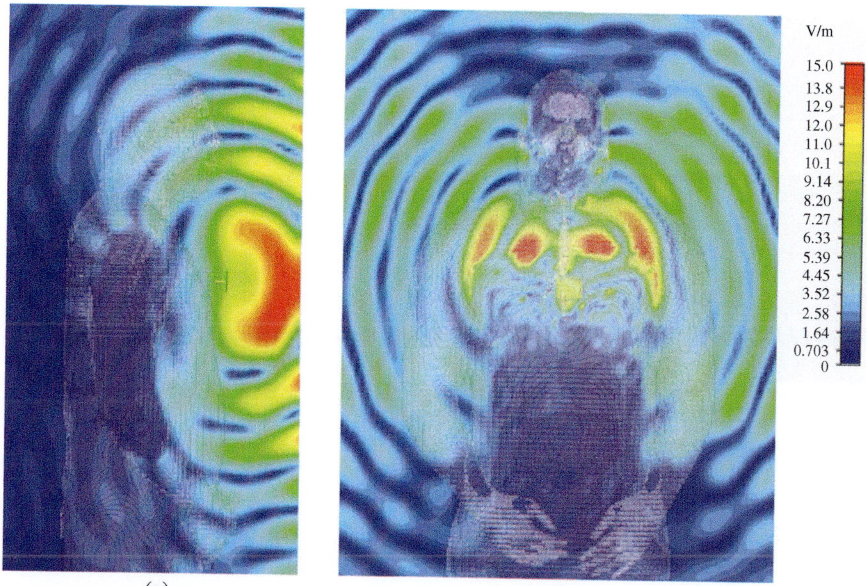

(a) (b)

FIGURE 13.3 Simulation results of the EM wave propagation alongside the trunk section of the Hugo model implemented in CST Microwave Studio. The radiating element is a microstrip patch antenna. (a) Antenna on-body; (b) antenna off-body.

Printed and bound by CPI Group (UK) Ltd, Croydon, CR0 4YY
17/09/2026
14964290-0001